A Survey of Modern Algebra

A Survey of Modern Algebra

Garrett Birkhoff
Harvard University

Saunders Mac Lane
The University of Chicago

CRC Press
Taylor & Francis Group
Boca Raton London New York

CRC Press is an imprint of the
Taylor & Francis Group, an **informa** business

AN AK PETERS BOOK

Reprinted 2010 by CRC Press
CRC Press
6000 Broken Sound Parkway, NW
Suite 300, Boca Raton, FL 33487
270 Madison Avenue
New York, NY 10016
2 Park Square, Milton Park
Abingdon, Oxon OX14 4RN, UK

Editorial, Sales, and Customer Service Office

A K Peters, Ltd.
5 Commonwealth Road, Suite 2C
Natick, MA 01760
www.akpeters.com

Copyright © 1997 by A K Peters, Ltd.

This fifth edition is a corrected reprint of the fourth edition by a new
publisher.

Originally printed in hardcover.
First paperback printing, 2008.
    ISBN-13: 978-1-56881-454-4

Library of Congress Cataloging-in-Publication Data
Birkhoff, Garrett, 1911-
    A survey of modern algebra / Garrett Birkhoff, Saunders Mac Lane.
        p. cm.   –   (AKP classics)
    Previously published: New York : Macmillan, c1977.
    "This fifth edition is a corrected reprint of the fourth edition by a new
publisher"–T.p. verso.
    Includes bibliographical references and index.
    ISBN 1-56881-068-7
    1. Algebra, Abstract. I. Mac Lane, Saunders, 1909-. II. Title. III. Series.

QA162.B57 1997
512'.02–dc21   97-372
                                                                CIP

Preface to the Fourth Edition

During the thirty-five years since the first edition of this book was written, courses in "modern algebra" have become a standard part of college curricula all over the world, and many books have been written for use in such courses. Nevertheless, it seems desirable to recall our basic philosophy, which remains that of the present book.

"We have tried throughout to express the conceptual background of the various definitions used. We have done this by illustrating each new term by as many familiar examples as possible. This seems especially important in an elementary text because it serves to emphasize the fact that the abstract concepts all arise from the analysis of concrete situations.

"To develop the student's power to think for himself in terms of the new concepts, we have included a wide variety of exercises on each topic. Some of these exercises are computational, some explore further examples of the new concepts, and others give additional theoretical developments. Exercises of the latter type serve the important function of familiarizing the student with the construction of a formal proof. The selection of exercises is sufficient to allow an instructor to adapt the text to students of quite varied degrees of maturity, of undergraduate or first year graduate level.

"Modern algebra also enables one to reinterpret the results of classical algebra, giving them far greater unity and generality. Therefore, instead of omitting these results, we have attempted to incorporate them systematically within the framework of the ideas of modern algebra.

"We have also tried not to lose sight of the fact that, for many students, the value of algebra lies in its applications to other fields: higher analysis, geometry, physics, and philosophy. This has influenced us in our emphasis on the real and complex fields, on groups of transformations as contrasted with abstract groups, on symmetric matrices and reduction to diagonal form, on the classification of quadratic forms under the orthogonal and Euclidean groups, and finally, in the inclusion of Boolean algebra, lattice theory, and transfinite numbers, all of which are important in mathematical logic and in the modern theory of real functions."

In detail, our Chapters 1–3 give an introduction to the theory of linear and polynomial equations in commutative rings. The familiar domain of integers and the rational field are emphasized, together with the rings of integers modulo n and associated polynomial rings. Chapters 4 and 5 develop the basic algebraic properties of the real and complex fields which are of such paramount importance for geometry and physics.

Chapter 6 introduces noncommutative algebra through its simplest and most fundamental concept: that of a group. The group concept is applied systematically in Chapters 7–10, on vector spaces and matrices. Here care is taken to keep in the foreground the fundamental role played by algebra in Euclidean, affine, and projective geometry. Dual spaces and tensor products are also discussed, but generalizations to modules over rings are not considered.

Chapter 11 includes a completely revised introduction to Boolean algebra and lattice theory. This is followed in Chapter 12 by a brief discussion of transfinite numbers. Finally, the last three chapters provide an introduction to general commutative algebra and arithmetic: ideals and quotient-rings, extensions of fields, algebraic numbers and their factorization, and Galois theory.

Many of the chapters are independent of one another; for example, the chapter on group theory may be introduced just after Chapter 1, while the material on ideals and fields (§§13.1 and 14.1) may be studied immediately after the chapter on vector spaces.

This independence is intended to make the book useful not only for a full-year course, assuming only high-school algebra, but also for various shorter courses. For example, a semester or quarter course covering linear algebra may be based on Chapters 6–10, the real and complex fields being emphasized. A semester course on abstract algebra could deal with Chapters 1–3, 6–8, 11, 13, and 14. Still other arrangements are possible.

We hope that our book will continue to serve not only as a text but also as a convenient reference for those wishing to apply the basic concepts of modern algebra to other branches of mathematics, including statistics and computing, and also to physics, chemistry, and engineering.

It is a pleasure to acknowledge our indebtedness to Clifford Bell, A. A. Bennett, E. Artin, F. A. Ficken, J. S. Frame, Nathan Jacobson, Walter Leighton, Gaylord Merriman, D. D. Miller, Ivan Niven, and many other friends and colleagues who assisted with helpful suggestions and improvements, and to Mrs. Saunders Mac Lane, who helped with the secretarial work in the first three editions.

Cambridge, Mass.
Chicago, Illinois

GARRETT BIRKHOFF
SAUNDERS MAC LANE

Contents

4 Real Numbers   94

5 Complex Numbers   107

6 Groups   124

7 Vectors and Vector Spaces   168

8 The Algebra of Matrices

9 Linear Groups

10 Determinants and Canonical Forms

15   Galois Theory   452

1

The Integers

1.1. Commutative Rings; Integral Domains

Modern algebra has exposed for the first time the full variety and richness of possible mathematical systems. We shall construct and examine many such systems, but the most fundamental of them all is the oldest mathematical system—that consisting of all the positive integers (whole numbers). A related but somewhat larger system is the collection **Z** of *all* integers 0, ±1, ±2, ±3, · · · . We begin our discussion with this system because it more closely resembles the other systems which arise in modern algebra.

The integers have many interesting algebraic properties. In this chapter, we will assume some especially obvious such properties as *postulates*, and deduce from them many other properties as logical consequences.

We first assume eight postulates for addition and multiplication. These postulates hold not only for the integers, but for many other systems of numbers, such as that of all rational numbers (fractions), all real numbers (unlimited decimals), and all complex numbers. They are also satisfied by polynomials, and by continuous real functions on any given interval. When these eight postulates hold for a system R, we shall say that R is a *commutative ring*.

Definition. *Let R be a set of elements a, b, c, · · · for which the sum $a + b$ and the product ab of any two elements a and b (distinct or not) of R are defined. Then R is called a* commutative ring *if the following postulates (i)–(viii) hold:*

   (i) *Closure.   If a and b are in R, then the sum $a + b$ and the product ab are in R.*

(ii) *Uniqueness.   If $a = a'$ and $b = b'$ in R, then*

$$a + b = a' + b' \quad and \quad ab = a'b'.$$

(iii) *Commutative laws.   For all a and b in R,*

$$a + b = b + a, \quad ab = ba.$$

(iv) *Associative laws.   For all a, b, and c in R,*

$$a + (b + c) = (a + b) + c, \quad a(bc) = (ab)c.$$

(v) *Distributive law.   For all a, b, and c in R,*

$$a(b + c) = ab + ac.$$

(vi) *Zero.   R contains an element 0 such that*

$$a + 0 = a \quad for \ all \ a \ in \ R.$$

(vii) *Unity.   R contains an element $1 \neq 0$ such that*

$$a1 = a \quad for \ all \ a \ in \ R.$$

(viii) *Additive inverse.   For each a in R, the equation $a + x = 0$ has a solution x in R.*

It is a familiar fact that the set **Z** of all integers satisfies these postulates. For example, the commutative and associative laws are so familiar that they are ordinarily used without explicit mention: thus $a + b + c$ customarily denotes the equal numbers $a + (b + c)$ and $(a + b) + c$. The property of zero stated in (vi) is the characteristic property of the number zero; and similarly, the property of 1 stated in (vii) is the characteristic property of the number one. Since these laws are formally analogous, we may say that 0 and 1 are the "identity elements" for addition and multiplication, respectively. The assumption $1 \neq 0$ in (vii) is included to eliminate trivial cases (otherwise the set consisting of the integer 0 alone would be a commutative ring).

The system **Z** of all integers has another property which cannot be deduced from the preceding postulates. Namely, if $c \neq 0$ and $ca = cb$ in **Z**, then necessarily $a = b$ (partial converse of (ii)). This property is not satisfied by real functions on a given interval, for example, though these form a commutative ring. The integers therefore constitute not only a commutative ring but also an *integral domain* in the sense of the following definition.

Definition. *An* integral domain *is a commutative ring in which the following additional postulate holds:*

(ix) *Cancellation law.   If $c \neq 0$ and $ca = cb$, then $a = b$.*

The domain $\mathbb{Z}[\sqrt{2}]$. An integral domain of interest for number theory consists of all numbers of the form $a + b\sqrt{2}$, where a and b are ordinary integers (in $\mathbb{Z}$). In $\mathbb{Z}[\sqrt{2}]$, $a + b\sqrt{2} = c + d\sqrt{2}$ if and only if $a = c$, $b = d$. Addition and multiplication are defined by

$$(a + b\sqrt{2}) + (c + d\sqrt{2}) = (a + c) + (b + d)\sqrt{2}$$
$$(a + b\sqrt{2})(c + d\sqrt{2}) = (ac + 2bd) + (ad + bc)\sqrt{2}.$$

Uniqueness and commutativity are easily verified for these operations, while $0 + 0\sqrt{2}$ acts as a zero and $1 + 0\sqrt{2}$ as a unity. The additive inverse of $a + b\sqrt{2}$ is $(-a) + (-b)\sqrt{2}$. The verification of the associative and distributive laws is a little more tedious, while that of the cancellation law will be deferred to the end of §1.2.

1.2. Elementary Properties of Commutative Rings

In elementary algebra one often takes the preceding postulates and their elementary consequences for granted. This seldom leads to serious errors, provided algebraic manipulations are checked against specific examples. However, much more care must be taken when one wishes to reach reliable conclusions about whole families of algebraic systems (e.g., valid for *all* integral domains generally). One must be sure that all proofs use only postulates listed explicitly and standard rules of logic.

Among the most fundamental rules of logic are the three basic laws for equality:

Reflexive law: $a = a$.
Symmetric law: If $a = b$,   then $b = a$.
Transitive law: If $a = b$ and $b = c$,   then $a = c$, valid for all $a, b,$ and c.

We now illustrate the idea of a formal proof for several rules valid in any commutative ring R.

RULE 1. $(a + b)c = ac + bc$, for all a, b, c in R.

This rule may be called the *right distributive law*, in contrast to postulate (v), which is the left distributive law.

Proof. For all a, b, and c in R:

1. $(a + b)c = c(a + b)$ (commutative law of mult.).
2. $c(a + b) = ca + cb$ (distributive law).
3. $(a + b)c = ca + cb$ (1, 2, transitive law).
4. $ca = ac, \ cb = bc$ (commutative law of mult.).
5. $ca + cb = ac + bc$ (4, uniqueness of addn.).
6. $(a + b)c = ac + bc$ (3, 5, transitive law).

RULE 2. For all a in R, $0 + a = a$ and $1 \cdot a = a$.

Proof. For all a in R:

1. $0 + a = a + 0$ (commutative law of addn.).
2. $a + 0 = a$ (zero).
3. $0 + a = a$ (1, 2, transitive law).

The proof for $1 \cdot a = a$ is similar.

RULE 3. If z in R has the property that $a + z = a$ for all a in R, then $z = 0$.

This rule states that R contains only one element 0 which can act as the identity element for addition.

Proof. Since $a + z = a$ holds for all a, it holds if a is 0.

1. $0 + z = 0$
2. $\quad 0 = 0 + z$ (1, symmetric law).
3. $0 + z = z$ (Rule 2 when a is z).
4. $\quad 0 = z$ (2, 3, transitive law).

In subsequent proofs such as this one, we shall condense the repeated use of the symmetric and transitive laws for equality.

RULE 4. For all a, b, c in R:

$$a + b = a + c \quad \text{implies} \quad b = c.$$

This rule is called the cancellation law for addition.

Proof. By postulate (viii) there is for the element a an element x with $a + x = 0$. Then

1. $x + a = a + x = 0$ (comm. law addn., trans. law).
2. $x = x, a + b = a + c$ (reflexive law, hypothesis).
3. $x + (a + b) = x + (a + c)$ (2, uniqueness of addn.).
4. $b = 0 + b = (x + a) + b$
   $= x + (a + b) = x + (a + c)$
   $= (x + a) + c = 0 + c = c.$

(Supply the reason for each step of 4!)

RULE 5.   For each a, R contains one and only one solution x of the equation $a + x = 0.$

This solution is denoted by $x = -a$, as usual. The rule may then be quoted as $a + (-a) = 0$. As customary, the symbol $a - b$ denotes $a + (-b)$.

Proof. By postulate (viii), there is a solution x. If y is a second solution, then $a + x = 0 = a + y$ by the transitive and symmetric laws. Hence by Rule 4, $x = y$.   Q.E.D.

RULE 6.   For given a and b in R, there is one and only one x in R with $a + x = b$.

This rule asserts that subtraction is possible and unique.
Proof. Take $x = (-a) + b$. Then (give reasons!)

$$a + x = a + ((-a) + b) = (a + (-a)) + b = 0 + b = b.$$

If y is a second solution, then $a + x = b = a + y$ by the transitive law; hence $x = y$ by Rule 4.   Q.E.D.

RULE 7.   For all a in R, $a \cdot 0 = 0 = 0 \cdot a$.

Proof.

1. $a = a, a + 0 = a$ (reflexive law, postulate (vi)).
1. $a(a + 0) = aa$ (1, uniqueness of mult.).
3. $aa + a \cdot 0 = a(a + 0) = aa$ (distributive law, etc.).
   $= aa + 0$
4. $a \cdot 0 = 0$ (3, Rule 4).
5. $0 \cdot a = a \cdot 0 = 0$ (comm. law mult., 4)

RULE 8.   If u in R has the property that $au = a$ for all a in R, then $u = 1$.

This rule asserts the uniqueness of the identity element 1 for multiplication. The proof, which resembles that of Rule 3, is left as an exercise.

RULE 9.   For all a and b in R, $(-a)(-b) = ab$.

A special case of this rule is the "mysterious" law $(-1)(-1) = 1$.
Proof.   Consider the triple sum (associative law!)

1. $[ab + a(-b)] + (-a)(-b) = ab + [a(-b) + (-a)(-b)]$.

By the distributive law, the definition of $-a$, Rule 7, and (vi),

2. $ab + [a(-b) + (-a)(-b)] = ab + [a + (-a)](-b)$
$= ab + 0(-b) = ab$.

For similar reasons,

3. $[ab + a(-b)] + (-a)(-b) = a[b + (-b)] + (-a)(-b)$
$= a \cdot 0 + (-a)(-b) = (-a)(-b)$.

The result then follows from 1, 2, and 3 by the transitive and symmetric laws for equality.   Q.E.D.

Various other simple and familiar rules are consequences of our postulates; some are stated in the exercises below.

Another basic algebraic law is the one used in the solution of quadratic equations, when it is argued that $(x + 2)(x - 3) = 0$ means either that $x + 2 = 0$ or that $x - 3 = 0$. The general law involved is the assertion

(1)   if $ab = 0$,   then either $a = 0$ or $b = 0$.

This assertion is not true in all commutative rings. But the proof is immediate in any integral domain D, by the cancellation law. For suppose that the first factor a is not zero. Then $ab = 0 = a \cdot 0$, and a may be cancelled; whence $b = 0$. Conversely, the cancellation law follows from this assertion (1) in any commutative ring R, for if $a \neq 0$, $ab = ac$ means that $ab - ac = a(b - c) = 0$, which by (1) makes $b - c = 0$. We therefore have

Theorem 1.   *The cancellation law of multiplication is equivalent in a commutative ring to the assertion that a product of nonzero factors is not zero.*

Nonzero elements a and b with a product $ab = 0$ are sometimes called "divisors of zero," so that the cancellation law in a commutative ring R is equivalent to the assumption that R contains no divisors of zero.

Theorem 1 can be used to prove the cancellation law for the domain $\mathbf{Z}[\sqrt{2}]$ defined at the end of §1.1, as follows. Suppose that $\mathbf{Z}[\sqrt{2}]$ included divisors of zero, with

$$(a + b\sqrt{2})(c + d\sqrt{2}) = (ac + 2bd) + (ad + bc)\sqrt{2} = 0.$$

By definition, this gives $ac + 2bd = 0$, $ad + bc = 0$. Multiply the first by d, the second by c, and subtract; this gives $b(2d^2 - c^2) = 0$, whence either $b = 0$ or $c^2 = 2d^2$. If $b = 0$, then the two preceding equations give $ac = ad = 0$, so either $a = 0$ or $c = d = 0$ by Theorem 1. But the first alternative, $a = 0$, would imply that $a + b\sqrt{2} = 0$ (since $b = 0$); the second that $c + d\sqrt{2} = 0$—in neither case do we have divisors of zero.

There remains the possibility $c^2 = 2d^2$; this would imply $\sqrt{2} = c/d$ rational, whose impossibility will be proved in Theorem 10, §3.7.

If one admits that $\sqrt{2}$ is a real number, and that the set of all real numbers forms an integral domain R, then one can very easily prove that $\mathbf{Z}[\sqrt{2}]$ is an integral domain, by appealing to the following concept of a subdomain.

Definition. *A subdomain of an integral domain D is a subset of D which is also an integral domain, for the same operations of addition and multiplication.*

It is obvious that such a subset S is a subdomain if and only if it contains 0 and 1, with any element a its additive inverse, and with any two elements a and b their sum $a + b$ and product ab.

Exercises

In each of Exercises 1–5 give complete proofs, supporting each step by a postulate, a previous step, one of the rules established in the text, or an already established exercise.

1. Prove that the following rules hold in any integral domain:
   (a) $(a + b)(c + d) = (ac + bc) + (ad + bd)$,
   (b) $a + [b + (c + d)] = (a + b) + (c + d) = [(a + b) + c] + d$,
   (c) $a + (b + c) = (c + a) + b$,
   (d) $a(bc) = c(ab)$,
   (e) $a(b + (c + d)) = (ab + ac) + ad$,
   (f) $a(b + c)d = (ab)d + a(cd)$.

2. (a) Prove Rule 8.   (b) Prove $1 \cdot 1 = 1$,
(c) Prove that the only "idempotents" (i.e., elements x satisfying $xx = x$) in an integral domain are 0 and 1.

3. Prove that the following rules hold for $-a$ in any integral domain:
(a) $-(-a) = a$,   (b) $-0 = 0$,
(c) $-(a + b) = (-a) + (-b)$,   (d) $-a = (-1)a$,
(e) $(-a)b = a(-b) = -(ab)$.

4. Prove Rule 9 from Ex. 3(d) and the special case $(-1)(-1) = 1$.

5. Prove that the following rules hold for the operation $a - b = a + (-b)$ in any integral domain:
(a) $(a - b) + (c - d) = (a + c) - (b + d)$,
(b) $(a - b) - (c - d) = (a + d) - (b + c)$,
(c) $(a - b)(c - d) = (ac + bd) - (ad + bc)$,
(d) $a - b = c - d$ if and only if $a + d = b + c$,
(e) $(a - b)c = ac - bc$.

6. Are the following sets of real numbers integral domains? Why?
(a) all even integers,   (b) all odd integers,   (c) all positive integers,
(d) all real numbers $a + b5^{1/4}$, where a and b are integers,
(e) all real numbers $a + b9^{1/4}$, where a and b are integers,
(f) all rational numbers whose denominators are 1 or a power of 2.

7. (a) Show that the system consisting of 0 and 1 alone, with addition and multiplication defined as usual, except that $1 + 1 = 0$ (instead of 2) is an integral domain.
(b) Show that the system which consists of 0 alone, with $0 + 0 = 0 \cdot 0 = 0$, satisfies all postulates for an integral domain except for the requirement $0 \neq 1$ in (vii).

8. (a) Show that if an algebraic system S satisfies all the postulates for an integral domain except possibly for the requirement $0 \neq 1$ in (vii), then S is either an integral domain or the system consisting of 0 alone, as described in Ex. 7(b).
(b) Is $0 \neq 1$ used in proving Rules 1–9?

9. Suppose that the sum of any two integers is defined as usual, but that the product of any two integers is defined to be zero. With this interpretation, which ones among the postulates for an integral domain are still satisfied?

10. Find two functions $f \not\equiv 0$ and $g \not\equiv 0$ such that $fg \equiv 0$.

1.3. Properties of Ordered Domains

Because the ring $\mathbf{Z}$ of all ordinary integers plays a unique role in mathematics, one should be aware of its special properties, of which the commutative and cancellation laws of multiplication are only two. Many other properties stem from the possibility of listing the integers in the usual order

$$\cdots -4, -3, -2, -1, 0, 1, 2, 3, 4, \cdots .$$

This order is customarily expressed in terms of the *relation* $a < b$, where the assertion $a < b$ (a is less than b) is taken to mean that the integer a stands to the left of the integer b in the list above. But the relation $a < b$ holds if and only if the difference $b - a$ is a positive integer. Consequently, every property of the relation $a < b$ can be derived from properties of the set of *positive* integers. We assume then as postulates the following three properties of the set of positive integers $1, 2, 3, \cdots$.

Addition: The sum of two positive integers is positive.

Multiplication: The product of two positive integers is positive.

Law of trichotomy: For a given integer a, one and only one of the following alternatives holds: either a is positive, or $a = 0$, or $-a$ is positive.

Incidentally, these properties are shared by the positive rational numbers and the positive real numbers; hence all the consequences of these properties are also shared. It is convenient to call an integral domain containing positive elements with these properties an *ordered domain*.

Definition. *An integral domain D is said to be* ordered *if there are certain elements of D, called the positive elements, which satisfy the addition, multiplication, and trichotomy laws stated above for integers.*

Theorem 2. *In any ordered domain, all squares of nonzero elements are positive.*

Proof. Let a^2 be given, with $a \neq 0$. By the law of trichotomy, either a or $-a$ is positive. In the first case, a^2 is positive by the multiplication law for positive elements; in the second, $-a$ is positive, and so $a^2 = (-a)^2 > 0$ by Rule 9 of §1.2.   Q.E.D.

It is a corollary that $1 = 1^2$ is always positive.

Definition. *In an ordered domain, the two equivalent statements $a < b$ (read "a is less than b") and $b > a$ ("b is greater than a") both mean that $b - a$ is positive. Also $a \leqq b$ means that either $a < b$ or $a = b$.*

According to this definition, the positive elements a can now be described as the elements a greater than zero. Elements $b < 0$ are called *negative*. One can deduce a number of familiar properties of the relation "less than" from its definition above.

Transitive law: If $a < b$ and $b < c$, then $a < c$.

Proof. By definition, the hypotheses $a < b$ and $b < c$ mean that $b - a$ and $c - b$ are positive. Hence by the addition principle, the sum $(b - a) + (c - b) = c - a$ is positive, which means that $a < c$.

The three basic postulates for positive elements are reflected by three corresponding properties of inequalities:

Addition to an inequality: If $a < b$, then $a + c < b + c$.

Multiplication of an inequality: If $a < b$ and $0 < c$, then $ac < bc$.

Law of trichotomy: For any a and b, one and only one of the relations $a < b$, $a = b$, or $a > b$ holds.

As an example, we prove the principle that an inequality may be multiplied by a positive number c. The conclusion requires us to prove that $bc - ac = (b - a)c$ is positive (cf. Ex. 5(e) of §1.2). But this is an immediate consequence of the multiplication postulate, for the factors $b - a$ and c are both positive by hypothesis. By a similar argument one may demonstrate that the multiplication of an inequality by a negative number inverts the sense of the inequality (see Ex. 1(c) below).

Definition. *In an ordered domain, the* absolute value $|a|$ *of a number is* 0 *if a is* 0, *and otherwise is the positive member of the couple a, $-a$.*

This definition might be restated as

(2) $|a| = +a$ if $a \geq 0$; $|a| = -a$ if $a < 0$.

By appropriate separate consideration of these two cases, one may prove the laws for absolute values of sums and products,

(3) $|ab| = |a|\,|b|$, $|a + b| \leq |a| + |b|$.

The sum law may also be obtained thus: by the definition, we have

$-|a| \leq a \leq |a|$ and $-|b| \leq b \leq |b|$; hence adding inequalities gives

$$-(|a| + |b|) \leq a + b \leq |a| + |b|.$$

This indicates at once that, whether $a + b$ is positive or negative, its absolute value cannot exceed $|a| + |b|$.

Exercises

1. Deduce from the postulates for an ordered domain the following rules:
    (a) if $a < b$, then $a + c < b + c$, and conversely,
    (b) $a - x < a - y$ if and only if $x > y$,

    (c) if $a < 0$, then $ax > ay$ if and only if $x < y$,
    (d) $0 < c$ and $ac < bc$ imply $a < b$,
    (e) $x + x + x + x = 0$ implies $x = 0$,
    (f) $a < b$ implies $a^3 < b^3$,
    (g) if $c \geq 0$, then $a \geq b$ implies $ac \geq bc$.
  2. Prove that the equation $x^2 + 1 = 0$ has no solution in an ordered domain.
  3. Prove as many laws on the relation $a \leq b$ as you can.
  4. Prove that $||a| - |b|| \leq |a - b|$ in any ordered domain.
 ★5. Prove that $a^7 = b^7$ implies $a = b$ in any ordered domain.
 ★6. In any ordered domain, show that $a^2 - ab + b^2 \geq 0$ for all a, b.
 ★7. Define "positive" element in the domain $\mathbf{Z}[\sqrt{2}]$, and show that the addition, multiplication, and trichotomy laws hold.
 ★8. Let D be an integral domain in which there is defined a relation $a < b$ which satisfies the transitive law, the principles for addition and multiplication of inequalities, and the law of trichotomy stated in the text. Prove that if a set of "positive" elements is suitably chosen, D is an ordered domain.
 ★9. Prove in detail that any subdomain of an ordered domain is an ordered domain.
★10. Let R be any commutative ring which contains a subset of "positive" elements satisfying the addition, multiplication, and trichotomy laws. Prove that R is an ordered domain. (*Hint*: Show that the cancellation law of multiplication holds, by considering separately the four cases $x > 0$ and $y > 0$, $x > 0$ and $-y > 0$, $-x > 0$ and $y > 0$, $-x > 0$ and $-y > 0$.)

1.4. Well-Ordering Principle

    A subset S of an ordered domain (such as the real number system) is called *well-ordered* if each nonempty subset of S contains a smallest member. In terms of this concept, one can formulate an important property of the integers, not characteristically algebraic and not shared by other number systems. This is the

    Well-ordering principle.   The positive integers are well-ordered.

    In other words, any nonempty collection C of positive integers must contain some smallest member m, such that whenever c is in C, $m \leq c$. For instance, the least positive even integer is 2.
    To illustrate the force of this principle, we prove

    Theorem 3. *There is no integer between* 0 *and* 1.

    This is immediately clear by a glance at the natural order of the integers, but we wish to show that this fact can also be *proved* from our

 * Here and subsequently exercises of greater difficulty are starred.

assumptions without "looking" at the integers. We give an indirect proof. If there is any integer c with $0 < c < 1$, then the set of all such integers is nonempty. By the well-ordering principle, there is a least integer m in this set, and $0 < m < 1$. If we multiply both sides of these inequalities by the positive number m, we have $0 < m^2 < m$. Thus m^2 is another integer in the set C, smaller than the supposedly minimum element m of C. This contradiction establishes Theorem 3.

Theorem 4. *A set S of positive integers which includes* 1, *and which includes* $n + 1$ *whenever it includes* n, *includes every positive integer.*

Proof. It is enough to show that the set S', consisting of those positive integers *not* included in S, is empty. Suppose S' were not empty; it would have to contain a least element m. But $m \neq 1$ by hypothesis; hence by Theorem 3, $m > 1$, and so $m - 1$ would be positive. But since $1 > 0$, $m - 1 < m$; hence by the choice of m, $m - 1$ would be in S. It follows by hypothesis that $(m - 1) + 1 = m$ would be in S. This contradiction establishes the theorem.

Exercises

1. Show that for any integer a, $a - 1$ is the greatest integer less than a.
2. Which of the following sets are well-ordered:
   (a) all odd positive integers,   (b) all even negative integers,
   (c) all integers greater than -7,   (d) all odd integers greater than 249?
3. Prove that any subset of a well-ordered set is well-ordered.
4. Prove that a set of integers which contains -1000, and contains $x + 1$ when it contains x, contains all the positive integers.
5. (a) A set S of integers is said to have the integer b as "lower bound" if $b \leq x$ for all x in S; b itself need not be in S. Show that any nonempty set S of integers having a lower bound has a least element.
   (b) Show that any nonempty set of integers having an "upper bound" has a greatest element.

1.5. Finite Induction; Laws of Exponents

We have now formulated a complete list of basic properties for the integers in terms of addition, multiplication, and order. Henceforth we assume that *the integers form an ordered integral domain* $\mathbf{Z}$ *in which the positive elements are well-ordered.* Every other mathematical property of the integers can be proved, by strictly logical processes, from those assumed. In particular, we can deduce the extremely important

Principle of Finite Induction. Let there be associated with each positive integer n a proposition $P(n)$ which is either true or false. If, first, $P(1)$ is true and, second, for all k, $P(k)$ implies $P(k + 1)$, then $P(n)$ is true for all positive integers n.

To deduce this principle from the well-ordering assumption, simply observe that the set of those positive integers k for which $P(k)$ is true satisfies the hypotheses and hence the conclusion of Theorem 4.

The method of proof by induction will now be used to prove various laws valid in any commutative ring. We first use it to establish formally the general distributive law for any number n of summands,

(4) $$a(b_1 + b_2 + \cdots + b_n) = ab_1 + ab_2 + \cdots + ab_n.$$

To be explicit, we define the repeated sum $b_1 + \cdots + b_n$ as follows:

$$b_1 + b_2 + b_3 = (b_1 + b_2) + b_3,$$
$$b_1 + b_2 + b_3 + b_4 = [(b_1 + b_2) + b_3] + b_4.$$

This convention can be stated in general as a recursive formula (for $k \geq 1$)

(5) $$b_1 + \cdots + b_k + b_{k+1} = (b_1 + \cdots + b_k) + b_{k+1},$$

which determines the arrangement of parentheses in $k + 1$ terms, given this arrangement for k terms.

The inductive proof of (4) requires first the proof for $n = 1$, which is immediate. Secondly, we assume the law (4) for $n = k$ and try to prove it for $n = k + 1$. By the definition (5) and the simple distributive law (v),

$$a(b_1 + \cdots + b_{k+1}) = a\,[(b_1 + \cdots + b_k) + b_{k+1}]$$
$$= a(b_1 + \cdots + b_k) + ab_{k+1}.$$

On the right, the first term can now be reduced by the assumed case of (4) for k summands, as

$$a(b_1 + \cdots + b_{k+1}) = (ab_1 + \cdots + ab_k) + ab_{k+1}.$$

Since the right-hand side is $ab_1 + \cdots + ab_{k+1}$, by the definition (5), we have completed the inductive proof of (4).

Similar but more complicated inductive arguments will yield the *general associative law*, which asserts that a sum $b_1 + \cdots + b_k$ or a product $b_1 \cdots b_k$ has the same value for any arrangement of parentheses (a special case appears in Ex. 9 below). Using this result and (4), one can then also establish the two-sided general distributive law

$$(a_1 + \cdots + a_m)(b_1 + \cdots + b_n)$$
$$= a_1 b_1 + \cdots + a_1 b_n + \cdots + a_m b_1 + \cdots + a_m b_n.$$

Note also the general associative and commutative law, according to which the sum of k given terms always has the same value, whatever the order or the grouping of the terms.

Positive integral exponents in any commutative ring **R** may also be treated by induction. If n is a positive integer, the power a^n stands for the product $a \cdot a \cdots a$, to n factors. This can also be stated as a "recursive" definition

(6) $a^1 = a,$ $a^{n+1} = a^n \cdot a$ (any a in **R**),

which makes it possible to compute any power a^{n+1} in terms of an already computed lower power a^n. From these definitions one may prove the usual laws, for any positive integral exponents m and n, as follows:

(7) $a^m a^n = a^{m+n},$

(8) $(a^m)^n = a^{mn},$ $(ab)^m = a^m b^m.$

For instance, the first law may be proved by induction on n. If $n = 1$, the law becomes $a^m \cdot a = a^{m+1}$, which is exactly the definition of a^{m+1}. Next assume that the law (7) is true for every m and for a given positive integer $n = k$, and consider the analogous expression $a^m a^{k+1}$ for the next larger exponent $k + 1$. One finds

$$a^m a^{k+1} = a^m(a^k a) = (a^m a^k)a = a^{m+k} a = a^{(m+k)+1} = a^{m+(k+1)},$$

by successive applications of the definition, the associative law, the induction assumption, and the definition again. This gives the law (7) for the case $n = k + 1$, and so completes the induction.

Finally, the binomial formula can be proved over any commutative ring R, as follows. First define the factorial function $n!$ on the nonnegative integers by recursion: $0! = 1$ and $(n + 1)! = (n!)(n + 1)$. Then define the binomial coefficients similarly for $n \geq 0$ in **Z** by

$$\binom{n}{0} = \binom{n}{n} = 1 \quad \text{and} \quad \binom{n+1}{k} = \binom{n}{k-1} + \binom{n}{k}$$

From these definitions it follows by induction on n that

(9) $(x + y)^n = x^n + nx^{n-1}y + \cdots + \binom{n}{k}x^{n-k}y^k + \cdots + y^n$

$$= \sum_{k=0}^{n} \binom{n}{k}x^{n-k}y^k$$

and that

(10) $(k!)(n-k)!\binom{n}{k} = n!$

(I.e., $\binom{n}{k} = (n!)/(k!)(n-k)!$ We leave the proof as an exercise.)

The Principle of Finite Induction permits one to assume the truth of $P(n)$ *gratis* in proving $P(n + 1)$. We shall now show that one can even assume the truth of $P(k)$ for all $k \leq n$. This is called the

Second Principle of Finite Induction. Let there be associated with each positive integer n a proposition $P(n)$. If, for each m, the assumption that $P(k)$ is true for all $k < m$ implies the conclusion that $P(m)$ is itself true, then $P(n)$ is true for all n.

Proof. Let S be the set of integers for which $P(n)$ is false. Unless S is empty, it will have a first member m. By choice of m, $P(k)$ will be true for all $k < m$; hence by hypothesis, $P(m)$ must itself be true, giving a contradiction. The only way out is to admit that S is empty.   Q.E.D.

Caution: In case $m = 1$, the set of all $k < 1$ is empty, so that one must implicitly include a proof of $P(1)$.

Exercises

1. Prove by induction that the following laws for positive exponents are valid in any integral domain:
   (a) $(a^m)^n = a^{mn}$,   (b) $(ab)^n = a^n b^n$,   (c) $1^n = 1$.
2. Prove by induction that $1 + 2 + \cdots + n = n(n + 1)/2$.
3. Prove formulas (9) and (10).
4. Prove by induction that $x_1^2 + \cdots + x_n^2 > 0$ unless $x_1 = \cdots = x_n = 0$.
5. Prove by induction the following summation formulas:
   (a) $1 + 4 + 9 + \cdots + n^2 = n(n + 1)(2n + 1)/6$,
   (b) $1 + 8 + 27 + \cdots + n^3 = [n(n + 1)/2]^2$.
6. In any ordered domain, show that every odd power of a negative element is negative.
7. Using induction, but not the well-ordering principle, prove Theorem 3. (*Hint:* Let $P(n)$ mean $n \geq 1$.)

★**8.** Using Ex. 7, prove the well-ordering principle from the Principle of Finite Induction. (*Hint:* Let $P(n)$ be the proposition that any class of positive integers containing a number $\leq n$ has a least member.)

9. Using the definition (5), prove the following associative law:

$$(a_1 + \cdots + a_m) + (b_1 + \cdots + b_n) = a_1 + \cdots + a_m + b_1 + \cdots + b_n.$$

10. Obtain a formula for the nth derivative of the product of two functions and prove the formula by induction on n.

★**11.** Prove that to any base $a > 1$, each positive integer m has a unique expression of the form

$$a^n r_n + a^{n-1} r_{n-1} + \cdots + a^2 r_2 + a r_1 + r_0,$$

where the integers r_k satisfy $0 \leq r_k < a$, $r_n \neq 0$.

★**12.** Illustrate Ex. 11 by converting the equation $63 \cdot 111 = 6993$ to the base 7, checking by multiplying out.

13. A druggist has only the five weights of 1, 3, 9, 27, and 81 ounces and a two-pan balance (weights may be placed in either pan). Show that he can weigh any amount up to 121 ounces.

14. Prove that the sum of the digits of any multiple of 9 is itself divisible by 9.

1.6. Divisibility

An equation $ax = b$ with integral coefficients does not always have an integral solution x. If there is an integral solution, b is said to be *divisible by a*; the investigation of this situation is the first problem of number theory.

An analogous concept of divisibility arises in every integral domain; it is defined as follows.

Definition. *In an integral domain D, an element b is divisible by an element a when $b = aq$ for some q in D. When b is divisible by a, we write $a \mid b$; we also call a a* factor *or* divisor *of b, and b a* multiple *of a. The divisors of 1 in D are called* units *or* invertibles *of D.*

Like the equality relation $a = b$, the relation $a \mid b$ is reflexive and transitive:

(11) $a \mid a$; $a \mid b$ and $b \mid c$ imply $a \mid c$.

The first law of (11) is trivial, since $a = a \cdot 1$ implies that $a \mid a$. To prove the second, recall that the hypotheses $a \mid b$ and $b \mid c$ are defined to mean

$b = ad_1$ and $c = bd_2$. for some integers d_1 and d_2. Substitution of the first equation in the second gives $c = a(d_1 d_2)$. Since $d_1 d_2$ is in D, this states according to the definition that $a \mid c$, as asserted in the conclusion of (11).

Theorem 5. *The only units of* **Z** *are* ± 1.

This theorem asserts, in effect, that for integers a and b, $ab = 1$ implies $a = \pm 1$ and $b = \pm 1$. But according to the rules for the absolute value of a product, $ab = 1$ gives $|ab| = |a| \cdot |b| = 1$. Since neither a nor b is zero, $|a|$ and $|b|$ are positive numbers. There are no positive integers between 0 and 1 (Theorem 3), so by the law of trichotomy $|a| \geqq 1$ and $|b| \geqq 1$. If either inequality held, the product $|a| \cdot |b|$ could not be 1. Therefore $|a| = |b| = 1$, so that $a = \pm 1$, $b = \pm 1$, as asserted.

Corollary. *If the integers a and b divide each other ($a \mid b$ and $b \mid a$), then $a = \pm b$.*

Proof. By hypothesis $a = bd_1$ and $b = ad_2$; hence $a = ad_2 d_1$. If $a = 0$, then $b = 0$, too. If $a \neq 0$, cancellation yields $1 = d_2 d_1$. Then $d_1 = \pm 1$ by the theorem, and hence again $a = \pm b$. Q.E.D.

Since $a = a \cdot 1 = (-a)(-1)$, any integer a is divisible by a, $-a$, $+1$, and -1.

Definition. *An integer p is a* prime *if p is not 0 or ± 1 and if p is divisible only by ± 1 and $\pm p$.*

The first few positive primes are

$$2, 3, 5, 7, 11, 13, 17, 19, 23, 29, 31.$$

Any positive integer which is not one or a prime can be factored into prime factors; thus

$$128 = 2^7; \quad 90 = 9 \cdot 10 = 3^2 \cdot 2 \cdot 5;$$
$$672 = 7 \cdot 96 = 7 \cdot 12 \cdot 8 = 7 \cdot 3 \cdot 2^5.$$

It is a matter of experience that we always get the same prime factors no matter how we proceed to obtain them. This uniqueness of the prime factorization can be proved by studying greatest common divisors, which we now do.

Exercises

1. Prove the following properties of units in any domain:
   (a) the product of two units is a unit,
   (b) a unit u of D divides every element of D,
   (c) if c divides every x in D, c is a unit.
2. Prove that if $a \mid b$ and $a \mid c$, then $a \mid (b + c)$.
3. Prove: If $b > 1$ is not prime, it has a positive prime divisor $d \leq \sqrt{b}$.
4. List all positive primes less than 100. (*Hint:* Throw away multiples of 2, 3, 5, 7, and use Ex. 3.)
5. If $a \mid b$, prove that $|a| \leq |b|$ when $b \neq 0$.

1.7. The Euclidean Algorithm

The ordinary process of dividing an integer a by b yields a quotient q and a remainder r. Formally, this amounts to the following assertion.

Division Algorithm. *For given integers a and b, with b > 0, there exist integers q and r such that*

(12) $$a = bq + r, \qquad 0 \leq r < b.$$

Geometric picture.   If we imagine the whole numbers displayed on the real axis, the possible multiples bq of b form a set of equally spaced division points on the line

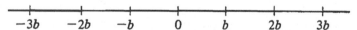

$$\begin{array}{ccccccc} -3b & -2b & -b & 0 & b & 2b & 3b \end{array}$$

The point representing a must fall in one of the intervals determined by these points, say in the interval between bq and $b(q + 1)$, exclusive of the right-hand end point. This means that $a - bq = r$, where r represents a length shorter than the whole length b of an interval. Hence $0 \leq r < b$, as asserted. This picture suggests the following proof based on our postulates.

Proof.   There certainly is some integral multiple of b not exceeding a; for instance, since $b > 0$, $b \geq 1$ by Theorem 3, so $(-|a|)b \leq -|a| \leq a$. Therefore the set of differences $a - bx$ contains at least one nonnegative integer, namely, $a - (-|a|)b$. Hence, by the well-ordering postulate, there is a *least* nonnegative $a - bx$, say $a - bq = r$. By construction, $r \geq 0$; while if $r \geq b$, then $a - b(q + 1) = r - b \geq 0$ would be less than $a - bq$, contrary to our choice of q. We conclude that $0 \leq r < b$, while $a = bq + (a - bq) = bq + r$.

Corollary 1. *For given integers a and b, the quotient q and the remainder r which satisfy (12) are uniquely determined.*

Proof. Suppose that $a = bq + r = bq' + r'$, where $0 \leq r < b$, $0 \leq r' < b$. Then $r - r' = b(q' - q)$ is numerically smaller than b, but is a multiple of b. It follows that $r - r'$ must be zero. Hence $r = r'$, $bq = bq'$, $q = q'$, which gives the uniqueness of q and r. Q.E.D.

Frequently, we have occasion to deal not with individual integers but with certain sets of integers, such as the set $\cdots -6, -3, 0, 3, 6, 9, \cdots$ which consists of all multiples of 3. This set has the important property that the sum or the difference of any two integers in the set is again an integer in the set. In general, a set S of integers is said to be *closed* under addition and subtraction if S contains the sum $a + b$ and the difference $a - b$ of any two integers a and b in S. All the even integers (positive, negative, and zero) form such a set. More generally, the set of all multiples xm of any fixed integer m is closed under addition and subtraction, for $xm \pm ym = (x \pm y)m$ is a multiple of m. We now prove that such sets of multiples are the only sets of integers with these properties.

Theorem 6. *Any nonvoid set of integers closed under addition and subtraction either consists of zero alone or else contains a least positive element and consists of all the multiples of this integer.*

Proof. Let such a set S contain an element $a \neq 0$. Then S contains the difference $a - a = 0$, and hence the difference $0 - a = -a$. Consequently, there is at least one positive element $|a| = \pm a$ in S. The well-ordering principle will provide a least positive element b in S.

The set S must contain all integral multiples of b. For one may first show by induction on n that any positive multiple nb is in S: if $n = 1$, b is in S; if kb is already known to lie in S, then $(k + 1)b = kb + b$ is a sum of two elements of S, hence is in S. Therefore, any negative multiple $(-n)b = 0 - (nb)$ is a difference of two elements of S, hence is in S.

The set S can contain nothing but the integral multiples of b. For if a is any element of S, the Division Algorithm may be applied to give a difference $a - bq = r$, which is also in S. The remainder r is nonnegative and less than b, while b is the smallest positive element in S. Therefore $r = 0$, and $a = bq$ is a multiple of b, as asserted. Q.E.D.

Definition. *An integer d is a greatest common divisor (g.c.d.) of the integers a and b if d is a common divisor of a and b which is a multiple of every other common divisor. In symbols, d must have the properties*

$$d \mid a; \quad d \mid b; \quad c \mid a \text{ and } c \mid b \text{ imply } c \mid d.$$

For example, both 3 and -3 are greatest common divisors of 6 and 9. According to the definition two different g.c.d.'s must divide each other, hence differ only in sign. Of the two possible g.c.d.'s $\pm d$ for a and b, the positive one is often denoted by the symbol (a, b). Note that the adjective "greatest" in the definition of a g.c.d. means not primarily that d has a greater *magnitude* than any other common divisor c, but that d is a *multiple* of any such c.

Theorem 7. *Any two integers $a \neq 0$ and $b \neq 0$ have a positive greatest common divisor (a, b). It can be expressed as a "linear combination" of a and b, with integral coefficients s and t, in the form*

(13) $(a, b) = sa + tb.$

Proof.   Consider the numbers of the form $sa + tb$. For any two such

$$(s_1 a + t_1 b) \pm (s_2 a + t_2 b) = (s_1 \pm s_2)a + (t_1 \pm t_2)b.$$

Therefore the set S of all integers $sa + tb$ is closed under addition and subtraction, so by Theorem 6 consists of all multiples of some minimum positive number $d = sa + tb$. From this formula it is clear that any factor of ca and b must be a factor of d. On the other hand, the original integers $a = 1 \cdot a + 0 \cdot b$ and $b = 0 \cdot a + 1 \cdot b$ both lie in the set S under consideration, and hence must be multiples of the minimum number d in this set. In other words, d is a common divisor. Hence it is the desired greatest common divisor.   Q.E.D.

Similarly, the set M of common *multiples* of a and b is closed under addition and subtraction. Its least positive member m will be a common multiple of a and b dividing every common multiple. Thus m is a "least common multiple" (or l.c.m.)..

Theorem 8. *Any two integers a and b have a least common multiple $m = [a, b]$ which is a divisor of every common multiple and which itself is a common multiple.*

To find explicitly the g.c.d. of two integers a and b, one may use the so-called *Euclidean algorithm.* We may suppose that a and b are both positive, since a negative integer b could be replaced by $-b$ without altering the g.c.d. $(a, b) = (a, -b)$. The Division Algorithm gives

(14) $a = bq_1 + r_1, \qquad 0 \leqq r_1 < b.$

Every integer which divides the terms a and b must divide the remainder r_1; conversely, every common divisor of b and r_1 is a divisor of a in (14).

Therefore the common divisors of a and b are the same as the common divisors of b and r_1, so the g.c.d.'s (a, b) and (b, r_1) are identical. This reduction can be repeated on b and r_1:

(15)
$$b = r_1 q_2 + r_2. \qquad 0 < r_2 < r_1;$$
$$r_1 = r_2 q_3 + r_3, \qquad 0 < r_3 < r_2;$$
$$\vdots \qquad\qquad \vdots$$
$$r_{n-2} = r_{n-1} q_n + r_n, \qquad 0 < r_n < r_{n-1};$$
$$r_{n-1} = r_n q_{n+1}.$$

Since the remainders continually decrease, there must ultimately† be a remainder r_{n+1} which is zero, as we have indicated in the last equation. The argument above shows that the desired greatest common divisor is

$$(a, b) = (b, r_1) = (r_1, r_2) = \cdots = (r_{n-1}, r_n).$$

But the last equation of (15) shows that r_n is itself a divisor of r_{n-1}, so that the last g.c.d. is just r_n itself. The g.c.d. of the given integers a and b is thus the last nonzero remainder r_n in the Euclidean algorithm (14) and (15).

The algorithm can also be used to represent the g.c.d. explicitly as a linear combination $sa + tb$. This can be done by expressing the successive remainders r_i in terms of a and b, as

$$r_1 = a - bq_1 = a + (-q_1)b,$$
$$r_2 = b - q_2 r_1 = (-q_2)a + (1 + q_1 q_2)b.$$

The form of these equations indicates that one would eventually obtain r_n as a linear combination of a and b with integral coefficients s and t which involve the quotients q_i.

The expression $(a, b) = sa + tb$ for the g.c.d. is of the greatest utility. One important consequence is the fact that a prime which divides a product of two numbers must always divide at least one of the factors:

Theorem 9. *If p is a prime, then $p \mid ab$ implies $p \mid a$ or $p \mid b$.*

Proof. By the definition of a prime, the only factors of p are ± 1 and $\pm p$. If the conclusion $p \mid a$ is false, the only common divisors of p and a are ± 1, so that 1 is a g.c.d. of a and p and can thus be expressed in the

† Why? Does a proof of this involve the well-ordering principle?

form $1 = sa + tp$. On multiplying through by b, we have

$$b = sab + tbp.$$

Both terms on the right are divisible by p, hence the left side b is divisible by p, as in the second alternative in the theorem. Q.E.D.

If $(a, b) = 1$, we call a and b *relatively prime*. In other words, two integers a and b are relatively prime if they have no common divisors except ± 1. The argument used to prove Theorem 9 will also prove the following generalization:

Theorem 10. *If $(c, a) = 1$ and $c \mid ab$, then $c \mid b$.*

One consequence may be drawn for an integer m which is a multiple of each of two relatively prime integers a and c. Such an m has the form $m = ad$ and is divisible by c, so by this theorem $c \mid d$, and $m = ad = a(cd')$. Therefore the product ac divides m. This argument proves

Theorem 11. *If $(a, c) = 1$, $a \mid m$, and $c \mid m$, then $ac \mid m$.*

Exercises

1. Use the Euclidean algorithm to find the g.c.d. of
   (a) $(14, 35)$,   (b) $(11, 15)$,   (c) $(180, 252)$,
   (d) $(2873, 6643)$,   (e) $(4148, 7684)$,   (f) $(1001, 7655)$.
2. Write (x, y) in the form $sx + ty$ (s, t integers) in Ex. 1(a)–(c).
3. Prove that $(0, a) = |a|$ for any integer a.
4. If $a > 0$, prove that $(ab, ac) = a(b, c)$.
5. Show that $b \mid c$ and $|c| < b$ imply $c = 0$. (This fact is used in proving Corollary 1.)
6. (a) Prove that any three integers a, b, c have a g.c.d. which can be expressed in the form $sa + tb + uc$.
   (b) Prove that $((a, b), c) = (a, (b, c)) = ((a, c), b)$.
7. Discuss Exs. 3–5 and 6(b) for the case of l.c.m.
8. Show that a set of integers closed under subtraction is necessarily also closed under addition.
9. Show that a set of integers closed under addition alone need not consist of all multiples of one fixed element.
10. In the Euclidean algorithm, show by induction on k that each remainder can be expressed in the form $r_k = s_k a + t_k b$, where s_k and t_k are integers.
11. Give a detailed proof of Theorem 10.
★12. Show that for any positive integers a, b the set of all $ma + nb$ (m, n positive integers) includes all multiples of (a, b) larger than ab.

13. If q is an integer such that for all integers a and b, $q \mid ab$ implies $q \mid a$ or $q \mid b$, prove that q is 0, ±1, or a prime (cf. Theorem 9).
14. (a) Prove that if $(a, m) = (b, m) = 1$, then $(ab, m) = 1$.
    (b) Prove that if $(a, c) = d$, $a \mid b$, and $c \mid b$, then $ac \mid bd$.
    (c) Prove that $[a, c] = ac/(a, c)$.

1.8. Fundamental Theorem of Arithmetic

It is now easy to prove the unique factorization theorem for integers, also called the fundamental theorem of arithmetic.

Theorem 12. *Any integer not zero can be expressed as a unit (±1) times a product of positive primes. This expression is unique except for the order in which the prime factors occur.*

Proof. That any integer a can be written as such a product may be proved by successively breaking a up into smaller factors. This process involves the second principle of finite induction and can be described as follows. It clearly suffices to consider only positive integers a.

Let $P(a)$ be the proposition that a can be factored as in Theorem 12. If $a = 1$ or if a is a prime, then $P(a)$ is trivially true. On the other hand, if a is composite, then it has a positive divisor b which is neither 1 nor a, so that $a = bc$, with $b < a$, $c < a$. But by the second induction principle, we can assume $P(b)$ and $P(c)$ to be true, so that b and c can be expressed as products of primes:

$$b = p_1 p_2 \cdots p_r, \qquad c = q_1 q_2 \cdots q_s,$$

yielding for a the composite expression

$$a = bc = p_1 p_2 \cdots p_r q_1 q_2 \cdots q_s,$$

which is of the desired form.

To prove the uniqueness, we have to consider two possible prime factorizations of an integer a,

$$a = (\pm 1) p_1 p_2 \cdots p_m = (\pm 1) q_1 q_2 \cdots q_n.$$

Since the primes p_i and q_j are all positive, the terms ±1 in the two decompositions must agree. The prime p_1 in the first factorization is a divisor of the product $a = \pm q_1 \cdots q_n$, so that repeated application of Theorem 9 insures that p_1 must divide at least one factor q_j of this product. Since $p_1 \mid q_j$ and both are positive primes, $p_1 = q_j$. Rearrange the factorization $q_1 q_2 \cdots q_i$ so that q_j appears first, then cancel p_1 against q_j, leaving

$$p_2 p_3 \cdots p_m = q_2' q_3' \cdots q_n',$$

where the accents denote the q's in their new order. Continue this process until no primes are left one one side of the resulting equation. There can then be no primes left on the other side, so that in the original factorization, $m = n$. We have caused the two factorizations to agree simply by rearranging the primes in the second factorization, as asserted in our uniqueness theorem.   Q.E.D.

In the factorization of a number the same prime p may occur several times. Collecting these occurrences, we may write the decomposition as

$$(16) \qquad a = \pm p_1^{e_1} p_2^{e_2} \cdots p_k^{e_k} \qquad (1 < p_1 < p_2 < \cdots < p_k).$$

Here our uniqueness theorem asserts that the exponent e_i to which each prime p_i occurs is uniquely determined by the given number a.

Exercises

1. Describe a systematic process for finding the g.c.d. and the l.c.m. of two integers whose prime-power decompositions (16) are known, illustrating with $a = 216$, $b = 360$, and $a = 144$, $b = 625$. (*Hint:* It is helpful to use "dummy" zero exponents for primes dividing one but not both of a or b.)

2. If $V_p(a)$ denotes the exponent of the highest power of the prime p dividing the nonzero integer a, prove the formulas
   (i) $V_p(a + b) \geqq \min\{V_p(a), V_p(b)\}$;
   (ii) $V_p((a, b)) = \min\{V_p(a), V_p(b)\}$;
   (iii) $V_p(ab) = V_p(a) + V_p(b)$;
   (iv) $V_p([a, b]) = \max\{V_p(a), V_p(b)\}$.

3. If $\|a\| = 2^{-V_p(a)}$, for V_p as in Ex. 2, prove that

$$\|ab\| = \|a\| \cdot \|b\| \qquad \text{and} \qquad \|a + b\| \leqq \max(\|a\|, \|b\|).$$

★4. Let $V(a)$ be a nonnegative function with integral values, defined for all nonzero integers a and having properties (i) and (iii) of Ex. 2. Prove that $V(a)$ is either identically 0 or a constant multiple of one of the functions $V_p(a)$ of Ex. 2. (*Hint:* First locate some p with $V(p) > 0$.)

5. Using the formulas of Ex. 2, show that for any positive integers a and b, $ab = (a, b)[a, b]$. (For a second proof, cf. Ex. 14(c), §1.7.)

6. Prove that the number of primes is infinite (Euclid). (*Hint:* If $p_1, \cdots, p_n$ are n primes, then the integer $p_1 p_2 \cdots p_n + 1$ is divisible by none of these primes.)

★7. Define the function $e(n)$ (n any positive integer) as the g.c.d. of the exponents occurring in the prime factorization of n. Prove that (a) for given r and n in **Z**, there is an integer x such that $x^r = n$ if and only if $r \mid e(n)$; (b) $e(n^r) = r \cdot e(n)$; (c) if $e(m) = e(n) = d$, then $d \mid e(mn)$.

8. If a product mn of positive integers is a square and if $(m, n) = 1$, show that both m and n are squares.

★9. The possible right trangles with sides measured by integers x, y, and z may be found as follows. Assume that x, y, and z have no common factors except ± 1.
   (a) If $x^2 + y^2 = z^2$, show that x and y cannot both be odd.
   (b) If y is even, apply Ex. 8, to show that $y = 2mn$, where m and n are integers with $x = m^2 - n^2$, $z = m^2 + n^2$. (*Hint:* Factor $z^2 - x^2$, and show $(z + x, z - x) = 2$.)

1.9. Congruences

In giving the time of day, it is customary to count only up to 12, and then to begin over again. This simple idea of throwing away the multiples of a fixed number 12 is the basis of the arithmetical notion of congruence. We call two integers *congruent* "modulo 12" if they differ only by an integral multiple of 12. For instance, 7 and 19 are so congruent, and we write $7 \equiv 19 \pmod{12}$.

Definition. $a \equiv b \pmod m$ *holds if and only if* $m \mid (a - b)$.

One might equally well say that $a \equiv b \pmod m$ means that the difference $a - b$ lies in the set of all multiples of m. There is still another alternative definition, based on the fact that each integer a on division by m leaves a unique remainder (Corollary 1 of §1.7). This alternative we state as follows:

Theorem 13. *Two integers a and b are congruent modulo m if and only if they leave the same remainder when divided by* $|m|$.

Since $a \equiv b \pmod m$ if and only if $a \equiv b \pmod{-m}$, it will suffice to prove this result for the case $m > 0$.

Proof. Suppose first that $a \equiv b \pmod m$ according to our definition. Then $a - b = cm$, a multiple of m. On division by m, b leaves a remainder $b - qm = r$, where $0 \leqq r < m$. Then

$$a = b + cm = (qm + r) + cm = (q + c)m + r.$$

This equation indicates that r is the unique remainder of a on division by m; hence a and b do have the same remainder.

Conversely, suppose that $a = qm + r$, $b = q'm + r$, with the same remainder r. Then $a - b = (q - q')m$ is divisible by m, so that $a \equiv b$ (mod m).   Q.E.D.

The relation of congruence for a fixed modulus m has for all integers a, b, and c the following properties, reminiscent of the laws of equality (§1.2):

Reflexive: $a \equiv a$

Symmetric: $a \equiv b$ implies $b \equiv a$ } all taken (mod m).

Transitive: $a \equiv b$ and $b \equiv c$ imply $a \equiv c$

Each of these laws may be proved by reversion to the definition of congruence. The symmetric law, so translated, requires that $m \mid (a - b)$ imply $m \mid (b - a)$. The hypothesis here is $a - b = dm$, which gives the conclusion $m \mid (b - a)$ in the form $b - a = (-d)m$.

The relation of congruence for a fixed modulus m has a further "substitution property," reminiscent of equality also: sums of congruent integers are congruent, and products of congruent integers are congruent.

Theorem 14. *If $a \equiv b$ (mod m), then for all integers x,*

$$a + x \equiv b + x \qquad ax \equiv bx, \qquad -a \equiv -b \qquad (all \bmod m).$$

Here again the proofs rest on an appeal to the definition. Thus the hypothesis becomes $a - b = km$ for some k; from this we may derive the conclusions in the form

$$m \mid (a + x - b - x), \qquad m \mid (ax - bx), \qquad m \mid (-a + b).$$

The law of cancellation which holds for equations need not hold for congruences. Thus $2 \cdot 7 \equiv 2 \cdot 1$ (mod 12) does not imply that $7 \equiv 1$ (mod 12). This inference fails because the 2 which was cancelled is a factor of the modulus. At best, a modified cancellation law can be found:

Theorem 15. *Whenever c is relatively prime to m,*

$$ca \equiv cb \ (mod \ m). \qquad implies \qquad a \equiv b \ (mod \ m).$$

Proof. By definition, the hypothesis states that $m \mid (ca - cb)$ or, in other words, that $m \mid c(a - b)$. But m is assumed relatively prime to the first factor c of this product, so Theorem 10 allows us to conclude that m divides the second factor $a - b$. This means that $a \equiv b$ (mod m), as asserted.

The study of linear equations may be extended to congruences.

Theorem 16. *If c is relatively prime to m, then the congruence* $cx \equiv b$ *(mod m) has an integral solution x. Any two solutions* x_1 *and* x_2 *are congruent, modulo m.*

Proof. By hypothesis, the g.c.d. (c, m) is 1, so $1 = sc + tm$ for suitable integers s and t. Multiplying by b, $b = bsc + btm$. The final term here is a multiple of m, so that $b \equiv (bs)c$ (mod m). This states that $x = bs$ is the required solution of $b \equiv xc$.

On the other hand, two solutions x_1 and x_2 of this congruence must satisfy $cx_1 \equiv cx_2$ because congruence is a transitive and symmetric relation. Since c is supposed prime to m, we can cancel the c here, as in Theorem 15, obtaining the desired conclusion $x_1 \equiv x_2$ (mod m). Q.E.D.

An important special case arises when the modulus m is a prime. In this case all integers not divisible by m are relatively prime to m. This fact gives the

Corollary. *If p is a prime and if* $c \not\equiv 0$ *(mod p), then* $cx \equiv b$ *(mod p) has a solution which is unique, modulo p.*

Simultaneous congruences can also be treated.

Theorem 17. *If the moduli* m_1 *and* m_2 *are relatively prime, then the congruences*

$$(17) \qquad\qquad x \equiv b_1 \,(mod\ m_1), \qquad x \equiv b_2 \,(mod\ m_2)$$

have a common solution x. Any two solutions are congruent modulo $m_1 m_2$.

Proof. For any integer y, $x = b_1 + ym_1$ is a solution of the first congruence. Such an x satisfies the second congruence also if and only if $b_1 + ym_1 \equiv b_2$ (mod m_2), or $ym_1 \equiv b_2 - b_1$ (mod m_2). Since m_1 is relatively prime to the modulus m_2, this congruence can be solved for y by Theorem 16.

Conversely, suppose that x and x' are two solutions of the given simultaneous congruences (17). Then $x - x' \equiv 0$ (mod m_1) and also (mod m_2). Since m_1 and m_2 are relatively prime, this implies that the difference $x - x'$ is divisible by the product modulus $m_1 m_2$, so that $x \equiv x'$ (mod $m_1 m_2$). Q.E.D.

The same methods of attack apply to two or more congruences of the form $a_i x \equiv b_i$ (mod m_i), with $(a_i, m_i) = 1$ and with the various moduli relatively prime in pairs.

Theorem 18 (Fermat). *If a is an integer and p is a prime, then*

$$a^p \equiv a \ (mod \ p).$$

Proof. For a fixed prime p, let $P(n)$ be the proposition that $n^p \equiv n$ (mod p). Then $P(0)$ and $P(1)$ are obvious. In the binomial expansion (9) for $(n + 1)^p$, every coefficient except the first and the last is divisible by p, hence $(n + 1)^p \equiv n^p + 1$ (mod p), whence $P(n)$ implies $(n + 1)^p \equiv n + 1 \ (mod \ p)$, which is the proposition $P(n + 1)$.

Exercises

1. Solve the following congruences:
   (a) $3x \equiv 2 \ (mod \ 5)$,   (b) $7x \equiv 4 \ (mod \ 10)$,
   (c) $243x + 17 \equiv 101 \ (mod \ 725)$,   (d) $4x + 3 \equiv 4 \ (mod \ 5)$,
   (e) $6x + 3 \equiv 4 \ (mod \ 10)$,   (f) $6x + 3 \equiv 1 \ (mod \ 10)$.
2. Prove that the relation $a \equiv b \ (mod \ m)$ is reflexive and transitive.
3. Prove directly that $a \equiv b \ (mod \ m)$ and $c \equiv d \ (mod \ m)$ imply $a + c \equiv b + d \ (mod \ m)$ *and* $ac \equiv bd \ (mod \ m)$.
★4. (a) Show that the congruence $ax \equiv b \ (mod \ m)$ has a solution if and only if $(a, m) \,|\, b$.
    (b) Show that if $(a, m) \,|\, b$, the congruence has exactly (a, m) incongruent solutions modulo m. (*Hint:* Divide a, b, and m by (a, m).)
5. If m is an integer, show that $m^2 \equiv 0, 1,$ or 4, modulo 8.
6. Prove $x^2 \equiv 35 \ (mod \ 100)$ has no solutions.
★7. Prove that if $x^2 \equiv n \ (mod \ 65)$ has a solution then so does $x^2 \equiv -n$ (mod 65).
8. If x is an odd number not divisible by 3, prove that $x^2 \equiv 1 \ (mod \ 24)$.
★9. (a) Show by tables that all numbers from 25 to 40 can be expressed as sums of four or fewer squares (the result is actually true for all positive numbers).
    (b) Prove that no integer $m \equiv 7$ (mod 8) can be expressed as a sum of three squares. (*Hint:* Use Ex. 5.)
10. Solve the simultaneous congruences:
    (a) $x \equiv 2 \ (mod \ 5)$, $2x \equiv 1 \ (mod \ 8)$,
    (b) $3x \equiv 2 \ (mod \ 5)$, $2x \equiv 1 \ (mod \ 3)$.
11. On a desert island, five men and a monkey gather coconuts all day, then sleep. The first man awakens and decides to take his share. He divides the coconuts into five equal shares, with one coconut left over. He gives the extra one to the monkey, hides his share, and goes to sleep. Later, the second man awakens and takes his fifth from the remaining pile; he too finds one extra and gives it to the monkey. Each of the remaining three men does likewise in turn. Find the minimum number of coconuts originally present (*Hint:* Try -4 coconuts.)
★12. Show by induction that Theorem 17 can be generalized to n congruences with moduli relatively prime in pairs.

★**13.** Prove that if $(m_1, m_2) = (a_1, m_1) = (a_2, m_2) = 1$, then the simultaneous congruences $a_i x \equiv b_i \pmod{m_i}$ $(i = 1, 2)$ have a common solution, and any two solutions are congruent modulo $m_1 m_2$.

★**14.** Generalize Ex. 13 to n simultaneous congruences.

15. For what positive integers m is it true that whenever $x^2 \equiv 0 \pmod{m}$ then also $x \equiv 0 \pmod{m}$?

16. If a and b are integers and p a prime, prove that $(a + b)^p \equiv a^p + b^p \pmod{p}$.

1.10. The Rings Z_n

From early antiquity, man has distinguished between the "even" integers $2, 4, 6, \cdots$ and the "odd" integers $1, 3, 5, \cdots$. The following laws for reckoning with even and odd integers are also familar:

(18)
$$\text{even} + \text{even} = \text{odd} + \text{odd} = \text{even}, \quad \text{even} + \text{odd} = \text{odd},$$
$$\text{even} \cdot \text{even} = \text{even} \cdot \text{odd} = \text{even}, \quad \text{odd} \cdot \text{odd} = \text{odd}.$$

These identities define a new integral domain Z_2, which consists of two elements 0 ("even") and 1 ("odd") alone, and having the addition and multiplication tables

$$0 + 0 = 1 + 1 = 0, \quad 0 + 1 = 1 + 0 = 1,$$
$$0 \cdot 0 = 0 \cdot 1 = 1 \cdot 0 = 0, \quad 1 \cdot 1 = 1.$$

We will now show that a similar construction can be applied to the remainders $0, 1, 2, \cdots, n - 1$ to any modulus n. Two such remainders can be added (or multiplied) by simply forming the sum (or product) in the ordinary sense (i.e., in Z), and then replacing the result by its remainder modulo n. Tables for the case $n = 5$ are

+	0	1	2	3	4			0	1	2	3	4
0	0	1	2	3	4		0	0	0	0	0	0
1	1	2	3	4	0		1	0	1	2	3	4
2	2	3	4	0	1		2	0	2	4	1	3
3	3	4	0	1	2		3	0	3	1	4	2
4	4	0	1	2	3		4	0	4	3	2	1

In every case the resulting system has properties (i)–(viii) of §1.1. That is, we have

Theorem 19. *Under addition and multiplication modulo any fixed $n \geq 2$, the set of integers $0, 1, \cdots, n - 1$ constitutes a commutative ring Z_n.*

Proof. In the last section, we saw that the relation $x \equiv y \pmod{n}$ is reflexive, symmetric, and transitive, like ordinary equality. In fact, by Theorem 14, $a \equiv b \pmod{n}$ and $c \equiv d \pmod{n}$ together imply

(19)　　　$a + c \equiv b + d \pmod{n}$,　　　$a \cdot c \equiv b \cdot d \pmod{n}$.

That is, postulates (i) and (ii) hold, provided "equality" in $\mathbf{Z}$ is reinterpreted to mean "congruent modulo n." Again, 0 and 1 in $\mathbf{Z}$ act in $\mathbf{Z}_n$ as identities for addition and multiplication, respectively, while $n - k$ is an additive inverse of k, modulo n.

It remains to verify postulates (iii)–(v); consider the distributive law. Since $a(b + c) = ab + ac$ for any integers, one must by (19) have $a(b + c) \equiv ab + ac \pmod{n}$ when remainders are taken mod n. This is the distributive law in $\mathbf{Z}_n$; the proofs of the commutative and associative laws are the same.　Q.E.D.

The only postulate for an integral domain not such an identity is the cancellation law of multiplication. According to Theorem 1, this law is equivalent to the assertion that there are no divisors of zero in $\mathbf{Z}_n$: $ab = 0$ implies $a = 0$ or $b = 0$. These equations in $\mathbf{Z}_n$ mean congruences for ordinary integers, so the law becomes the statement: $ab \equiv 0 \pmod{n}$ implies $a \equiv 0 \pmod{n}$ or $b \equiv 0 \pmod{n}$. This is equivalent to the assertion that $n \mid ab$ implies $n \mid a$ or $n \mid b$. This is true if n is a prime (Theorem 9). If n is not prime, n has a nontrivial factorization $n = ab$, so $n \mid ab$ although neither $n \mid a$ nor $n \mid b$, and $\mathbf{Z}_n$ has zero-divisors. This proves

Theorem 20. *The ring $\mathbf{Z}_n$ of integers modulo n is an integral domain if and only if n is a prime.*

There are other, more systematic ways to construct the algebra of integers modulo n. The device of replacing congruence by equality means essentially that all the integers which leave the same remainder on division by n are grouped together to make one new "number." Each such group of integers is called a "residue class." For the modulus 5 there are five such classes, corresponding to the possible remainders, 0, 1, 2, 3, and 4; some of these classes are

$$1_5 = \{\cdots, -14, -9, -4, 1, 6, 11, 16, \cdots\},$$
$$2_5 = \{\cdots, -13, -8, -3, 2, 7, 12, 17, \cdots\},$$
$$3_5 = \{\cdots, -12, -7, -2, 3, 8, 13, 18, \cdots\}.$$

For any modulus n the residue class r_n determined by a remainder r with $0 \leq r < n$ consists of all integers a which leave the remainder r on division by n. Each integer belongs to one and only one residue class, and two integers will belong to the same residue class if and only if they are congruent (Theorem 13). There are n residue classes: 0_n, 1_n, $\cdots$, $(n-1)_n$.

The algebraic operations of Z_n can be carried out directly on these classes. For suppose that two residues r and s give in Z_n a remainder t as sum, $r + s \equiv t \pmod{n}$. The answer would be obtained if one used instead of the residues r and s any other elements in the corresponding classes. If a is in r_n, b in s_n, then $a + b$ is in the class t_n belonging to the sum t, for $a \equiv r$ and $b \equiv s$ give $a + b \equiv r + s \equiv t \pmod{n}$. In general, the algebra Z_n could be defined as the algebra of these residue classes: to add (or multiply) two classes, pick any representatives a and b of these classes, and find the residue class containing the sum (or the product) of these representatives. If a_n denotes the residue class which contains a, this rule may be stated as

$$(20) \qquad (a + b)_n = a_n + b_n, \qquad (ab)_n = a_n b_n.$$

For instance, the sum $1_5 + 2_5 = 3_5$ of the classes listed above may be found by adding any chosen representatives $6 + (-13)$ to get a result -7 which lies in the sum class 3_5. Other choices $-9 + (-3) = -12$, $11 + 7 = 18$, $-14 + 17 = 3$, all give the *same* sum, 3_5.

The residue classes which we have defined in terms of remainders may also be defined directly in terms of congruences, by a general method to be discussed in §6.13.

Exercises

1. Construct addition and multiplication tables for Z_3 and Z_4.

2. Compute in Z_7:　$(3 \cdot 4) \cdot 5$,　$3 \cdot (4 \cdot 5)$,　$3 \cdot (4 + 5)$,　$3 \cdot 4 + 3 \cdot 5$.

3. Find all divisors of zero in Z_{26}, Z_{24}.

4. Determine the exact set of all sums $x + y$ and that of all products xy for x in 4_8, y in 4_8. How are these related to the sets $4_8 + 4_8$ and $4_8 \cdot 4_8$?

5. Verify the associative law for the addition of residue classes, as in the proof of Theorem 19.

6. For real numbers x and y, let $x \equiv y \pmod{2\pi}$ mean that $x = y + 2n\pi$ for some integer n. Show that addition of residue classes can then be defined as in (20), whereas multiplication of residue classes cannot be so defined.

★7. Show that in Z_n any element c which is not a unit is a zero-divisor.

★8. (a) Enumerate the units of Z_{15}.

(b) Show that if $n = 2m + 1$ is odd, then the number of units of Z_n is even.

★9. Show that k is a unit of Z_n if and only if $(k, n) = 1$ in Z.

1.11. Sets, Functions, and Relations

At this point, we pause to discuss briefly the fundamental notions of set, function, binary operation, and relation.

A *set* is a quite arbitrary collection of mathematical objects: for example, the set of all odd numbers or the set of all points in the plane equidistant from two given points. If A is a set, we write $x \in A$ to signify that the object x is an element of the set A, and $x \notin A$ when x is not an element of A. A finite set A can be specified by listing its elements; for example, $\{0, 2, 4\}$ denotes the set whose (only) elements are the numbers 0, 2, and 4. More generally, any set is determined by its elements, in the sense that two sets A and B are equal (the "same") if and only if they have the same elements. This principle (called the *axiom of extensionality*) can also be stated symbolically: $A = B$ means that for all x, $x \in A$ if and only if $x \in B$. The resulting equality of sets is clearly a reflexive, symmetric, and transitive relation, as required in §1.2 for any equality.

A set S is called a *subset* of a set A if and only if every element x of S is also in A; the symbol $S \subset A$ indicates that S is a subset of A. If both $T \subset S$ and $S \subset A$, then clearly $T \subset A$, so the relation "subset of" is transitive. Likewise, the condition for the equality of sets becomes the statement that $A = B$ if and only if both $A \subset B$ and $B \subset A$. Moreover, the *empty* set $\varnothing$ (the set with no members) is a subset of every set.

Starting with any set, such as the set of all integers, we can pick out various subsets: the set of all positive integers, the set of all odd positive integers, the set of all integers greater than 18, and so on. These examples illustrate the principle that any property determines a subset; more exactly, given any set A and a property P, one may form the subset

(21) $$S = \{x \mid x \in A \quad \text{and} \quad x \text{ has } P\}$$

of all those elements of A which have the property P.

Generally, if A and B are sets, a *function* $\phi: A \to B$ on A to B is a rule which assigns to each element a in A an element $a\phi$ in B. We will write this $a \mapsto a\phi$. Thus $x \mapsto x^2$ is a function ϕ on the set $A = \mathbf{Q}$ of all rational numbers to the set B of all nonnegative rationals (it can also be considered as a function $\phi: \mathbf{Q} \to \mathbf{Q}$). Likewise, the operation "add one" sends each integer n to another, by $n \mapsto n + 1$; hence it is a function $\phi: \mathbf{Z} \to \mathbf{Z}$. In any ordered domain D, the process of taking the absolute value, $a \mapsto |a|$, is similarly a function on the set D to the set of nonnegative elements in D. Taking the negative, $a \mapsto -a$, is still another function on D to D.

The relation $a \mapsto a\phi$ is sometimes written $a \mapsto \phi a$ or $a \mapsto \phi(a)$, with the symbol ϕ for the function in front. A function $\phi: A \to B$ is also called a *mapping*, a *transformation*, or a *correspondence* from A to B. The set A is called the *domain* of the function ϕ, and B its *codomain*. For example, the usual telephone dial

A B C	D E F	G H I	J K L	M N O	P R S	T U V	W X Y	Z
\\|/	\\|/	\\|/	\\|/	\\|/	\\|/	\\|/	\\|/	|
2	3	4	5	6	7	8	9	0

defines a function on a set A of 25 letters (the alphabet, Q omitted) to the set $\{0, 1, \cdots, 9\}$ of all ten digits.

The *image* (or "range") of a function $\phi: A \to B$ is the set of all the "values" of the function; that is, all $a\phi$ for a in A. The image is a subset of the codomain B, but need not be all of B. For example, the image of the telephone-dial function is the subset $\{0, 2, \cdots, 9\}$, with 1 omitted.

A function $\phi: A \to B$ is called *surjective* (or *onto*) when every element $b \in B$ is in the image—that is, when the image is the whole codomain. For example, absolute value $a \mapsto |a|$ for integers is a function $\mathbf{Z} \to \mathbf{Z}$, but is *not* surjective because the image is the (proper) subset $\mathbf{N} \subset \mathbf{Z}$ of all nonnegative integers. However, the rule $a \mapsto |a|$ also defines a function $\mathbf{Z} \to \mathbf{N}$ that *is* surjective. To decide whether or not a function is *onto*, we must know the intended codomain.

A function $\phi: A \to B$ is an *injection* (or *one-one into*) when different elements of A always have different images—in other words, when $a\phi = a'\phi$ always implies $a = a'$. For example, $x \to 2x$ is an injection $\mathbf{Z} \to \mathbf{Z}$ (but is not surjection).

A function $\phi: A \to B$ is a *bijection* (or *bijective*, or *one-one onto*) when it is both injective and surjective; that is, when to each element $b \in B$ there is one and only one $a \in A$ which has image b, with $a\phi = b$. For example, $n \mapsto n + 1$ is a bijection $\mathbf{Z} \leftrightarrow \mathbf{Z}$ and, for any domain D, $a \mapsto a$ is a bijection $D \to D$. Bijections $\phi: A \to B$ are also called one-one correspondences (of A onto B), while not necessarily injective correspondences have been called many-one correspondences.

Binary Operations. Operations on pairs of numbers arise in many contexts—the addition of two integers, the addition of two residue classes in $\mathbf{Z}_n$, the multiplication of two real numbers, the subtraction of one integer from another, and the like. In such cases we speak of a *binary operation*. In general, a binary operation "$\circ$" on a set S of elements $a, b, c, \cdots$ is a rule which assigns to each ordered pair of elements a and b from S a uniquely defined third element $c = a \circ b$ in the same set S. Here by "uniquely" we mean the substitution property

(22) $\qquad a = a'$ and $b = b'$ imply $a \circ b = a' \circ b'$,

as in the uniqueness postulate for a commutative ring.

It is convenient to write $S \times T$ for the set of all ordered pairs of elements (a, b) with $a \in S, b \in T$; this is called the *Cartesian product* (or simply "product") of S and T. One also writes S^2 for the product $S \times S$ of a set with itself; a binary operation is then the same thing as a function $\circ: S^2 \to S$.

Two given integers may be "related" to each other in many ways, such as "$a = b$," "$a < b$," "$a \equiv b \pmod{7}$," or "$a \mid b$." Each of these phrases is said to express a certain "binary relation" between a and b. One may readily mention many other relations between other types of mathematical objects; there are also nonmathematical relations, such as the relation "is a brother of" between people. To discuss relations in general we introduce a symbol R to stand for any relation ("R" stands for "$<$," "$\equiv$," or "$\mid$," etc.). Formally, "R" denotes a *binary relation* on a given set S of objects if, given two elements a and b in the set S, either a stands in the relation R to b (in symbols, aRb), or a does not stand in the relation R to b (in symbols, $aR'b$).

Especially important in mathematics are the relations R on a set S which, like congruence and equality, satisfy the following laws:

Reflexive: aRa for all a in S.
Symmetric: aRb implies bRa for all a, b in S.
Transitive: aRb and bRc imply aRc for all a, b, c in S.

Reflexive, symmetric, and transitive relations are known as *equivalence relations*. For example, the relation of congruence between triangles in the plane is such an equivalence relation.

Exercises

1. Which of the following binary operations $a \circ b$ on integers a and b are associative, and which ones are commutative?

$$a - b, \quad a^2 + b^2, \quad 2(a + b), \quad -a - b.$$

2. Which of the three properties "reflexive," "symmetric," and "transitive" apply to each of the following relations between integers a and b?

$$a \leqq b, \quad a < b, \quad a \mid b, \quad a^2 + a = b^2 + b, \quad a < |b|.$$

3. Do the same for the following relations on the class of all people: "is a father of," "is a brother of," "is a friend of," "is an uncle of," "is a descendant of." Would any of your answers be changed if these relations are restricted to apply only to the class of all men?

★4. How is the relation "is an uncle of" connected with the relations "is a brother of" and "is a parent of"? Can you state any similar general rule for making a new relation out of two given ones?

5. A relation R is called "circular" if aRb and bRc imply cRa. Show that a relation is reflexive and circular if and only if it is reflexive, symmetric, and transitive.

★6. What is wrong with the following "proof" that the symmetric and transitive laws for a relation R imply the reflexive law? "By the symmetric law, aRb implies bRa; by the transitive laws, aRb and bRa imply aRa."

7. Each of the following rules defines a function $f: \mathbf{Z} \to \mathbf{Z}$. In each case specify the image and whether or not the function is injective.

   (a) $a \mapsto |a| + 1$,   (b) $a \mapsto a^2$,
   (c) $a \mapsto 2a + 5$,   (d) $a \mapsto$ g.c.d. $(a, 6)$.

8. Do Ex. 7, replacing $\mathbf{Z}$ by the class $\mathbf{Z}^+$ of positive integers.

9. For what integers n is the function $x \mapsto 6x + 7$ bijective on $\mathbf{Z}_n$? surjective on $\mathbf{Z}_n$?

10. Show that any relation R on a set S can be regarded as a function $f: S^2 \to \{0, 1\}$.

1.12. Isomorphisms and Automorphisms

One of the most important concepts of modern algebra is that of *isomorphism*. We now define this concept for commutative rings as follows:

Definition. *An* isomorphism *between two commutative rings R and R' is a one-one correspondence $a \leftrightarrow a'$ of the elements a of R with the elements a' of R', which satisfies for all elements a and b the conditions*

$$(23) \qquad (a + b)' = a' + b', \qquad (ab)' = a'b'.$$

The rings R and R' are called isomorphic if there exists such a correspondence.

On account of the laws (23) one may say that the isomorphism $a \leftrightarrow a'$ "preserves sums and products." Loosely speaking, two commutative rings are isomorphic when they differ only in the notation for their elements. An appropriate example is the algebra of "even" and "odd" as compared with the integral domain $\mathbf{Z}_2$, as discussed in §1.10. The one-one correspondence

$$\text{even} \leftrightarrow 0 \qquad \text{odd} \leftrightarrow 1$$

is an isomorphism between these domains because corresponding elements are added and multiplied according to the same rules (cf. formula (18)).

Many integral domains have important isomorphisms with themselves. Such isomorphisms are called *automorphisms*; they are analogous to symmetries of geometrical figures (see §6.1). Consider, for example, the domain $\mathbf{Z}[\sqrt{2}]$ described in §1.1 as the set of all numbers $m + n\sqrt{2}$ for m and n in the domain $\mathbf{Z}$ of integers; it is isomorphic to itself under the nontrivial correspondence $m + n\sqrt{2} \leftrightarrow m - n\sqrt{2}$. This correspondence is an isomorphism, since for any $a = m + n\sqrt{2}$ and $b = m_1 + n_1\sqrt{2}$, we have

$$
\begin{aligned}
(ab)' &= [(m + n\sqrt{2})(m_1 + n_1\sqrt{2})]' \\
&= [(mm_1 + 2nn_1) + (mn_1 + m_1 n)\sqrt{2}]' \\
&= (mm_1 + 2nn_1) - (mn_1 + m_1 n)\sqrt{2}, \\
a'b' &= (m - n\sqrt{2})(m_1 - n_1\sqrt{2}) \\
&= (mm_1 + 2nn_1) - (mn_1 + m_1 n)\sqrt{2}
\end{aligned}
$$

and, similarly, $(a + b)' = a' + b'$.

Any isomorphism $a \leftrightarrow a'$ preserves not only sums and products, but also differences. By definition, $a - b$ is the solution of the equation $b + x = a$, so that $b + (a - b) = a$. Since the correspondence preserves sums, $b' + (a - b)' = a'$; this asserts that $(a - b)'$ is the (unique) solution of the equation $b' + x = a'$, or that

$$
(a - b)' = a' - b'.
$$

Other rules are

$$
(24) \qquad 0' = 0, \qquad 1' = 1, \qquad (-a)' = -(a').
$$

In words: the zero (unity) of R corresponds to the zero (unity) of R'.

We shall see later that the idea of isomorphism applies to algebraic systems in general. One may even describe abstract algebra as the study of those properties of algebraic systems which are preserved under isomorphism.

In describing the system of integers as an ordered domain in which each set of positive integers has a least element, we claimed that these postulates completely describe the integers for all mathematical purposes. We can now state this more precisely (it will be proved in §2.6). Any ordered domain in which the positive elements are well-ordered is isomorphic to the domain $\mathbf{Z}$ of integers. Such a characterization of $\mathbf{Z}$ "up to isomorphism" is the most that could be achieved with any postulate system of the type we have used, for it is clear, in general, that if a system S satisfies such a system of postulates, and if S' is another system isomorphic to S, then S' must also satisfy the postulates. Thus if S satisfies a commutative law for addition, then $a + b = b + a$ for all a and b in S. The corresponding elements in the given isomorphism must be equal, so $(a + b)' = (b + a)'$. Since the isomorphism preserves sums, $a' + b' = b' + a'$. This asserts that the commutative law also holds in S'. This argument is of a general character and applies to all our postulates.

Exercises

1. Prove that the properties (24) hold for any isomorphism.
2. Let $\mathbf{Z}[\sqrt{3}]$ be the domain of all numbers $m + n\sqrt{3}$ for $m, n \in \mathbf{Z}$. Exhibit a nontrivial isomorphism of $\mathbf{Z}[\sqrt{3}]$ with itself.
3. Prove that the correspondence $m + n\sqrt{2} \leftrightarrow m + n\sqrt{3}$ is not an isomorphism between the domains $\mathbf{Z}[\sqrt{2}]$ and $\mathbf{Z}[\sqrt{3}]$.
4. (a) Prove that under any isomorphism an element x satisfying an equation $x^2 = 1 + 1$ must correspond to an element $y = x'$ satisfying the equation $y^2 = 1' + 1'$.
   (b) Use (a) to show that no isomorphism is possible between $\mathbf{Z}[\sqrt{2}]$ and $\mathbf{Z}[\sqrt{3}]$.
5. Show that the domain $\mathbf{Z}$ of integers has no nontrivial isomorphisms with itself.
★6. Prove that an integral domain with exactly three elements is necessarily isomorphic to $\mathbf{Z}_3$.
7. Prove that isomorphism is an "equivalence relation" (i.e., a reflexive, symmetric, and transitive relation).

2

Rational Numbers and Fields

2.1. Definition of a Field

Both the integral domain $\mathbf{Q}$ of all rational numbers and the integral domain $\mathbf{R}$ of all real numbers have an essential algebraic advantage over the domain $\mathbf{Z}$ of integers: any equation $ax = b$ $(a \neq 0)$ can be solved in them. Commutative rings with this property are called *fields*; we now show that division is possible and has its familiar properties in any commutative ring where all nonzero elements have nonmultiplicative inverses.

Definition. *A field F is a commutative ring which contains for each element $a \neq 0$ an "inverse" element a^{-1} satisfying the equation $a^{-1}a = 1$.*

It is easy to show that the cancellation law (ix) of §1.1 holds in any field, for if $c \neq 0$ and $ca = cb$, then

$$a = 1a = (c^{-1}c)a = c^{-1}(ca) = c^{-1}(cb) = (c^{-1}c)b = 1b = b.$$

In other words, every field is an integral domain; more generally, so is every *subdomain* of a field (and for the same reason). Conversely, in this section and the next we will show that any integral domain can be extended to a field in one and only one minimal way. The method of extension is illustrated by the standard representation of fractions as quotients of integers.

Theorem 1. *Division (except by zero) is possible and is unique in any field.*

Proof. We have to show that for given $a \neq 0$ and b in a field F the equation $ax = b$ has one and only one solution x in F. If $a \neq 0$, the inverse a^{-1} may be used to construct an element $x = a^{-1}b$ which on substitution proves to be a solution of $ax = b$. It is the only solution, for by the cancellation law proved above, $ax = b$ and $ay = b$ together imply $x = y$ if $a \neq 0$. Q.E.D.

The solution of $ax = b$ is denoted by b/a (the quotient of b by a). In particular, $1/a = a^{-1}$.

All the rules for algebraic manipulation listed in §1.2 are satisfied in fields, considered as integral domains. The usual rules for the manipulation of quotients can also be proved from the postulates for a field.

Theorem 2. *In any field, quotients obey the following laws (where $b \neq 0$ and $d \neq 0$),*

(i)	$(a/b) = (c/d)$	*if and only if*	$ad = bc$,
(ii)	$(a/b) \pm (c/d) = (ad \pm bc)/(bd)$,		
(iii)	$(a/b)(c/d) = (ac/bd)$,		
(iv)	$(a/b) + (-a/b) = 0$,		
(v)	$(a/b)(b/a) = 1$	*if* $(a/b) \neq 0$.	

Proof of (i). The hypothesis $(a/b) = (c/d)$ means $ab^{-1} = cd^{-1}$. This gives $ad = a(b^{-1}b)d = cd^{-1}(bd) = cd^{-1}db = bc$. Conversely, if $ad = bc$, then $a/b = b^{-1}a = b^{-1}add^{-1} = b^{-1}bcd^{-1} = cd^{-1} = c/d$, as desired.

Proof of (ii). Observe that $x = a/b$ and $y = c/d$ denote the solutions of $bx = a$ and $dy = c$. These equations may be combined to give

$$dbx = da, \qquad bdy = bc, \qquad bd(x \pm y) = ad \pm bc.$$

Thus $x \pm y$ is the unique solution $z = (ad \pm bc)/bd$ of the equation $bdz = ad \pm bc$.

Proof of (iii). As above, the equations $bx = a$ and $dy = c$ can be combined to give

$$(bd)(xy) = (bx)(dy) = ac,$$

whence
$$xy = (ac)/(bd).$$

Proof of (iv). Substituting in (ii), we have

$$(a/b) + (-a/b) = (ab - ba)/b^2 = 0/b^2 = 0 \cdot (b^2)^{-1} = 0.$$

Proof of (v). Substituting in (iii), we have $(a/b)(b/a) = ab/ba$. But ab/ba is the unique solution of the equation $bax = ab$. Clearly, $x = 1$ satisfies this equation; hence $ab/ba = 1$. Q.E.D.

Arguments similar to those just employed can be used to prove such other familiar laws as the following:

(1) $\quad (bd)^{-1} = d^{-1}b^{-1}, \qquad (-b)^{-1} = -(b^{-1}) \qquad$ if $\qquad b, d \neq 0.$

(2) $\quad a \pm (b/c) = (ac \pm b)/c, \qquad a(b/c) = ab/c, \qquad\qquad c \neq 0.$

(3) $(a/b)/(c/d) = ad/bc, \qquad (a/b)/c = a/bc, \qquad a/1 = a; \quad b, c, d \neq 0.$

(4) $\quad -(a/b) = (-a)/b = a/(-b), \qquad (-a)/(-b) = a/b, \qquad b \neq 0.$

The proofs will be left to the reader as exercises.

Fields exist in great variety. Thus, for any prime p, the integral domain $\mathbf{Z}_p$ constructed in §1.10 is a field. This follows from the corollary of Theorem 16, §1.9. Again, if one assumes that the real numbers form a field, one can easily construct other examples of fields by using the notion of a *subfield*.

Definition. *A subfield of a given field F is a subset of F which is itself a field under the operations of addition and multiplication in F.*

All identities (viz., the commutative, associative, and distributive laws) which hold in F hold *a fortiori* in any subset of F, provided the operations in question can be performed. In testing a subset S of F for being a subfield, one can therefore ignore the postulates which are identities and test only those which involve some "existence" assertion, such as the existence of an inverse. This gives the following result:

Theorem 3. *A subset S of a field F is a subfield if S contains the zero and unity of F, if S is closed under addition and multiplication, and if each a of S has its negative and (provided a $\neq$ 0) its inverse a^{-1} in S.*

Theorem 3 may now be applied to show that the set of all real numbers of the form $a + b\sqrt{2}$, with rational coefficients a and b, is a subfield of the field of all real numbers. This subfield is customarily denoted by $\mathbf{Q}(\sqrt{2})$, where $\mathbf{Q}$ designates the field of rationals. Theorem 3 does apply, for the sum of any two numbers of $\mathbf{Q}(\sqrt{2})$ is another one of the same sort, and similarly the product is

$$(a + b\sqrt{2})(c + d\sqrt{2}) = (ac + 2bd) + (bc + ad)\sqrt{2}.$$

Again, $\mathbf{Q}(\sqrt{2})$ contains $0 = 0 + 0\sqrt{2}$, $1 = 1 + 0\sqrt{2}$, and $-(a + b\sqrt{2}) = -a - b\sqrt{2}$ if it contains $a + b\sqrt{2}$. Finally, an inverse $(a + b\sqrt{2})^{-1}$ of any nonzero element may be found by "rationalizing the denominator,"

$$\frac{1}{a + b\sqrt{2}} = \frac{1}{a + b\sqrt{2}}\left(\frac{a - b\sqrt{2}}{a - b\sqrt{2}}\right) = \left(\frac{a}{a^2 - 2b^2}\right) - \left(\frac{b}{a^2 - 2b^2}\right)\sqrt{2}.$$

The new denominator $a^2 - 2b^2$ is never zero (as is proved in §3.6), and the resulting inverse does have the required form $a' + b'\sqrt{2}$ with rational coefficients $a' = a/(a^2 - 2b^2)$, $b' = -b/(a^2 - 2b^2)$. One may easily verify that this inverse does indeed satisfy the equation

$$(a' + b'\sqrt{2})(a + b\sqrt{2}) = 1.$$

Similarly, the set $\mathbf{Q}(\sqrt[3]{5})$ of all real numbers $a + b\sqrt[3]{5} + c\sqrt[3]{25}$ with rational a, b, c is a field. Addition, subtraction, and multiplication are performed within this set much as in $\mathbf{Q}(\sqrt{2})$, using this time the fact that $(\sqrt[3]{5})^3 = 5$ is a rational number. Finally, $(a + b\sqrt[3]{5} + c\sqrt[3]{25})^{-1}$ may be computed by showing that the equation

$$(a + b\sqrt[3]{5} + c\sqrt[3]{25})(x + y\sqrt[3]{5} + z\sqrt[3]{25}) = 1 + 0 \cdot \sqrt[3]{5} + 0 \cdot \sqrt[3]{25}$$

is equivalent to a system of simultaneous linear equations. These equations can always be solved for x, y, and z, unless $a = b = c = 0$.

We may construct still other subfields if we assume that there is a field of *complex* numbers $a + bi$, where $i = \sqrt{-1}$ and a and b are real. The quadratic equation

$$\omega^2 + \omega + 1 = 0$$

will have a root $\omega = (-1 + \sqrt{-3})/2 = -1/2 + (\sqrt{3}/2)i$ in the field. (Note that since $\omega^3 - 1 = (\omega - 1)(\omega^2 + \omega + 1) = 0$, ω is an "imaginary" cube root of unity!) All $a + b\omega$ (a, b rational) form a subfield $\mathbf{Q}(\omega)$ of the field of all complex numbers, for

$$(a + b\omega) + (c + d\omega) = (a + c) + (b + d)\omega,$$
$$(a + b\omega)(c + d\omega) = ac + (bc + ad)\omega + bd\omega^2$$
$$= (ac - bd) + (bc + ad - bd)\omega,$$

where the equation $\omega^2 = -\omega - 1$ has been used to get rid of the term in ω^2. Furthermore, any $a + b\omega \neq 0$ has an inverse in the set, for

$$(a + b\omega)\left[\frac{-(b - a + b\omega)}{a^2 - ab + b^2}\right] = \frac{a^2 - ab + b^2}{a^2 - ab + b^2} = 1.$$

The denominator $a^2 - ab + b^2$ appearing in this inverse is never zero, for $a^2 - ab + b^2 = (a^2 + b^2)/2 + (a - b)^2/2$ is certainly positive unless $a = b = 0$.

Exercises

1. Prove formulas (1)–(4) from the postulates for a field.

2. Make a table which exhibits c^{-1} for each $c \neq 0$ in Z_{11}.

3. If the set of real numbers is assumed to be a field, which of the following subsets of reals are fields? (a) all positive integers, (b) all numbers $a + b\sqrt{3}$, with a, b rational, (c) all numbers $a + b\sqrt[3]{5}$, with a, b rational, (d) all rational numbers which are not integers, (e) all numbers $a + b\sqrt{5}$, with a and b rational.

4. Show that in Theorem 3 the conditions $0 \in S$ and $1 \in S$ can be replaced by the condition "S contains at least two elements." (*Hint:* Consider $ax = a$.)

★5. Show that the law $a + b = b + a$ is implied by postulates (i), (ii), and (iv)–(vii) of §1.1, together with
  (viii') For each a in R, the equations $a + x = 0$ and $y + a = 0$ have solutions x and y in R.

6. Is every integral domain isomorphic to a field itself a field? Why?

7. Prove that the only subfield of the field Q of rational numbers is Q itself.

8. State and prove an analogue of Theorem 3 for subdomains.

9. Show that a subfield of $Q(\sqrt{2})$ is either Q itself or the whole field $Q(\sqrt{2})$.

10. If S and S' are two subfields of a given field F, show that the set of elements common to S and S' is also a subfield.

11. Can you state a general theorem on the possible subdomains of Z? of Z_n?

★12. Construct addition and multiplication tables for a field of four elements, assuming that $1 + 1 = 0$ (addition is mod 2) and that there is an element x such that $x^2 = x + 1$.

★13. Find all subfields of the field of Ex. 12.

2.2. Construction of the Rationals

We will now prove rigorously that the (ordered) field Q of rational numbers can be constructed from the well-ordered domain Z of all integers, whose existence was postulated in Chap. 1. Indeed, we will prove more: that a similar construction can be applied to *any* integral domain.

The integers alone do not form a field; the construction of the rational numbers from the integers is essentially just the construction of a field which will contain the integers. Clearly, this field must also contain solutions for all equations $bx = a$ with integral coefficients a and $b \neq 0$.

To construct abstractly the "rational numbers" which solve these equations, we simply introduce certain new symbols (or couples) $r = (a, b)$, each of which is intended to stand for a solution of an equation $bx = a$. To realize this intention we must specify that these new objects shall be added, multiplied, and equated exactly as are the quotients a/b in a field (Theorem 2, (i)–(iii)).

The preceding specification makes good sense whether we start with the domain of integers $\mathbf{Z}$, or from some other integral domain D. It can be formulated precisely as follows.

Definition. *Let D be any integral domain. The field of quotients $Q(D)$ of D consists of all couples (a, b) with $a, b \in D$ and $b \neq 0$. The "equality" of such couples is governed by the convention that*

(5) $$(a, b) \equiv (a', b') \quad \textit{if and only if} \quad ab' = a'b,$$

while sums and products are defined, respectively, by

(6) $$(a, b) + (a', b') = (ab' + a'b, bb'),$$

(7) $$(a, b) \cdot (a', b') = (aa', bb').$$

Note that since D contains no "divisors of zero" (§1.2, Theorem 1), the product $bb' \neq 0$ in (6) and (7), and so $Q(D)$ is closed under addition and multiplication.

We wish to regard the relation " $\equiv$ " of "congruence" between couples as an equality. Since this relation is not formal identity ((a, b) identical to (a', b') would mean $a = a'$ and $b = b'$), we must prove that this congruence has the properties of equality listed in §1.2 (for formal identity these properties would have been trivial). In the first place, we may check by straightforward argument that " $\equiv$ " is reflexive, symmetric, and transitive. And then, the sum and product are uniquely determined in the sense of this congruence. For instance, $(a, b) \equiv (a', b')$ implies $(a, b) + (a'', b'') \equiv (a', b') + (a'', b'')$. For each sum in the conclusion is given by a formula like (6), and these two results are congruent in the sense (5) if and only if $(ab'' + a''b)b'b'' = (a'b'' + a''b')bb''$. But this equation follows from the hypothesis $(a, b) \equiv (a', b')$ (i.e., $ab' = a'b$). A similar uniqueness assertion holds for the product. We conclude that the equality defined by (5) has the desired properties.

Various algebraic laws in $Q(D)$ may now be checked. Thus, for the distributive law one can reduce each side of the law systematically, according to definitions (6) and (7), in the following way, where r, r', and

r'' are any three couples:

$$r(r' + r'') \qquad\qquad\qquad r r' + r r''$$
$$(a, b)[(a', b') + (a'', b'')] \qquad (a, b)(a', b') + (a, b)(a'', b'')$$
$$(a, b)(a'b'' + a''b', b'b'') \qquad (aa', bb') + (aa'', bb'')$$
$$(aa'b'' + aa''b', bb'b'') \qquad (aa'bb'' + aa''bb', bb'bb'').$$

These two results give equal couples in the sense of (5), as the second result differs from the first only in the presence of an extra nonzero factor b in all terms. Such an extra factor in a couple always gives an equal couple, $(bx, by) \equiv (x, y)$, for by (5) this equality amounts simply to the identity $bxy = byx$.

This explicit proof of the distributive law in $Q(D)$ is but an illustration. By the same straightforward use of the definitions and the laws for D, one proves the associative and commutative laws. An identity element for addition (a zero) is the couple $(0, 1)$, for

$$(0, 1) + (a, b) = (0 \cdot b + 1 \cdot a, 1 \cdot b) = (a, b).$$

The cancellation law holds, and the couple $(1, 1)$ is an identity for multiplication. The negative of (a, b) is $-(a, b) = (-a, b)$. This verifies all the postulates listed in §1.1 for an integral domain.

Theorem 4. *The field of quotients $Q(D)$ is a field for any integral domain D.*

Proof. It remains only to prove that every equation $rx = 1$ with $r \neq 0$ has a solution x in $Q(D)$—that is, the existence for every $r \neq 0$ in $Q(D)$ of an inverse for r. But this is easy; more generally, any equation

(8) $\qquad\qquad (a, b)(x, y) \equiv (c, d) \qquad$ with $\qquad (a, b) \not\equiv (0, 1)$

has a solution suggested by (3), namely,

(8') $\qquad\qquad\qquad\qquad (x, y) = (bc, ad).$

For by direct substitution $(a, b)(bc, ad) = (abc, bad)$, and $(abc, bad) \equiv (c, d)$ because $abcd = badc$. The hypothesis $(a, b) \not\equiv (0, 1)$ insures that $a \neq 0$, hence that (x, y) has a second term ad not zero, as required by our definition of a rational number. Q.E.D.

We now wish to show that $Q(D)$ actually contains our original integral domain D as a subdomain—in other words, that $Q(D)$ is actually an *extension* of D. This is not strictly possible, since a couple (a, b) can't be the same thing as an element of D. However, we can associate with each $a \in D$ a couple $(a, 1)$ which behaves under equality, addition, and multiplication exactly like a itself, as shown by

$$(a, 1) + (b, 1) = (a \cdot 1 + b \cdot 1, 1 \cdot 1) = (a + b, 1),$$

$$(a, 1) \cdot (b, 1) = (ab, 1 \cdot 1) = (ab, 1),$$

$$(a, 1) \equiv (b, 1) \qquad \text{if and only if} \qquad a = b.$$

One may conclude that the one-one correspondence $a \leftrightarrow (a, 1)$ is an isomorphism of the given integral domain D to a subdomain of the field $Q(D) = F$. Moreover, equations (8) and (8') show that any couple $r = (a, b) \in Q(D)$ is the solution of an equation $(b, 1)r = (a, 1)$, or $br = a$; hence $r = (a, b)$ is the quotient a/b. This proves

Theorem 5. *Any integral domain D can be embedded isomorphically in a field $Q(D)$, each element of which is a quotient of two elements of D.*

Theorem 5 applies in particular to the domain $\mathbf{Z}$; indeed it is suggestive to follow through the preceding arguments thinking of the special case that $D = \mathbf{Z}$, so that $Q(D) = Q(\mathbf{Z})$ is the set of all ordinary fractions. Hence we have the

Corollary. *The integral domain $\mathbf{Z}$ can be embedded as a subdomain in a field $\mathbf{Q} = Q(\mathbf{Z})$, each element of which is a quotient a/b of integers, $b \neq 0$.*

We now show that the rational field $\mathbf{Q} = Q(\mathbf{Z})$ is in fact exactly characterized (up to isomorphism) by the preceding statement. Since $\mathbf{Z}$ is defined by its postulates only up to an isomorphism, this is as complete a characterization as we can hope for. We will, in fact, prove the analogous result for any domain D.

Theorem 6. *Let an integral domain D be contained as a subdomain in any field F. Then the set of all those elements of F of the form a/b, $a, b \in D$, $b \neq 0$, is a subfield S of F; moreover, this subfield S is isomorphic to $Q(D)$ under the correspondence $a/b \leftrightarrow (a, b)$.*

Note. An isomorphism between two fields F and F' means an isomorphism between F and F' regarded as commutative rings. Specifically, it is a one-one correspondence between F and F' such that if

$x \leftrightarrow x'$ and $y \leftrightarrow y'$, then

$$(x + y) \leftrightarrow (x' + y') \quad \text{and} \quad (xy) \leftrightarrow (x'y').$$

Proof. The field F contains quotients a/b which are solutions of equations $bx = a$ with coefficients a and $b \neq 0$ in D. The set S of all these quotients contains all the integers $a/1 = a$; by the laws of Theorem 2, S is closed under addition, subtraction, multiplication, and division, so that S might be described as the closure of D under these operations in F. In any event, S is a field (Theorem 3).

The way in which these quotients a/b add, multiply, and become equal is described by (i)–(iii) of Theorem 2. Exactly the same rules are used for the couples (a, b). Hence the correspondence $a/b \leftrightarrow (a, b)$ is an isomorphism of the closure S of D onto $Q(D)$.   Q.E.D.

Observe, in particular, that this correspondence maps each a in D onto $a/1 \leftrightarrow (a, 1) = a$.

Combining Theorem 6 with the preceding corollary, we get

Theorem 7. *The integral domain* $\mathbf{Z}$ *can be embedded in one and only one way in a field* $\mathbf{Q} = Q(\mathbf{Z})$ *so that each element of* $\mathbf{Q}$ *is a quotient of integers.*

This completes the construction of the rational field $\mathbf{Q}$ from the integers.

Exercises

1. Prove in detail the commutative and the associative laws for multiplication of couples.
2. Prove that the "equality" relation defined by (5) is reflexive, symmetric, and transitive.
3. Let $\mathbf{Z}[i]$ be the set of all complex numbers $a + bi$, where a and b are integers and $i^2 = -1$. (a) State explicitly how to add and multiply two such numbers. (b) Prove that they form an integral domain. (c) Describe its quotient field.
4. Can the ring $\mathbf{Z}_6$ of integers modulo 6 be embedded in a field? Why?
5. Describe the field of quotients of the ring $\mathbf{Z}_5$ of integers modulo 5.
6. What is the field of quotients of the field $\mathbf{Q}$? Generalize.
7. Show that under any isomorphism $F \leftrightarrow F'$ between two fields, $a \leftrightarrow a'$, $b \leftrightarrow b'$, and $c \leftrightarrow c'$ imply $c^{-1} \leftrightarrow c'^{-1}$ and $(a - b)/c \leftrightarrow (a' - b')/c'$, provided $c \neq 0$. (Cf. Ex. 1 of §1.12.)
8. Prove that the correspondence $a + b\sqrt{7} \leftrightarrow a + b\sqrt{11}$ (a, b rational) is *not* an isomorphism.

★**9.** Prove that there is no isomorphism between the field $\mathbf{Q}(\sqrt{7})$ of numbers of the form $a + b\sqrt{7}$ and that of numbers of the form $a + b\sqrt{11}$ $(a, b$ rational). (*Hint:* Show that nothing can correspond to $\sqrt{7}$.)

10. What can one say about the fields of quotients a/b and a'/b' from isomorphic integral domains D and D'? Prove your statements.

★**11.** Prove that any rational number not 0 or ± 1 can be expressed uniquely in the form $(\pm 1)p_1{}^{e_1} \cdots p_r{}^{e_r}$, where the p_i are positive primes with $p_1 < p_2 < \cdots < p$, and the exponents e_i are positive or negative integers.

★**12.** Prove that any rational number $r/s \neq 0$ can be expressed uniquely in the form $r/s = b_1 + b_2/2! + b_3/3! + \cdots + b_n/n!$, where n is a suitable integer, and each b_k is an integer, with $0 \leqq b_k < k$ if $k > 1$, and $b_n \neq 0$.

13. For a fixed prime p show that the set $\mathbf{Z}_{(p)}$ of all rationals m/n with n prime to p is an integral domain. Identify its field of quotients.

14. Find the smallest subdomain of $\mathbf{Q}$ containing the rational numbers $1/6$ and $1/5$.

★**15.** Describe all possible integral domains which are subdomains of $\mathbf{Q}$.

16. Show that any field with exactly two elements is isomorphic to $\mathbf{Z}_2$.

17. Show that the integral domain $\mathbf{Z}[\sqrt{3}]$, consisting of all $a + b\sqrt{3}$ for integers a and b, has a field of quotients isomorphic to the set of all real numbers of the form $r + s\sqrt{3}$, r and s rational, and obtain an explicit isomorphism.

2.3. Simultaneous Linear Equations

A field need not consist of ordinary "numbers"; for instance, if p is a prime, the integers modulo p form a field containing only a finite number of distinct (i.e., incongruent) elements. The fact that the domain $\mathbf{Z}_p$ is a field is a corollary of

Theorem 8. *Any finite integral domain D is a field.*

Proof. The assumption that D is finite means that the elements of D can be completely enumerated in a list $b_1, b_2, \cdots, b_n$, where n is some positive integer (a discussion of finite sets in general appears in Chap. 12). To prove D a field, we need only provide an inverse for any specified element $a \neq 0$ in D. Try *all* the products

(9) $ab_1, ab_2, \cdots, ab_n$ $(b_1, \cdots, b_n$ the elements of D).

This gives n elements in D which are all distinct, because $ab_i = ab_j$ for $i \neq j$ would by the cancellation law entail $b_i = b_j$, counter to the assumption that the b's are distinct. Since this list (9) exhausts all of D, the unity element 1 of D must somewhere appear in the list as $1 = ab_i$. The corresponding element b_i is the desired inverse of a.   Q.E.D.

To actually find the inverse in $\mathbf{Z}_p$ by the proof, one proceeds by trial of all possible numbers b_i in $\mathbf{Z}_p$. Inverses can also be computed directly, for the equation $ax = 1$ with $a \neq 0$ in $\mathbf{Z}_p$ is simply another form of the congruence $ax \equiv 1 \pmod{p}$ with $a \neq 0$, and the latter can be solved for the integer x by the Euclidean algorithm methods, as in Theorem 16 of §1.9.

It is a remarkable fact that the entire theory of simultaneous linear equations applies to fields in general. Thus, consider the two simultaneous equations

$$(10) \qquad ax + by = e, \qquad cx + dy = f,$$

where the letters $a, \cdots, f$ stand for arbitrary elements of the field F. Multiplying the first equation by d, the second by b, and subtracting, we get $(ad - bc)x = de - bf$; multiplying the second equation by a, the first by c, and subtracting, we get $(ad - bc)y = af - ce$. Hence, if we define the *determinant* of the coefficients of (10) as (cf. Chap. 10)

$$\Delta = \begin{vmatrix} a & b \\ c & d \end{vmatrix} = ad - bc,$$

and if $\Delta \neq 0$, then equations (10) have the solution

$$(10') \qquad x = \frac{de - bf}{\Delta}, \qquad y = \frac{af - ce}{\Delta} \qquad (\Delta = ad - bc),$$

and no other solution. Whereas if $\Delta = 0$, then equations (10) have either no solution or many solutions (the latter eventuality arises when $c = ka$, $d = kb$, $f = ke$, so that the two equations are "proportional").

Gauss Elimination. The preceding device of *elimination* can be extended to m simultaneous linear equations in n unknowns $x_1, \cdots, x_n$, of the form

$$
\begin{aligned}
a_{11}x_1 + a_{12}x_2 + \cdots + a_{1n}x_n &= b_1, \\
a_{21}x_1 + a_{22}x_2 + \cdots + a_{2n}x_n &= b_2, \\
&\vdots \\
a_{m1}x_1 + a_{m2}x_2 + \cdots + a_{mn}x_n &= b_m.
\end{aligned}
$$

(11)

Here both the known coefficients a_{ij}, b_i and the unknowns x_j are restricted to a specified field F. We will now describe a general process, known as *Gauss elimination*, for finding all solutions of the given set (system) of equations. The idea is to replace the given system by a simpler system, which is *equivalent* to the given system in the sense of having precisely the same solutions. (Thus, the degenerate equation $0 \cdot x_1 + \cdots + 0 \cdot x_n = b_i$ is "equivalent" to $0 = b_i$, which cannot be satisfied.)

In a more compact notation, we write down only the ith equation, indicating its form by a sample term $a_{ij}x_j$ and the statement that the equation is to be summed over $j = 1, \cdots, n$ by writing

(11') $\displaystyle\sum_{j=1}^{n} a_{ij}x_j = b_i$ for $i = 1, \cdots, m$;   all $a_{ij} \in F$.

We argue by induction on n, the number of unknowns, distinguishing two cases.

Case 1. Every $a_{i1} = 0$. Then, trivially, the system (11') is equivalent to a "smaller" system of m equations in the $n - 1$ unknowns $x_2, \cdots, x_n$; x_1 is arbitrary for any solution of the smaller system.

Case 2. Some $a_{i1} \neq 0$. By interchanging two equations (if necessary), we get an equivalent system with $a_{11} \neq 0$. Multiplying the first equation by a_{11}^{-1}, we then get an equivalent system in which $a_{11} = 1$. Then subtracting a_{i1} times the new first equation from each ith equation in turn ($i = 2, \cdots, m$), we get an equivalent system of the form

$$x_1 + a_{12}'x_2 + a_{13}'x_3 + \cdots + a_{1n}'x_n = b_1'$$
$$a_{22}'x_2 + a_{22}'x_3 + \cdots + a_{2n}'x_n = b_2'$$
(12) $$\vdots$$
$$a_{m2}'x_2 + a_{m3}'x_3 + \cdots + a_{mn}'x_n = b_n'$$

For example, over the field $\mathbf{Z}_{11}$ this would reduce

$$3x + 5y + 7z \equiv 6 \qquad\qquad x + 9y + 6z \equiv 2$$
$$5x + 9y + 6z \equiv 7 \quad \text{to} \quad 8y + 9z \equiv 8$$
$$2x + y + 4z \equiv 3 \qquad\qquad 5y + 3z \equiv 10,$$

where all equations are understood to be modulo 11.

Proceeding by induction on m, we obtain

Theorem 9. *Any system (11) of m simultaneous linear equations in n unknowns can be reduced to an equivalent system whose ith equation has the form*

(13) $x_i + c_{i,i+1}x_{i+1} + c_{i,i+2}x_{i+2} + \cdots + c_{in}x_n = d_i,$

for some subset of r of the integers i = 1, $\cdots$, m, plus m − r equations of the form 0 = d_k.

Proof. If Case 2 always arises, we get m equations of the form (12), and the given system is said to be *compatible.* If Case 1 arises, then we may get degenerate equations of the form $0 = d_k$. If all $d_k = 0$, these can be ignored; if one $d_k \neq 0$, the original system (11) is *incompatible* (has no solutions). Q.E.D.

Written out in full, the system (13) looks like the display written below

$$x_1 + c_{12}x_2 + c_{13}x_3 + \cdots \quad \cdots + c_{1n}x_n = d_1,$$

$$x_2 + c_{23}x_3 + \cdots \quad \cdots + c_{2n}x_n = d_2,$$

$$x_3 + \cdots \quad \cdots + c_{3n}x_n = d_3,$$

$$\vdots$$

$$x_r + \cdots + c_{rn}x_n = d_r \qquad (r \leq m),$$

which is said to be in *echelon* form.

Solutions of any system of the echelon form (13) are easily described. Consider $x_n, x_{n-1}, x_{n-2}, \cdots, x_1$ in succession. If a given x_i in this sequence is the first variable in an equation of (13), then it is determined by $x_n, \cdots, x_{i+1}$ from the relation

(13') $$x_i = d_i - c_{i,i+1} - c_{i,i+2} - \cdots - c_{in}x_n.$$

If it is not, then this x_i can be chosen arbitrarily. This proves the

Corollary. *In the compatible case of Theorem 9, the set of all solutions of (11) is determined as follows. The m − r variables x_k not occurring in (13) can be chosen arbitrarily (they are free parameters). For any choice of these x_k, the remaining x_i can be computed recursively by substituting in (13').*

In the numerical example displayed, $8y + 9z \equiv 8 \pmod{11}$ would first be reduced to $y + 8z \equiv 1 \pmod{11}$. Subtracting five times this equation from $5y + 3z \equiv 10 \pmod{11}$, we get $7z \equiv 5 \pmod{11}$, whence $z \equiv 7 \pmod{11}$. The echelon form of the given system is thus

$$\left. \begin{array}{r} x + 9y + 6z \equiv 2 \\ y + 8z \equiv 1 \\ z \equiv 7 \end{array} \right\} \pmod{11}.$$

Solving, we get $y \equiv 1 - 8z \equiv 0 \pmod{11}$, and $x \equiv 2 - 9y - 6z \equiv 4 \pmod{11}$. The solution $x = 4, y = 0, z = 7$ can be checked by substituting into the original equation.

A system of equations (11) is *homogeneous* if the constants b_i on the right are all zero. Such a system always has a (trivial) solution $x_1 = x_2 = \cdots = x_n = 0$. There may be no further solutions, but if the number of variables exceeds the number of equations, the last equation of (12, will always contain an extra variable which can be chosen at will. Furthermore, the possible inconsistent equations $0 = d_i$ can never arise for homogeneous equations. Hence,

Theorem 10. *A system of m homogeneous linear equations in n variables, with m < n, always has a solution in which not all the unknowns are zero.*

Exercises

1. Solve the following simultaneous congruences:
   - (a) $3x + 2y \equiv 1 \pmod{7}$, $4x + 6y \equiv 3 \pmod{7}$;
   - (b) $2x + 7y \equiv 3 \pmod{11}$, $3x + 4z \equiv 6 \pmod{11}$,
     $$4x + 7y + z \equiv 0 \pmod{11};$$
   - (c) $x - 2y + z \equiv 5 \pmod{13}$, $2x + 2y \equiv 7 \pmod{13}$,
     $$5x - 3y + 4z \equiv 1 \pmod{13}.$$

2. Solve equations (a) and (b) in Ex. 1, with moduli deleted, in the field **Q** of rational numbers.

3. Solve in $\mathbf{Q}(\sqrt{2})$ the simultaneous equations
   $$(1 + \sqrt{2})x + (1 - \sqrt{2})y = 2, \qquad (2 - \sqrt{2})x + (3 - \sqrt{2})y = 1.$$

4. Find all incongruent solutions of the simultaneous congruences
   $$x + y + z \equiv 0 \pmod{5}, \qquad 3x + 2y + 4z \equiv 0 \pmod{5}.$$

5. Find all incongruent solutions of the simultaneous congruences:
   - (a) $x + 2y - z + 5t \equiv 4$, $2x + 5y + z + 2t \equiv 1$,
     $$x + 3y + 2s + 6t \equiv 2, \text{ all mod 7};$$
   - (b) $x + y + z \equiv 1 \pmod{5}$, $3x + 3y + 3z \equiv 4 \pmod{5}$.

6. Prove that two equations $a_1 x_1 + \cdots + a_n x_n = c$, $b_1 x_1 + \cdots + b_n x_n = d$ always have a solution for coefficients in a given field, provided there are no constants $k \neq 0$ and $m \neq 0$ with $k a_i = m b_i$ for $i = 1, \cdots, n$.

7. Prove that if $(x_1, \cdots, x_n)$ is any solution of a system of homogeneous linear equations, then $(-x_1, \cdots, -x_n)$ is another solution. What can be said about the sum of two solutions?

★8. (a) Prove that the *three* simultaneous equations

$$ax + by + cz = d, \qquad a'x + b'y + c'z = d', \qquad a''x + b''y + c''z = d'',$$

have one and only one solution in any field F if the 3×3 determinant

$$\Delta = ab'c'' + a'b''c + a''bc' - a''b'c - a'bc'' - ab''c' \neq 0.$$

(b) Compute a formula for x in (a), and use it to show that $x = 4$ for the three simultaneous linear equations over $\mathbf{Z}_{11}$ displayed below (12).

2.4. Ordered Fields

A field F is said to be *ordered* if it contains a set P of "positive" elements with the additive, multiplicative, and trichotomic properties listed in §1.3; in other words, a field is ordered if, when considered as a domain, it is an ordered integral domain. We know by experience that the rational numbers do constitute such an ordered field; we shall now prove this from our construction of rationals as couples of integers, and shall show further that the "natural" method of ordering is the only way of making the rational numbers into an ordered field.

First recall that in any ordered domain a nonzero square b^2 is always positive. If a quotient a/b is positive, the product $(a/b)b^2 = ab$ must therefore also be positive, and conversely. Hence in any ordered field,

$$(14) \qquad\qquad a/b > 0 \qquad \text{if and only if} \qquad ab > 0.$$

But the rational number (a, b) was intended to represent the quotient a/b. Hence we define a rational number (a, b) to be *positive* if and only if the product ab is positive in $\mathbf{Z}$.

Theorem 11. *The rational numbers form an ordered field if $(a, b) > 0$ is defined to mean that the integer ab is positive.*

Proof. Since we have defined equality by convention, we must prove that equals of positive elements are positive: $(a, b) > 0$ and $(a, b) \equiv (c, d)$ imply $(c, d) > 0$. This is true, since cd has the same sign as b^2cd, ab the same sign as abd^2, and since $abd^2 = b^2cd$ in virtue of the hypothesis $ad = bc$. Positiveness also has the requisite additive, multiplicative, and trichotomic properties. For instance, the sum of two positive couples (a, b) and (c, d) is positive, since $ab > 0$ and $cd > 0$ imply $d^2ab > 0$ and $b^2cd > 0$, whence

$$bd(ad + bc) = d^2ab + b^2cd > 0,$$

which is to say that the sum $(ad + bc, bd)$ is positive. Finally, the definition of "positive" for fractions agrees with the natural order of the special fractions $(a, 1)$ which represent integers, for $(a, 1)$ is positive by the definition (14) only if $1 \cdot a > 0$.   Q.E.D.

Since the proof of Theorem 11 involves only the assumption that the integers are an ordered domain, it in fact establishes the following more general result.

Theorem 12. *The field Q of quotients of an ordered integral domain D may be ordered by the stipulation that a quotient a/b of elements a and b of D is positive if and only if ab is positive. This is the only way in which the order of D may be extended to make Q an ordered field.*

There are many other ordered fields: the field of real numbers, the field $\mathbf{Q}(\sqrt{2})$ of numbers $a + b\sqrt{2}$ (see §2.1), and other subfields of the real number field. In any such field an absolute value can be introduced as in §1.3, and the properties of inequalities established there will hold. In any ordered field, in addition to the rules valid in any ordered domain, one may prove

(15) $\qquad\qquad 0 < 1/a \qquad$ if and only if $\qquad a > 0,$

(16) $\qquad\qquad a/b < c/d \qquad$ if and only if $\qquad abd^2 < b^2cd,$

(17) $\qquad\qquad 0 < a < b \qquad$ implies $\qquad 0 < 1/b < 1/a,$

(18) $\qquad\qquad a < b < 0 \qquad$ implies $\qquad 0 > 1/a > 1/b,$

(19) $\qquad\qquad\qquad a_1^2 + a_2^2 + \cdots + a_n^2 \geqq 0.$

The two rules (17) and (18) are the usual ones for the division of inequalities. The rule (19) that a sum of squares is never negative (Theorem 2, §1.3) is especially useful. For instance, if $a \neq b$, then $(a - b)^2 > 0$, so $a^2 - 2ab + b^2 > 0$, which gives $a^2 + b^2 > 2ab$. In this, set $x = a^2$ and $y = b^2$ and divide by 2. Then

$$(x + y)/2 > \sqrt{xy} \qquad (x \neq y).$$

This states that the arithmetic mean of two distinct positive real numbers exceeds the geometric mean $\sqrt{xy}$.

Exercises

1. Assuming that the integers form an ordered domain, prove that the product of two positive rational numbers is positive.

2. Prove similarly that if $(a, b) \neq 0$, then just one of the two alternatives $(a, b) > 0$ and $-(a, b) > 0$ holds in $Q(D)$, D an ordered domain.

3. Prove $|xx' + yy'| \leq \sqrt{(x^2 + y^2)(x'^2 + y'^2)}$ in any ordered field in which all positive numbers have square roots. (*Hint:* Square both sides.)

4. Prove formulas (15)–(19) of the text.

5. If n is a positive integer and a and b positive rational numbers, prove that $(a^n + b^n)/2 \geq ((a + b)/2)^n$. (*Hint:* Set $(a + b)/2 = r, a = r + d, b = r - d$.)

6. (a) Prove: any subfield of an ordered field is an ordered field.
(b) Is any subdomain of an ordered field an ordered domain?

7. For the rational numbers (or, more generally, in any ordered field), prove that if $a < b$, there are infinitely many x satisfying $a < x < b$.

8. Prove that in no ordered field do the positive elements form a well-ordered set.

9. A common mistake in arithmetic is the assumption that $a/b + a/c = a/(b + c)$.
(a) Show that in any field, $a/b + a/c = a/(b + c)$ implies $a = 0$ or $b^2 + bc + c^2 = 0$.
(b) Show that in an ordered field, it implies $a = 0$.

★2.5. Postulates for the Positive Integers

Although we have used the domain **Z** of all integers as the starting point for our review of the basic number systems of mathematics, this procedure is really quite sophisticated because it assumes that negative numbers exist. In the rest of this chapter we shall show how this assumption can be avoided, by showing how to derive the negative integers and their properties from familiar facts about positive integers alone.

For consistency, we begin by listing some basic properties of the system **Z**$^+$ of all positive integers that follow easily from the results of Chap. 1.

Theorem 13. *The system* **Z**$^+$ *of all positive integers in* **Z** *has the following properties:*

(i) *It is closed under uniquely defined binary operations of addition and multiplication, which are associative, commutative, and distributive.*

(ii) *There exists a multiplicative identity 1 in* **Z**$^+$, *such that* $m \cdot 1 = m$ *for all m in* **Z**$^+$.

(iii) *Furthermore, the following cancellation law holds in* **Z**$^+$:

(20) 　　　　　　　　*if* $mx = nx$, 　　*then*　　$m = n$.

★ Sections which are starred may be omitted without loss of continuity.

(iv) *Again, for any two elements m and n of* $\mathbf{Z}^+$, *exactly one of the following alternatives holds: m = n, or m + x = n has a solution x in* $\mathbf{Z}^+$, *or m = n + y has a solution y in* $\mathbf{Z}^+$.

(v) *Finally, the Principle of Finite Induction holds in* $\mathbf{Z}^+$: *any subset of* $\mathbf{Z}^+$ *which contains 1, and n + 1 whenever it contains n, contains every element of* $\mathbf{Z}^+$.

We leave the proof of these properties of $\mathbf{Z}^+$ as an exercise.

Conversely, if the properties (i)–(v) stated in this theorem are viewed as postulates, they do completely characterize the positive integers, in the sense that the positive integers as we previously defined them do have these properties and that any other system satisfying these postulates can be proved to be isomorphic to this system of positive integers. Note in particular that if $m + x = n$ in $\mathbf{Z}^+$, then

$$n + z = (m + x) + z = m + (x + z) = m + (z + x) = (m + z) + x,$$

whence $m + z = n + z$ is impossible, by (iv). Similarly, $n = m + y$ is incompatible with $m + z = n + z$. Therefore, appealing a third time to property (iv), we obtain

(21)   if $m + z = n + z,$ then $m = n.$

Furthermore, the three alternatives about the equations $m + x = n$ take the place of some of the order properties of the positive integers.

Starting with the positive integers, as given by these postulates, one may reconstruct the system $\mathbf{Z}$ of all integers. The object of this construction is to get a system larger than $\mathbf{Z}^+$ in which subtraction will always be possible. Hence we introduce as new elements certain couples (m, n) of positive integers, where each couple is to behave as if it were the solution of the equation $n + x = m$. The details of this construction resemble the construction of the rationals from the integers (§2.2).

Definition. *An integer is a couple* (m, n) *of positive integers m and n. "Equality" of couples is defined by the convention*

(22) $(m, n) \equiv (r, s)$ means $m + s = n + r,$

while sums and products are defined by

(23) $(m, n) + (r, s) = (m + r, n + s),$

(24) $(m, n) \cdot (r, s) = (mr + ns, ms + nr).$

Finally, (m, n) *is "positive" if and only if* $n + x = m$ *for some positive integer x.*

The couples as introduced by these definitions do in fact satisfy all the postulates we have given for the integers. One must first verify that the equality introduced by (22) is reflexive, symmetric, and transitive, and that sums and products as given by (23) and (24) are uniquely determined in the sense of this equality. The various formal laws for an integral domain then follow by a systematic application of the definitions (23) and (24) to these laws, much as in the discussion of rational numbers. In particular, $(2, 1)$ is a unity and $(1, 1)$ a zero for the system just defined. Additive inverses exist since

$$(m, n) + (n, m) \equiv (1, 1) \qquad \text{for all} \quad (m, n).$$

The cancellation law for the multiplication of couples is harder to prove; one proof uses condition (iv) of Theorem 13. With this proved, we know that the couples form an integral domain.

By the postulate (iv) in Theorem 13, every couple can be written in just one of the three forms (m, m), $(n + x, n)$, $(m, m + x)$. Those of the first form are equal to the zero $(1, 1)$; those of the second form $(n + x, n)$ are the positive couples, and may be shown to have the additive, multiplicative, and trichotomic properties required in the definition of an ordered integral domain (§1.3). Moreover, $(m + x, m) \equiv (n + y, n)$ if and only if $x = y$. Hence, if "congruent" couples are actually identified, the correspondence $x \mapsto (n + x, n)$ is an injection from the given positive integers x and the new positive couples $(n + x, n)$. It is even a homomorphism, since by definitions (23) and (24),

$$(m + x, m) + (n + y, n) = (m + n + x + y, m + n),$$
$$(m + x, m)(n + y, n) = (mn + my + nx + mn + xy,$$
$$mn + nx + mn + my).$$

Hence the new "positive" couples satisfy the law of finite induction. We have thus sketched a proof of the following result:

Theorem 14. *The system* $\mathbf{Z}^+$ *of positive integers can be embedded in a larger system* $\mathbf{Z}$ *in which subtraction is possible, in such a way that any element of* $\mathbf{Z}$ *is a difference of two positive integers from* $\mathbf{Z}^+$. *The system* $\mathbf{Z}$ *thus constructed is an ordered domain whose positive elements satisfy the Principle of Finite Induction.*

By §1.5, Ex. 8, this result implies the well-ordering principle. It should be noticed that the proof just sketched involves only our postulates on $\mathbf{Z}^+$. Conversely, in any integral domain containing $\mathbf{Z}^+$, the differences $(a - b)$ of elements of $\mathbf{Z}^+$ must satisfy definitions (22)–(24). (Cf. §1.2, Ex.

5.) This proves

Theorem 15. *Any integral domain containing the system* $\mathbf{Z}^+$ *contains a subdomain isomorphic to the domain* $\mathbf{Z}$ *of all integers.*

Exercises

1. Prove the relation defined by (22) is reflexive, symmetric, and transitive.
2. Prove that if $(m, n) \equiv (m', n')$, then $(m, n) + (r, s) \equiv (m', n') + (r, s)$ and $(m, n) \cdot (r, s) \equiv (m', n') \cdot (r, s)$ for all (r, s).
3. Prove that the "addition" defined by (23) is commutative and associative.
4. Prove the same for the "multiplication" defined by (24).
5. Prove that (m, m) is the same for all m, and is an additive zero. Show that the first statement follows from the second.
6. Prove that $(m + 1, m)$ is a multiplicative identity.
7. Prove the distributive law.
8. Prove the cancellation law for multiplication.
9. What properties of $\mathbf{Z}^+$ have been used in Exs. 1–8? State a theorem bearing the same relation to Theorems 14 and 15 that Theorem 7 bears to Theorem 5.
10. Show that Theorem 14 would not hold for any definition of "positiveness" of couples (m, n) other than that stated after (24).
11. Prove Theorem 13 in detail.
★12. Show that postulate (iv) of Theorem 13 may be replaced by the requirement that $m + 1 \neq 1$ for every m of $\mathbf{Z}^+$. (This is essentially Peano's postulate (iii), as stated in Theorem 16.)
13. In $\mathbf{Z}^+$, define $m < n$ to means that $m + x = n$ for some $x \in \mathbf{Z}^+$. Prove (a) $m < n$ and $n < r$ imply $m < r$, (b) $m < m$ for no m; (c) $m < n$ implies $m + r < n + r$ for all r; (d) $m < n$ implies $mr < nr$ for all r.
★14. Show that conditions (c) and (d) of Ex. 13 may be used to replace the cancellation laws (20) and (21) in the list of postulates for $\mathbf{Z}^+$.
15. Show that repetition of the process used to obtain $\mathbf{Z}$ from $\mathbf{Z}^+$ yields no new extension of $\mathbf{Z}$. Can you generalize this result?

★2.6. Peano Postulates

Instead of regarding addition and multiplication as undefined operations on the set $P = \mathbf{Z}^+$ of positive integers, one can define them in terms of the successor function

(25) $$S(n) = n + 1.$$

Theorem 16. *The set P of positive integers and the successor function S have the following properties:*
(i) *$1 \in P$;*
(ii) *if $n \in P$, then $S(n) \in P$;*
(iii) *for no n in P is $S(n) = 1$;*
(iv) *for n and m in P, $S(n) = S(m)$ implies $n = m$;*
(v) *a subset of P which contains 1, and which contains $S(n)$ whenever it contains n, must equal P.*

Proof. The cited properties are immediate from Theorem 13. Note, in particular, that (v) is the Principle of Finite Induction.   Q.E.D.

The properties (i)–(v) are known as the *Peano postulates* for the positive integers. They suffice, as will be shown below, to prove all the properties of the positive integers. We shall now use them to show that our original postulates for the integers determine the integers up to isomorphism.

Theorem 17. *In any ordered domain D, there is a unique subset P' which satisfies the Peano postulates with respect to the unity $1'$ and the successor function $S'(a) = a + 1'$.*

Remark. Intuitively, it is clear that the sequence $1', 2', 3', \cdots$ defined by $2' = 1' + 1'$, $3' = 1' + 1' + 1'$, etc., is such a set P'. But we wish a formal proof, based on our postulates for ordered domains.

Proof. The set D^+ of all positive elements of D clearly contains $1'$ and satisfies (i) and (ii). Now let Σ be the class of all subsets T of D^+ which have the properties (i) and (ii) of P; we define P' to be the intersection of all these sets T; i.e., $a \in P'$ if and only if a is in every such set T.

By definition, (i) and (ii) hold for P'. Since P' consists only of positive elements, (iii) holds; since $a + 1' = b + 1'$ implies $a = b$, (iv) holds. To prove (v), let A be a subset of P' which contains $1'$ and contains $S'(a)$ whenever it contains a. Then A is one of the sets T used above, hence P' is contained in A, and therefore $P' = A$. This proves (v) for P', and (v) shows that P' is the only possible such set, since P' satisfies (i) and (ii).

Theorem 18. *The subset P' of Theorem 17 is isomorphic to the set P of positive integers with respect to addition, multiplication, and order.*

Remark. Informally, it is clear that $1 \mapsto 1'$, $2 \mapsto 2', \cdots$ should yield the required isomorphism. Since $1' < 1' + 1' < 1' + 1' + 1' < \cdots$, this correspondence should preserve order.

Proof. First let $Q(n)$ be the proposition that there is a unique correspondence $x \mapsto \phi_n(x)$ between the integers $1 \leq x \leq n$ in P and elements $\phi_n(x)$ in P' under which

(26) $\qquad \phi_n(1) = 1', \qquad \phi_n(S(x)) = S'(\phi_n(x)) \qquad$ for $1 \leqq x < n$.

Clearly $Q(1)$ holds. Given $Q(n)$ and hence a ϕ_n, we can construct a unique ϕ_{n+1} by setting $\phi_{n+1}(x) = \phi_n(x)$ for $1 \leqq x \leqq n$ and $\phi_{n+1}(n+1) = S'(\phi_n(n))$. Hence $Q(n)$ implies $Q(n+1)$. This proves $Q(n)$ by induction.

Again, if $1 \leqq x \leqq n < m$, one can prove by induction on x that $\phi_n(x) = \phi_m(x)$; hence $\phi_n(x)$ is independent of n, provided that $x \leqq n$. Let $\phi(x)$ denote this element of P'. This gives a correspondence $x \mapsto \phi(x)$ of P to P' with the properties

(27) $\qquad\qquad\qquad \phi(1) = 1', \qquad \phi(S(x)) = S'(\phi(x)).$

Every element of P' is the correspondent $\phi(x)$ of some $x \in P$, for the set of elements $\phi(x)$ includes $1'$, and includes with any $\phi(x)$ its successor; hence the set is all of P' by property (v) of P'.

Both in P and in P' we have

(28) $\qquad\qquad n + 1 = S(n) \qquad n + S(m) = S(n + m),$

(29) $\qquad\qquad n \cdot 1 = n \qquad\quad n \cdot S(m) = n \cdot m + n.$

From these equations and (27), one can easily prove by induction on m that $\phi(n + m) = \phi(n) + \phi(m)$ and $\phi(nm) = \phi(n)\phi(m)$; in other words, ϕ is an isomorphism with respect to addition and multiplication.

Next, ϕ preserves order; that is, $m < n$ implies $\phi(m) < \phi(n)$. Indeed, by the definition, $m < n$ means that $n - m$ is positive; that is,

(30) $\quad m < n \qquad$ if and only if $\qquad n = m + k \qquad$ for some k in P.

Hence $m < n$ yields $n = m + k$, hence $\phi(n) = \phi(m) + \phi(k)$; since $\phi(k)$ is positive in D, this proves $\phi(m) < \phi(n)$, as required.

Finally, ϕ is a bijection of P to P'; since we already know that $\phi(x)$ includes all of P', we need only show that $n \neq m$ implies $\phi(n) \neq \phi(m)$. But $n \neq m$ means, say, that $m < n$, hence $\phi(m) < \phi(n)$ and therefore $\phi(n) \neq \phi(m)$.

To summarize our conclusions, we define an *order-isomorphism* between two ordered domains to be an isomorphism which preserves order. In view of Theorem 15, we get from Theorem 18 the following corollary:

Corollary 1. *Any ordered domain D contains a subdomain order-isomorphic with* **Z.**

Combining this result with Theorems 6 and 7, we have

Corollary 2. *Any ordered field contains a subfield order-isomorphic with the field* **Q** *of rational numbers.*

This result gives an abstract characterization of the rational field as the *smallest ordered field*.

Finally, in case the positive elements in D are well-ordered, the set P' of Theorem 17 can easily be shown to consist of *all* the positive elements of D. This proves:

Corollary 3. *There is, up to order-isomorphism, only one ordered domain* **Z** *whose positive elements form a well-ordered set.*

This shows that the postulates we have used for the integers determine the integers uniquely, up to isomorphism.

The treatment of the integers can begin, not with the postulates for a well-ordered domain, but with the Peano postulates. The essential point is the observation that the recursive equations (28) and (29) can be used to define complete addition and multiplication tables. Formally, one can establish, much as in the proof of Theorem 15, that there is one and only one binary operation + satisfying (28), and similarly for multiplication. The various properties listed in Theorem 13 can then be established by induction, and the construction of couples given in §2.5 then yields the integers from the Peano postulates.

Exercises

In the following exercises, assume only the Peano postulates and that addition and multiplication are defined by (28) and (29).

1. Show by induction that $n + 1 = 1 + n$.
2. Using Ex. 1, show that addition is commutative.
3. Prove that addition is associative.
4. Prove that multiplication is associative.
5. Prove the distributive law.

3

Polynomials

3.1. Polynomial Forms

Let D be any integral domain, and let x be any element of a larger integral domain E which contains D as a subdomain. In E one can form sums, differences, and products of x with the elements of D and with itself.

By performing these operations repeatedly, one evidently gets all expressions of the form

$$(1) \quad a_0 + a_1x + \cdots + a_nx^n \quad (a_0, \cdots, a_n \in D; \quad a_n \neq 0 \text{ if } n > 0),$$

where x^n (n any positive integer) is defined as $xx \cdots x$ to n factors. But conversely, using only the postulates for an integral domain, one can add, subtract, or multiply any two expressions of the form (1), obtaining a third such expression. For example, if D is the domain of integers,

$$
\begin{aligned}
f(x) &= (0 + 1 \cdot x + (-2)x^2)(2 + 3 \cdot x) \\
&= 0 \cdot 2 + 0 \cdot 3 \cdot x + 1 \cdot x \cdot 2 + 1 \cdot x \cdot 3 \cdot x + (-2)x^2 \cdot 2 \\
&\qquad\qquad\qquad\qquad\qquad\qquad\qquad + (-2)x^2 \cdot 3 \cdot x \\
&= 0 + 0 \cdot x + 2x + 3x^2 + (-4)x^2 + (-6)x^3 \\
&= 0 + (0 + 2)x + (3 + (-4))x^2 + (-6)x^3 \\
&= 0 + 2x + (-1)x^2 + (-6)x^3,
\end{aligned}
$$

by the generalized distributive law, the commutative and associative law, and finally the distributive law.

This argument can be generalized. Indeed, let

$$p(x) = a_0 + a_1 x + \cdots + a_m x^m$$
$$\text{and} \quad q(x) = b_0 + b_1 x + \cdots + b_n x^n$$

be any two expressions of the form (1). If $m > n$, then we have

(2) $p(x) \pm q(x) = (a_0 \pm b_0) + \cdots + (a_n \pm b_n)x^n + a_{n+1}x^{n+1}$
$$+ \cdots + a_m x^m.$$

A similar formula holds if $m \leq n$. Again, by the distributive law,

$$p(x)q(x) = \sum_{i=0}^{m} \sum_{j=0}^{n} a_i b_j x^{i+j}.$$

Collecting terms with the same exponent and adding coefficients, we have

(3) $p(x)q(x) = a_0 b_0 + (a_0 b_1 + a_1 b_0)x + \cdots + a_m b_n x^{m+n}.$

In this formula, the coefficient of x^k is clearly a sum

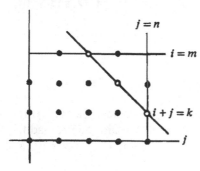

Figure 1

$$\sum_i a_i b_{k-i}$$

for all i with $0 \leq i \leq m$ and $0 \leq k - i \leq n$. See Figure 1.

We have thus proved the following result:

Theorem 1. *Assume there exists an integral domain E containing a subdomain isomorphic with the given domain D, and an element x not in D. Then the polynomials (1) in this element x are added, subtracted, and multiplied by formulas (2) and (3), and so form a subdomain of E.*

In order to prove that there always does exist such an integral domain *E*, one wants the following definition.

Definition. *By a* polynomial *in x over an integral domain D is meant an expression of the form (1). The integer n is called the* degree *of the form (1). Two polynomials are called equal if they have the same degree and if corresponding coefficients are equal.*

Since nothing is assumed known about the symbol x, the expression (1) is also often called a polynomial *form* (to distinguish it from a polynomial function; see §3.2), and the symbol x itself is called an *indeterminate*.

Theorem 2. *If addition and multiplication are defined by formulas (2) and (3), then the different polynomial forms in x over any integral domain D form a new integral domain D[x] containing D.*

Proof. The absence of zero-divisors (cancellation law of multiplication) follows from (3), since the leading coefficient $a_m b_n$ of the product of two nonzero polynomial forms is the (nonzero) product of the nonzero leading coefficients a_m and b_n of its factors. The properties of 0 and 1 and the existence of additive inverses follow readily from (2) and (3).

To prove the commutative, associative, and distributive laws it is convenient to introduce "dummy" zero coefficients. This changes (2) and (3) to the simpler forms

(2')
$$\sum_{k=0}^{\infty} a_k x^k + \sum_{k=0}^{\infty} b_k x^k = \sum_{k=0}^{\infty} (a_k + b_k) x^k,$$

(3')
$$\left(\sum_{k=0}^{\infty} a_k x^k\right)\left(\sum_{k=0}^{\infty} b_k x^k\right) = \sum_{k=0}^{\infty} \left(\sum_{i+j=k} a_i b_j x^k\right),$$

where all but a finite number of coefficients are zero. Any law such as the distributive law may then be verified simply by multiplying out both sides of the law by rules (2') and (3'), as

$$\left(\sum_k a_k x^k\right)\left(\sum_k b_k x^k + \sum_k c_k x^k\right) = \sum_k \left[\sum_{i+j=k} a_i(b_j + c_j)\right]x^k,$$

$$\left(\sum_k a_k x^k\right)\left(\sum_k b_k x^k\right) + \left(\sum_k a_k x^k\right)\left(\sum_k c_k x^k\right) = \sum_k \left[\left(\sum_{i+j=k} a_i b_j\right) + \left(\sum_{i+j=k} a_i c_j\right)\right]x^k,$$

and showing that the coefficient of each power x^k of x is the same in both expressions. By the distributive law in the domain D, the coefficient of the k-th power of x is the same in both expressions. Similar arguments complete the proof of Theorem 2.

Now recalling Theorem 7 of §2.2, we see that if we define a *rational form* in the indeterminate x over D as a formal *quotient*

$$\frac{p(x)}{q(x)} = \frac{a_0 + a_1 x + \cdots + a_m x^m}{b_0 + b_1 + \cdots + b_n x^n} \qquad (a_i, b_j \text{ in } D; a_m \neq 0 \text{ if } m > 0; b_n \not\equiv 0)$$

of polynomial forms with non-zero denominator, and define equality, addition, and multiplication by (5), (6), and (7) of the Definition of §2.2, we get a field.

Corollary. *The rational forms in an indeterminate x over any integral domain D constitute a field. This field is denoted by $D(x)$.*

Exercises

1. Reduce to the form (1): $x^2 - 5x(3x + 7)^2$,
   $(x^2 + 5x - 4)(x^2 - 2x + 3)$, $(3x^2 + 7x - 1/2)(x^3 - x/2 + 1)$.
2. Compute similarly $(3x^3 + 5x - 4)(4x^3 - x + 3)$, where the coefficients are the integers mod 7.
3. Is $x^3 + 5x - 4$ of the form (1)? Reduce it to this form. Reduce $(1 + x + 2x^2 + 3x^3) - (0 + x + x^2 + 3x^3)$ to the form (1), stating which postulates are used at each step.
4. (a) Is $1/2 + 3 \cdot x^{1/2} + 5x$ a polynomial form over the rational field?
   (b) Why is $x^3 \cdot x^4$ not equal to x^2 in the domain of polynomial forms with coefficients in $\mathbf{Z}_5$?
5. Discuss the following statements:
   (a) The degree of the product of two polynomial forms is the sum of the degrees of the factors.
   (b) The degree of the sum of two polynomials is the larger of the degrees of the summands.
6. Prove that the associative laws for addition and multiplication hold in $D[x]$.
7. The "formal derivative" of $p(x) = a_0 + a_1x + \cdots + a_nx^n$ is defined as $p'(x) = a_1 + 2a_2x + \cdots + na_nx^{n-1}$. Prove, over any integral domain:
   (a) $(cp)' = cp'$,   (b) $(p + q)' = p' + q'$,
   (c) $(pq)' = pq' + p'q$,   (d) $(p^n)' = np^{n-1}p'$.
★8. If $p(y)$ and $q(x)$ are polynomial forms in indeterminates y and x, show that the substitution $y = q(x)$ yields a polynomial $p(q(x))$. For the formal derivative of Ex. 7, prove that $[p(q(x))]' = p'(q(x)) \cdot q'(x)$.
★9. For given D show how to construct an integral domain $D\{t\}$ consisting of all "formal" infinite power series $a_0 + a_1t + a_2t^2 + \cdots$ in a symbol t, with coefficients a_i in D.
★10. (a) If D is an ordered domain, show that the polynomial forms (1) constitute an ordered domain $D[x]$ if $p(x) > 0$ is defined to mean that the first nonzero coefficient a_k in $p(x)$ is positive in D.
   (b) Show that $D[x]$ is also an ordered domain if we define $p(x) > 0$ to mean that $a_n > 0$ in (1).
★11. Setting $D = \mathbf{Z}$ in Ex. 10(b), show that 1 is the least "positive" polynomial in $\mathbf{Z}[x]$, although $\mathbf{Z}[x]$ fails to satisfy the well-ordering principle.

3.2. Polynomial Functions

As before, let D be any integral domain, and let

$$f(x) = a_0 + a_1 x + \cdots + a_m x^m$$

be any polynomial form in x over D. If the indeterminate x is replaced by an element $c \in D$, $f(x)$ no longer remains an empty expression: it can be evaluated as a definite member $a_0 + a_1 c + \cdots + a_m c^m$ of D. In other words, if x is regarded as an independent *variable* in the sense of the calculus, instead of as an abstract symbol outside of D, $f(x)$ becomes an ordinary *function*: "If x is given (as c), then $f(x)$ is determined (as $f(c)$)." By abstraction, we shall define generally a "function" f of a variable on D as a rule assigning to each element x of D a "value" $f(x)$, also in D. We shall define two such functions to be *equal* (in symbols, $f = g$) if and only if $f(x) = g(x)$ for all x. The *sum* $h = f + g$, the *difference* $q = f - g$, and the *product* $p = fg$ of two functions are defined by the rules $h(x) = f(x) + g(x)$, $q(x) = f(x) - g(x)$, and $p(x) = f(x)g(x)$ for all x. A *constant* function is one whose value b is independent of x; the *identity* function is the function j with $j(x) = x$ for all x.

Definition. *A polynomial function is a function which can be written in the form (1).*

Since the only rules used in deriving formulas (2) and (3) are valid in any integral domain, they hold no matter what value c (in D) is assigned to the indeterminate† x. That is, they are *identities*, and therefore sums and products of polynomial functions can also be computed by formulas (2) and (3). As will be explained in §3.3, it follows that the polynomial functions over D constitute a commutative ring in the sense defined in §1.1.

By definition, each form (1) determines a unique polynomial function, and each polynomial function is determined by at least one such form. Therefore there is certainly a mapping which preserves sums and products, from the polynomial forms to the polynomial functions over any given integral domain D. (Such correspondences are called homomorphisms onto, or epimorphisms; see §3.3.)

If we could be certain that the mapping was *one-one*, we would know that it was an *isomorphism*. Hence, from the point of view of abstract algebra, it would be permissible to forget the distinction between polynomial forms and polynomial functions. Unfortunately, such is not the case.

---

† Indeed, this is the secret of solving equations by letting "x be the unknown quantity": every manipulation allowed on x must be true for every possible value of x.

Indeed, over the field $\mathbf{Z}_3$ of integers mod 3, the distinct forms $f(x) = x^3 - x$ and $g(x) = 0$ determine the same function—the function which is identically zero. By Fermat's theorem (§1.9, Theorem 18), the same is true over $\mathbf{Z}_p$ for $x^p - x$ and 0. Hence, over any $\mathbf{Z}_p$, equality has an effectively different meaning for functions than it does for forms.

We shall now show that it is no accident that the domain of coefficients in the preceding example is finite. We could not construct such an example over the field of rationals. But before doing this, we recall some elementary definitions. By the *degree* of a nonzero form (1), we mean n, its biggest exponent. The term $a_n x^n$ of biggest degree is called its *leading term*, a_n its *leading coefficient*, and if $a_n = 1$, the polynomial is termed *monic*.

Theorem 3. *A polynomial form $r(x)$ over a domain D is divisible by $x - a$ if and only if $r(a) = 0$.*

Here the statement "$r(x)$ is divisible by $x - a$" means that $r(x) = (x - a)s(x)$ for some polynomial form over D.

Proof. Set $r(x) = c_0 + c_1 x + \cdots + c_n x^n$ $(c_n \neq 0)$. For every a, we have, by high school algebra,

$$\sum_{k=0}^{n} c_k x^k - \sum_{k=0}^{n} c_k a^k = \sum_{k=0}^{n} c_k (x^k - a^k)$$

$$= \sum_{k=1}^{n} c_k [(x - a)(x^{k-1} + x^{k-2} a + \cdots + a^{k-1})].$$

Therefore $r(x) - r(a) = (x - a)s(x)$, where $s(x)$ is a polynomial form of degree $n - 1$. Conversely, if $r(x) = (x - a)s(x)$, substituting a for x gives $r(a) = 0$.

Corollary. *A polynomial form $r(x)$ of degree n over an integral domain D has at most n zeros in D.*

(By a zero of $r(x)$ is meant a root of the equation $r(x) = 0$; that is, an element $a \in D$ such that $r(a) = 0$.)

Proof. If a is a zero, then by the theorem, $r(x) = (x - a)s(x)$, where $s(x)$ has degree $n - 1$. By induction, $s(x)$ has at most $n - 1$ zeros, but $r(x) = 0$ by Theorem 1 of §1.2 if and only if $x = a$ or $s(x) = 0$. Hence $r(x) = 0$ has at most n zeros.

Theorem 4. *If an integral domain D is infinite, then two polynomial forms over D which define the same function have identical coefficients.*

Proof. As in (1), let $p(x)$ and $q(x)$ be two given forms in the indeterminate x. If they determine the same function, then $p(a) = q(a)$ for every element a chosen from D; the desired conclusion is then that $p(x)$ and $q(x)$ have the same degree and have corresponding coefficients equal. In terms of the difference $r(x) = p(x) - q(x)$, this is to say that $r(a) = c_0 + c_1 a + \cdots + c_n a^n = 0$ for all a in D implies that $c_0 = c_1 = \cdots = c_n = 0$. This conclusion follows by Theorem 3, for unless the coefficients c_i are all zero, the polynomial $r(x)$ is zero for at most n values of x—whence, since D is infinite, there will be remaining values of x on which $r(x) \neq 0$.

Thus, if D is infinite, the concepts of polynomial function and polynomial form are equivalent (technically, the ring of polynomial functions is *isomorphic* to that of polynomial forms).

On the other hand, Theorem 4 never holds if D is a *finite* integral domain, with elements $a_1, \cdots, a_n$. For example, the monic polynomial form $(x - a_1)(x - a_2) \cdots (x - a_n)$ of degree n determines the same function as the form 0, in this case.

Since any system isomorphic with an integral domain is itself an integral domain, Theorem 4 implies the following corollary:

Corollary. *The polynomial functions on any infinite integral domain themselves form an integral domain.*

If D is an *infinite field*, distinct rational forms define distinct rational functions, and the rational *functions* on D form a field. (*Caution:* A rational function is not defined at all points, but only where the denominator is not zero. Thus it is defined, if D is a field, at all but a finite number of points.)

It is often desired to find a polynomial $p(x)$ of minimum degree which assumes given values $y_0, y_1, \cdots, y_n$ in a field F at $n + 1$ given points $a_0, a_1, \cdots, a_n \in F$, so that

$$(4) \qquad p(a_i) = y_i \qquad (i = 0, 1, \cdots, n; \quad a_i \neq a_j \text{ if } i \neq j).$$

This is called the problem of polynomial interpolation.

To solve this problem, consider the polynomials

$$q_i(x) = \prod_{j \neq i} (x - a_j) = (x - a_0) \cdots (x - a_{i-1})(x - a_{i+1}) \cdots (x - a_n).$$

Evidently, $q_i(a_j) = 0$ if $j \neq i$, while

$$C_i = q_i(a_i) = \prod_{j \neq i} (a_i - a_j) \neq 0.$$

Hence C_i^{-1} exists, and the following polynomial of degree n or less

$$(5) \qquad p(x) = \sum_{i=0}^{n} C_i^{-1} y_i q_i(x) = \sum_{i=0}^{n} \frac{y_i \prod_{j \neq i} (x - a_j)}{\prod_{j \neq i} (a_i - a_j)}$$

satisfies equations (4). Formula (5) is called Lagrange's interpolation formula.

In view of Theorem 3, at *most* one polynomial of degree n or less can satisfy equations (4): the difference of two such polynomials would have $n + 1$ zeros, and so would be the polynomial form zero. This proves the following result.

Theorem 5. *There is exactly one polynomial form of degree n or less which assumes given values at $n + 1$ distinct points.*

Exercises

1. In the domain Z_5, find a second polynomial form determining the same function as $x^2 - x + 1$.
2. Show that $x^2 - 1$ has four zeros over Z_{15}. Why doesn't this contradict the corollary of Theorem 3?
3. Show that if $a_0 = a_1 - h$, $a_2 = a_1 + h$, and $1 + 1 \neq 0$, then (4) can be solved for $n = 2$ by the parabolic interpolation formula

$$p(x) = y_1 + \frac{1}{2}(y_2 - y_0)\left(\frac{x - a_1}{h}\right) + \frac{1}{2}(y_2 - 2y_1 + y_0)\left(\frac{x - a_1}{h}\right)^2.$$

4. Find a cubic polynomial $f(x) = a + bx + cx^2 + dx^3$ satisfying $f(0) = 0$, $f(1) = 1$, $f(2) = 0$, $f(3) = 1$, by treating a, b, c, d as unknowns in four equations, of which the last is $a + 3b + 9c + 27d = 1$. (This is the method of undetermined coefficients.)
5. Use the interpolation formula (5) to show that every function on any finite field (such as Z_p) is equal to some polynomial function.
★6. Let D be a finite integral domain with n elements $a_1, \cdots, a_n$. Let $m(x)$ denote the fixed polynomial form $(x - a_1) \cdots (x - a_n)$.
    (a) Show that if two polynomial forms $f(x)$ and $g(x)$ determine the same function, then $m(x)$ is a divisor of the form $f(x) - g(x)$.
    (b) Compute $m(x)$ for the domains Z_3 and Z_5.
    (c) Show that $m(x) = x^p - x$ in case $D = Z_p$. (*Hint:* Use Fermat's theorem.)
7. Prove that over an infinite field, distinct rational forms which determine the same functions are formally equal in the sense of §2.2.

8. (a) If D and D' are isomorphic domains, prove that $D[x]$ is isomorphic to $D'[y]$, where $D[x]$ and $D'[y]$ are the domains of polynomial forms in indeterminates x and y over D and D', respectively.
   (b) How about $D(x)$ and $D'(y)$?
9. If Q is the field of quotients of a domain D (Theorem 4, §2.2), prove that the field $D(x)$ is isomorphic to the field $Q(x)$.

3.3. Homomorphisms of Commutative Rings

Let D be any given integral domain, and let $D\langle x \rangle$ denote the system of polynomial functions over D. For all $x \in D$, $f(x) + g(x) = g(x) + f(x)$, $0 + f(x) = f(x)$, $1 \cdot f(x) = f(x)$, and so on. Hence addition and multiplication are commutative, associative, and distributive; identity elements exist for addition and multiplication; and inverses exist for addition. In summary, $D\langle x \rangle$ satisfies all the postulates for an integral domain *except* the cancellation law of multiplication. This breaks down when D is finite because there exists a zero product $(x - a_1)(x - a_2) \cdots (x - a_n)$ of nonzero factors.

In other words, $D\langle x \rangle$ is a *commutative ring* in the sense defined in §1.1. For convenience, we recapitulate this definition here.

Definition. *A commutative ring is a set closed under two binary, commutative, and associative operations, called addition and multiplication, and in which further:*
   (i) *multiplication is distributive over addition;*
   (ii) *an additive identity (zero) 0 and additive inverses exist;*
   (iii) *a multiplicative identity (unity) 1 exists.*†

It will be recalled that Rules 1–9 in §1.2 were proved to be valid in any commutative ring. Also, an interesting family of *finite* commutative rings $\mathbf{Z}_m$ was constructed in §1.10, Theorem 19.

Another instance of a commutative ring is furnished by the system D^* of *all* functions on any integral domain D, where addition and multiplication are defined as in §3.2. There are zero-divisors in the domain D^* of all functions even on infinite integral domains D. Thus, if D is any *ordered* domain, and if we define $f(x) = |x| + x$ and $g(x) = |x| - x$, then $fg = h$ is $h(x) = |x|^2 - x^2 = 0$ for all x, yet $f \neq 0$, $g \neq 0$. On the other hand, D^* has every other defining property of an integral domain. One can prove each law for D^* from the corresponding law for D by the simple

---

† Some authors omit condition (iii) in defining commutative rings. *Noncommutative* rings will be considered in Chap. 13.

device of writing "for all x" in the right places. Thus $f(x) + g(x) = g(x) + f(x)$ for all x implies $f + g = g + f$. Again, if we define e as the constant function $e(x) = 1$ for all x, then $e(x)f(x) = 1 \cdot f(x) = f(x)$ for all x and f, whence $ef = f$ for all f, so that e is a multiplicative identity (unity) of D^*. (See why the cancellation law of multiplication cannot be proved in this way.) Since the cancellation law for multiplication was nowhere used in the above, we may assert:

Lemma 1. *The functions on any commutative ring A themselves form a commutative ring.*

Now let us define (by analogy with "subdomain") a *subring* of a commutative ring A as a subset of A which contains, with any two elements f and g, also $f \pm g$ and fg, and which also contains the unity of A.

By Theorem 1, the set $D\langle x \rangle$ of polynomial functions on any integral domain D (1) is a subring of the ring D^* of all functions on D, (2) contains all constant functions and the identity function, and (3) is contained in any other such subring. In this sense $D\langle x \rangle$ is the subring of D^* *generated* by the constant functions and the identity function. This gives a simple algebraic characterization of the concept of a polynomial function.

Deeper insight into commutative rings can be gained by generalizing the notion of isomorphism as follows.

Definition. *A function $\phi: a \mapsto a\phi$ from a commutative ring R into a commutative ring R' is called a* homomorphism *if and only if it satisfies, for all $a, b \in R$,*

$$(6) \qquad\qquad (a + b)\phi = a\phi + b\phi,$$

$$(7) \qquad\qquad (ab)\phi = (a\phi)(b\phi),$$

and carries the unity of R into the unity of R'.

These conditions state that the homomorphism preserves addition and multiplication. They have been written in the compact notation of §§1.11–1.12, whereby $a\phi$ signifies the transform of a by ϕ. If we write $\phi(a)$ instead of $a\phi$, they become $\phi(a + b) = \phi(a) + \phi(b)$ and $\phi(ab) = \phi(a)\phi(b)$ instead. Evidently, an isomorphism is just a homomorphism which is bijective (one-one and onto).

One easily verifies that the function from n to the residue class containing n, for any fixed modulus m, is a homomorphism $\mathbf{Z} \to \mathbf{Z}_m$

mapping the domain of integers onto the ring $\mathbf{Z}_m$ of §1.10, Theorem 19. We now prove another easy result.

Lemma 2. *Let ϕ be a homomorphism from a commutative ring R into a commutative ring R'. Then 0ϕ is the zero of R', and $(a - b)\phi = a\phi - b\phi$ for all $a, b \in R$.*

Proof. By (6), $0\phi = (0 + 0)\phi = 0\phi + 0\phi$, which proves that 0ϕ is the zero of R'. Likewise, if $x = a - b$ in R, then $b + x = a$ and so $a\phi = (b + x)\phi = b\phi + x\phi$, whence $x\phi = a\phi - b\phi$ in R'.

Theorem 6. *The correspondence $p(x) \mapsto f(x)$ from the domain $D[x]$ of polynomial forms over any integral domain D to the ring $D\langle x\rangle$ of polynomial functions over D is a homomorphism.*

Proof. For any element x in D, the addition and multiplication of the numbers $p(x)$ and $q(x)$ in D must conform to identities (2) and (3), since the derivation of these identities in §3.1 used only the postulates for an integral domain.

The result of Theorem 4 states that if D is infinite, then the homomorphism of Theorem 6 is an isomorphism.

Exercises

1. (a) Show that there are only four different functions on the field $\mathbf{Z}_2$, and write out addition and multiplication tables for this ring of functions.
   (b) Express each of these functions as a polynomial function.
   (c) Is this ring of functions isomorphic with the ring of integers modulo 4?
2. How many different functions are there on the ring $\mathbf{Z}_n$ of integers modulo n?
3. Are the following sets of functions commutative rings with unity?
   (a) all functions f on a domain D for which $f(0) = 0$,
   (b) all functions f on D with $f(0) = f(1)$,
   (c) all functions f on D with $f(0) \neq 0$,
   (d) all functions f on $\mathbf{Q}$ (the rational field) with $-7 \leq f(x) \leq 7$ for all x,
   (e) all f on $\mathbf{Q}$ with $f(x + 1) = f(x)$ for all x (such an f is periodic).
4. Construct two commutative rings of functions not included in the examples of Ex. 3.
5. Let D^* be defined as in the text. Prove the associative law for sums and products in D^*.
6. (a) If D and D' are isomorphic domains, prove that $D\langle x\rangle$ and $D'\langle x\rangle$ are isomorphic.
   (b) How about D^* and $(D')^*$?
7. Show that one cannot embed in a field the ring $\mathbf{Z}_p\langle x\rangle$ of all polynomial functions over $\mathbf{Z}_p$.

8. Show that if a homomorphism maps a commutative ring R *onto* a commutative ring R', then the unity of R is carried by ϕ into the unity of R'.
9. Show that if $\phi: R \to R'$ is any homomorphism of rings, then the set K of those elements in R which are mapped onto 0 in R' is a subring.

★3.4. Polynomials in Several Variables

The discussion of §§3.1–3.3 dealt with polynomials in a *single* variable (indeterminate) x. But most of the results extend without difficulty to the case of *several* variables (or indeterminates) $x_1, \cdots, x_n$.

Definition. *A polynomial form over D in indeterminates $x_1, \cdots, x_n$ is defined recursively as a form in x_n over the domain $D[x_1, \cdots, x_{n-1}]$ of polynomial forms in $x_1, \cdots, x_{n-1}$ (in short, $D[x_1, \cdots, x_n] = D[x_1, \cdots, x_{n-1}][x_n]$). A polynomial function of variables $x_1, \cdots, x_n$ on an integral domain D is one which can be built up by addition, subtraction, and multiplication from the constant function $f(x_1, \cdots, x_n) = c$ and the n identity functions $f_i(x_1, \cdots, x_n) = x_i$ $(i = 1, \cdots, n)$.*

Thus, in the case of two variables x, y, one such form would be $p(x, y) = (3 + x^2) + 0 \cdot y + (2x - x^3)y^2$—usually written in the more flexible form $3 + x^2 + 2xy^2 - x^3y^2$.

A corollary of Theorem 4 and induction on n is

Theorem 7. *Each polynomial function in $x_1, \cdots, x_n$ can be expressed in one and only one way as a polynomial form if D is infinite. Whether D is infinite or not, $D[x_1, \cdots, x_n]$ is an integral domain.*

It is obvious from the definition that every permutation of the subscripts induces a natural automorphism of the commutative ring $D\langle x_1, \cdots, x_n \rangle$ of polynomial functions of n variables. It follows by Theorem 7 that if D is infinite, the same is true of polynomial forms (whose definition is not symmetrical in the variables). We shall now show that this result is true for *any* integral domain D.

Theorem 8. *Every permutation of the subscripts induces a different automorphism on $D[x_1, \cdots, x_n]$.*

Proof. Consider the case of two indeterminates x, y. Each form

$$p(y, x) = \sum_i \left(\sum_j a_{ij} y^j \right) x^i$$

of $D[y, x]$ can be rearranged by the distributive, commutative, and associative laws in $D[y, x]$ to give an expression of the form

$$p(y, x) = \sum_j \left(\sum_i a_{ij} x^i \right) y^j.$$

This result has the proper form, and can be interpreted as if it were a polynomial $p'(x, y)$ in the domain $D[x, y]$ (x first, then y). The correspondence $p(y, x) \mapsto p'(x, y)$ thus set up is one-one—every finite set of non-zero coefficients a_{ij} corresponds to just one element of $D[y, x]$ and just one of $D[x, y]$. Finally, since rules (2) and (3) for addition and multiplication can be deduced from the postulates for an integral domain, which both $D[y, x]$ and $D[x, y]$ are, we see that the correspondence preserves sums and products.

The case of n indeterminates can be treated similarly with a more elaborate general notation—or deduced by induction from the case of two variables.

Thus $D[x_1, \cdots, x_n]$ in fact depends *symmetrically* on $x_1, \cdots, x_n$. This suggests framing a definition of $D[x_1, \cdots, x_n]$ from which this symmetry is immediately apparent. This may be done in the case $n = 2$, for the domain $D'' = D[x, y]$, roughly as follows. Firstly, D'' is generated by x, y, and elements of D (every element of D'' may be obtained from x, y, and D by repeated sums and products); in the second place, the generators x and y are *simultaneous indeterminates* over D or are *algebraically independent* over D). By this we mean that a finite sum

$$\sum_{i,j} a_{ij} x^i y^j$$

with coefficients a_{ij} in D can be zero if and only if all coefficients a_{ij} are zero. These two properties uniquely determine the domain $D[x, y]$ in a symmetrical manner (see Ex. 9 below).

Exercises

1. Represent as polynomials in y with coefficients in $D[x]$:
   (a) $p(x, y) = y^3 x + (x^2 - xy)^2$,
   (b) $q(x, y) = (x + y)^3 - 3yx(x^2 + x - 1)$.
2. Compute the number of possible functions of two variables x, y on the domain $\mathbf{Z}_2$.
3. Rearrange the following expression as a polynomial in x with coefficients which are polynomials in y (as in the proof of Theorem 8):

$$(3x^2 + 2x + 1)y^3 + (x^4 + 2)y^2 + (2x - 3)y + x^4 - 3x^2 + 2x.$$

4. Let D be any integral domain. Prove that the correspondence which carries each $p(x)$ into $p(-x)$ is an automorphism of $D[x]$. Is it also one of $D\langle x\rangle$?
5. Is the correspondence $p(x) \mapsto p(x + c)$, where c is a constant, an automorphism of $D[x]$? Illustrate by numerical examples.
6. If F is a field, show that the correspondence $p(x) \mapsto p(ax)$ is an automorphism of $F[x]$ for any constant $a \neq 0$.
7. Exhibit automorphisms of $D[x, y]$ other than those described in Theorem 8.
8. Prove Theorem 7, (a) for $n = 2$; (b) for any n.
9. (a) Prove in detail that the domain $D[x, y]$ (first x, then y) is indeed generated over D by two "simultaneous indeterminates" x and y.
    (b) Let D' and D'' be two domains each generated over D by two simultaneous indeterminates x', y' and x'', y'', respectively. Prove that D' is isomorphic to D'' under a correspondence which maps x' on x'', y' on y'', and each element of D on itself.
    (c) Use parts (a) and (b) to give another proof of Theorem 8, for $n = 2$.

3.5. The Division Algorithm

The Division Algorithm for polynomials (sometimes called "polynomial long division") provides a standard scheme for dividing one polynomial $b(x)$ by a second one $a(x)$ so as to get a quotient $q(x)$ and remainder $r(x)$ of degree less than that of the divisor $a(x)$. We shall now show that this Division Algorithm, although usually carried out with rational coefficients, is actually possible for polynomials with coefficients in any field.

Theorem 9. *If F is any field, and $a(x) \neq 0$ and $b(x)$ are any polynomials over F, then we can find polynomials $q(x)$ and $r(x)$ over F so that*

$$(8) \qquad b(x) = q(x)a(x) + r(x),$$

where $r(x)$ is either zero or has a degree less than that of $a(x)$.

Informal proof. Eliminate successively the highest terms of the dividend $b(x)$ by subtracting from it products of the divisor $a(x)$ by suitable monomials cx^k. If $a(x) = a_0 + a_1x + \cdots + a_mx^m$ ($a_m \neq 0$) and $b(x) = b_0 + b_1x + \cdots + b_nx^n$ ($b_n \neq 0$), and if the degree n of $b(x)$ is not already less than that m of $a(x)$, we can form the difference

$$(9) \qquad b_1(x) = b(x) - (b_n/a_m)x^{n-m}a(x)$$
$$= 0 \cdot x^n + (b_{n-1} - a_{m-1}b_n/a_m)x^{n-1} + \cdots,$$

which will be of degree less than n, or zero. We can then repeat this process until the degree of the remainder is less than m.

A formal proof for this Division Algorithm can be based on the Second Induction Principle, as formulated in §1.5. Let m be the degree of $a(x)$. Any polynomial $b(x)$ of degree $n < m$ then has a representation $b(x) = 0 \cdot a(x) + b(x)$, with a quotient $q(x) = 0$. For a polynomial $b(x)$ of degree $n \geq m$, transposition of (9) gives

$$(10) \qquad\qquad b(x) = b_1(x) + (b_n/a_m)x^{n-m}a(x),$$

where the degree k of $b_1(x)$ is less than n unless $b_1(x) = 0$. By the second induction principle, we can assume the expansion (8) to be possible for all $b(x)$ of degree $k < n$, so that we have

$$(11) \qquad\qquad b_1(x) = q_1(x)a(x) + r(x),$$

where the degree of $r(x)$ is less than m, unless $r(x) = 0$. Substituting (11) in (10), we get the desired equation (8), as

$$b(x) = [q_1(x) + (b_n/a_m)x^{n-m}]a(x) + r(x).$$

In particular, if the polynomial $a(x) = x - c$ is monic and linear, then the remainder $r(x)$ in (8) is a constant $r = b(x) - (x - c)q(x)$. If we set $x = c$, this equation gives $r = b(c) - 0q(c) = b(c)$. Hence we have

Corollary 1. *The remainder of a polynomial $p(x)$, when divided by $x - c$, is $p(c)$* (*Remainder Theorem*).

When the remainder $r(x)$ in (8) is zero, we say that $b(x)$ is divisible by $a(x)$. More exactly, if $a(x)$ and $b(x)$ are two polynomial forms over an integral domain D, then $b(x)$ is *divisible* by $a(x)$ *over D* or *in $D[x]$* if and only if $b(x) = q(x)a(x)$ for some polynomial form $q(x) \in D[x]$.

Exercises

1. Show that $q(x)$ and $r(x)$ are unique for given $a(x)$ and $b(x)$ in (8).
2. Compute $q(x)$, $r(x)$ if $b(x) = x^5 - x^3 + 3x - 5$ and $a(x) = x^2 + 7$.
3. The same as Ex. 2 if $a(x)$ is respectively $x - 2, x + 2, x^3 + x - 1$.
4. (a) Do Ex. 2 for the field Z_5.
   (b) Do Ex. 3 for the field Z_3.
5. Given distinct numbers $a_0, a_1, \cdots, a_n$ in a field F, let $a(x) = \prod_{j=0}^{n}(x - a_j)$.
   Show that the remainder $r(x)$ of any polynomial $f(x)$ over F upon division by $a(x)$ is precisely the Lagrange interpolant to $f(x)$ at these points.
6. Is $x^3 + x^2 + x + 1$ divisible by $x^2 + 3x + 2$ over any of the domains Z_3, Z_5, Z_7?

7. Find all possible rings $\mathbf{Z}_n$ over which $x^5 - 10x + 12$ is divisible by $x^2 + 2$.
8. (a) If a polynomial $f(x)$ over any domain has $f(a) = 0 = f(b)$, where $a \neq b$, show that $f(x)$ is divisible by $(x - a)(x - b)$.
   (b) Generalize this result.
9. In the application of the Second Induction Principle to the Division Algorithm, what specifically is $P(n)$ (see §1.5)?

3.6. Units and Associates

One can get a complete analogue for polynomials of the fundamental theorem of arithmetic. In this analogue, the role of prime numbers is played by "irreducible" polynomials, defined as follows.

Definition. *A polynomial form is called* reducible *over a field F if it can be factored into polynomials of lower degree with coefficients in F; otherwise, it is called* irreducible *over F.*

Thus the polynomial $x^2 + 4$ is irreducible over the field of rationals. For suppose instead $x^2 + 4 = (x + a)(x + b)$. Substituting $x = -b$ this gives $(-b)^2 + 4 = (-b + a)(-b + b) = 0$, hence $(-b)^2 = -4$. This is clearly impossible, as a square cannot be negative. Since the same reasoning holds in any ordered field, we conclude that $x^2 + 4$ is also irreducible over the real field or any other ordered field.

To clarify the analogy between irreducible polynomials and prime numbers, we now define certain divisibility concepts for an arbitrary integral domain D, be it the polynomial domain $\mathbf{Q}[x]$, the domain $\mathbf{Z}$ of integers, or something else.

An element a of D is divisible by b (in symbols, $b \mid a$) if there exists some c in D such that $a = cb$. Two elements a and b are *associates* if both $b \mid a$ and $a \mid b$. An associate of the unity element 1 is called a *unit*. Since $1 \mid a$ for all a, an element u is a unit in D if and only if it has in D a multiplicative inverse u^{-1} with $1 = uu^{-1}$. Elements with this property are also called *invertible*.

If a and b are associates, $a = cb$ and $b = c'a$, hence $a = cc'a$. The cancellation law gives $1 = cc'$, so both c and c' are units. Conversely, $a = ub$ is an associate of b if u is a unit. Hence two elements are associates if and only if each may be obtained from the other by introducing a unit factor.

Example 1. In a field, every $a \neq 0$ is a unit.

Example 2. In the domain $\mathbf{Z}$ of integers, the units are ± 1; hence the associates of any a are $\pm a$.

EXAMPLE 3. In a polynomial domain $D[x]$ in an indeterminate x, the degree of a product $f(x) \cdot g(x)$ is the sum of the degrees of the factors. Hence any element $b(x)$ with a polynomial inverse $a(x)b(x) = 1$ must be a polynomial $b(x) = b$ of degree zero. Such a constant polynomial b has an inverse only if b already has an inverse in D. Therefore the units of $D[x]$ are the units of D.

If F is a field, the units of the polynomial domain $F[x]$ are thus exactly the nonzero constants of F, so that two polynomials $f(x)$ and $g(x)$ are associates in $F[x]$ if and only if each is a constant multiple of the other

EXAMPLE 4. In the domain $\mathbf{Z}[\sqrt{2}]$ of all numbers $a + b\sqrt{2}$ (a, b integers), $(a + b\sqrt{2})(x + y\sqrt{2}) = 1$ implies $x = a/(a^2 - 2b^2)$, $y = -b/(a^2 - 2b^2)$—and these are integers if and only if $a^2 - 2b^2 = \pm 1$. Thus, $1 \pm \sqrt{2}$ and $3 \pm 2\sqrt{2}$ are units, whereas $2 + \sqrt{2}$ is not a unit in $\mathbf{Z}[\sqrt{2}]$.

An element b of an arbitrary integral domain D is divisible by all its associates and by all units. These are called "improper" divisors of b. An element not a unit with no proper divisors is called *prime* or *irreducible* in D.

EXAMPLE 5. Over any field F, a linear polynomial $ax + b$ with $a \neq 0$ is irreducible, for its only factors are constants (units) or constant multiples of itself (associates).

EXAMPLE 6. Consider the domain $\mathbf{Z}[\sqrt{-1}]$ of "Gaussian integers" of the form $a + b\sqrt{-1}$, with $a, b \in \mathbf{Z}$. If $a + b\sqrt{-1}$ is a unit, then, for some $c + d\sqrt{-1}$, we have

$$1 = (a + b\sqrt{-1})(c + d\sqrt{-1}) = (ac - bd) + (ad + bc)\sqrt{-1}.$$

Hence $ac - bd = 1$, $ad + bc = 0$, and

$$1 = (ac - bd)^2 + (ad + bc)^2 = (a^2 + b^2)(c^2 + d^2),$$

as can easily be checked. Since $a^2 + b^2$, $c^2 + d^2$ are nonnegative integers, we infer $a^2 + b^2 = c^2 + d^2 = 1$; the only possibilities are thus $1, -1, \sqrt{-1}$, and $-\sqrt{-1}$, giving four units.

Lemma. *In any integral domain D, the relation "a and b are associates" is an equivalence relation.*

The proof will be left to the reader. (See also Exs. 1–3 below.)

Exercises

1. In any integral domain D, prove that
   (a) the relation "$b \mid a$" is reflexive and transitive,
   (b) if $c \neq 0$, then $b \mid a$ if and only if $bc \mid ac$,
   (c) any two elements have a common divisor and a common multiple,
   (d) if $a \mid b$ and $a \mid c$, then $a \mid (b \pm c)$.
2. Prove that the units of Z_m are the integers relatively prime to m.
3. In any integral domain, let "$a \sim b$" mean "a is associate to b." Prove that
   (a) if $a \sim b$, then $c \mid a$ if and only if $c \mid b$,
   (b) if $a \sim b$, then $a \mid c$ if and only if $b \mid c$,
   (c) if $a \mid c$ if and only if $b \mid c$, then $a \sim b$,
   (d) if p is prime and $p \sim q$, then q is prime.
4. Show that if $a \sim a'$ and $b \sim b'$, then $ab \sim a'b'$—whereas in general $a + b \sim a' + b'$ fails.
5. Prove the "generalized law of cancellation": If $ax \sim by$, $a \sim b$, and $a \neq 0$, then $x \sim y$.
6. List all associates of $x^2 + 2x - 1$ in $Z_5[x]$.
7. Find all units in the domain $D[x, y]$ of polynomials in two indeterminates.
8. For which elements a of an integral domain D is the correspondence $p(x) \rightarrow p(ax)$ an automorphism of $D[x]$?
9. Find all the units in the domain D which consists of all rational numbers m/n with m and n integers such that n is not divisible by 7.
10. Where $\alpha = a + b\sqrt{3}$, define $N(\alpha) = a^2 - 3b^2$. Prove
    (a) $N(\alpha\alpha') = N(\alpha)N(\alpha')$,
    (b) that if α is a unit in $Z[\sqrt{3}]$, then $N(\alpha) = \pm 1$.
11. Let $Z[\sqrt{5}]$ be the domain of all numbers $\alpha = a + b\sqrt{5}$ (a, b integers), and set $N(\alpha) = a^2 - 5b^2$.
    (a) Prove that $9 + 4\sqrt{5}$ is a unit in this domain. (Cf. Ex. 10.)
    (b) Show that $1 - \sqrt{5}$ and $3 + \sqrt{5}$ are associates, but are not units.
    (c) Show generally that α is a unit if and only if $N(\alpha) = \pm 1$.
    (d) If $N(\alpha)$ is a prime in Z, show that α is a prime in $Z[\sqrt{5}]$.
    (e) Show that $4 + \sqrt{5}$ and $4 - \sqrt{5}$ are primes.
    (f) Show that 2 and $3 + \sqrt{5}$ are primes. (*Hint:* $x^2 \equiv 2 \pmod 5$ is impossible for $x \in Z$.)
    (g) Use $2 \cdot 2 = (3 + \sqrt{5})(3 - \sqrt{5})$ to show that $Z[\sqrt{5}]$ is not a unique factorization domain (§3.9).
12. Prove in detail the lemma of the text.

3.7. Irreducible Polynomials

A basic problem in polynomial algebra consists in finding effective tests for the irreducibility of polynomials over a given field. The nature of such tests depends entirely on the field F in question. Thus over the complex field **C**, the polynomial $x^2 + 1$ can be factored as $x^2 + 1 = (x + \sqrt{-1})(x - \sqrt{-1})$. In fact, as will be shown in §5.3, the only irreducible polynomials of **C**$[x]$ are linear. Yet $x^2 + 1$ is irreducible over the real field **R**.

Again, since $x^2 - 28 = (x - \sqrt{28})(x + \sqrt{28})$, the polynomial $x^2 - 28$ is reducible over the real field. The same polynomial is irreducible over the rational field, as we shall now prove rigorously.

Lemma. *A quadratic or cubic polynomial $p(x)$ is irreducible over a field, F, unless $p(c) = 0$ for some $c \in F$.*

Proof. In any factorization of $p(x)$ into polynomials of lower degree, one factor must be linear, since the degree of a product of polynomials is the sum of the degrees of the factors.

Theorem 10. *Let $p(x) = a_0 x^n + a_1 x^{n-1} + \cdots + a_n$ be a polynomial with integral coefficients. Any rational root of the equation $p(x) = 0$ must have the form r/s, where $r \mid a_n$ and $s \mid a_0$.*

Proof. Suppose $p(x) = 0$ for some fraction $x = b/c$. By dividing out the g.c.d. of b and c, one can express b/c in "lowest terms" as a quotient r/s of relatively prime integers r and s. Substitution of this value in $p(x)$ gives

$$(12) \qquad 0 = s^n p(r/s) = a_0 r^n + a_1 r^{n-1} s + \cdots + a_n s^n,$$

whence

$$-a_0 r^n = s(a_1 r^{n-1} + a_2 r^{n-2} s + \cdots + a_n s^{n-1}), \qquad \text{and} \qquad s \mid a_0 r^n.$$

But $(s, r) = 1$; hence, by successive applications of Theorem 10 of §1.7, $s \mid a_0 r^{n-1}, \cdots, s \mid a_0$. Similarly, as $-a_n s^n = r(a_0 r^{n-1} + \cdots + a_{n-1} s^{n-1})$, $r \mid a_n$.

Corollary. *Any rational root of a monic polynomial having integral coefficients is an integer.*

It is now easy to prove that $x^2 - 28$ is irreducible over **Q**. By the Corollary, $x^2 = 28$ implies that $x = r/s$ is an integer. But $x^2 - 28 > 0$ if $|x| \geq 6$, and $x^2 - 28 < 0$ if $|x| \leq 5$. Hence no integer can be a root of $x^2 - 28 = 0$, and (by the Lemma) $x^2 - 28$ is irreducible over the rational field.

There is no easy general test for irreducibility of polynomials over the rational field **Q** (but see §3.10).

Exercises

1. Test the following equations for rational roots:
   (a) $3x^3 - 7x = 5$,   (b) $5x^3 + x^2 + x = 4$,
   (c) $8x^5 + 3x^2 = 17$,   (d) $6x^3 - 3x = 18$.
2. Prove that $30x^n = 91$ has a rational root for no integer $n > 1$. (*Hint:* Use the fundamental theorem of arithmetic.)
3. For which rational numbers x is $3x^2 - 7x$ an integer? Find necessary and sufficient conditions.
4. For what integers a between 0 and 250 does $30x^n = a$ have a rational root for some $n > 1$?
5. Is $x^2 + 1$ irreducible over $\mathbf{Z}_3$? over $\mathbf{Z}_5$? How about $x^3 + x + 2$?
6. Find a finite field over which $x^2 - 2$ is (a) reducible, (b) irreducible.
7. Find all monic irreducible quadratic polynomials over the field $\mathbf{Z}_5$.
8. Find all monic irreducible cubic polynomials over $\mathbf{Z}_3$.
9. Prove that if $a_0 + a_1x + a_2x^2 + \cdots + a_nx^n$ is irreducible, then so is $a_n + a_{n-1}x + a_{n-2}x^2 + \cdots + a_0x^n$.
10. Decompose into irreducible factors the polynomial $x^4 - 5x^2 + 6$ over the field of rationals, over the field $\mathbf{Q}(\sqrt{2})$ of §2.1, and over the field of reals.
★11. Show that if $4ac > b^2$, then $ax^2 + bx + c$ is irreducible over any ordered field.

3.8. Unique Factorization Theorem

Throughout this section we shall be considering factorization in the domain $F[x]$ of polynomial forms in one indeterminate x over a *field F*. The main result is that factorization into irreducible (prime) factors is unique, the proof being a virtual repetition of that of the analogous fundamental theorem of arithmetic (Chap. 1). The analogy involves the following fundamental notion, which will be considered systematically in Chap. 13.

Definition. *A nonvoid subset C of a commutative ring R is called an* ideal *when $a \in C$ and $b \in C$ imply $(a \pm b) \in C$, and $a \in C$, $r \in R$ imply $ra \in C$.*

Remark. For any $a \in R$, the set of all multiples ra of a is an ideal, since

$$ra \pm sa = (r \pm s)a \quad \text{and} \quad s(ra) = (sr)a, \quad s, r \in R.$$

Such an ideal is called a *principal ideal*. We will now show that all ideals in any $F[x]$ are principal.

Theorem 11. *Over any field F, any ideal C of F[x] consists either* (i) *of 0 alone, or* (ii) *of the set of multiples $q(x)a(x)$ of any nonzero member $a(x)$ of least degree.*

Proof. Unless $C = 0$, it contains a nonzero polynomial $a(x)$ of least degree $d(a)$, and, with $a(x)$, all its multiples $q(x)a(x)$. In this case, if $b(x)$ is *any* polynomial of C, by Theorem 9 some $r(x) = b(x) - q(x)a(x)$ has degree less than $d(a)$. But by hypothesis C contains $r(x)$, and by construction it contains no nonzero polynomial of degree less than $d(a)$. Hence $r(x) = 0$ and $b(x) = q(x)a(x)$, proving the theorem.

Now let $a(x)$ and $b(x)$ be any two polynomials, and consider the set C of all the "linear combinations" $s(x)a(x) + t(x)b(x)$ which can be formed from them with any polynomial coefficients $s(x)$ and $t(x)$. This set C is obviously nonvoid, and contains any sum, difference, or multiple of its members, since (in abbreviated notation)

$$(sa + tb) \pm (s'a + t'b) = (s \pm s')a + (t \pm t')b,$$

$$q(sa + tb) = (qs)a + (qt)b.$$

Hence the set C is an ideal and so, by Theorem 11, consists of the multiples of some polynomial $d(x)$ of least degree.

This polynomial $d(x)$ will divide both $a(x) = 1 \cdot a(x) + 0 \cdot b(x)$ and $b(x) = 0 \cdot a(x) + 1 \cdot b(x)$, and will be divisible by any common divisor of $a(x)$ and $b(x)$, since $d(x) = s_0(x)a(x) + t_0(x)b(x)$. Our conclusion is

Theorem 12. *In F[x], any two polynomials a and b have a "greatest common divisor" d satisfying* (i) $d \mid a$ *and* $d \mid b$, (i') $c \mid a$ *and* $c \mid b$ *imply* $c \mid d$. *Moreover,* (ii) *d is a "linear combination"* $d = sa + tb$ *of a and b.*

We remark that the Euclidean algorithm, described in detail in §1.7, can be used to compute d explicitly from a and b. (This is because our Division Algorithm allows us to compute remainders of polynomials explicitly.)

Also, if d satisfies (i), (i'), and (ii), then so do all associates of d. Incidentally, (i) and (ii) imply (i').

The g.c.d. $d(x)$ is unique except for unit factors, for if d and d' are two greatest common divisors of the same polynomials a and b, then by (i) and (i'), $d \mid d'$ and $d' \mid d$, so that d and d' are indeed associates. Conversely, if d is a g.c.d., so is every associate of d. It is sometimes convenient to speak of the unique monic polynomial associate to d as "the" g.c.d.

Two polynomials $a(x)$ and $b(x)$ are said to be *relatively prime* if their greatest common divisors are unity and its associates. This means that polynomials are relatively prime if and only if their only common factors are the nonzero constants of F (the units of the domain $F[x]$).

Theorem 13. *If $p(x)$ is irreducible, then $p(x)|a(x)b(x)$ implies that $p(x)|a(x)$ or $p(x)|b(x)$.*

Proof. Because $p(x)$ is irreducible, the g.c.d. of $p(x)$ and $a(x)$ is either $p(x)$ or the unity 1. In the former case, $p(x)|a(x)$; in the latter case, we can write $1 = s(x)p(x) + t(x)a(x)$ and so

$$b(x) = 1 \cdot b(x) = s(x)p(x)b(x) + t(x)[a(x)b(x)].$$

Since $p(x)$ divides the product $a(x)b(x)$, it divides both terms on the right, hence does divide $b(x)$, as required for the theorem.

Theorem 14. *Any nonconstant polynomial $a(x)$ in $F[x]$ can be expressed as a constant c times a product of monic irreducible polynomials. This expression is unique except for the order in which the factors occur.*

First, such a factorization is possible. If $a(x)$ is a constant or irreducible, this is trivial. Otherwise, $a(x)$ is the product $a(x) = b(x)b'(x)$ of factors of lower degree. By the Second Induction Principle, we can assume

$$b(x) = cp_1(x) \cdots p_m(x), \qquad b'(x) = c'p_1{}'(x) \cdots p_n{}'(x),$$

whence $a(x) = (cc')p_1(x) \cdots p_m(x)p_1{}'(x) \cdots p_n{}'(x)$, where cc' is a constant and the $p_i(x)$ and $p_j{}'(x)$ are irreducible and monic polynomials.

To prove the uniqueness, suppose $a(x)$ has two possible such "prime" factorizations,

$$a(x) = cp_1(x) \cdots p_m(x) = c'q_1(x) \cdots q_n(x).$$

Clearly, $c = c'$ will be the leading coefficient of $a(x)$ (since the latter is the product of the leading coefficients of its factors). Again, since $p_1(x)$ divides $c'q_1(x) \cdots q_n(x) = a(x)$, it must by Theorem 13 divide some (nonconstant) factor $q_i(x)$; since $q_i(x)$ is irreducible, the quotient $q_i(x)/p_1(x)$ must be a constant; and since $p_1(x)$ and $q_i(x)$ are both monic, it must be 1. Hence $p_1(x) = q_i(x)$. Cancelling, $p_2(x) \cdots p_m(x)$ equals the product of the $q_k(x)$ $[k \neq i]$, and has a lower degree than $a(x)$. Therefore, again by the Second Induction Principle, the $p_j(x)$ $[j \neq 1]$ and $q_k(x)$ $[k \neq i]$ are equal in pairs, completing the proof.

It is a corollary (cf. §1.8, last paragraph) that the exponent e_i to which each (monic) irreducible polynomial $p_i(x)$ occurs as a factor of

$a(x)$ is uniquely determined by $a(x)$, and is the biggest e such that $p_i(x)^e | a(x)$.

If a polynomial $a(x)$ is decomposed into irreducible factors $p_i(x)$ which are not necessarily monic, the factors are no longer absolutely unique, as in Theorem 14. However, each factor $p_i(x)$ divided by its leading coefficient gives a (unique) monic irreducible factor, and therefore is associate to this irreducible in $F[x]$. Hence any two such factorizations can be made to agree with each other simply by reordering terms and replacing each factor by a suitable associate factor. This situation is summarized by the statement that the decomposition of a polynomial in $F[x]$ is unique to *within order and unit factors* (or to within order and replacement of factors by associates).

Exercises

1. Show that if ϕ is any homomorphism from a commutative ring R to a commutative ring R', then the antecedents of the additive zero of R' form an ideal in R.
2. (a) Find the g.c.d. of $x^3 - 1$ and $x^4 + x^3 + 2x^2 + x + 1$ over $\mathbf{Q}$.
   (b) Express this g.c.d. as a linear combination $d(x) = s(x)a(x) + t(x)b(x)$ of the given polynomials. (*Caution:* The coefficients need not be integers.)
   (c) The same for $x^{18} - 1$, $x^{33} - 1$.
3. Find the g.c.d. of $2x^3 + 6x^2 - x - 3$, $x^4 + 4x^3 + 3x^2 + x + 1$ over $\mathbf{Q}$.
4. Do Ex. 3, assuming that the polynomials have coefficients in $\mathbf{Z}_3$.
5. Show that $x^3 + x + 1$ is irreducible modulo 5.
6. Factor the following polynomials in $\mathbf{Z}_3$:
   (a) $x^2 + x + 1$, (b) $x^3 + x + 2$, (c) $2x^3 + 2x^2 + x + 1$,
   (d) $x^4 + x^3 + x + 1$, ★(e) $x^4 + x^3 + x + 2$.
7. List (to within associates) all divisors of $x^4 - 1$ in the domain of polynomials with rational coefficients, proving that every divisor of $x^4 - 1$ is associate to one on your list.
8. Do the same for $x^6 - 1$, $x^8 - 1$.
9. Prove that two polynomial forms $q(x)$ and $r(x)$ over $\mathbf{Z}$ represent the same function on $\mathbf{Z}_p$ if and only if $(x^p - x) | [q(x) - r(x)]$. (*Hint:* Use Ex. 6 of §3.2.)
10. Prove that any finite set of polynomials over a field has a g.c.d., which is a linear combination of the given polynomials.
11. (a) Prove that the set of all common multiples of any two given polynomials over a field is an ideal.
    (b) Infer that the polynomials have a l.c.m.; illustrate by finding the l.c.m. of $x^2 + 3x + 2$ and $(x + 1)^2$.
12. If a given polynomial $p(x)$ over F has the property that $p(x) | a(x)b(x)$ always implies either $p(x) | a(x)$ or $p(x) | b(x)$, prove $p(x)$ irreducible over F.

13. If $p(x)$ is a given polynomial such that any other polynomial is either relatively prime to $p(x)$ or divisible by $p(x)$, prove $p(x)$ irreducible.

14. If $m(x)$ is a power of an irreducible polynomial, show that $m(x)|a(x)b(x)$ implies either $m(x)|a(x)$ or $m(x)|b(x))^e$ for some e.

15. If $h(x)$ is relatively prime to both $f(x)$ and $g(x)$, prove $h(x)$ relatively prime to $f(x)g(x)$.

16. If $h(x)|f(x)g(x)$ and $h(x)$ is relatively prime to $f(x)$, prove that $h(x)|g(x)$.

17. If $f(x)$ and $g(x)$ are relatively prime polynomials in $F[x]$, and if F is a subfield of K, prove that $f(x)$ and $g(x)$ are relatively prime also in $K[x]$.

★18. If two polynomials with rational coefficients have a real root in common, prove that they have a common divisor with rational coefficients which is not a constant.

19. The following descriptions give certain sets of polynomials with rational coefficients. Which of these sets are ideals? When the set is an ideal, find in it a polynomial of least degree.
    (a) all $b(x)$ with $b(3) = b(5) = 0$,
    (b) all $b(x)$ with $b(3) \neq 0$ and $b(2) = 0$,
    (c) all $b(x)$ with $b(3) = 0$, $b(6) = b(7)$,
    (d) all $b(x)$ such that some power of $b(x)$ is divisible by $(x + 1)^4(x + 2)$.

20. Let S be any set of polynomials over F which contains the difference of any two of its members and contains with any $b(x)$ both $xb(x)$ and $ab(x)$ for each constant a in F. Show that S is an ideal.

★3.9. Other Domains with Unique Factorization

Consider the domain $\mathbf{Q}[x, y]$ of polynomial forms in two indeterminates over the rational field $\mathbf{Q}$. The only common divisors of $a(x, y) = x$ and $b(x, y) = y^2 + x$ are 1 and its associates, yet there are no polynomials $s(x, y)$ and $t(x, y)$ such that $xs(x, y) + (y^2 + x)t(x, y) = 1$, since the polynomial $xs + (y^2 + x)t$ would never have a constant term not zero, whatever the choice of s and t. Similarly, in the domain $\mathbf{Z}[x]$ of polynomials with integral coefficients, g.c.d. $(2, x) = 1$, yet $s(x) \cdot 2 + t(x) \cdot x = 1$ has no solution. Thus Theorem 12 does not hold in either domain.

Nevertheless, one can show that in both cases, factorization into primes is possible and unique (Theorem 14 holds).

Definition. *By a unique factorization domain (sometimes called a "Gaussian domain") is meant an integral domain in which*
   (i) *any element not a unit can be factored into primes;*
   (ii) *this factorization is unique to within order and unit factors.*

Our main result will be that if G is any unique factorization domain, then so is any domain $G[x_1, \cdots, x_n]$ of polynomial forms over G. Using induction on n, one can evidently reduce the problem to the case $G[x]$ of a single indeterminate, and it is this case which we shall consider.

First, we shall embed G in the *field* $F = Q(G)$ of its formal quotients (§2.2, Theorem 4), and we shall consider $F[x]$ along with $G[x]$. We may typically imagine G as the domain of the integers and F correspondingly as that of the rationals.

Second, we shall call a polynomial of $F[x]$ *primitive* when its coefficients (i) are in G ("integers") and (ii) have no common divisors except units in G. Thus $3 - 5x^2$ is primitive, $3 - 6x^2$ is not.

Lemma 1 (Gauss). *The product of any two primitive polynomials is itself primitive.*

Proof. Write

$$\sum_k c_k x^k = \sum_i a_i x^i \cdot \sum_j b_j x^j;$$

if this is not primitive, then some prime $p \in G$ will divide every c_k. But let a_m and b_n be the first coefficients *not* divisible by p in $\sum_i a_i x^i$ and $\sum_j b_j x^j$, respectively (they certainly exist, since the polynomials are primitive). Then the formula (3) for the coefficient c_{m+n} in the product gives

$$a_m b_n = c_{m+n} - [a_0 b_{m+n} + \cdots + a_{m-1} b_{n+1} + a_{m+1} b_{n-1} + \cdots + a_{m+n} b_0],$$

so that the product $a_m b_n$ is divisible by p, since all the terms on the right are so divisible. This means that the prime p must appear in the unique decomposition of one of the factors a_m or b_n, in contradiction to the choice of a_m and b_n as not divisible by p.

Lemma 2. *Any nonzero polynomial $f(x)$ of $F[x]$ can be written as $f(x) = c_f f^*(x)$, where c_f is in F and $f^*(x)$ is primitive. Moreover, for a given $f(x)$, the constant c_f and the primitive polynomial $f^*(x)$ are unique except for a possible unit factor from G.*

Proof. First write $f(x) = (b_0/a_0) + (b_1/a_1)x + \cdots + (b_n/a_n)x^n$, $a_i, b_i \in G$ ("integers"). If $c = 1/a_0 a_1 \cdots a_n$, we have $f(x) = cg(x)$, where $g(x)$ has coefficients in G. Now let c' be a greatest common divisor of the coefficients of $g(x)$ (this exists, since the unique factorization theorem holds in G). Clearly, $f^*(x) = g(x)/c'$ is primitive, and $f(x) = (cc')f^*(x)$. This is the first result, with $c_f = cc'$.

To prove the uniqueness of c_f and f^*, it suffices to show that f^* is unique to within units in G. To this end, suppose $f^*(x) = cg^*(x)$, where

$f^*(x)$ and $g^*(x)$ are primitive and $c \in F$. Write $c = u/v$, where $u, v \in G$ are relatively prime, so that $ug^*(x) = vf^*(x)$. The coefficients of $ug^*(x)$ will then have v as a common factor, whence, since u and v are relatively prime, v divides every coefficient of $g^*(x)$. But $g^*(x)$ is primitive, hence v is a unit in G. By symmetry, u is a unit, and so u/v is a unit in G. This completes the proof.

The constant c_f of Lemma 2 is called the *content* of $f(x)$; it is unique to within associateness in G.

Lemma 3. *If* $f(x) = g(x)h(x)$ *in* $G[x]$ *or even* $F[x]$, *then* $c_f \sim c_g c_h$ *and* $f^*(x) \sim g^*(x)h^*(x)$, *where* "$\sim$" *denotes the relation of being associate in* $G[x]$.

Proof. By Lemma 1, $g^*(x)h^*(x)$ is primitive; it is also clearly a constant multiple of $f^*(x)$; hence by Lemma 2 the two differ by a unit factor u in G (are associate); hence $c_f = u^{-1}c_g c_h$. Q.E.D.

It is a corollary that if $f(x)$ is in $G[x]$ and reducible in $F[x]$, then $f(x) = uc_f g^*(x)h^*(x)$. This gives the following generalization of the Corollary of Theorem 10.

Theorem 15. *A polynomial with integral coefficients which can be factored into polynomials with rational coefficients can already be factored into polynomials of the same degrees with integral coefficients.*

What is more important, by Lemma 3 the factorization of any $f(x)$ in $G[x]$ splits into independent parts: the factorization of its "content" c_f and that of its "primitive part" $f^*(x)$. The former takes place in G and so by hypothesis is possible and unique. By Lemma 3, the latter is essentially equivalent to factorization in $F[x]$, which is possible and unique by Theorem 14. This suggests

Lemma 4. *If* G *is a unique factorization domain, so is* $G[x]$.

Proof. By Lemma 2, any polynomial $f(x)$ has a factorization $f(x) = c_f f^*(x)$, hence a prime element $f(x)$ in $G[x]$ must have one of these factors c_f or f^* a unit of $G[x]$. Therefore the primes of $G[x]$ are of two types: the primes p of G, and the primitive polynomials $q(x)$ which are irreducible, both in $G[x]$ and (Theorem 15) in $F[x]$.

Now consider any polynomial $f(x)$ in $G[x]$. It has a factorization in $F[x]$, and hence is associate to a product of primitive irreducibles of $G[x]$, as $f(x) \sim q_1(x) \cdots q_m(x)$. Thus $f(x) = dq_1(x) \cdots q_m(x)$, where the element d of G can be factored into irreducibles p_i of G. All told, $f(x)$

has the decomposition

$$f(x) = p_1 \cdots p_r q_1(x) \cdots q_m(x),$$

where each p_i is a prime of G, each $q_j(x)$ a primitive irreducible of $G[x]$.

In this factorization the polynomials $q_j(x)$ which appear are uniquely determined, to within units in G, as the primitive parts of the unique irreducible factors of $f(x)$ in $F[x]$. Since the $q_j(x)$ are primitive, the product $p_1 \cdots p_r$ is the essentially unique content c_f of $f(x)$. Therefore the p_i are the (essentially) unique factors of c_f in the given domain G. This shows that $G[x]$ is a unique factorization domain.   Q.E.D.

Fom Lemma 4 and an induction on n one concludes

Theorem 16. *If G is any unique factorization domain, so is every polynomial domain $G[x_1, \cdots, x_n]$ over G.*

In §14.10, we shall exhibit an integral domain which is *not* a unique factorization domain, in which neither Theorem 12 nor Theorem 14 holds (cf. §3.6, Ex. 11(g)).

Exercises

1. Represent each of the following as a product of a constant by a primitive polynomial of $\mathbf{Z}[x]$: $3x^2 + 6x + 9$, $x^2/2 + x/3 + 7$.
2. List *all* the divisors of $6x^2 + 3x - 3$ in $\mathbf{Z}[x]$.
★3. Describe a systematic method for finding all linear factors $ax + b$ of a polynomial $f(x)$ in $\mathbf{Z}[x]$.
4. For what integers n is $2x^2 + nx - 7$ reducible in $\mathbf{Q}[x]$?
5. Find the prime factors of the following polynomials in $\mathbf{Q}[x]$:

$$x^3 - 1001x^2 - 1, \qquad x^4 + 50x^2 + 2.$$

6. Prove that two elements a and b in a unique factorization domain always have a g.c.d. (a, b) and an l.c.m. $[a, b]$.
7. Prove that $ab \sim (a, b)[a, b]$ in any unique factorization domain.
8. Do the properties of "relatively prime" elements as stated in Exs. 15 and 16 of §3.8 hold in every unique factorization domain?
9. In the notation of the text, show directly
   (a) that $c_f f^*(x) \mid c_g g^*(x)$ in $G[x]$ if and only if $c_f \mid c_g$ in G and $f^*(x) \mid g^*(x)$ in $F[x]$;
   (b) using (a), that a "prime" of $G[x]$ which divides a product $a(x)b(x)$ must divide $a(x)$ or $b(x)$.
10. If $f(x)$ and $g(x)$ are relatively prime in $F[x]$, prove that $yf(x) + g(x)$ is irreducible in $F[x, y]$.

11. Decompose each of the following into irreducible factors in $Q[x, y]$, and prove that your factors are actually irreducible:
    (a) $x^3 - y^3$, (b) $x^4 - y^2$, (c) $x^6 - y^6$, (d) $x^7 + 2x^3y + 3x^2 + 9y$.

12. Find all irreducible polynomials of degree 2 or less in $Z_2[x, y]$.

13. Show that there exist in $Q[x, y]$ no polynomial solutions for the equation
    $1 = s(x, y)(x - 2) + t(x, y)(x + y - 3)$.

14. Show that the polynomial $f(x, y)$ is irreducible in $F[x, y]$ if there is a substitution $x \to t'$, $y \to t^s$ which yields a polynomial $f(t', t^s)$ irreducible in $F[t]$, provided the degree of $f(t', t^s)$ is the maximum of the integers mr and ns for all pairs m, n appearing as the exponents of some term $x^m y^n$ of f.

★15. (Kronecker.) If $f(x)|g(x)$ in $Z[x]$, prove that $f(c)|g(c)$ for each c in Z. Develop from this fact (and the interpolation formula (5) of §3.2) a systematic method of finding in a finite number of steps all factors of given degree of any $f(x)$ of $Z[x]$.

16. Let D be the set of all rational numbers which can be written as fractions a/b with a denominator b relatively prime to 6. Prove that D is a unique factorization domain.

★3.10. Eisenstein's Irreducibility Criterion

It is obvious that the equation $x^n = 1$, n odd, has no rational root except $x = 1$. It follows that $x^n - 1$ has no monic linear factors over Q except $x - 1$. But this does not show that the quotient

(13)
$$\phi(x) = \frac{x^n - 1}{x - 1} = x^{n-1} + x^{n-2} + \cdots + x + 1$$

is irreducible. Indeed, this polynomial is reducible unless n is a prime.

We now show that if $n = p$ is a prime, then the cyclotomic polynomial $\phi(x)$ defined by (13) is irreducible, so that $x^p - 1 = (x - 1)\phi(x)$ gives the (unique) factorization of $x^p - 1$ into monic irreducible factors. This result will be deduced from the following sufficient condition for irreducibility due to Eisenstein:

Theorem 17. *For a given prime p, let $a(x) = a_n x^n + a_{n-1}x^{n-1} + \cdots + a_0$ be a polynomial with integral coefficients, such that $a_n \not\equiv 0 \pmod p$, $a_{n-1} \equiv a_{n-2} \equiv \cdots \equiv a_0 \equiv 0 \pmod p$, $a_0 \not\equiv 0 \pmod{p^2}$. Then $a(x)$ is irreducible over the field of rationals.*

Proof. In any possible factorization $(n = m + k)$

$$a(x) = (b_m x^m + b_{m-1}x^{m-1} + \cdots + b_0)(c_k x^k + c_{k-1}x^{k-1} + \cdots + c_0)$$

we may assume by Theorem 15 that both factors have integral coefficients b_i and c_j. Since $a_0 = b_0 c_0$, the third hypothesis $a_0 \not\equiv 0 \pmod{p^2}$ means that not both b_0 and c_0 are divisible by p. To fix our ideas, suppose that $b_0 \not\equiv 0 \pmod p$, while $c_0 \equiv 0 \pmod p$. But $b_m c_k = a_n \not\equiv 0 \pmod p$, so $c_k \not\equiv 0 \pmod p$. Pick the smallest index $r \leqq k$ for which $c_r \not\equiv 0 \pmod p$, with $c_{r-1} \equiv \cdots \equiv c_0 \equiv 0 \pmod p$. Then

$$a_r = b_0 c_r + b_1 c_{r-1} + \cdots + b_r c_0 \equiv b_0 c_r \pmod p.$$

But $b_0 \not\equiv 0$ and $c_r \not\equiv 0$, give $a_r \not\equiv 0$, for p is a prime. By the hypothesis, the only coefficient a_r for which this is possible is a_n, so $r = n$: the degree of the second of the proposed factors must be n, so that the polynomial $f(x)$ is indeed irreducible.   Q.E.D.

This criterion may be applied to the polynomial (13) when $n = p$; it gives the *cyclotomic* polynomial

(13') $\phi(x) = (x^p - 1)/(x - 1) = x^{p-1} + x^{p-2} + \cdots + x + 1.$

The Eisenstein criterion does not apply to (13') as it stands, but a simple change of variable $y = x - 1$ works, for the binomial expansion gives

$$(x^p - 1)/(x - 1) = [(y + 1)^p - 1]/y$$
$$= y^{p-1} + py^{p-2} + \frac{p(p - 1)}{1 \cdot 2} y^{p-3} + \cdots + p.$$

The binomial coefficients which appear on the right are all integers divisible by the prime p, for p occurs in each numerator as a factor, and can never be cancelled out by the (smaller) integers in the denominator. The polynomial in y thus satisfies the hypotheses of the Eisenstein criterion, hence is irreducible; this entails the irreducibility of the original cyclotomic polynomial $\phi(x)$ of (13').

Exercises

1. Which of the following polynomials are irreducible over the field of rationals?

     $x^3 + 2x^2 + 4x + 2, \quad x^3 + 2x^2 + 2x + 4, \quad x^7 - 47, \quad x^4 + 15.$

2. Use Eisenstein's criterion to show $x^2 + 1$ irreducible over the rationals.
3. If $f(x)$ is irreducible over a field F, show that $f(x + a)$ also is, for any a in F.
4. If a polynomial $f(x)$ of degree $n > k$ satisfies the hypotheses $a_n \not\equiv 0, a_k \not\equiv 0, a_{k-1} \equiv \cdots \equiv a_0 \equiv 0 \pmod p$ and $a_0 \not\equiv 0 \pmod{p^2}$, show that $f(x)$ has an irreducible factor of degree at least k.

★5. Show that the irreducibility of a polynomial of odd degree $2n + 1$ is enforced by the conditions $a_{2n+1} \not\equiv 0 \pmod p$, $a_{2n} \equiv \cdots \equiv a_{n+1} \equiv 0$ $(\mathrm{mod}\, p)$, $a_n \equiv a_{n-1} \equiv \cdots \equiv a_0 \equiv 0 \pmod{p^2}$, $a_0 \not\equiv 0 \pmod{p^3}$.

6. (a) If $f(x)$ is a monic polynomial with integral coefficients, show that the irreducibility of $f(x)$ modulo p implies its irreducibility over **Q**.
   (b) Show that every factor of $f(x)$ over **Z** must reduce modulo p to a factor of the same degree over $\mathbf{Z}_p$.
   (c) Use this to test (using small p) the irreducibility over **Q** of

$$x^3 + 6x^2 + 5x + 25, \quad x^2 + 6x^2 + 11x + 8, \quad x^4 + 8x^3 + x^2 + 2x + 5.$$

7. (a) Let $F[t]$ be the domain of all polynomials in an indeterminate t. State and prove an analogue of the Eisenstein Theorem for polynomials $f(x)$ with *coefficients* in $F[t]$. (*Hint:* Use t in place of p.)
   (b) Use this to prove that $x^3 + 3t^2x^2 + 2tx^2 + t^4x + 7t + t^2$ is irreducible in the domain $F[t, x]$.

★3.11. Partial Fractions

The unique decomposition theorem for polynomials can be applied to rational functions to obtain certain simplified representations, like the partial fraction decomposition used in integral calculus. This we now discuss, assuming throughout that the polynomials and rational forms used have coefficients in some fixed field F.

Consider first a rational form $b(x)/a(x)$ in which the denominator has a factorization $a(x) = c(x)d(x)$ with relatively prime factors $c(x)$ and $d(x)$. Theorem 12 gives polynomials $s(x)$ and $t(x)$ with $1 = sc + td$; hence

(14) $\qquad b(x)/[c(x)d(x)] = [s(x)b(x)]/d(x) + [t(x)b(x)]/c(x).$

The result in words is

Lemma 1. *A rational form in which the denominator is the product of relatively prime polynomials $c(x)$ and $d(x)$ can be expressed as a sum of two quotients with denominators $c(x)$ and $d(x)$, respectively.*

If the denominator $a(x)$ is a power $a(x) = [c(x)]^m$, $m > 1$, this process does not apply directly. Instead, divide the numerator by $c(x)$ as in the division algorithm, $b(x) = q_0(x)c(x) + r_0(x)$, then divide the quotient $q_0(x)$ again by $c(x)$, to get $q_0(x) = q_1(x)c(x) + r_1(x)$. Combined, these give

$$b(x) = q_1(x)[c(x)]^2 + r_1(x)c(x) + r_0(x).$$

Repeating this process (this phrase disguises an induction; remove the camouflage!), one finds, in abbreviated notation, that †

$$(15) \qquad b(x) = q_{m-1}c^m + r_{m-1}c^{m-1} + \cdots + r_1c + r_0,$$

where each polynomial $r_i = r_i(x)$, if not zero, has degree less than that of $c(x)$. The rational form $b(x)/a(x)$ now becomes

$$(16) \quad b/c^m = q_{m-1} + r_{m-1}/c + r_{m-2}/c^2 + \cdots + r_1/c^{m-1} + r_0/c^m.$$

This proves

Lemma 2. *A rational form with a power $[c(x)]^m$ as denominator can be expressed as a polynomial plus a sum of rational forms with denominators which are powers of $c(x)$ and numerators which have degrees less than that of $c(x)$.*

To combine these results, decompose an arbitrary given denominator $a(x)$ into a product of monic irreducibles. If equal irreducibles are grouped together, one has

$$(17) \qquad a(x) = a_0[p_1(x)]^{m_1}[p_2(x)]^{m_2} \cdots [p_k(x)]^{m_k},$$

with integral exponents m_i. Any two distinct monic irreducibles $p_1(x)$ and $p_2(x)$ are certainly relatively prime, so that the powers $[p_1(x)]^{m_1}$ and $[p_2(x)]^{m_2}$ have no common factors except units, hence are relatively prime. Lemma 1 can therefore be applied to that factorization of the denominator in which one factor is $c_1(x) = [p_1(x)]^{m_1}$, while the other factor is all the rest of (17). Repetition gives b/a as a sum of fractions, each with a denominator $[p_i(x)]^{m_i}$. To these denominators the reduction of (16) may be applied.

Theorem 18. *Any rational form $b(x)/a(x)$ can be expressed as a polynomial in x plus a sum of ("partial") fractions of the form $r(x)/[p(x)]^m$, where $p(x)$ is irreducible and $r(x)$ has degree less than that of $p(x)$. The denominators $[p(x)]^m$ which occur are all factors of the original denominator $a(x)$.*

If the explicit partial fraction decomposition of a given rational function $b(x)/a(x)$ is to be found, the successive steps of the proof of Theorem 18 may be carried out to get the explicit result. Such a proof,

---

† This is the analogue of the decimal expansion of an integer presented in Ex. 11, §1.5.

which can always be used for actual computation of the objects concerned, is known as a "constructive" proof.

For example, consider, over the field $\mathbf{Q}$, $(x + 1)/(x^3 - 1)$. The denominator is $(x - 1)(x^2 + x + 1)$, and the second factor is irreducible. The Division Algorithm gives $x^2 + x + 1 = (x + 2)(x - 1) + 3$. Multiplying this equation by the numerator $x + 1$ of the original equation, we get

$$3(x + 1) = (x + 1)(x^2 + x + 1) - (x^2 + 3x + 2)(x - 1);$$

$$\frac{3(x + 1)}{x^3 - 1} = \frac{x + 1}{x - 1} - \frac{x^2 + 3x + 2}{x^2 + x + 1}.$$

Each of the resulting fractions may be simplified by a further long division,† to give

$$\frac{3(x + 1)}{x^3 - 1} = \frac{2}{x - 1} - \frac{2x + 1}{x^2 + x + 1}$$

Over the field $\mathbf{R}$ of real numbers the only irreducible polynomials are the linear ones and the quadratic polynomials $ax^2 + bx + c$ with $b^2 - 4ac < 0$. (This statement will be proved in §5.5, Theorem 7.) Therefore over $\mathbf{R}$ any rational function can be expressed as a sum of terms with denominators which are powers of linear and quadratic expressions. This fact is used in calculus to prove that the indefinite integral of any rational function can be expressed in terms of "elementary functions" (i.e., algebraic, trigonometric, and exponential functions, and their inverses). By Theorem 18, the rational form to be integrated is essentially a sum of terms of the types $c(x + a)^{-m}$ and $c(x + d)(x^2 + ax + b)^{-m}$. Hence the proposition on integrals will be proved if one can integrate these two types by elementary functions (which can be done).

Exercises

1. Decompose into partial fractions (over the rational field):

(a) $\dfrac{3x + 4}{x^2 + 3x + 2}$, (b) $\dfrac{1}{x^2 - a^2}$, (c) $\dfrac{1}{x^3 + x}$,

(d) $\dfrac{a^2}{x^3 - a^3}$, (e) $\dfrac{3}{x^4 + 5x^2 + 4}$, (f) $\dfrac{3x - 7}{(x - 2)^2}$.

† Compare the directness of this method with that often used in texts on calculus, where one must solve for the "unknown" coefficients A, B, C which occur in the terms $A/(x - 1)$ and $(Bx + C)/(x^2 + x + 1)$.

2. Decompose $(4x + 2)/(x^3 + 2x^2 + 4x + 8)$ over (a) the field $\mathbf{Z}_5$ of integers mod 5, (b) the field $\mathbf{Q}$ of rational numbers.

3. If $a_0, a_1, \cdots, a_n$ are distinct, prove that

$$\frac{1}{\prod_i (x - a_i)} = \sum_i \frac{C_i^{-1}}{x - a_i}, \qquad \text{where } C_i = \prod_{j \neq i} (a_i - a_j).$$

   (*Hint:* Expand $p(a_i) = 1$ by Lagrange's interpolation formula.)

4. Prove equation (13) by induction on m.

5. Give a detailed proof by induction of Theorem 18.

6. (a) Prove that any rational form not a polynomial can be represented as a polynomial plus another rational form in which the numerator is 0 or has lower degree than the denominator.

   (b) Is this representation unique?

7. If all fractions (including the partial fractions) are restricted to have numerators of degrees lower than the respective denominators, show that the representation is unique (a) in Lemma 1, (b) in Lemma 2, (c) in Theorem 18.

8. (a) If $(x - a)$ is not a factor of $f(x)$, prove that

$$\frac{1}{(x - a)^r f(x)} = \frac{C}{(x - a)^r} + \frac{g(x)}{(x - a)^{r-1} f(x)},$$

   where $C = 1/f(a)$ and $g(x)$ is a suitable polynomial.

   ★(b) Using Ex. 8(a) or Ex. 3, deduce a canonical form for those rational functions whose denominators can be factored into linear factors.

9. (a) If $p(x)$ is irreducible, prove that any representation of a fraction $b(x)/p(x)$ (with b relatively prime to p) as a sum of fractions must involve at least one fraction with a denominator divisible by $p(x)$. (This means that further partial fraction decompositions of $b(x)/p(x)$ are out of the question.)

   (b) Can the same be said for $b(x)/[p(x)]^m$?

★10. Find the sum of $[(x + 1)(x + 2)]^{-1} + 2[(x + 2)(x + 4)]^{-1} + \cdots + 2^n[(x + 2^n)(x + 2^{n+1})]^{-1}$.

★11. Develop a method of representing any rational number as a sum of "partial fractions" of the special form a/p^n (p prime, $0 \leq a < p$). For example, $1/6 = 1/2 - 1/3$.

★12. Assuming Theorem 6 of §5.3, show that the indefinite integral of any complex rational function is the sum of a rational function and a linear combination of complex logarithms $\log(z + a_i) = \int dz/(z + a_i)$.

★13. Show that over any ordered domain D, the polynomial domain $D[x]$ becomes ordered if we choose for "positive" polynomials those having a positive leading coefficient, so that $a_n > 0$ in (1).

4

Real Numbers

4.1. Dilemma of Pythagoras

Although "modern" algebra properly stresses the wealth of properties holding in general fields and integral domains, the real and complex fields are indispensable for describing quantitatively the world in which we live. For example, these two fields are crucial in the relation of algebra to geometry, both in elementary analytic geometry and in the further development of vectors and vector analysis (Chap. 7). Moreover, they also have unique algebraic properties, which will be exploited in later chapters of this book. Especially important are the *order* completeness of the real field **R** and the *algebraic* completeness of the complex field **C**. We shall devote the next two chapters to these completeness properties and their algebraic implications.

A completely geometric approach to real numbers was used by the Greeks. For them, a number was simply a ratio $(a:b)$ between two line segments a and b. They gave direct geometric constructions for equality between ratios and for addition, multiplication, subtraction, and division of ratios. The postulates stating that the real numbers form an *ordered field* (§2.4) appeared to the Greeks as a series of geometric theorems, to be proved from postulates for plane geometry (including the parallel postulate).

The ancient Greek philosopher Pythagoras knew that the ratio $r = d/s$ between the length d of a diagonal of a square and the length s of its side must satisfy the equation

$$(1) \qquad d^2 = (rs)^2 = r^2 s^2 = s^2 + s^2 \qquad \text{(Pythagorean theorem)}.$$

So, he reasoned, there is a "number" r satisfying $r^2 = 1 + 1 = 2$.

On the other hand, he found r could not be represented as a quotient $r = a/b$ of integers, for $(a/b)^2 = 2$ would imply $a^2 = 2b^2$. By the prime factorization theorem, 2 divides a^2 just twice as often as it divides a—hence an even number of times; similarly, it divides $2b^2$ an odd number of times. Therefore, $a^2 = 2b^2$ has no solution in integers.

From this "dilemma of Pythagoras" one can escape only by creating *irrational* numbers: numbers which are not quotients of integers.

Similar arguments show that both the ratio $\sqrt{3}$ of the length of a diagonal of a cube C to the length of its side, and the ratio $\sqrt[3]{2}$ of the length of a side of C to the side of a cube having half as much volume, are irrational numbers. These results are special cases of Theorem 10 of §3.7.

Further irrational numbers are π (which thus cannot be exactly $\frac{22}{7}$ or even 3.1416), e, and many others. In Chap. 14 we shall prove that the vast majority of real numbers not only are irrational, but also (unlike $\sqrt{2}$) even fail to satisfy any algebraic equation. To answer the fundamental question *"what is a real number?"* we shall need to use entirely new ideas.

One such idea is that of *continuity*—the idea that if the real axis is divided into two segments, then these segments must touch at a common frontier point. A second such idea is that the ordered field **Q** of rational numbers is *dense* in the real field, so that every real number is a *limit* of one or more sequences of rational numbers (e.g., of finite decimal approximations correct to n places). This idea can also be expressed in the statement

(2)   If $x < y$, then there exists $m/n \in Q$ such that $x < m/n < y$.

This property of real numbers was first recognized by the Greek mathematician Eudoxus. Thinking of $x = a:b$ and $y = c:d$ as ratios of lengths of line segments, integral multiples $n \cdot a$ of which could be formed geometrically, Eudoxus stipulated that $(a:b) = (c:d)$ if and only if, for all positive integers m and n,

(3) $na > mb$ implies $nc > md$, $na < mb$ implies $nc < md$.

The two preceding ideas can be combined into a single postulate of *completeness*, which also permits one to construct the real field as a natural extension of the ordered field **Q**. This "completeness" postulate is analogous to the well-ordering postulate for the integers (§1.4): both deal with properties of infinite sets, and so are *nonalgebraic*. As we shall see, this completeness postulate is needed to establish certain essential algebraic properties of the real field (e.g., that every positive number has a square root).

Exercises

1. Give a direct proof that $\sqrt{3}$ is irrational.
2. Prove that $\sqrt[n]{a}$ is irrational unless the integer a is the nth power of some integer.
3. Prove that $\log_{10} 3$ is irrational (*Hint:* Use the definition of the logarithm.)
4. Show that, if $a \neq 0$ and b are rational, then $au + b$ is rational if and only if u is rational.
★5. Prove that $\sqrt{2} + \sqrt{5}$ is irrational. (*Hint:* Find a polynomial equation for $x = \sqrt{2} + \sqrt{5}$, starting by squaring both sides of $x - \sqrt{2} = \sqrt{5}$.)
★6. Prove that e, as defined by the convergent series $\sum\limits_{k=0}^{\infty} 1/k!$, is irrational. (*Hint:* If e were rational, then $(n!)e$ would be an integer for some n.)

4.2. Upper and Lower Bounds

The real field can be most simply characterized as an ordered field in which arbitrary bounded sets have greatest lower and least upper bounds. We now define these two notions, which are analogous to the concepts of greatest common divisor and least common multiple in the theory of divisibility.

Definition. *By an* upper bound *to a set S of elements of an ordered domain D is meant an element b (which need not itself be in S) such that* $b \geqq x$ *for every x in S. An upper bound b of S is a* least upper bound *if no smaller element of D is an upper bound for S, that is, if for any $b' < b$ there is an x in S with $b' < x$.*

The concepts of *lower bound* and *greatest lower bound* of S are defined dually, by interchanging $>$ with $<$ throughout, in the above definition.

It follows directly from the definition that a subset S of D has at most one least upper bound and at most one greatest lower bound (why?).

Intuitively, think of the real numbers as the points of a continuous line (the x-axis), and imagine the rational numbers as sprinkled densely on this line in their natural positions. From this picture, one readily concludes that every real number a can be characterized as the least upper bound of the set S of all rational numbers $r = m/n$ $(n > 0)$ such that $r < a$. For example, $\sqrt{2}$ is the least real number greater than all ratios m/n $(m > 0, n > 0)$ such that $m^2 < 2n^2$. That is, the number $\sqrt{2}$ is the least upper bound of the set of positive rational numbers m/n such that $m^2 < 2n^2$.

The concept of real numbers as least upper bounds of sets of rationals is directly involved in the familiar representation of real numbers by unlimited decimals. Thus we can write $\sqrt{2}$ as both a least upper bound (l.u.b.) and a greatest lower bound (g.l.b.),

(4)
$$\sqrt{2} = \text{l.u.b.} \, (1.4, 1.41, 1.414, 1.4142, \cdots)$$
$$= \text{g.l.b.} \, (1.5, 1.42, 1.415, 1.4143, \cdots).$$

Assuming the familiar properties of this decimal representation, it is very easy to "see" that every nonempty set T of positive real numbers has a greatest lower bound, as follows.

Consider the n-place decimals which express members of T to the first n places: there will be a least among them, because there are only a finite number of nonnegative n-place decimals less than any given member of T. Let this least n-place decimal be $k + 0.d_1 d_2 \cdots d_n$, where k is some integer and each d_i is a digit. The least $(n + 1)$-st place decimal coincides with this through the first n places, so has the form $k + 0.d_1 d_2 \cdots d_n d_{n+1}$, with one added digit. Our construction hence defines a certain unlimited decimal $c = k + 0.d_1 d_2 d_3 \cdots$. By construction, this is a lower bound to T (since its decimal expansion is greater than that of no x in T), and even a *greatest* lower bound (any bigger decimal would lose this property).

However, if the real numbers are defined as unlimited decimals, it is very hard to prove what is implicitly assumed in high-school algebra: that the system of unlimited decimals is an ordered field.†

Exercises

1. Prove that $x = .12437437437 \cdots$ represents a rational number. (*Hint:* Compute $1000x - x$.)
2. Do the same for $y = 1.23672367 \cdots$.
★3. Prove that *any* "repeating decimal," like those of Exs. 1 and 2, represents a rational number. Define your terms carefully.
★4. Prove conversely that the decimal expansion of any rational number is "repeating." (*Suggestion:* Show that if the same remainder occurs after m as after $m - k$ divisions by 10, the block of k digits between gets repeated indefinitely.)
★5. Does the result of Ex. 4 holds in the duodecimal scale?
6. Find three successive approximations to $\sqrt{2}$ in the domain of all rational numbers with denominators powers of 3.

† For details, see J. F. Ritt, *Theory of Functions* (New York: Kings Crown Press, 1947). The difficulty begins with equations like $.19999 \cdots = .20000 \cdots$ between different decimals.

7. Describe two different sets of rational numbers which both have the same l.u.b. 2.

★8. Define the sequence $(2, 3/2, 17/12, 577/408, \cdots)$ recursively by $x_1 = 2$, $x_{k+1} = (x_k/2) + (1/x_k)$.
   (a) Show that for $k > 1$, $x_k = m_k/n_k$, where $m_k^2 = 2n_k^2 + 1$.
   (b) Defining $\epsilon_k = x_k - \sqrt{2}$, show that $0 < \epsilon_{k+1} < \epsilon_k^2/2\sqrt{2}$.
   (c) Show that g.l.b. $(2, 3/2, 17/12, \cdots) = \sqrt{2}$.

4.3. Postulates for Real Numbers

We shall now describe the real numbers by a brief set of postulates. Subsequently we shall see (Theorem 6) that these postulates determine the real numbers uniquely, up to an isomorphism.

Definition. *An ordered domain D is* complete *if and only if every nonempty set S of positive elements of D has a greatest lower bound in D.*

Postulate for the real numbers. The real numbers form a complete ordered field **R**.

From the properties of the real numbers given by this postulate, one can actually deduce all the known properties of the real numbers, including such a result as Rolle's theorem, which is known to be fundamental in the proof of Taylor's theorem and elsewhere in the calculus.

However, we shall confine our attention to a few simple applications.

Theorem 1. *In the field* **R** *of real numbers, every nonempty subset S which has a lower bound has a greatest lower bound, and, dually, every nonempty subset T which has an upper bound has a least upper bound.*

Proof. Suppose S has a lower bound b. If $1 - b$ is added to each number x of S, there results a set S' of positive numbers $x - b + 1$. By our postulate, this set S' has a g.l.b. c'. Consequently, the number $c = c' + b - 1$ is then a g.l.b. for the original set S, as may be readily verified.

Dually, if the set T has an upper bound a, the set of all negatives $-y$ of elements of T has a lower bound $-a$. Hence, by the previous proof, the set has a greatest lower bound b^*. The number $a^* = -b^*$ then proves to be a least upper bound of the given set T.   Q.E.D.

Our postulates make the real numbers an ordered field **R**, so Corollary 2 of Theorem 18 in §2.6 shows that **R** must contain a subfield isomorphic to the field **Q** of rationals. Since **Q** is defined in Chap. 2 only up to isomorphism, we can just as well assume that the field **R** of real numbers does contain all the rationals and hence all the integers. This convention adjusts our postulates to fit ordinary usage, and enables us to prove the following property of the reals (often called the Archimedean law).

Theorem 2. *For any two numbers $a > 0$ and $b > 0$ in the field* **R** *of all real numbers (as defined by our postulates), there exists an integer n for which $na > b$.*

Proof. Suppose the conclusion false for two particular real numbers a and b, so that, for every n, $b \geqq na$. The set S of all the multiples na then has the upper bound b so that it has also a least upper bound b^*. Therefore $b^* \geqq na$ for every n, so that also $b^* \geqq (m + 1)a$ for every m. This implies $b^* - a \geqq ma$ so that $b^* - a$ is an upper bound for the set S of all multiples of a, although it is smaller than the given least upper bound, a contradiction.

Corollary. *Given real numbers a and b, with $b > 0$, there exists an integer q such that $a = bq + r$, $0 \leqq r < b$.*

The proof of this extension of the Division Algorithm will be left to the reader.

The so-established "Archimedean property" may be used to justify the condition of Eudoxus (cf. §4.1, (3)).

Theorem 3. *Between any two real numbers $c > d$, there exists a rational number m/n such that $c > m/n > d$.*

As before, this is to be proved simply from the postulate that the reals form a complete ordered field. By hypothesis, $c - d > 0$, so the Archimedean law yields a positive integer n such that $n(c - d) > 1$, or $1/n < c - d$. Now let m be the smallest integer such that $m > nd$; then $(m - 1)/n \leqq d$, so that

$$m/n = (m - 1)/n + 1/n < d + (c - d) = c.$$

Since $m/n > d$, this completes the proof.

We can visualize the above proof as follows. The various fractions 0, $\pm 1/n$, $\pm 2/n, \cdots$ with a fixed denominator n are spaced along the real axis at intervals of length $1/n$. To be sure that one such point falls between c and d, we need only make the spacing $1/n$ less than the given difference $c - d$.

This theorem may be used to substantiate formally the idea used intuitively in a representation like (4) of a real number as a l.u.b. of rationals.

Corollary. *Every real number is the l.u.b. of a set of rationals.*

Proof. For a given real number c, let S denote the set of all rationals $m/n \leqq c$. Then c is an upper bound of S; by the theorem no smaller real number d could be an upper bound of S, hence c is the *least* upper bound of S.

Exercises

1. Prove that there is no ordered domain D in which *every* nonempty set has a l.u.b. (*Hint:* Show that D itself can have no upper bound.)

★2. Show that the ordered domain **Z** is complete.

3. State in geometrical language a postulate on points of the real axis which asserts that bounded sets have l.u.b. and g.l.b. (use the words "left" and "right").

4. Exhibit the l.u.b. of each of the following sets of rational numbers:
   (a) $1/3, 4/9, 13/27, 40/81, \cdots$;   (b) $1/2, 3/4, 7/8, 15/16, \cdots$.

5. Let a set S have a l.u.b. a^* and a g.l.b. b^*.
   (a) Show in detail why the set of all numbers $-3x$, for x in S, has the l.u.b. $-3b^*$ and the g.l.b. $-3a^*$.
   (b) In the same way, find the l.u.b. and the g.l.b. of the set of all numbers $x + 5$, for x in S.

6. In Ex. 5, what is the l.u.b. (a) of the set of all numbers $7x + 2$ for x in S, (b) of the set of all numbers $1/x$ for $x \neq 0$ in S, if $b^* > 0$?

7. Let S_1 and S_2 be sets of real numbers with the respective least upper bounds b_1 and b_2. What is the least upper bound (a) of the set $S_1 + S_2$ of all sums $s_1 + s_2$ (for s_1 in S_1 and s_2 in S_2), (b) of the set of all elements belonging either to S_1 or to S_2?

8. Collect in one list a complete set of postulates for the real numbers.

★9. Construct a system of postulates for the positive real numbers. (*Hint:* Cf. §2.5.)

10. Show that an element a^* in an ordered field is a least upper bound for a set S if and only if (i) $x \leqq a^*$ for all $x \in S$ and (ii) for each positive e in the field, there is an x in S with $|x - a^*| < e$.

11. Show that between any two real numbers $c < d$, there exists a rational cube $(m/n)^3$ such that $c < (m/n^3) < d$. Is this true for rational squares?

12. If $h > 1$ is an integer, prove that between any two real numbers $c > d$, there lies a rational number of the form m/h^k, where m and k are suitable integers.

13. Let a, b, c, and d be positive elements of a complete ordered field. Show that $a/b = c/d$ if and only if the condition (3) of Eudoxus is satisfied.

14. Prove in detail the corollary to Theorem 2.

4.4. Roots of Polynomial Equations

We shall now show how to use the existence of least upper bounds to prove various properties of the real number system **R**, including first the existence of solutions for equations such as $x^2 = 2$.

Theorem 4. *If $p(x)$ is a polynomial with real coefficients, if $a < b$, and if $p(a) < p(b)$, then for every constant C satisfying $p(a) < C < p(b)$, the equation $p(x) = C$ has a root between a and b.*

Geometrically, the hypothesis means that the graph of $y = p(x)$ meets the horizontal line $y = p(a)$ at $x = a$ and the line $y = p(b)$ at $x = b$; the conclusion asserts that the graph must also meet each intermediate horizontal line† $y = C$ at some point with an x-coordinate between a and b.

The proof depends upon two lemmas.

Lemma 1. *For any real x and h, we have $p(x + h) - p(x) = hg(x, h)$, where $g(x, h)$ is a polynomial depending only on $p(x)$.*

Proof. (Cf. Theorem 3, §3.2.) For each monomial term $a_k x^k$ of $p(x)$, this is true by the binomial theorem. Now summing over k and taking out the common factor h, we get the desired result.

Lemma 2. *For given a, b, and $p(x)$, there exists a real constant M such that $|p(x + h) - p(x)| \leqq Mh$ for all x and all positive h satisfying $a \leqq x \leqq b$, $a \leqq x + h \leqq b$.*

Proof. By Lemma 1, it suffices to show that $|g(x, h)| \leqq M$ whenever $|x| \leqq |a| + |b|$, $|h| \leqq |b - a|$. But if we replace each term in $g(x, h)$ by its absolute value, we increase $|g(x, h)|$ or leave it undiminished, by formula (3) of §1.3 and induction. We do the same again if we replace $|x|$ and $|h|$ by $|a| + |b|$ and $|b - a|$, respectively. This substitution gives us, however, a real constant M, depending only on the coefficients of $p(x)$ and the interval $a \leqq x \leqq b$.

Having Lemma 2 at our disposal, we are ready to prove Theorem 4. Let S denote the set of real numbers between a and b satisfying $p(x) \leqq C$. Since $p(a) < C$, S is nonvoid, and it has b as upper bound; hence it has a real *least* upper bound c. We shall show that $p(c) = C$.

For this purpose, it is clearly sufficient to exclude the possibilities $p(c) < C$ and $p(c) > C$. But $p(c) < C$ would imply that $p(c + h) \leqq C$

---

† There is a general theorem of analysis which asserts this conclusion, not only for polynomial functions $p(x)$, but for any continuous function.

for $h = [C - p(c)]/M$, by Lemma 2, whence $(c + h) \in S$. This would contradict our definition of c, as an upper bound to S. (Lemma 2 applies because $c + h \geqq b$ is evidently also impossible.)

There remains the possibility $p(c) > C$. But in this case, again by Lemma 2, $p(c - h) > C$ for all positive $h \leqq [p(c) - C]/(2M)$. This would contradict our definition of c as the *least* upper bound of S: $c - [p(c) - C]/(2M)$ would give a smaller upper bound. There remains only the possibility $p(c) = C$.   Q.E.D.

From the theorem one readily proves:

Corollary 1. *If $p(x)$ is a polynomial with positive coefficients and no constant term, and if $C > 0$, then $p(x) = C$ has a positive real root.*

Corollary 2. *If $p(x)$ is of odd degree, then $p(x) = C$ has a real root for every real number C.*

Theorem 4 does not give a construction for actually computing a root of $p(x) = C$ in decimal form, but this is easy to do. For example, one can let $c_1 = (a + b)/2$; then $p(c_1) = C$ or $p(c_1) > C$ or $p(c_1) < C$. In the first case the root is found; in the second and third cases there is a root in an interval (either $a \leqq x \leqq c_1$ or $c_1 \leqq x \leqq b$) half as long as before. By repeating this construction, a root of $p(x) = C$ can be found to any desired approximation.

Convergence would be much faster if one used linear interpolation and set

$$c_1 = a + [C - p(a)][b - a][p(b) - p(a)]^{-1}.$$

Other efficient methods of calculating roots of equations are studied in analysis. For example, if $|x| < 1$, one may use the infinite series

$$(5) \qquad \sqrt{1 + x} = 1 + \frac{1}{2}x + \frac{1}{2}\left(-\frac{1}{2}\right)\frac{x^2}{2!} + \frac{1}{2}\left(-\frac{1}{2}\right)\left(-\frac{3}{2}\right)\frac{x^3}{3!} + \cdots.$$

Appendix. Trigonometric Solution of Cubic. In the case of a cubic equation

$$(6) \qquad a_3 x^3 + a_2 x^2 + a_1 x + a_0 = 0, \qquad a_3 \neq 0,$$

the real roots can be found as follows. Dividing through by a_3, we reduce (6) to the case $a_3 = 1$. Now, by making the substitution $x = y - a_2/3$ and transposing the constant term, we reduce (6) to

$$(7) \qquad y^3 + py = q.$$

If $p = 0$, the solution is immediate.

Otherwise, setting $y = hz$ and multiplying (7) through by k, where $h = \sqrt{4|p|/3}$, $k = 3/(h|p|)$, we can reduce it to one of the forms

(8) $$4z^3 + 3z = C \quad \text{or} \quad 4z^3 - 3z = C.$$

To solve the first equation, one can use the familiar trigonometric identity $\sinh 3\theta = 4 \sinh^3 \theta + 3 \sinh \theta$, whence

(9a) $$z = \sinh [(1/3) \sinh^{-1} C].$$

To solve the second equation, if $C \geq 1$, we use the analogous formula $\cosh 3\theta = 4 \cosh^3 \theta - 3 \cosh \theta$ to get

(9b) $$z = \cosh [(1/3) \cosh^{-1} C].$$

If $C \leq -1$, the same method applies after changing the sign of z. To solve the second equation when $|C| < 1$ (this is the so-called irreducible case of §15.8), use similarly $\cos 3\theta = 4 \cos^3 \theta - 3 \cos \theta$, to get

(9c) $$z = \cos [(1/3) \cos^{-1} C].$$

In this case z assumes *three* values because $\cos^{-1} C$ has three values, differing by multiples of 120°.

Exercises

1. Prove that every positive real number has a real square root.
2. Show that for any positive real number a and any integer n, the equation $x^n = a$ has one and only one positive real root $\sqrt[n]{a}$.
3. Show that $x^4 - x = C$ has two real roots for every $C > -3/8$.
4. Find $\sqrt{5}$ to four decimal places, using (5) and $(\sqrt{5}/2)^2 = 1 + 1/4$.
5. Find $\sqrt{2}$ to six places using (5) and $(5\sqrt{2}/7)^2 = 1 + 1/49$.
6. Show that a monic polynomial of even degree assumes a least value K, and every value $C > K$ twice.
7. (a) If a and b are positive reals, show that $ax^{n+1} > bx^n$ for all sufficiently large positive values of x.
   (b) Given a polynomial $p(x)$ with a positive leading coefficient, find a real number m such that $p(x) > 0$ for all $x > M$.
8. Prove Corollary 1.
9. Prove Corollary 2.
10. Find to three decimal places the real roots of
    (a) $3x^3 - x = 1/9$,   (b) $x^3 - 3x^2 + 6x = 7$,   (c) $x^3 + 3x^2 + 2 = 0$.

★4.5. Dedekind Cuts

Imagine the rational numbers sprinkled in their natural position on the x-axis. But cutting the x-axis (say with scissors), one divides the rational numbers into two classes, L on the left and U on the right. Every rational number falls into one of these two classes, while a rational number m/n is in both only if the axis is cut exactly at the point $x = m/n$. Observe especially that if x is in L, then $x \leqq y$ for every y of U; conversely, if $x \leqq y$ for all y in U, x must lie in L. This leads to the idea of a *Dedekind cut*.

Formally, let F be any ordered field. By a "Dedekind cut" in F, we mean a pair of nonvoid subsets L and U such that

(i) L is the set of all lower bounds to the elements of U, and

(ii) U is the set of all upper bounds to the elements of L.

Lemma 1. *The lower and upper halves of a Dedekind cut taken together include all elements; they have at most one element in common.*

Proof. Let $x \in F$ be given. If $x \leqq a$ for some $a \in L$, then $x \leqq a \leqq y$ for all $y \in U$, whence $x \in L$. Otherwise, by the trichotomy law, $x > a$ for all $a \in L$, and so $x \in U$, which proves the first assertion: every element of F is in either L or U. Again, let a and b each be both in L and in U. Then $a \geqq b$ (since $a \in U$, $b \in L$) and $a \leqq b$ (since $a \in L$, $b \in U$), whence $a = b$, proving the second assertion.

If L and U have an element a in common, the cut will be said to go *through a*. Clearly, there is a cut (L_a, U_a) through every a, if L_a is the set of $x \leqq a$, and U_a the set of $x \geqq a$.

Dedekind Cut Axiom (*on an ordered field F*). *Every cut goes through some element a.*

Theorem 5. *The Dedekind cut axiom holds in an ordered field F if and only if F is a complete ordered field.*

Proof. Let (L, U) be any cut. If the existence of least upper bounds is given, L has a *least* upper bound a. Since a is an upper bound of L, it must lie in U; since it is a *least* upper bound, it is a lower bound for all the upper bounds, and so for all the elements of U. By the definition of a cut this means that a lies in L, so the given cut does go through the element a.

Conversely, suppose the Dedekind axiom holds, and that S is a nonempty bounded set. Let U be the set of all upper bounds of S, and L the set of all lower bounds of U (clearly, L contains S). To prove (L, U) a cut, one need only establish that U is the set of all upper bounds of L. But by the construction of L, every element of U is an upper bound of L ($x \leqq y$ for all $x \in L$, $y \in U$); while since L contains S, U includes *all* such upper bounds. Now by the Dedekind axiom, the cut (L, U) goes through some element a, which is an upper bound to S *qua* an element of U, and a *least* upper bound (i.e., $a \leqq x$ for all $x \in U$) *qua* an element of L. This completes the proof.

We shall now sketch a proof of the categorical nature of our postulate (§4.3) that the real number system is a complete ordered field.

Theorem 6. *Any two complete ordered fields are isomorphic.*

Proof. Let F' and F'' be any two such fields; by Corollary 2 of Theorem 18 in §2.6, they will contain isomorphic "rational" subfields $\mathbf{Q}'$ and $\mathbf{Q}''$. We shall extend the isomorphism between $\mathbf{Q}'$ and $\mathbf{Q}''$ (an isomorphism which preserves order as well as sums and products) to an isomorphism between F' and F''.

Indeed, every $a' \in F'$ defines a cut in F', and thereby a cut in $\mathbf{Q}'$ (the subfield of rationals). But by Theorem 3, a' is determined by this cut in $\mathbf{Q}'$—and every cut (L_R, U_R) in $\mathbf{Q}'$ determines an $a' = $ l.u.b. $L_R = $ g.l.b. U_R in this way. Cuts in $\mathbf{Q}''$ behave similarly, whence the elements of F' and F'' are bijective to the cuts in $\mathbf{Q}'$ and $\mathbf{Q}''$, respectively. This bijection clearly preserves order.

Finally, the *operations* in F' and F'' can be defined from those of $\mathbf{Q}'$ and $\mathbf{Q}''$ so as to extend the isomorphism. More precisely, let a and b correspond to cuts (L_a, U_a) and (L_b, U_b) in $\mathbf{Q}'$. Then $a + b$ corresponds to the cut† $(L_a + L_b, U_a + U_b)$—where $L_a + L_b$ is the set of sums $x + y$ ($x \in L_a$, $y \in L_b$), while $U_a + U_b$ is similarly described. To multiply *positive* elements a and b, form similar cuts in the system of positive rationals. Then ab corresponds to the cut $(L_a L_b, U_a U_b)$—where $L_a L_b$ is the set of products xy ($x \in L_a$, $y \in L_b$), and similarly for $U_a U_b$. Since $(-a)b = a(-b) = -ab$ and $(-a)(-b) = ab$, this extends to all products. We omit the details.

Conversely, one may use the cuts to "construct" the real numbers from the integers or positive integers. One first proves that the rationals form an ordered field $\mathbf{Q}$ having the Archimedean property stated in Theorem 2. By defining the addition and multiplication of cuts in $\mathbf{Q}$ in the way sketched in the previous paragraph, one can show that the cuts in $\mathbf{Q}$ form an ordered field satisfying the Dedekind cut axiom—hence giving a

---

† In certain cases, $(L_a + L_b, U_a + U_b)$ fails to be a cut because the number $a + b$ appears in neither half; but one then obtains a cut if the missing number is adjoined to both halves. A similar remark applies to $L_a L_b$ below.

complete ordered field. But the proof is long, and would lead us far afield, so that we shall just state the result.

Theorem 7. *There is one and (except for isomorphic fields) only one complete ordered field.*

Instead of using Dedekind cuts, it is also possible to construct the real numbers from the rationals as limits of sequences of rationals.†

Exercises

1. Show that if (L, U) and (L', U') are cuts in the rational field, every rational number with one exception at most can be written either as $x + y$ ($x \in L$, $y \in L'$) or as $u + v$ ($u \in U, v \in U'$).
2. State and prove an analogous theorem for the *positive* rational numbers under multiplication.
3. Why does this theorem fail for negative rationals?
4. Show that for every $\epsilon > 0$ there is an n so large that $10^{-n} < \epsilon$.
5. A Dedekind cut in an ordered field F is sometimes defined as a pair of subsets L' and U' of F such that every element of F lies either in L' or in U' and such that $x < y$ whenever $x \in L'$ and $y \in U'$. By adding and deleting suitable single numbers, show that every cut (L', U') of this type gives a cut (L, U) in the sense of the text, and conversely.
6. If t is an element in an ordered domain D with $0 < t < 1$, show that $s = 2 - t$ has the properties $s > 1$, $st \leqq 1$.
7. Let D be a "complete" ordered domain not isomorphic to $\mathbf{Z}$. Show that D contains an element t with $0 < t < 1$. If b and c are any positive elements of D, show that $t^n b < c$ for some n.
★8. Use Exs. 6 and 7 to show that any "complete" ordered domain is isomorphic either to $\mathbf{Z}$ or to $\mathbf{R}$. (*Hint:* To find the inverse of $b > 1$, consider all x with $xb \leqq 1$.)
9. (a) Prove that any isomorphism of $\mathbf{R}$ with itself preserves the relation $x \leqq y$. (*Hint:* $x \leqq y$ if and only if $z^2 = y - x$ has a root.)
   (b) Using (a), prove that the only isomorphism of $\mathbf{R}$ with itself is the trivial isomorphism $x \mapsto x$.
★10. Show that if $D = F$ is an ordered field and if, for each rational function,

$$R(x) = \frac{b_0 + b_1 x + \cdots + b_r x^r}{a_0 + a_1 x + \cdots + a_n x^n} \neq 0, \qquad a_n b_r \neq 0,$$

we define $R(x) > 0$ to mean that $a_n b_r > 0$, then $F(x)$ becomes an ordered field.
★11. Show that in Ex. 10 $R(x) > 0$ if and only if $R(t) > 0$ for all sufficiently large t in F.

† See the treatment in Chapter VI of C. C. MacDuffee, *Introduction to Abstract Algebra* (New York: Wiley, 1940).

5

Complex Numbers

5.1. Definition

Especially in algebra, but also in the theory of analytic functions and differential equations, many algebraic theorems have much simpler statements if one extends the real number system **R** to a larger field **C** of "complex" numbers. This we shall now define, and show that it is what one gets from the real field if one desires to make every polynomial equation have a root.

Definition. *A* complex number *is a couple* (x, y) *of real numbers—x being called the real and y the imaginary component of* (x, y). *Complex numbers are added and multiplied by the rules*:

(1) $$(x, y) + (x', y') = (x + x', y + y'),$$

(2) $$(x, y) \cdot (x', y') = (xx' - yy', xy' + yx').$$

The system of complex numbers so defined is denoted by **C**.

We owe the above definition not to divine revelation, but to simple algebraic experimentation. First, it was observed that the equation $x^2 = -1$ had no real root (x^2 being never negative). This suggested inventing an *imaginary* number i, satisfying $i^2 = -1$, and otherwise satisfying the ordinary laws of algebra. Stated in precise language, it suggested the plausible hypothesis that there was an integral domain D containing such an element i and the real field **R** as well.

In D, any expression of the form $x + yi$ (x, y real numbers) would represent an element. Moreover, by the definition of an integral domain

107

(laws of ordinary algebra),

$$(1') \qquad (x + yi) \pm (x' + y'i) = (x \pm x') + (y \pm y')i,$$

$$(2') \qquad (x + yi) \cdot (x' + y'i) = xx' + (xy' + yx')i + yy'i^2.$$

Since $i^2 = -1$, we get from (2')

$$(2'') \qquad (x + yi) \cdot (x' + y'i) = (xx' - yy') + (xy' + yx')i.$$

It is a corollary that the subdomain of D generated by $\mathbf{R}$ and i contains all elements of the form $x + yi$ and no others.

Again, $(x + yi) = (x' + y'i)$ implies $(x - x') = (y' - y)i$; hence squaring both sides, $(x - x')^2 = -(y' - y)^2$. And since $(x - x')^2 \geqq 0$, $-(y' - y)^2 \leqq 0$, this is impossible unless $x = x'$, $y = y'$. In summary, distinct couples (x, y) of real numbers determine distinct elements $x + yi$ of D. This establishes a one-one correspondence of the form $(x, y) \leftrightarrow x + yi$ between the elements of $\mathbf{C}$ and those of the subdomain of D generated by $\mathbf{R}$ and i. Finally, comparing formulas (1')–(2'') with (1)–(2), we see that the correspondence preserves sums and products, hence is an isomorphism. This proves

Theorem 1. *Let D be any integral domain containing the real number system $\mathbf{R}$ and a square root i of -1. Then the subdomain of D generated by $\mathbf{R}$ and i is isomorphic with $\mathbf{C}$.*

We now prove our conjecture that there does indeed exist an integral domain D which contains the real numbers and a square root of -1.

Theorem 2. *The complex number system, as defined above, is a field containing a subfield isomorphic to $\mathbf{R}$ and a root of $x^2 + 1 = 0$.*

Proof. For the couples (x, y), the commutative and associative laws of addition, the fact that $(0, 0)$ is an additive identity, and the fact that $(-x, -y)$ is an additive inverse of (x, y) are immediate consequences of the fact that *real* and *imaginary* components are added independently, while the corresponding laws hold for them.

Similarly, the commutative and associative laws of multiplication, the facts that $(1, 0)$ is a multiplicative identity and that every $(x, y) \neq (0, 0)$ has a multiplicative inverse

$$(3) \qquad (x, y)^{-1} = (x/(x^2 + y^2), -y/(x^2 + y^2))$$

follow from the fact to be established in §5.2, that "arguments" and "absolute values" of complex numbers combine independently under multiplication and themselves satisfy the same laws. But at the present stage, it is preferable to check these laws by direct substitution in the definition (2)—only the calculation for the associative law is long-winded. We omit the details.

Finally, we can check the distributive law by similar direct substitution. Thus let $z = (x, y)$, $z' = (x', y')$, $z'' = (x'', y'')$. Then substituting in (1) and (2),

$$z(z' + z'') = (x, y)(x' + x'', y' + y'')$$
$$= (x(x' + x'') - y(y' + y''), x(y' + y'') + y(x' + x'')),$$

$$zz' + zz'' = (xx' - yy', xy' + yx') + (xx'' - yy'', xy'' + yx'')$$
$$= (xx' - yy' + xx'' - yy'', xy' + yx' + xy'' + yx'');$$

from this, $z(z' + z'') = zz' + zz''$ can be checked directly.

In this field **C** of couples of numbers one may find a subfield of real numbers by exploiting the correspondence $(x, y) \leftrightarrow x + yi$, used in Theorem 1, in which the real numbers x correspond to couples with second term zero and the couple $(0, 1)$ to i. Specifically, if the second components y and y' in the definitions (1) and (2) are both zero, then the first components x and x' add and multiply just as do the real numbers x and x'. This is just the recognition that the correspondence $x \leftrightarrow (x, 0)$ is an isomorphism of the field **R** of reals to a subset of **C**. We agree, as in previous cases, that each such special complex number $(x, 0)$ is simply to be identified with the corresponding real number x.

Finally, the desired square root of -1 is presumably the couple $(0, 1)$; and in fact, a special case of the definition (2) shows that $(0, 1)^2 = (-1, 0) = -1$. Hence we *define* i to be the couple $(0, 1)$. Any couple (x, y) then has the form

(4) $(x, y) = (x, 0) + (0, y) = (x, 0) + (y, 0)(0, 1) = x + yi$

The notation $x + yi$ is so suggestive that we shall usually employ it instead of (x, y) in the sequel. For brevity, we shall also often write $z = (x, y) = x + yi$, $w = (u, v) = u + vi$, $c = (a, b) = a + bi$, and so on—in other words, we use a single letter to denote a complex number, and the two immediately preceding letters of the alphabet for its real and imaginary components.

Exercises

1. Check that complex multiplication is commutative and associative.
2. Check that $(x, y)(x, y)^{-1} = (1, 0)$ holds if formula (3) is used.

3. Solve $(1, 1)(x, y) = (2, 1)$
   (a) as a pair of simultaneous linear equations in x and y,
   (b) using (3).

4. Find complex numbers $z = x + yi$ and $w = u + vi$ which satisfy
   (a) $z + iw = 1$, $iz + w = 1 + i$,
   (b) $(1 + i)z - iw = 3 + i$, $(2 + i)z + (2 - i)w = 2i$.

5. Find all complex roots of $z^2 = -a$, where a is any positive real number. Justify your answer.

6. Describe the subfield of $\mathbf{C}$ which is generated by i and the *rational* numbers.

7. Is Theorem 1 still true if D is a commutative ring? Give details.

8. (a) Show that $z^2 = a + ib$ has solutions $z = x + iy$ with

$$x = [(1/2)(a + \sqrt{a^2 + b^2})]^{1/2}, \qquad y = b/(2x).$$

   (b) Show also that $y = [(1/2)(\sqrt{a^2 + b^2} - a)]^{1/2}$, $x = b/(2y)$. (Note that these formulas are more accurate for numerical computation when a is negative and b/a small.)

9. The equation $z^3 + 3iz = 3 + i$ has $-i$ for one root. Compute one other root in decimal form.

★10. Show that if F is any ordered field, then there exists a larger field F^* containing a subfield isomorphic to F and a square root of -1.

★11. Using the methods of Theorems 1 and 2, show without recourse to the real numbers that the rational field $\mathbf{Q}$ can be extended to a larger field $\mathbf{Q}(\sqrt{2})$ containing $\mathbf{Q}$ and a square root of 2.

12. Show that there is no possible definition of "positive complex number" which would make $\mathbf{C}$ an ordered field.

5.2. The Complex Plane

There is a fundamental one-one mapping of the complex numbers onto the points of a Cartesian plane. Namely, each complex number $z = x + iy$ is mapped onto the point $P = (x, y)$ with the real component x of z as abscissa and the imaginary component y as ordinate.

Polar coordinates may be used in this plane. We may recall that each point P of the plane and hence each complex number z is uniquely determined by the two polar coordinates r and θ, where r is the (non-negative) length of the segment $\overline{Oz}$ joining the point P to the origin, while θ is the angle from the x-axis to this segment (Figure 1), so

(5) $\qquad |z| = r = (x^2 + y^2)^{1/2}, \qquad \arg z = \theta = \tan^{-1} y/x.$

One calls r the *absolute value* of the complex number z and θ the *argument* of z. They determine x and y by

(6) $\qquad x = r \cos \theta, \qquad y = r \sin \theta, \qquad z = r(\cos \theta + i \sin \theta),$

the usual laws for the transformation from polar to rectangular coordinates. One also writes (6) in the form $z = re^{i\theta}$, since the usual Taylor series expansion gives

$$e^{i\theta} = 1 + i\theta + \frac{(-1)\theta^2}{2!} + \frac{(-i)\theta^2}{3!} + \cdots$$
$$= \cos\theta + i\sin\theta.$$

The importance of the absolute values and arguments rests largely on de Moivre's formulas, which may be stated as follows:

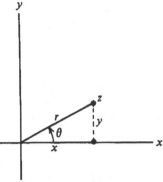

Figure 1

Theorem 3. *The absolute value of a product of complex numbers is the product of the absolute values of the factors; the argument is the sum of the arguments of the factors; in other words,*

(7) $$|zz'| = |z| \cdot |z'|, \qquad \arg zz' = \arg z + \arg z'.$$

Proof. As in (6), $z = r(\cos\theta + i\sin\theta)$, $z' = r'(\cos\theta' + i\sin\theta')$. Substituting in the definition (2), we get

$$zz' = rr'[(\cos\theta\cos\theta' - \sin\theta\sin\theta') + i(\cos\theta\sin\theta' + \sin\theta\cos\theta')];$$

by well-known trigonometric formulas, this is equivalent to

$$zz' = rr'[\cos(\theta + \theta') + i\sin(\theta + \theta')].$$

This gives the result (7).

Not only the multiplicative, but the additive properties of (inequalities on) absolute values are valid for complex as well as real numbers. That is,

(8) $$|z| > 0 \quad \text{unless } z = 0, \quad |0| = 0;$$

(9) $$|z + z'| \leq |z| + |z'|.$$

To prove these, note that formula (1) means that the sum $z + z'$ may be found by drawing (Figure 2) the parallelogram with three vertices at z, 0, and z'; the fourth vertex will be $z + z'$. Formulas (8) and (9) now follow from the identity between absolute values and geometrical lengths.

Complex nth roots of unity may be found using trigonometry. From the de Moivre formulas (7) one sees immediately that

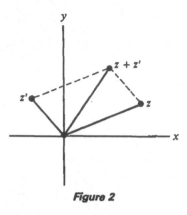

Figure 2

$$[r(\cos \theta + i \sin \theta)]^{-1}$$
$$= (1/r)[\cos (-\theta) + i \sin (-\theta)].$$

Further, one sees that $z^n = 1$ if and only if $|z|^n = 1$ and $n \cdot \arg z$ is an integral multiple $2k\pi$ of 2π. Since $|z| \geqq 0$, $|z| = 1$. Since $\arg z$ is single-valued on $0 \leqq \theta < 2\pi$, there are thus precisely n solutions of $z^n = 1$; in rectangular coordinates they are 1, $\cos 2\pi/n + i \sin 2\pi/n$, $\cdots$, $\cos 2\pi(n - 1)/n + i \sin 2\pi(n - 1)/n$. If we denote $\cos 2\pi/n + i \sin 2\pi/n$ by ω, we obtain another representation of these nth roots of unity as 1, ω, $\omega^2, \cdots, \omega^{n-1}$. Geometrically stated, this is

Theorem 4. *The complex nth roots of unity are the vertices of a regular polygon of n sides inscribed in the unit circle $|z| = 1$.*

Consider more generally the equation $z^n = c$, where $c \neq 0$ is any complex number. In polar coordinates, one solution of this is

$$z_0 = |c|^{1/n}(\cos \theta + i \sin \theta), \quad \text{with} \quad \theta = (1/n) \arg c.$$

Moreover, wz_0 is a root of $x^n = c$ if and only if $c = (wz_0)^n = w^n z_0^n = w^n c$, whence $w^n = 1$. Thus the nth roots of c are $z_0, \omega z_0, \omega^2 z_0, \cdots, \omega^{n-1} z_0$, where ω is as defined above. In particular, they are also represented by the vertices of a regular polygon.

One can easily compute the nth roots $z_0, \omega z_0, \cdots, \omega^{n-1} z_0$ of $c = a + bi$ numerically, with the aid of logarithmic and trigonometric tables. From the identity

$$\log |z_0| = \log |c|^{1/n} = (1/n) \log (a^2 + b^2)^{1/2} = (1/2n) \log (a^2 + b^2)$$

one can compute $|z_0|$. By de Moivre's formulas (7), $\arg z_0$ equals $(1/n) \tan^{-1} (b/a)$, and $\arg \omega^k z_0 = (1/n) \tan^{-1} (b/a) + 360k/n$ in degrees. The computation is completed by the formula

$$z = r(\cos \theta + i \sin \theta) = |z| \cos (\arg z) + i |z| \sin (\arg z).$$

Each complex nth root of unity ω satisfies a polynomial equation with rational coefficients irreducible over the field of rationals. These equations, known as the "cyclotomic" equations, play an important role in the theory of equations.

By definition, every nth root of unity satisfies $z^n - 1 = 0$; moreover, all except $z = 1$ satisfy

(10) $q_n(z) = (z^n - 1)/(z - 1) = z^{n-1} + z^{n-2} + \cdots + z + 1 = 0.$

In §3.10 Eisenstein's criterion was used to show that $q_p(z)$ is irreducible if $n = p$ is a prime.

If n is not a prime, the facts become less simple. Thus, if $n = 4$, $z^3 + z^2 + z + 1 = (z + 1)(z^2 + 1)$ is reducible. In general, we can factor out from (10) the cyclotomic polynomials satisfied by kth roots of unity, where k runs through the proper divisors of n. The nth roots of unity which are not also kth roots of unity for some $k < n$ are called *primitive* nth roots of unity. (Thus, the primitive fourth roots of unity are i and $-i$.) They are the ω^m with m relatively prime to n, and they all satisfy the same irreducible equation over the rational field. But the proof of this result, and the computation of the degree of this equation, involve more number theory than is desirable here.

Exercises

1. Prove the commutative and associative laws of multiplication and the existence of multiplicative inverses from de Moivre's formulas.
2. Describe geometrically the correspondence $z \mapsto zi$.
3. Find to 4 decimal places (using trigonometric tables) the real and imaginary components of the cube roots and the fifth roots of unity.
4. Find to 4 decimal places the cube and fourth roots of $2 + 2i$.
5. List the primitive twelfth roots of unity, and plot them on graph paper, drawing a large "unit circle."
6. Describe geometrically the effect of transformations $z \mapsto cz + d$ ($c, d \in \mathbf{C}$, $c \neq 0$). What if $|c| = 1$? (*Hint:* Use the words "translation," "rotation," and "expansion.")
7. Find the irreducible factors of $z^6 - 1$ over $\mathbf{Q}$ (the rationals).
8. (a) Prove that $\omega = \cos(2\pi/n) + i \sin(2\pi/n)$ is a primitive nth root of unity.
   (b) Prove that ω^m is a primitive nth root of unity if and only if m is relatively prime to n.

5.3. Fundamental Theorem of Algebra

We saw in §5.1 that the complex number system is obtained by adjoining to the real number system $\mathbf{R}$ an imaginary root i of the equation $z^2 + 1 = 0$. But why stop here? Why not try to add "imaginary" roots of other polynomial equations so as to get still larger fields?

The answer is contained in the so-called Fundamental Theorem of Algebra: as soon as i is adjoined, every polynomial equation *has* actual (complex) roots, so that one does not need to invent imaginary ones to solve equations.

Theorem 5 (Euler–Gauss). *Every polynomial p(z) of positive degree with complex coefficients has a complex root.*

Many proofs of this celebrated theorem are known.[†] All proofs involve nonalgebraic concepts like those introduced in Chap 4; we have selected one whose nonalgebraic part is especially plausible intuitively. We do not prove the nonalgebraic part in detail from the relevant axioms of Chap. 4.

Proof.   Since $p(z) = a_m z^m + a_{m-1} z^{m-1} + \cdots + a_0$, with $a_m \neq 0$, has the same roots as

$$q(z) = z^m + (a_{m-1}/a_m)z^{m-1} + \cdots + (a_0/a_m)$$
$$= z^m + c_{m-1} z^{m-1} + \cdots + c_0,$$

only the case where the leading coefficient is unity need be discussed.

In this case let us picture *two* complex planes, labeling one the "z-plane," and the other the "w-plane." The given function $q(z)$ maps each point $z_0 = (x_0, y_0)$ of the z-plane onto a point $w_0 = q(z_0)$ of the w-plane. Moreover, if z describes a continuous curve on the z-plane, then $q(z)$ (being differentiable) will describe a continuous curve on the w-plane. Our object is to show that the origin 0 of the w-plane is the "image" $q(z)$ of some z on the z-plane—or, what is the same thing, that the image of some circle on the z-plane passes through 0.

For each fixed $r > 0$, the function $w = q(re^{i\theta})$ defines a closed curve γ_r' in the w-plane: the image of the circle γ_r: $|z| = r$ ($z = re^{i\theta}$) of radius r and center 0 in the z-plane. For each fixed r, consider the line integral.[‡]

$$\phi(r, \theta) = \int_0^\theta d(\arg w) = \int_0^\theta (u\,dv - v\,du)/(u^2 + v^2);$$

this is defined for any γ_r' *not* passing through the origin $w = 0$. (If γ_r'

[†] Cf., for example, L. E. Dickson, *New First Course in the Theory of Equations* (New York: Wiley, 1939), Appendix, or L. Weisner, *Introduction to the Theory of Equations* (New York: Macmillan, 1938), p. 145.
[‡] In proving the existence of line integrals, essential use is made of the *completeness* of **R**. The identity $(d \arg w) = (u\,dv - v\,du)/(u^2 + v^2)$ holds, since $\arg w = \arctan(v/u)$.

passes through $w = 0$, the conclusion of
Theorem 5 is immediate.) It is geometrically
obvious that $\phi(r, 2\pi) = 2\pi n(r)$, where the
winding number $n(r)$ is the number of times
that γ_r' winds counterclockwise around the
origin. Thus $n(r) = 1$ in the imaginary example depicted in Figure 3.

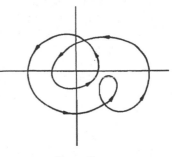

Figure 3

Now consider the variation of $n(r)$ with r.
Since $q(re^{i\theta})$ is a continuous function, $n(r)$
varies continuously with r except when γ_r'
passes through the origin. Again, $n(0) = 0$
(unless $c_0 = 0$ in which case 0 is a root). Now
assume $c_0 \neq 0$. We shall now show that if r is large enough, $n(r)$ is the
degree m of $q(z)$. Indeed, let

$$q(z) = z^m + c_{m-1}z^{m-1} + \cdots + c_1 z + c_0 = z^m\left(1 + \sum_{k=1}^{m} c_{m-k}z^{-k}\right).$$

By de Moivre's formulas (7),

$$\arg q(z) = m \arg z + \arg\left(1 + \sum_{k=1}^{m} c_{m-k}z^{-k}\right).$$

Hence, as z describes the circle γ_r counterclockwise, the net change in
$\arg q(z)$ is the sum of m times the change in $\arg z$ (which is $m \cdot 2\pi$) plus
the change in

$$\arg\left(1 + \sum_{k} c_{m-k}z^{-k}\right).$$

But if $|z| = r$ is sufficiently large, by formulas (8) and (9)

$$1 + \sum_{k} c_{m-k}z^{-k} = u$$

stays in the circle $|u - 1| < 1/2$, and so goes around the origin zero
times (make a figure to illustrate this).

We conclude that if r is large enough, $n(r) = m$: the total change in
$\arg q(z)$ is $2\pi m$. But as r changes, γ_r' is deformed continuously (since
$q(z)$ is continuous). It is geometrically evident,[†] however, that a curve

---

[†] This is proved as a theorem in plane topology; cf., for example, S. Lefschetz,
Introduction to Topology (Princeton University Press, 1949), p. 127.

which winds around the origin $n \neq 0$ times cannot be continuously deformed into a point without being made to pass through the origin at some stage of the deformation. It follows that, for some r, γ_r' must pass through the origin; where this happens, $q(z) = 0$! Q.E.D.

As a corollary, we note that if $p(z_1) = 0$, then by the Remainder Theorem (§3.5) we can write $p(z) = (z - z_1)r(z)$. If the degree m of $p(z)$ exceeds 1, the quotient $r(z)$ has positive degree, hence also has a complex root $z = z_2$. Proceeding thus, we find m linear factors for $p(z)$, as

$$(11) \qquad p(z) = c(z - z_1)(z - z_2) \cdots (z - z_m).$$

It follows that *the only irreducible polynomials over* **C** *are linear.* A corollary of this and the unique factorization theorem of Chap. 3 is

Theorem 6. *Any polynomial with complex coefficients can be written in one and only one way in the form (11).*

The roots of $p(z)$ are evidently the z_i in (11)—since a product vanishes if and only if one of its factors is zero. If a factor $(z - z_i)$ occurs repeatedly, the number of its occurrences is called the *multiplicity* of the root z_i. It can also be defined, using the calculus, as the "order" to which $p(z)$ vanishes at z_i: the greatest integer ν such that $p(z)$ and its first $(\nu - 1)$ derivatives all vanish at z_i.

Exercises

1. Prove the uniqueness of the decomposition (11) without using the general uniqueness theorem of §3.8.
2. Prove that any rational complex function which is finite for all z is a polynomial.
3. Do couples (w, z) of complex numbers when added and multiplied by rules (1) and (2) form a commutative ring with unity? a field?
4. Show that any quadratic polynomial can be brought to one of the forms $cz(z - 1)$ or cz^2 by a suitable automorphism of **C**[z].
5. (a) Using the MacLaurin series, show formally that $e^{ix} = \cos x + i \sin x$.
   (b) Show that every complex number can be written as $re^{i\theta}$.
   (c) Derive the identities $\cos z = (e^{iz} + e^{-iz})/2$, $\sin z = (e^{iz} - e^{-iz})/2i$.
6. Use partial fractions to show that any rational function over the field **C** can be written as a sum of a polynomial plus rational functions in which each numerator is a constant and each denominator a power of a linear function.
7. Factor $z^2 + z + 1 + i$.

5.4. Conjugate Numbers and Real Polynomials

In the complex field $\mathbf{C}$, the equation $z^2 = -1$ has two roots i and $-i = 0 + (-1)i$. The correspondence $x + yi \mapsto x + y(-i) = x - yi$ carries the first of these roots into the second and conversely, while leaving all real numbers unchanged. Furthermore, this correspondence carries sums into sums and products into products, as may be checked either by direct substitution in formulas (1) and (2) or by application of Theorem 1. In other words, the correspondence is an *automorphism* of $\mathbf{C}$ (an isomorphism of $\mathbf{C}$ with itself).

We can state this more compactly as follows. By the "conjugate" z^{*1} of a complex number $z = x + yi$, we mean the number $x - yi$. The correspondence $z \mapsto z^*$ is an automorphism of period two of $\mathbf{C}$, in the sense that

(12) $(z_1 + z_2)^* = z_1^* + z_2^*$, $(z_1 z_2)^* = z_1^* z_2^*$, $(z^*)^* = z$.

It amounts geometrically to a reflection of the complex plane in the x-axis; the only numbers which are equal to their conjugates are the real numbers.

Conjugate complex numbers are very useful in mathematics and physics (especially in wave mechanics). In using them, it is convenient to memorize such simple formulas as

$$|z|^2 = zz^*, \qquad z^{-1} = z^*/|z|^2.$$

Their use enables one to derive the factorization theory of real polynomials easily out of Theorem 6.

Lemma. *The nonreal complex roots of a polynomial equation with real coefficients occur in conjugate pairs.*

This generalizes the well-known fact that a quadratic $ax^2 + bx + c$ with *discriminant* $b^2 - 4ac < 0$ has two roots $x = (-b \pm \sqrt{b^2 - 4ac})/2a$ which are complex conjugates.

Proof. Let $p(z)$ be the given polynomial; we can write it in the form (11), where the z_i are complex (not usually real). Since the correspondence $z_i \mapsto z_i^*$ applied to these roots z_i is an automorphism, it carries $p(z)$ into another polynomial $p^*(z) = c^*(z - z_1^*)(z - z_2^*) \cdots (z - z_n^*)$ in which each coefficient is the conjugate of the corresponding coefficient of $p(z)$. But since the coefficients of $p(z)$ are real, $p(z) = p^*(z)$. Hence, the factorization (11) being unique, $c = c^*$ is real, and the z_i are also real or complex conjugate in pairs.

[1]Now commonly written as $\bar{z}$.

Theorem 7. *Any polynomial with real coefficients can be factored into* (real) *linear polynomials and* (real) *quadratic polynomials with negative discriminant.*

Proof. The real z_i in the lemma give (real) linear factors $(z - z_i)$. A pair of conjugate complex roots $a + bi$ and $a - bi$ with $b \neq 0$ may be combined as in

$$(z - (a + bi))(z - (a - bi)) = z^2 - 2az + (a^2 + b^2)$$

to give a quadratic factor of $p(z)$ with real coefficients and with a real discriminant $4a^2 - 4(a^2 + b^2) = -4b^2 < 0$. Q.E.D.

Conversely, linear polynomials and quadratic polynomials with negative discriminant are irreducible over the real field (the latter since they have only complex roots, and hence no linear factors). It is a corollary that the factorization described in Theorem 7 is unique.

Exercises

1. Solve: (a) $(1 + i)z + 3iz^* = 2 + i$,
   (b) $zz^* + 2z = 3 + i$,   (c) $zz^* + 3(z - z^*) = 4 - 3i$.
2. Solve: $zz^* + 3(z + z^*) = 7$, $zz^* + 3(z + z^*) = 3i$.
3. Solve simultaneously: $iz + (1 + i)w = 3 + i$, $(1 + i)z^* - (6 + i)w^* = 4$.
4. Give an independent proof of Corollary 2 of Theorem 4 (§4.4).
5. Show that if one adjoins to the real number system an imaginary root of *any* irreducible nonlinear real polynomial, one gets a field isomorphic with **C**.
6. Show that over any ordered field $ax^2 + bx + c$ is irreducible if $b^2 - 4ac < 0$.
7. Show that every automorphism of **C** in which the real numbers are all left fixed is either the identity automorphism $(z \mapsto z)$ or the automorphism $z \mapsto z^*$.

★5.5. Quadratic and Cubic Equations

In §5.3 we proved the existence of roots for any polynomial equation with complex coefficients, but did not show how to calculate roots effectively. We shall show, in §§5.5–5.6, how to do this for polynomials of degrees two, three, and four. The procedures will involve only the four rational operations (addition, multiplication, subtraction, and division) and the extraction of nth roots. We showed how to perform these operations on complex numbers in §§5.1–5.2; the procedure to be used now will also apply to any other field in which nth roots of arbitrary numbers can be constructed and in which $1 + 1 \neq 0$ and $1 + 1 + 1 \neq 0$.

Quadratic equations can be solved by "completing the square" as in high-school algebra. Such an equation

$$(13) \qquad az^2 + bz + c = 0 \qquad (a \neq 0),$$

is equivalent to (has the same roots as) the simpler equation

$$(14) \qquad z^2 + Bz + C = 0 \qquad (B = b/a, \quad C = c/a).$$

If one sets $w = z + B/2$ (i.e., $z = w - B/2$), so as to complete the square, one sees that (14) is equivalent to

$$(15) \qquad w^2 = B^2/4 - C.$$

Substituting back z, a, b, c for w, B, C, this gives

$$(16) \qquad z = w - B/2 = (-b + \sqrt{b^2 - 4ac})/(2a)$$

and so yields two solutions, by §5.2.

Cubic equations can be solved similarly. First reduce the cubic, as in §4.4, to the form

$$(17) \qquad z^3 + pz + q = 0.$$

Then make Vieta's substitution $z = w - p/(3w)$. The result (after cancellation) is

$$(18) \qquad w^3 - p^3/(27w^3) + q = 0.$$

Multiplying through by w^3, we get a quadratic in w^3, which can be solved by (16), giving

$$(19) \qquad w^3 = -q/2 + \sqrt{q^2/4 + p^3/27} \qquad \text{(two values)}.$$

This gives *six* solutions for w in the form of cube roots. Substituting these in the formula $z = w - p/(3w)$, we get three pairs of solutions for z, paired solutions being equal.

It is interesting to relate the preceding formulas to Theorem 6. Thus, in the quadratic case, writing $z^2 + Bz + C = (z - z_1)(z - z_2)$, we have

$$(20) \quad z_1 + z_2 = -B, \quad z_1 z_2 = C, \quad \text{whence} \quad (z_1 - z_2)^2 = B^2 - 4C.$$

The quantity $B^2 - 4C = D$ is the discriminant of (14). In terms of the original coefficients of (13), $D = (b^2 - 4ac)/a^2$.

Similarly, if z_1, z_2, z_3 are the roots of the reduced cubic equation (17),

$$(21) \quad z_1 + z_2 + z_3 = 0, \quad\quad z_1 z_2 + z_2 z_3 + z_3 z_1 = p, \quad\quad z_1 z_2 z_3 = -q.$$

Combining the first two relations, we get the formulas

$$(22) \quad p = z_1 z_2 - z_3{}^2, \quad (z_1 - z_2)^2 = -4p - 3z_3{}^2, \quad z_1{}^2 + z_2{}^2 + z_3{}^2 = -2p.$$

We now define the *discriminant* of a cubic equation by

$$(23) \quad D = \prod (z_i - z_j)^2 = P^2, \quad \text{where} \quad P = (z_1 - z_2)(z_2 - z_3)(z_1 - z_3).$$

Squaring P and using the second relation of (22), we get after some calculation

$$(24) \quad\quad\quad\quad\quad\quad D = -4p^3 - 27q^2,$$

which can be used to simplify (19) to $w = -q/2 + \sqrt{-D}/6$.

Theorem 8. *A quadratic or cubic equation with real coefficients has real roots if its discriminant is nonnegative, and two imaginary roots if its discriminant is negative.*

Proof. By the Corollary of Theorem 7, either all roots are real or there are two conjugate imaginary roots $z_1 = x_1 + iy$, and $z_2 = x_1 - iy$. If all roots are real, $(z_i - z_j)^2 \geq 0$ for all $i \neq j$, and so $D \geq 0$. In the opposite case $(z_1 - z_2)^2 = -4y^2 < 0$, and since $z_3 = x_3$ is real, $(z_1 - z_3)(z_2 - z_3) = (x_1 - x_3)^2 + y^2 > 0$, so that $D < 0$. Q.E.D.

By (23), the condition $D = 0$ gives a simple test for multiple roots.

Unfortunately, precisely in the case $D > 0$ that $z^3 + pz + q = 0$ has all real roots, formula (19) expresses them in terms of complex numbers. We shall show in §15.6 that this cannot be helped!

Exercises

1. Prove that for any (complex) y, p there exists a z satisfying $y = z - p/3z$. How many exist?

2. Solve in radicals

    (a) $z^2 + iz = 2$,   (b) $z^3 + 3iz = 1 + i$,   (c) $z^3 + 3iz^2 = 10i$.

3. Convert one root in each of Exs. 2(a)–(c) into decimal form.

4. (a) Prove (22). (b) Prove (24).

★5. (a) Show that $\sinh 3\gamma = \sinh(3\gamma + 2\pi i)$.
   (b) Using formula (9a) in §4.4, show that $4z^3 + 3z = C$ has, in addition to the real root $\sinh[(1/3)\sinh^{-1} C] = \sinh\gamma$, also the complex roots $-(1/3)\cosh\gamma \pm i(\sqrt{3}/2)\sinh\gamma$.

6. Let $\omega = c^{2\pi i/5}$ be a primitive fifth root of unity, and let $\zeta = \omega + 1/\omega$.
   (a) Show that $\zeta^2 + \zeta = 1$.
   (b) Infer that in a regular pentagon with center at $(0,0)$ and one vertex at $(1,0)$, the x-coordinate of either adjacent vertex is $(\sqrt{5} - 1)/4$.

7. Using the formula $\cos\theta = (e^{i\theta} + e^{-i\theta})/2$, show that $\cos n\theta = T_n(\cos\theta)$ for a suitable polynomial T_n of degree n, and compute T_1, T_2, T_3, T_4.

★5.6. Solution of Quartic by Radicals

Any method which reduces the solution of an algebraic equation to a sequence of rational operations and extractions of nth roots of quantities already known is called a "solution by radicals."

Theorem 9. *Any polynomial equation of degree $n \leqq 4$ with real or complex coefficients is solvable by radicals.*

Proof. Since the case $n = 1$ is solvable over any field, while the cases $n = 2, 3$ were treated in §5.5, we need only consider

$$ax^4 + bx^3 + cx^2 + dx + e = 0 \ (a \neq 0).$$

Again, dividing through by a, and replacing x by $z = x + b/4a$ (so as to "complete" the quartic), we get the equation

$$(25) \qquad z^4 + pz^2 + qz + r = 0,$$

whose roots differ from those of the original equation by $b/4a$. But for all u, (25) is equivalent to

$$(26) \qquad z^4 + z^2 u + u^2/4 - z^2 u - u^2/4 + pz^2 + qz + r = 0$$

$$\text{or} \quad (z^2 + u/2)^2 - [(u - p)z^2 - qz + (u^2/4 - r)] = 0.$$

The first term is a perfect square P^2, with $P = z^2 + \frac{1}{2}u$. The term in square brackets is a perfect square Q^2 for those u such that (equating the discriminant to zero)

$$(27) \qquad q^2 = 4(u - p)(u^2/4 - r).$$

This cubic equation in u can be solved by radicals, using Theorem 8. If the coefficients of (25) are real, one can even show that at least one real number $u_1 \geqq p$ satisfies (27), for the right side of (27) is zero if $u = p$ and becomes larger than q^2, or any other preassigned constant, when u is sufficiently large and positive. Hence, by Theorem 4 of §4.4, (27) has the desired real root u_1.

Substituting this constant u_1 into (26), the left side of (25) assumes the form $P^2 - Q^2 = (P + Q)(P - Q)$, or

$$(28) \qquad (z^2 + u_1/2 + Q)(z^2 + u_1/2 - Q),$$

where

$$(29) \qquad Q = Az - B, \qquad A = \sqrt{u_1 - p}, \qquad B = q/2A.$$

The roots of (25) are clearly those of the two quadratic factors of (28), which can be found by (16). Note that these factors are real if the coefficients a, b, c, d, e of the original equation were real.

It is interesting to recall the history of the solution of equations by radicals. The solution of the quadratic was known to the Hindus and in its geometric form (§4.1) to the Greeks. The cubic and quartic were solved by the Renaissance Italian mathematicians Scipio del Ferro (1515) and Ferrari (1545). However, not until the nineteenth century did Abel and Galois prove the impossibility of solving all polynomial equations of degree $n \geqq 5$ in the same way (§15.9).

Exercises

1. Solve by radicals: $z^4 - 4z^3 + (1 + i)z = 3i$.
2. Prove, without using the Fundamental Theorem of Algebra, that every real polynomial of degree $n < 6$ has a complex root.
3. Solve the simultaneous equations: $zw = 1 + i$, $z^2 + w^2 = 3 - i$.

★5.7. Equations of Stable Type

Many physical systems are stable if and only if all roots of an appropriate polynomial equation have negative real parts. Hence equations with this property may be called "of stable type."

In the case of real quadratic equations $z^2 + Bz + C = 0$, it is easy to test for stability. If $4C \leqq B^2$, both roots are real. They have the same sign if and only if $z_1z_2 = C > 0$, the sign being negative if and only if $B = -(z_1 + z_2) > 0$. If $4C > B^2$, the roots are two conjugate complex numbers. They both have negative real parts $x_1 = x_2$ if and only if $B = -2x_1 = -2x_2 > 0$; in this case also $C > B^2/4 > 0$. Hence in both cases the condition for "stability" is $B > 0$, $C > 0$.

In the case of real cubic equations $z^2 + Az^2 + Bz + C = 0$, conditions for stability are also not hard to find. (It is not, of course, sufficient to consider the reduced form (17).) Indeed, if all roots have negative real parts, then, since one root $z = -a$ is real, we have a factorization

$$(30) \qquad z^3 + Az^2 + Bz + C = (z + a)(z^2 + bz + c).$$

Here $a > 0$, and by the previous case $b > 0$ and $c > 0$. Therefore $A = a + b > 0$, $B = (ab + c) > 0$, and $C = ac > 0$ are necessary for stability. Furthermore, $AB - C = b(a^2 + ab + c) > 0$.

Conversely, suppose that $A > 0, B > 0, C > 0$, and consider the real factorization (30), which always exists by Theorem 7. Since $ac = C > 0$, a and c have the same sign. But, if they were both negative, then b would have to be negative to make $ab + c > 0$, and so $A = a + b < 0$, contrary to hypothesis. Hence $a > 0$ and $c > 0$, implying $a^2 + ab + c = a(a + b) + c > 0$. But this implies $b = (AB - C)/(a^2 + ab + c) > 0$, whence both factors of (30) are "stable." Hence we have proved the following result.

Theorem 10. *The real quadratic equation* $z^2 + Bz + C = 0$ *is of stable type if and only if* $B > 0$ *and* $C > 0$. *The real cubic equation* $z^3 + Az^2 + Bz + C = 0$ *is of stable type if and only if* $A > 0, B > 0$, $C > 0$, *and* $AB > C$.

Exercises

1. Test the following polynomials for stability:
   (a) $z^3 + z^2 + 2z + 1$,   (b) $z^3 + z^2 + 2z + 2$.
2. Show that for monic real polynomial of degree n to be of stable type, all its coefficients must be positive.
★3. Show that $z^4 + Az^3 + Bz^2 + Cz + D$ with real coefficients is of stable type if and only if all its coefficients are positive, and $ABC > A^2D + C^2$.
★4. Assuming Ex. 3, obtain necessary and sufficient conditions for a complex quadratic equation $z^2 + Bz + C = 0$ to be of stable type. (*Hint:* Consider $(z^2 + Bz + C)(z^2 + B^*z + C^*) = 0$.)

6

Groups

6.1. Symmetries of the Square

The idea of "symmetry" is familiar to every educated person. But fewer people realize that there is a consequential algebra of symmetry. This algebra will now be introduced in the concrete case of the symmetries of the square.

Imagine a cardboard square laid on a plane with fixed axes, so that the center of the square falls on the origin of coordinates, and one side is horizontal. It is clear that the square has rotational symmetry: it is carried into itself by the following rigid motions.

R: a 90° rotation clockwise around its center O.
R', R'': similar rotations through 180° and 270°.

The square also has reflective symmetry; it can be carried into itself by the following rigid reflections.

H: a reflection in the horizontal axis through O.
V: a reflection in the vertical axis through O.
D: a reflection in the diagonal in quadrants I and III.
D': a reflection in the diagonal in quadrants II and IV.

Our list thus includes seven symmetries so far.

The algebra of symmetries has its genesis in the fact that we can *multiply* two motions by performing them in succession. Thus, the product HR is obtained by first reflecting the square in a horizontal axis, then rotating clockwise through 90°. By experimenting with a square

124

piece of cardboard, one can verify that this has the same net effect as D', reflection about the diagonal from the upper left- to the lower right-hand corner. Alternatively, the equation $HR = D'$ can be checked by noting that both sides have the same effect on each vertex of the square. Thus, in Figure 1, HR sends 1 into 4 by H and then 4 into 3 by R—hence 1 into 3, just as does D'.

Similarly, RH is defined as a result of a clockwise rotation through 90° followed by reflection in a horizontal axis. (*Caution:* The plane of Figure 1, which contains the axes of reflection, is not imagined as rotated with the square.)

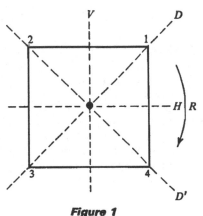

Figure 1

A computation shows that $RH = D \neq HR$, from which we conclude incidentally that our "multiplication" is not in general commutative! It is, however, associative, as we shall see in §6.2.

The reader will find it instructive to compute other products of symmetries of the square (a complete list is given in Table 1, §6.4). If he does this, he will discover one exception to the principle that successive applications of any two symmetries yield a third symmetry. If, for example, he multiplies R with R'', he will see that their product is a motion which leaves every point fixed: it is the so-called "identity" motion I. This is not usually considered a symmetry by nonmathematicians; nevertheless, we shall consider it a (degenerate) symmetry, in order to be able to multiply all pairs of symmetries.

In general, a symmetry of a geometrical figure is, by definition, a one-one transformation of its points which preserves distance. It can be readily seen that any symmetry of the square must carry the vertex 1 into one of the four possible vertices, and that for each such choice there are exactly two symmetries. Thus all told there are only eight symmetries, which are those we have listed.

Not only the square, but every regular polygon and regular solid (e.g., the cube and regular icosahedron) has an interesting group of symmetries, which may be found by the elementary method sketched above.

Similarly, many ornaments have interesting symmetries. Thus consider the infinite ornamental pattern

in which the arrowheads are spaced uniformly one inch apart along a line. Three simple symmetries of this figure are T, a translation to the right by one inch, T', a translation to the left by one inch, and H, a reflection in the horizontal axis of the figure. Others (in fact, all others!) may be found by multiplying these together repeatedly.

Exercises

1. Compute HV, HD', $D'H$, $R'D'$, $D'R'$, $R'R''$.
2. Describe TH and HT in the ornamental "arrowhead" pattern.
3. List the symmetries of an equilateral triangle, and compute five typical products.
4. List the symmetries of a general rectangle, and compute all their products.
★5. How many symmetries are possessed by the regular tetrahedron? by the regular octahedron? Draw figures.
★6. Show that any symmetry of the ornamental pattern of the text can be obtained by repeatedly multiplying H, T, and T'.

6.2. Groups of Transformations

The algebra of symmetry can be extended to one-one transformations of any set S of elements whatever. Although it is often suggestive to think of the set S as a "space" (e.g., a plane or a sphere), its elements as "points," and the bijections as "symmetries" of S with respect to suitable properties, the bijections of S satisfy some nontrivial algebraic laws in any case.

To understand these laws, one must have clearly in mind the definitions of function, injection, surjection, and bijection made in §1.11. To illustrate these afresh, we give some new examples; as in §1.11, we will usually abbreviate $f(x)$ to xf (read "the transform of x by f"), $g(x)$ to xg, etc.

The function $f(x) = e^{2\pi i x}$ maps the field $\mathbf{R}$ of all real numbers *into* the field $\mathbf{C}$ of all complex numbers; its range (image) is the unit circle. Similarly, $g(z) = |z|$ is a function $g : \mathbf{C} \to \mathbf{R}$ whose image is the set of all nonnegative real numbers.

Again, consider the following functions $\phi_0 : \mathbf{Z} \to \mathbf{Z}$ and $\psi_0 : \mathbf{Z} \to \mathbf{Z}$ on the domain $\mathbf{Z}$ of all integers to itself:

$$n\phi_0 = 2n, \quad \text{and} \quad m\psi_0 = \begin{cases} m/2 & \text{if } m \text{ is even,} \\ 0 & \text{if } m \text{ is odd.} \end{cases}$$

By the cancellation law of multiplication ϕ_0 is one-one; yet its range consists only of even integers, so that ϕ_0 does not transform $\mathbf{Z}$ onto $\mathbf{Z}$. On the other hand, ψ_0 is not one-one, since all odd integers are mapped onto zero, but it does map $\mathbf{Z}$ onto $\mathbf{Z}$; thus ψ_0 is surjective but not injective.

We turn now to the algebra of transformations. Two transformations $\phi: S \to T$ and $\phi': S \to T$ with the same domain S and the same codomain T are called *equal* if they have the same effect upon every point of S; that is,

(1) $\phi = \phi'$ means that $p\phi = p\phi'$ for every $p \in S$.

The *product* or *composite* $\phi\psi$ of two transformations is again defined as the result of performing them in succession; first ϕ, then ψ, provided however that the codomain of ϕ is the domain of ψ. In other words, if

$$\phi: S \to T, \qquad \psi: T \to U,$$

then $\phi\psi$ is the transformation of S into U given by the equation

(2) $$p(\phi\psi) = (p\phi)\psi,$$

which defines the effect of $\phi\psi$ upon any point $p \in S$. In particular, the product of two transformations of S (into itself) is always defined. We shall now restrict our attention to this case, although almost all the identities proved below apply also to the general case, provided that the products involved are defined.

Multiplication of transformations conforms to the

Associative law: $(\phi\psi)\theta = \phi(\psi\theta),$

whenever the products involved are defined. This is obvious intuitively: both $(\phi\psi)\theta$ and $\phi(\psi\theta)$ amount to performing first ϕ, then ψ, and finally θ, in that order. Formally, we have for each $p \in S$,

$$p[\phi(\psi\theta)] \underset{\phi(\psi\theta)}{=} (p\phi)(\psi\theta) \underset{\psi\theta}{=} [(p\phi)\psi]\theta \underset{\phi\psi}{=} [p(\phi\psi)]\theta \underset{(\phi\psi)\theta}{=} p[(\phi\psi)\theta],$$

where each step depends on applying the definition (2) of multiplication to the product indicated below the equality symbol for that step. By the definition (1) of equality for transformations, this proves the associative law $\phi(\psi\theta) = (\phi\psi)\theta$.

The *identity* transformation $I = I_S$ on the set S is that transformation $I: S \to S$ which leaves every point of S fixed. This is stated algebraically in the identity

(3) $pI = p$ for every $p \in S$.

From the above definitions there follows directly the

Identity law: $I\phi = \phi I = \phi$ for all ϕ.

To see this, note that $p(I\phi) = (pI)\phi = p\phi$ for all p and, similarly, that $p(\phi I) = (p\phi)I = p\phi$.

Return now to the special transformations ϕ_0 and ψ_0 defined above on the set $\mathbf{Z}$, and compute their products. Clearly, $m\psi_0\phi_0 = m$ if m is even, and 0 if m is odd; hence $\psi_0\phi_0 \neq I$. On the other hand, $m\phi_0\psi_0 = m$ for all $m \in \mathbf{Z}$, hence $\phi_0\psi_0 = I$. We may thus call ψ_0 a right-inverse (but not a left-inverse) of ϕ_0.

In general, if the transformations $\phi: S \to S$ and $\psi: S \to S$ have the product $\phi\psi = I: S \to S$, then ϕ is called a *left-inverse* of ψ, and ψ a *right-inverse* of ϕ. These definitions are closely related to the concepts of being "one-one" (injective) and "onto" (surjective), as defined earlier.

Theorem 1. *A transformation $\phi: S \to S$ is one-one if and only if it has a right-inverse; it is onto if and only if it has a left-inverse.*

Proof. If ϕ has a right-inverse ψ, $\phi\psi = I$ and $p\phi = p'\phi$ imply

$$p = p(\phi\psi) = (p\phi)\psi = (p'\phi)\psi = p'(\phi\psi) = p'.$$

Thus $p\phi = p'\phi$ implies $p = p'$, so that ϕ is one-one. Similarly, if ϕ has a left-inverse ψ', then $\psi'\phi = I$. Hence, any q in S can be written $q = qI = q(\psi'\phi) = (q\psi')\phi$, as the ϕ-image of a suitable point $p = q\psi'$. Therefore ϕ is onto.

Conversely, given *any* $\phi: S \to S$, we first construct a second transformation $\psi: S \to S$, as follows. For each q in S which is the image under ϕ of one or more points p of S, choose† as image $q\psi$ any one of these points p. Then $q(\psi\phi) = (q\psi)\phi = p\phi = q$ for any q of the form $p\phi$. Let ψ map the remaining points q of S in any way whatever, say on some fixed point of the (nonempty) set S.

Now if ϕ is onto, every q has the form $p\phi$, and hence $\psi\phi = I$, so that ϕ has ψ as left-inverse. On the other hand, if ϕ is one-one, then, for each p, $(p\phi)\psi$ must be the unique antecedent p of $q = p\phi$; hence $\phi\psi = I$ and ψ is a right-inverse of ϕ as asserted.

---

† In case the set of such points q is infinite, the Axiom of Choice (cf. §12.2) asserts the possibility of making such an infinite number of choices of p, one for each q.

Remark. The functional notation $y = \phi(x)$ of the calculus suggests writing $y = \phi x$ where we have written $y = x\phi$ above. In this notation, the composite of ϕ and $z = \psi(y)$ is naturally written $z = (\psi\phi)x$, as an abbreviation for $z = \psi(\phi(x))$, instead of $z = x\phi\psi$. Hence $\psi\phi$ means "perform first ϕ, then ψ," and the notions of right- and left-inverse become interchanged. Either notation by itself is satisfactory, but confusion between them must be avoided. The meaning of two-sided inverse stays the same, however, as do the following corollaries.

Corollary 1. *A transformation* $\phi: S \to S$ *is a bijection if and only if it has both a right-inverse and a left-inverse. When this is the case, any right-inverse of* ϕ *is equal to any left-inverse of* ϕ.

Indeed, if ϕ has a right-inverse θ and a left-inverse ψ, then

$$\theta = I\theta = (\psi\phi)\theta = \psi(\phi\theta) = \psi I = \psi.$$

Define a (two-sided) inverse of $\phi: S \to S$ as any transformation ϕ^{-1} which satisfies the

Inverse law: $\phi\phi^{-1} = \phi^{-1}\phi = I.$

These equations also state that ϕ^{-1} is a two-sided inverse of ϕ, hence the further corollary:

Corollary 2. *A transformation* $\phi: S \to S$ *is bijective if and only if* ϕ *has a (two-sided) inverse* ϕ^{-1}. *When this is the case, any two inverses of* ϕ *are equal, and*

(4) $(\phi^{-1})^{-1} = \phi.$

This corollary is what will be used below; it has an immediate direct proof, for ϕ^{-1} is simply that transformation of S which takes each point $q = p\phi$ back into its unique antecedent p. In the special case when S is finite, ϕ is one-one if and only if it is onto, so that the more elaborate discussion of left- and right-inverses is pointless in this case.

Theorem 1 and its corollaries also hold, together with their proofs, for functions $\phi: S \to T$ on a set into another set T. One need only observe that a left-inverse ψ or a right-inverse θ is a transformation of the second set T into S and that

$$\psi\phi = I_T: \ T \to T, \qquad \phi\theta = I_S: \ S \to S.$$

Here I_S and I_T are the identity transformations on S and T, respectively.

We are now ready to define the important concept of a group of transformations. By a *group of transformations* on a "space" S is meant any set G of one-one transformations ϕ of S onto S such that (i) the identity transformation of S is in G; (ii) if ϕ is in G, so is its inverse; (iii) if ϕ and ψ are in G, so is their product $\phi\psi$.

Theorem 2. *The set G of all bijections of any space S onto itself is a group of transformations.*

Proof. Since $II = I$, the identity I of S is bijective, hence is in the set G, as required by condition (i) above. If ϕ is in G, Corollary 2 above shows that ϕ^{-1} is also one-one onto, hence is likewise in G, as in (ii). Finally, the product of any two one-one transformations ϕ and ψ of S onto S has an inverse, for by hypothesis

$$(\phi\psi)(\psi^{-1}\phi^{-1}) = \phi(\psi\psi^{-1})\phi^{-1} = \phi I\phi^{-1} = \phi\phi^{-1} = I,$$

$$(\psi^{-1}\phi^{-1})(\phi\psi) = \psi^{-1}(\phi^{-1}\phi)\psi = \psi^{-1}I\psi = \psi^{-1}\psi = I.$$

Therefore $\phi\psi$ is also bijective (one-one onto), and has as inverse

$$(5) \qquad\qquad (\phi\psi)^{-1} = \psi^{-1}\phi^{-1}.$$

In words, the inverse of a product is the product of the inverses, taken in the opposite order.   Q.E.D.

A bijection of a finite set S to itself is usually called a *permutation* of S. The group of all permutations of n elements is called the *symmetric group* of degree n; it evidently contains $n!$ permutations, for the image k_1 of the first element can be chosen in n ways, that of the second element can then be chosen in $n - 1$ ways from the elements not k_1, and so on.

Exercises

1. Compute VD, $(VD)R''$, DR'', $V(DR'')$ in the group of the square.
2. Compute similarly HR, $R'(HR)$, $R'H$, $(R'H)R$.
3. Let S consist of all real numbers (or all points x on a line), while the transformations considered have the form $x\phi = ax + b$. In each of the following cases, find when the set of all possible ϕ's with coefficients a and b of the type indicated is a group of transformations. Give reasons.
   (a) a and b rational numbers;   (b) $a = 1$, b an odd integer;
   (c) $a = 1$, b a positive integer or 0;   (d) $a = 1$, b an even integer;
   (e) a an integer, $b = 0$;   (f) $a \neq 0$, a and b real numbers;
   (g) $a \neq 0$, a an integer, b a real number;

(h) $a \neq 0, a$ a real number, b an integer;

(i) $a \neq 0, a$ an integer, b an irrational number;

(j) $a \neq 0, a$ a rational number, b a real number.

In which of these groups is "multiplication" commutative?

4. Find all the transformations on a "space" S of exactly three "points." How many are there? How many of these are one-one?

5. Show that the transformation $n \mapsto n^2$ on the set of positive integers has no left-inverse, and exhibit explicitly two right-inverses.

6. Exhibit two distinct left-inverses of the transformation $\psi_0 : \mathbf{Z} \to \mathbf{Z}$ defined in the text, and two right-inverses of ϕ_0.

7. Show that if ϕ and ψ both have right-inverses, then so does $\phi\psi$.

8. Compute $(R^{-1}(VR))^{-1}((R^{-1}D)R)$ for the group of the square.

9. Solve the equation $RXR' = D$ for the group of the square.

10. Check that, in the group of the square, $(RH)^{-1} = H^{-1}R^{-1} \neq R^{-1}H^{-1}$.

11. Find the inverse of every symmetry of the rectangle, and test the rule (5).

12. If $\phi_1, \cdots, \phi_n$ are one-one, prove that so is $\phi_1\phi_2 \cdots \phi_n$, with

$$(\phi_1\phi_2 \cdots \phi_n)^{-1} = \phi_n^{-1}\phi_{n-1}^{-1}, \cdots, \phi_1^{-1}.$$

13. Show that for any $\phi : S \to S$, the transformation ψ constructed in the second part of the proof of Theorem 1 satisfies $\phi\psi\phi = \phi$.

★14. Show that a transformation $\phi : S \to S$ which has a unique right-inverse or a unique left-inverse is necessarily a one-one transformation of S onto S.

6.3. Further Examples

The symmetries of a cube form another interesting group. Geometrically speaking, these symmetries are the one-one transformations which preserve distances on the cube. They are known as "isometries," and are 48 in number. To see this, note that any initial vertex can be carried into any one of the eight vertices. After the transform of any one vertex has been fixed, the three adjacent vertices can be permuted in any of six ways, giving $6 \cdot 8 = 48$ possibilities. When one vertex and the three adjacent vertices occupy known positions, every point of the cube is in fixed position, so the whole symmetry is known. Hence the cube has exactly 48 symmetries. Many of them have special geometrical properties, such as the one which reflects each point into the diametrically opposite point.

A familiar group containing an infinite number of transformations is the so-called Euclidean group. This consists of the "isometries" of the plane—or, in the language of elementary geometry, of the transformations under which the plane is congruent to itself. It is made up of products of translations, rigid rotations, and reflections; it will be discussed in greater detail in Chap. 9.

Another group consists of the "similarity" transformations of space—those one-one transformations which multiply all distances by a  constant factor $k > 0$ (a factor of proportionality). The rigid motions of the surface of any sphere into itself again constitute a group.

The isometries of the plane leaving invariant a regular hexagonal network (Figure 2) form another interesting group.

Again, a rubber band held in a straight line between fixed endpoints P and Q may be deformed in many ways along this line. All such deformations form a group (the group of so-called "homeomorphisms" of the segment PQ).

Figure 2

Generally speaking, those one-one transformations of any set of elements which preserve any given property or properties of these elements form a group. Felix Klein (*Erlanger Programm,* 1872) has eloquently described how the different branches of geometry can be regarded as the study of those properties of suitable spaces which are preserved under appropriate groups of transformations. Thus Euclidean geometry deals with those properties of space preserved under all isometries, and topology with those which are preserved under all homeomorphisms. Similarly, "projective" and "affine" geometry deals with the properties which are preserved under the "projective" and "affine" groups to be defined in Chap. 9.

Exercises

1. Describe all the symmetries of a wheel with six equally spaced radial spokes.
2. Describe the six symmetries of a cube with one vertex held fixed.
3. Let S, T be reflections of a cube in planes parallel to distinct faces. Describe ST geometrically.
4. Describe some isometries of the plane which carry the hexagonal network of Figure 2 onto itself.
5. Do the same for a network of squares. Can you enumerate all such transformations (this is difficult)?
6. Do the same for the network of equilateral triangles, and relate this to the group of Ex. 1.
7. Do the same for an infinite cylinder, for a finite cylinder, for a helix wound around the cylinder making a constant angle with the axis of the cylinder.
★8. Show that the transformations $x \mapsto x' = (ax + b)/(cx + d)$, with $ad - bc = 1$ and with coefficients in any field F, constitute a group acting on the set consisting of the elements of F and a symbolic element ∞.

6.4. Abstract Groups

Groups of transformations are by no means the only systems with a multiplication which satisfies the associative, identity, and inverse laws of §6.2. For instance, the nonzero numbers of any field (e.g., of the rational, real, or complex field) satisfy them. The product of any two nonzero numbers is a nonzero number; the associative law holds; the unit 1 of the field satisfies the identity law, and $1/x = x^{-1}$ satisfies the inverse law.

Similarly, the elements (including zero, this time) of any integral domain satisfy our laws when combined under addition. Thus, any two elements have a uniquely determinate sum; addition is associative; while zero satisfies the identity law, and $-x$ the inverse law, relative to addition. In other words, the elements of any integral domain form a group under addition.

It is convenient to introduce the abstract concept of a group to include these and other instances.

Definition. *A* group *G is a system of elements with a binary operation which* (i) *is associative,* (ii) *admits an identity satisfying the identity law, and* (iii) *admits for each element a an element* a^{-1} *(called its* inverse*) satisfying the inverse law.*

Groups can be defined abstractly, without reference to transformations, in many ways; groups so defined are often called *abstract groups.*

In discussing abstract groups, elements will be denoted by small Latin letters $a, b, c, \cdots$. The product notation "ab" will ordinarily be used to denote the result of applying the group operation to two elements a and b of G—but other notations, such as "$a + b$" and "$a \circ b$," are equally valid. In the product notation, with "e" for the identity, the three laws defining groups become

Associative law:	$a(bc) = (ab)c$	for all a, b, c.
Identity law:	$ae = ea = a$	for all a.
Inverse law:	$aa^{-1} = a^{-1}a = e$	for each a and some a^{-1}.

A group whose operation satisfies the commutative law is called a "commutative" or "Abelian" group. Using this concept, we can simplify the definition of a field as follows.

Definition. *A field is a system F of elements closed under two uniquely defined binary operations, addition and multiplication, such that*
   (i) *under addition, F is a commutative group with identity* 0;

(ii) *under multiplication, the nonzero elements form a commutative group;*
(iii) *both distributive laws hold:* $a(b + c) = ab + ac$.

To see that this definition is equivalent to that given in §2.1, observe that the postulates just given include all those previously stated for a field, except for the associative law for products with a factor 0; this can be verified in detail.

Some of the results of the first sections of Chaps. 1 and 2 will now appear as corollaries of the following theorem on groups.

Theorem 3. *In any group,* $xa = b$ *and* $ay = b$ *have the unique solutions* $x = ba^{-1}$ *and* $y = a^{-1}b$. *Hence* $ca = da$ *implies* $c = d$, *and so does* $ac = ad$ *(cancellation law).*

Proof. If a^{-1} is the element specified in the inverse law, clearly, $(ba^{-1})a = b(a^{-1}a) = be = b$, and, similarly, $a(a^{-1}b) = b$. Conversely, $xa = b$ implies $x = xe = xaa^{-1} = ba^{-1}$, and, similarly, $ay = b$ implies $y = a^{-1}b$.

Note that in this proof a^{-1} is not assumed to be the only element satisfying $xa = e$. But it is, since if $xa = e$, then

$$x = xe = x(aa^{-1}) = (xa)a^{-1} = ea^{-1} = a^{-1}.$$

Similarly, a^{-1} is the only element such that $ay = e$.

Since in any group G the equations $ex = e$ and $ay = e$ have by Theorem 3 the unique solutions $x = e$ and $y = a^{-1}$, we get the

Corollary. *A group has only one identity element, and only one inverse a^{-1} for each element a.*

Theorem 4. *In the preceding definition of a group, the identity and inverse laws can be replaced by the weaker laws,*

Left-identity: For some e, $ea = a$ for all a.
Left-inverse: Given a, $a^{-1}a = e$ for some a^{-1}.

Proof. Given these weaker laws, cancellation on the left is possible; that is, $ca = cb$ implies $a = b$, for we need only to premultiply each side of $ca = cb$ by c^{-1} and apply the associative law to get $(c^{-1}c)a = (c^{-1}c)b$, which is $ea = eb$, and gives $a = b$.

The given left-identity is also a right-identity, for

$$a^{-1}ae = ee = e = a^{-1}a,$$

whence, by left-cancellation, $ae = a$ for all a. Finally, left-inverses are also right-inverses, for

$$a^{-1}(aa^{-1}) = (a^{-1}a)a^{-1} = ea^{-1} = a^{-1} = a^{-1}e,$$

since the left-identity is a right-identity. Left-cancellation now gives $aa^{-1} = e$. This completes our proof.

There are many other postulate systems for groups. A useful one may be set up in terms of the possibility of division, as follows:

Theorem 5. *If G is a nonvoid system closed under an associative multiplication for which all equations $xa = b$ and $ay = b$ have solutions x and y in G, then G is a group.*

The proof is left as an exercise (Ex. 12).

Besides systematizing the algebraic laws governing multiplication in any group G, we may list in a "multiplication table" the special rules for forming the product of any two elements of G, provided the number of elements in G is finite. This is a square array of entries, headed both to the left and above by a list of the elements of the group. The entry opposite a on the left and headed above by b is the product ab (in that order).

In Table 1 we have tabulated for illustration the multiplication table

Table 1. Group of the Square

	I	R	R'	R''	H	V	D	D'
I	I	R	R'	R''	H	V	D	D'
R	R	R'	R''	I	D	D'	V	H
R'	R'	R''	I	R	V	H	D'	D
R''	R''	I	R	R'	D'	D	H	V
H	H	D'	V	D	I	R'	R''	R
V	V	D	H	D'	R'	I	R	R''
D	D	H	D'	V	R	R''	I	R'
D'	D'	V	D	H	R''	R	R'	I

for the group of symmetries of the square. The computations can be modeled on those made in §6.1 in proving that $HR = D'$ and $RH = D$. Another method is described in §6.6.

Most of the group properties can be read directly from the table. Thus, the existence of an identity states that some row and the corresponding column must be replicas of the top heading and of the left heading, respectively. The possibility of solving the equation $ay = b$ means that the row opposite a must contain the entry b; since the solution is unique, b can occur only once in this row. The group is commutative if and only if its table is symmetric about the principal diagonal (which extends from upper left to lower right). Unfortunately, the associative law cannot be easily visualized in the table.

Exercises

1. Let a, b, c be fixed elements of a group. Prove the equation $xaxba = xbc$ has one and only one solution.
2. In a group of $2n$ elements, prove there is an element besides the identity which is its own inverse.
3. Do the positive real numbers form a group under addition? under multiplication? Do the even integers form one under addition? Do the odd ones? Why?
4. In the field $\mathbf{Z}_{11}$ of integers modulo 11, which of the following sets are groups under multiplication?
   (a) $(1, 3, 4, 5, 9)$;   (b) $(1, 3, 5, 7, 8)$;   (c) $(1, 8)$;   (d) $(1, 10)$.
5. Prove that a group with 4 or fewer elements is necessarily Abelian. (*Hint:* ba is one of e, b, a, ab, except in trivial cases.)
6. Prove that if $xx = x$ in a group, then $x = e$.
7. Do the following multiplication tables describe groups?

	a	b	c	d
a	b	d	a	c
b	d	c	b	a
c	a	b	c	d
d	c	a	d	b

	a	b	c	d
a	a	b	c	d
b	b	a	d	c
c	c	d	a	a
d	d	c	b	b

8. Prove that Rules 2, 4, and 6 of §1.2 are valid in any commutative group.

9. Which of the following sets of numbers are groups? Why?
   (a) All rational numbers under addition; under multiplication.
   (b) All irrational numbers under multiplication.
   (c) All complex numbers of absolute value 1, under multiplication.
   (d) All complex numbers z with $|z| = 1$, under the operation $z \circ z' = |z| \cdot z'$.
   (e) All integers under the operation of subtraction.
   (f) The "units" (§3.6) of any integral domain under multiplication.
10. Prove that the following postulates describe an Abelian group: (i) $(ab)c = a(cb)$ for all a, b, c; (ii) the "left-identity" postulate of Theorem 4; (iii) the "left-inverse" postulate of Theorem 4.
★11. Prove that if $x^2 = e$ for all elements of a group G, then G is commutative.
★12. Prove Theorem 5. (*Hint:* If $ax = a$, then x is a right-identity, and any right-identity equals any left-identity.)
★13. Let S be a nonvoid set closed under a multiplication such that $ab = ba$, $a(bc) = (ab)c$, and $ax = ay$ implies $x = y$.
   (a) If S is finite, prove that S is a group.
   (b) If S is finite or infinite, prove that S can be embedded in a group.

6.5. Isomorphism

Consider the transformation $x \mapsto \log x$ on the domain of positive reals. It is well known that as x increases in the interval $0 < x < +\infty$, $\log x$ increases continuously in the interval $-\infty < y < +\infty$; that is, the correspondence is one-one between the system of positive real numbers and the system of all real numbers (the inverse transformation being $y \mapsto e^y$). Moreover $\log(xy) = \log x + \log y$ for all x, y: we can replace computations of products by parallel computations with sums. This is indeed the main practical use of logarithms!

Next, let $\mathbf{Z}_3$ be the field of the integers mod 3 (§1.10), and let G be the group of the rigid rotations of an equilateral triangle into itself. If I, R, and R' are the rotations through 0°, 120°, and 240°, respectively, the bijection $0 \leftrightarrow I$, $1 \leftrightarrow R$, $2 \leftrightarrow R'$ associating integers with rotations is one which carries sums in $\mathbf{Z}_3$ into products of the corresponding rotations. For instance, consider the correspondences

$$1 + 2 \equiv 0 \,(\text{mod } 3), \qquad RR' = I,$$
$$2 + 2 \equiv 1 \,(\text{mod } 3), \qquad R'R' = R.$$

These are instances of the general concept of "isomorphism" mentioned in §1.12. This concept is simpler and also more important for groups than for integral domains.

Definition. *By an* isomorphism *between two groups G and G' is meant a bijection a ↔ a' between their elements which preserves group multiplication—i.e., which is such that if a ↔ a' and b ↔ b', then ab ↔ a'b'.*

Thus, in the first example, we have described an isomorphism between the group of positive real numbers under multiplication, and that of all real numbers under addition. In the second, we have pointed out the isomorphism of the additive group of the integers mod 3 with the group of rotational symmetries of the equilateral triangle.

Similarly, the mapping $0 \mapsto 1$, $1 \mapsto 2$, $2 \mapsto 4$, $3 \mapsto 3$ is an isomorphism from the group of integers under addition mod 4 to the group of nonzero integers mod 5 under multiplication. It is convenient to check this result by comparing the group table for the integers under addition mod 4 with that for the nonzero elements of Z_5 under multiplication. See Tables 2 and 3.

In turn, the group of the integers under addition modulo 4 is isomorphic with the group of rotational symmetries of the square. That the bijection $0 \leftrightarrow I$, $1 \leftrightarrow R$, $2 \leftrightarrow R'$, $3 \leftrightarrow R''$ is an isomorphism can be checked by comparing Tables 2 and 3 with part of Table 1 (§6.4).

Table 2

+	0	1	2	3
0	0	1	2	3
1	1	2	3	0
2	2	3	0	1
3	3	0	1	2

Table 3

×	1	2	4	3
1	1	2	4	3
2	2	4	3	1
4	4	3	1	2
3	3	1	2	4

The notion of isomorphism is technically valuable because it gives form to the recognition that the same abstract group–theoretic situation can arise in entirely different contexts. The fact that isomorphic groups are abstractly the same (and differ only in the notation for their elements) can be seen in a number of ways.

Thus, by definition, two finite groups G and G' are isomorphic if and only if every group table for G yields a group table for G', by appropriate substitution. It follows from the next to the last sentence of §6.4 that G' is Abelian if and only if G is; that is, any isomorphic image of a finite Abelian group is Abelian. Again, isomorphism behaves like equality in another respect.

Theorem 6. *The relation "G is isomorphic to G'" is a reflexive, symmetric, and transitive relation between groups.*

Proof. The reflexive property is trivial (every group is isomorphic to itself by the identity transformation). As for the symmetric property, let $a \leftrightarrow aT$ be any isomorphic correspondence between G and G'; since T is bijective, it has an inverse T^{-1}, which is an isomorphism of G' onto G. Finally, if T maps G isomorphically on G', while T' maps G' isomorphically on G'', then TT' is an isomorphism of G with G''.   Q.E.D.

It is worth observing that Theorem 6 and its proof hold equally for isomorphisms between integral domains, and indeed for isomorphisms between algebraic systems of any kind whatever.

Theorem 7. *Under an isomorphism between two groups, the identity elements correspond and the inverses of corresponding elements correspond.*

Proof. The unique solution e of $ax = a$ goes into the unique solution e' of $a'x = a'$; hence the identities correspond. Consequently, the unique solution a^{-1} of the equation $ax = e$ in G goes into the unique solution a'^{-1} of $a'x = e'$ in G'; this completes the proof.

We shall finally prove a remarkable result of Cayley, which can be interpreted as demonstrating the completeness of our postulates on the multiplication of transformations.

Theorem 8. *Any abstract group G is isomorphic with a group of transformations.*

Proof. Associate with each element $a \in G$ the transformation $\phi_a : x \to xa = x\phi_a$ on the "space" of all elements x of G. Since $e\phi_a = e\phi_b$ implies $a = ea = eb = b$, distinct elements of G correspond to distinct transformations. Since

$$(6) \qquad\qquad x(\phi_a\phi_b) = (x\phi_a)\phi_b = (xa)b = x(ab) = x\phi_{ab}$$

holds for all x, the product $\phi_a\phi_b$ is ϕ_{ab}, and the set G' of all ϕ_a contains, with any two transformations, their product. Again, since $x\phi_e = xe = x$ for all x, G' contains the identity. One can similarly show that $(\phi_a)^{-1}$ exists and is in G' for all a, being in fact $\phi_{a^{-1}}$. Hence G' is a group of transformations, which is by (6) isomorphic with G.

Exercises

1. Are any two of the following groups isomorphic: (a) the group of symmetries of an equilateral triangle; (b) the group of symmetries of a square; (c) the group of rotations of a regular hexagon; (d) the additive group of integers mod 6?

2. The same question for (a) the group of rotations of a square; (b) the group of symmetries of a rectangle; (c) the group of symmetries of a rhombus (equilateral parallelogram); (d) the multiplicative group of 1, 5, 8, 12, mod 13; (e) the multiplicative group of 1, 5, 7, 11, mod 12.
3. (a) Prove that the additive group of "Gaussian" integers $m + n\sqrt{-1}$ $(m, n \in \mathbf{Z})$ is isomorphic to the multiplicative group of rational fractions of the form $2^n 3^m$ $(m, n \in \mathbf{Z})$.
   (b) Show that both are isomorphic to the group of all translations of a rectangular network.
★4. Is the multiplicative group of nonzero real numbers isomorphic with the additive group of all real numbers?
5. Determine all the isomorphisms between the additive group of $\mathbf{Z}_4$ and the group of rotations of the square.
6. (a) Exhibit an isomorphism between the group of the square and a group of transformations on the four vertices 1, 2, 3, 4 of the square.
   (b) Show explicitly how inverses correspond under this isomorphism, as in Theorem 7.
7. Do the same for the group of all rotations of a regular hexagon.
8. Illustrate Theorem 8 by exhibiting a group of transformations isomorphic with each of the following groups:
   (a) the additive group of all real numbers,
   (b) the multiplicative group of all nonzero real. numbers,
   (c) the additive group of integers mod 8.

6.6. Cyclic Groups

In any group, the integral *powers a^m* of the group element a can be defined separately for positive, zero, and negative exponents. If $m > 0$, we define

(7) $a^m = a \cdot a \cdots a$ (to m factors), $a^0 = e$, $a^{-m} = (a^{-1})^m$.

Two of the usual laws of exponents hold,

(8) $a^r a^s = a^{r+s}$, $(a^r)^s = a^{rs}$.

On the other hand, $(ab)^r \neq a^r b^r$ in general (cf. Ex. 2).

If both exponents r and s are positive, the laws (8) follow directly† from the definition (7) (cf. §1.5). In the other cases for the first law of (8), one of r or s may be zero, in which case (8) is immediate, or both r and s may be negative, in which case the result comes directly from the last part

---

† Thus r factors "a" followed by s factors "a" give all told $r + s$ factors. Again, s sets of r factors "a" each give all told sr factors.

of the definition (7). There remains the case when one exponent is negative and one positive, say $r = -m$ and $s = n$, with $m > 0$ and $n > 0$. Then

$$a^{-m}a^n = (a^{-1})^m a^n = (a^{-1} \cdots a^{-1})(a \cdots a).$$

By the associative law we can cancel successive a's against the inverses a^{-1}. In case $n \geq m$ we have left a^{n-m}, while if $n < m$, we have some inverses left, $(a^{-1})^{m-n}$ or $a^{-(m-n)}$. In both cases we have the desired law $a^{-m}a^n = a^{n+(-m)}$.

The second half of (8) can be established even more simply. If s is positive, then by the first half of (8),

$$a^r a^r \cdots a^r \text{ (to } s \text{ factors)} = a^{r+r+\cdots+r} = a^{rs}.$$

If s is negative, we can make a similar expansion, noting that $(a^r)^{-1} = a^{-r}$ whether r is positive, zero, or negative. If s is zero, the result is immediate.

Definition. *The order of an element a in a group is the least positive integer† m such that $a^m = e$; if no positive power of a equals the identity e, a has order infinity. The group G is cyclic if it contains some one element x whose powers exhaust G; this element is said to generate the group.*

For example, the group of all rotations of a square into itself consists of the four powers R, R^2, R^3, and $R^4 = I$ of the clockwise rotation R of $90°$. This group might equally well have been generated by R^3, which is a counterclockwise rotation of $90°$, since $R^2 = (R^3)^2$, $R = (R^3)^3$, and $I = (R^3)^4$ with R^3 exhaust the group.

Theorem 9. *If an element a generates the cyclic group G, then the order of a determines G to within isomorphism. In fact, if the order of a is infinite, G is isomorphic with the additive group of the integers; if the order of a is some finite integer n, G is isomorphic with the additive group of the integers modulo n.*

Proof. First, $a^r = a^s$ if and only if

(9) $$e = a^r(a^s)^{-1} = a^r a^{-s} = a^{r-s}, \quad \text{by (8)}.$$

Again, if $r \neq s$, either $r > s$, or $s > r$; hence if the order of a is infinite, so that $a^{r-s} = e$ for no $r > s$, no two powers of a are equal. Moreover,

† The well-ordering principle of §1.4 guarantees the existence of this m.

by (8) $a^s a^t = a^{s+t}$; therefore, the correspondence $a^s \mapsto s$ makes G isomorphic with the additive group of the integers, proving our first assertion.

If the order of a is finite, then the set of those integers t with $a^t = e$ contains 0, and by (8) contains the sum and difference of any two of its members. Hence, by Theorem 6 of §1.7, $a^t = e$ if and only if t is a multiple of the order n of a—and so by (9), $a^r = a^s$ if and only if $n \mid (r - s)$; that is, $a^r = a^s$ if and only if $r \equiv s \pmod{n}$. Finally, by (8) again, $a^r a^s = a^{r+s}$; consequently, the function $a^r \mapsto r$ is an isomorphism of G to the additive group of the integers modulo n.   Q.E.D.

It is a corollary that the number of elements in any cyclic group G is equal to the order of any generator of G, and that any two cyclic groups of the same order are isomorphic.

The group of the square is not cyclic, but is generated by the two elements R and H; indeed Table 1 (§6.4) shows that

$$R^0 = I, \qquad R = R, \qquad R^2 = R', \qquad R^3 = R'';$$
$$H = H, \qquad HR = D', \qquad HR^{2} = V, \qquad HR^3 = D.$$

The elements of the group are thus represented uniquely as $H^i R^j$ with $i = 0, 1$ and $j = 0, 1, 2, 3$. Furthermore, H and R satisfy

$$R^4 = I, \qquad H^2 = I, \qquad RH = HR^3.$$

These are called "defining relations" because they suffice to put the product of any two elements $H^i R^j$ ($i = 0, 1$) into the same form. For example,

$$D'V = HRHR^2 = HHR^3R^2 = IR = R;$$

similar calculations will give the whole multiplication table for the group of the square (Table 1).

Exercises

1. Using the definitions $a^1 = a$, $a^{m+1} = a^m a$, prove laws (8), for positive exponents, by induction.
2. Prove that if $(ab)^n = a^n b^n$ for all a and b in G and all positive integers n, then G is commutative, and conversely.
3. How many different generators has a cyclic group of order 6?
4. Show that if a commutative group with 6 elements contains an element of order 3, it is cyclic.

5. Is the multiplicative group of 1, 2, $\cdots$, 6 mod 7 cyclic? of 1, 3, 5, 7 mod 8? of 1, 2, 4, 5, 7, 8 mod 9?

6. If a cyclic group G is generated by a of order m, prove that a^k generates G if and only if g.c.d. $(k, m) = 1$.

7. Under the hypotheses of Ex. 6, find the order of any element a^k of G.

8. Find the order of every element in the group of the square.

9. Give the elements and the multiplication table of the group generated by two elements x and y subject to the defining relations $x^2 = y^2 = e, xy = yx$.

10. The dihedral group D_n is the group of all symmetries of a regular polygon of n sides (if $n = 4$, D_n is the group of the square). Show that D_n contains $2n$ elements and is generated by two elements R and H with $R^n = I$, $H^2 = I$, and $RH = HR^{n-1}$.

★11. Obtain generators and defining relations for the groups of symmetries of the three infinite patterns

imagined as extended to infinity in both directions. Are any two of these three groups isomorphic?

★12. Make similar studies for the groups described in Exs. 1, 2, 4, and 5 of §6.3.

6.7. Subgroups

Many groups are contained in larger groups. Thus, the group of rotations of the square is a part of the group of all symmetries of the square. Again, the group of the eight permutations of the vertices of the square induced by symmetries is a part of the group of all $4! = 24$ permutations of these vertices. The group of the even integers under addition is a part of the group of all integers under addition.

These examples suggest the concept of a subgroup. A subset S of a group G is called a *subgroup* of G if S is itself a group with respect to the binary operation (multiplication) of G.

In any group G, the set consisting of the identity e alone is a subgroup. The whole group G is also a subgroup of itself. Subgroups of G other than the trivial ("improper") subgroups e and G are called *proper* subgroups.

Theorem 10. *A nonvoid subset S of a group G is a subgroup if and only if (i) a and b in S imply ab in S and (ii) a in S implies a^{-1} in S.*

Proof. Under these hypotheses, clearly S is a subgroup: the associativity is trivial; the identity $e = aa^{-1}$ of G is in S, for there is at least one element a in S; the other group postulates are assumed. Conversely, we must prove that (i) and (ii) hold in any subgroup. The identity $x = e'$ of any subgroup of G satisfies $xx = x$, and so is the identity of G (Ex. 6, §6.4). Consequently, since G has but one inverse for any a, the inverse of any element a in the subgroup is the same as its inverse in G, so (ii) holds. Condition (i) is obvious.

For elements a of finite order m, clearly $a^{m-1}a = a^m = e$, and so $a^{-1} = a^{m-1}$. Hence one has the following simplified condition.

Theorem 11. *A nonvoid subset S of a finite group G is a subgroup of G if and only if the product of any two elements in S is itself in S.*

Among the subgroups of a given non-Abelian group G, one of the most important is its *center*. This is defined as the set of all elements $a \in G$ such that $ax = xa$ for all $x \in G$. We leave to the reader the verification that the center is, in fact, always a subgroup of G.

The problem of determining all subgroups of a specified group G is in general very difficult. We shall now solve it in the case that G is a cyclic group.

Theorem 12. *Any subgroup S of a cyclic group G is itself cyclic.*

Proof. Let G consist of the powers of an element a. If a^s and a^t are in S, then $a^{s+t} = a^s a^t$ and $a^{s-t} = a^s(a^t)^{-1}$ are in S, by Theorem 10. The set of integers s for which a^s is in S is therefore closed under addition and subtraction, so† consists of the multiples of some least positive exponent r (Theorem 6, §1.7). Therefore S itself consists of the powers $a^{kr} = (a^r)^k$, hence is cyclic with generator a^r. Q.E.D.

In case G is infinite, every $r > 0$ determines a different subgroup. If G has n elements, then since $a^n = e$ is surely in S, only those $r > 0$ which are divisors of n determine subgroups in this manner—but again these subgroups are all distinct.

To obtain material for further development, we now enumerate all the subgroups of the group of the square. By examining the definitions given in §6.1 for the operations of this group, one finds the (proper) subgroups leaving invariant each of the following configurations:

A diagonal	*An axis*	*A face*	*An axis and a diagonal*
$[I, D, D', R']$	$[I, H, V, R']$	$[I, R, R', R'']$	$[I, R']$

Vertex 1 (or 3)	*Vertex 2 (or 4)*	*A vertical side*	*A horizontal side*
$[I, D]$	$[I, D']$	$[I, H]$	$[I, V]$

† The conclusion, with $r = 0$, also holds if the set S consists of 0 alone.

By the transformations leaving a face invariant, we understand those which do not turn the square over. All these subgroups may be displayed in their relation to each other in a table, where each group is joined to all of its subgroups by a descending line or sequence of lines, as shown in Figure 3.

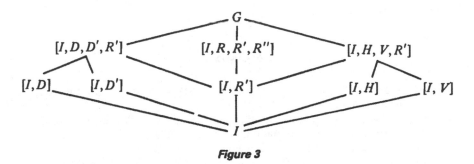

Figure 3

Without using geometry we could still find all these subgroups. Indeed, the determination of all the subgroups of a specified finite group G is most efficiently handled by considering the group elements purely abstractly, as follows.

Observe first that if a subgroup S of G contains an element a, it also contains the "cyclic" subgroup $\{a\}$ (prove it is a subgroup!) consisting of all the powers of a. In the present case, this gives us all but the first two of the subgroups listed. Next, observe that any other must contain not only two cyclic subgroups $\{a\}$ and $\{b\}$, but also the set $\{a, b\}$ of all products, such as $a^2b^{-3}a$, of powers of a and b. (Prove, using Theorem 11, that these form a subgroup!) In the present case, this procedure gives us the remaining subgroups. (We shall see in §6.8 why they all contain either 2 or 4 elements.) In general, we may have to test further for subgroups $\{a, b, c\}$ generated by three or more elements, but this can never happen unless the number of elements in the group is a product of at least four primes.

The *intersection* $S \cap T$ of two subgroups (indeed, of any two sets!) S and T is the set of all elements which belong both to S and to T.

Theorem 13. *The intersection $S \cap T$ of two subgroups S and T of a group G is a subgroup of G.*

Proof. By Theorem 10, a in $S \cap T$ implies a in S, hence a^{-1} in S; likewise it implies a^{-1} in T, and so a^{-1} in $S \cap T$. Similarly, a and b in $S \cap T$ implies ab in S, ab in T, and so ab in $S \cap T$. Hence by Theorem

10, $S \cap T$ is a subgroup. Also, $S \cap T$ contains e, and so is nonvoid. Q.E.D.

Clearly, $S \cap T$ is the largest subgroup contained in both S and T; dually, there exists a least subgroup containing both S and T. It consists of all the products of positive and negative powers of elements of S and T—it is called the *join* of S and T, and denoted $S \vee T$. We shall return to these concepts in Chap. 11.

Exercises

1. In the group of symmetries of the regular hexagon, what is the subgroup leaving a diagonal fixed?
2. If T is a subgroup of S, and S a subgroup of G, prove T is a subgroup of G.
3. In the group of all permutations ϕ of four digits 1, 2, 3, 4, find the following subgroups: (a) all ϕ carrying the set $\{1, 2\}$ into the set $\{1, 2\}$; (b) all ϕ such that $a \equiv b \pmod{2}$ implies $a\phi \equiv b\phi \pmod{2}$ for any digits a, b of the set 1, 2, 3, 4.
4. Prove that Theorem 11 still holds if G is infinite, but all elements of G have a finite order. Show that the additive group of $\mathbf{Z}_p[x]$ is such a group.
5. Tabulate all the subgroups of the following groups: (a) the additive group mod 12; (b) the group of a regular pentagon; (c) the group of a regular hexagon; ★(d) the group of all permutations of four letters.
★6. Let $a \leftrightarrow a'$ be an isomorphism between two groups G and G' of permutations, and let S consist of those permutations of G leaving one letter fixed. Does the set S' of all elements of G' corresponding to a's in S necessarily form a subgroup of G'? Must the set S' leave a letter fixed? Illustrate.
7. Prove that the center of any group G is a subgroup of G.
8. Find the center of the group (a) of the square, (b) of the equilateral triangle.
★9. Do the same for the group of a regular polygon of n sides.
★10. Show that the elements of finite order in any commutative group G form a subgroup.

6.8. Lagrange's Theorem

We now come to a far-reaching concept of abstract group theory: the idea that any subgroup S of a group G decomposes G into cosets.

Definition. *By the* order *of a group or subgroup is meant the number of its elements. By a* right coset *(left coset) of a subgroup S of a group G is meant any set Sa (or aS) consisting of all the right-multiples sa (left-multiples as) of the elements s of S by a fixed element a in G. The number of distinct right cosets is called the "index" of S in G.*

Since $Se = S$, S is a right coset of itself. Moreover, one has

Lemma 1. *If S is finite, each right coset Sa of S has exactly as many elements as S does.*

For, the transformation $s \mapsto sa$ is bijective: each element $t = sa$ of the coset Sa is the image of one and only one element $s = ta^{-1}$ of S. (Cf. also Theorem 8.)

Lemma 2. *Two right cosets Sa and Sb of S are either identical or without common elements.*

For, suppose Sa and Sb have an element $c = s'a = s''b$ (s', s'' in S) in common. Then Sb contains every element $sa = ss'^{-1}s'a = (ss'^{-1}s'')b$ of Sa, and similarly Sa contains every element of Sb. Consequently, $Sa = Sb$.

It is easy to illustrate these results. Thus, if G is the group of symmetries of the square, the subgroup $S = [I, H]$ has the four right cosets

$$[I, H]I = [I, H], \qquad [I, H]R = [R, HR] = [R, D'],$$
$$[I, H]R' = [R', HR'] = [R', V], \qquad [I, H]R'' = [R'', HR''] = [R'', D].$$

Each coset has two elements, and every element of the group falls into one of the four right cosets.

Again, if G is the additive group of the integers, the subgroup of multiples $\pm 5n$ of 5 has for right cosets the different residue classes modulo 5. Finally, let G be the symmetric group of all permutations of the symbols $1, \cdots, 6$, while S is the subgroup leaving the symbol 1 fixed. Then $1\phi = k$ implies for all $\psi \in S$ that $1(\psi\phi) = (1\psi)\phi = 1\phi = k$. Hence the coset $S\phi$ contains only (and so by Lemma 1 all) the 5! permutations carrying $1 \to k$. Therefore the right cosets of S are the subsets carrying $1 \to 1, 1 \to 2, \cdots, 1 \to 6$, respectively.

From the preceding lemmas, we obtain a classic result which is of fundamental importance for the theory of finite groups. Since any right coset Sa always contains $a = ea$, any group G is exhausted by its right cosets. Therefore G is decomposed by S into nonoverlapping subsets, each of which has exactly as many elements as S. If G is finite,† the conclusion is:

Theorem 14 (Lagrange). *The order of a finite group G is a multiple of the order of every one of its subgroups.*

† The extension to the infinite case follows immediately from the discussion of Chap. 12—but the importance of the result disappears.

Each element a of G generates a cyclic subgroup, whose order is (Theorem 9) simply the order of a. Therefore we have

Corollary 1. *Every element of a finite group G has as order a divisor of the order of G.*

Corollary 2. *Every group G of prime order p is cyclic.*

For, the cyclic subgroup A generated by any element $a \neq e$ in such a group has an order $n > 1$ dividing p. But this implies $n = p$, and so $G = A$ is cyclic.

More generally, Lagrange's theorem can be applied to determine (up to isomorphism) all abstract groups of any low order. As an example, define the *four group* as the group with four commuting elements: e (the identity) and $a, b, c = ab$, the latter each of order two. It will be shown in §6.9 that this group is isomorphic to the group of symmetries of a rectangle. We now prove

Corollary 3. *The only abstract groups of order four are the cyclic group of that order and the four group.*

In other words, every group of order four is isomorphic to either the cyclic group of that order or the four group.

Proof. If a group G of order 4 contains an element of order 4, it is cyclic. Otherwise, by Corollary 1, all elements of G except e must have order 2. Call them a, b, c. By the cancellation law, ab cannot be $ae = a$, $eb = b$, or $aa = e$; hence $ab = c$. Similarly, $ba = c$, $ac = ca = b$, $bc = cb = a$. But these, together with $a^2 = b^2 = c^2 = e$, and $ex = xe = x$ for all x, give the multiplication table of the four group.

Lagrange's theorem can also be applied to number theory.

Corollary 4 (Fermat). *If a is an integer and p a prime, then $a^p \equiv a$ (mod p).*

Proof. The multiplication group mod p (excluding zero) has $p - 1$ elements. The order of any element a of this group is then a divisor of $p - 1$, by Corollary 1, and so $a^{p-1} \equiv 1 \pmod{p}$ whenever $a \not\equiv 0 \pmod{p}$. If we multiply by a on both sides, we obtain the desired congruence, except for the case $a \equiv 0 \pmod{p}$, for which the conclusion is trivially true. (This is a new proof of Theorem 18 in Chapter 1).

Exercises

1. Check Fermat's theorem for $p = 7$ and $a = 2, 3, 6$.
2. (a) Enumerate the subgroups of the dihedral group (§6.6, Ex. 10) of order 26. How many are there?
   (b) Generalize your result.
3. Prove: the number of right cosets of any subgroup of a finite group equals the number of its left cosets. (*Hint:* Use the correspondence $x \mapsto x^{-1}$.)
4. Determine the cosets of the subgroup $[I, D]$ of the group of the square.
5. If S is any subgroup of a group G, let SaS denote the set of all products sas' for s, s' in S. Prove that for any $a, b \in G$, either $SaS \cap SbS$ is void or $SaS = SbS$.
6. For a subgroup S, let $x \equiv y \pmod{S}$ be defined to mean $xy^{-1} \in S$.
   (a) Prove that this relation is reflexive, symmetric, and transitive, and show that $x \equiv y \pmod{S}$ if and only if x and y lie in the same right coset of S.
   (b) Show that $x \equiv y \pmod{S}$ implies $xa \equiv ya \pmod{S}$ for all a.
7. Let G be the group of a regular hexagon, S the subgroup leaving one vertex fixed. Find the right and left cosets of S.
8. Prove that a group of order p^m, where p is a prime, must contain a subgroup of order p.
9. (a) If G is the group of all transformations $x \mapsto ax + b$ of $\mathbf{R}$, where $a \neq 0$ and b are real, while S is the subgroup of all such transformations with $a = 1$, describe the right and left cosets of S in G.
   (b) Do the same for the subgroup T of all transformations with $b = 0$.
★10. (a) Show that in any commutative ring R, the units (those elements with multiplicative inverses) form a group G.
   (b) Show that if $R = \mathbf{Z}_n$, then G consists of the positive integers $k < n$ relatively prime to n.
   (c) The *order* of G, in case $R = \mathbf{Z}_n$, is denoted $\phi(n)$ and called Euler's ϕ-function. Show that $\phi(p) = p - 1$ if $n = p$ is a prime, and compute $\phi(12)$, $\phi(16)$, $\phi(30)$.
   (d) Using Lagrange's theorem, infer that if $(k, n) = 1$, then $k^{\phi(n)} \equiv 1 \pmod{n}$.
★11. If S and T are subgroups of orders s and t of a group G, and if u and v are the orders of $S \cap T$ and $S \cup T$, prove that $st \leq uv$.
★12. Prove that the only abstract groups of order 6 are the cyclic group and the symmetric group on three letters.
★13. Let $2^h + 1$ be a prime p.
   (a) Prove that in the multiplicative group mod p, the order of 2 is $2h$.
   (b) Using Fermat's theorem, infer that $2h$ divides $p - 1 = 2^h$.
   (c) Conclude that h is a power of 2.

6.9. Permutation Groups

A *permutation* is a one-one transformation of a finite set into itself. For instance, the set might consist of the five digits 1, 2, 3, 4, 5. One permutation might be the transformation ϕ,

(10) $1\phi = 2,$ $2\phi = 3,$ $3\phi = 4,$ $4\phi = 5,$ $5\phi = 1.$

Another might be the transformation ϕ' with

(11) $1\phi' = 2,$ $2\phi' = 3,$ $3\phi' = 1,$ $4\phi' = 5,$ $5\phi' = 4.$

The reader will find it instructive to compute $\phi\phi'$, $\phi'\phi$, and to note that $\phi\phi' \neq \phi'\phi$.

Permutations which, like the permutation ϕ defined above, give a circular rearrangement of the symbols permuted (Figure 4) are called

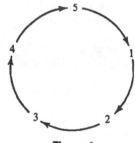

cyclic permutations or *cycles*. There is a suggestive notation for cyclic permutations—simply write down inside parentheses first any letter involved, then its transform, $\cdots$, and finally the letter transformed into the original letter. Thus, the permutation ϕ of (10) might be written in any one of the equivalent forms (12345), (23451), (34512), (45123), or (51234).

Figure 4

Theorem 15. *A cyclic permutation of n symbols has order n.*

Proof. The cyclic permutation $\gamma = (a_1 a_2 \cdots a_n)$ carries a_i into a_{i+1}. Hence γ^2 has the doubled effect of carrying each a_i into a_{i+2}, and generally γ^k carries a_i into a_{i+k}, where all subscripts are to be reduced modulo n. We have in γ^k the identity I if and only if a_{i+k} equals a_i; that is, if and only if $k \equiv 0 \pmod{n}$. The smallest k with $\gamma^k = I$ is then n itself, so γ does have the order n (see the definition in §6.6). The cycle γ is said to have length n.

The notation for a cyclic permutation can be extended to any permutation. For example, the permutation ϕ' in (11) cyclically permutes the digits 1, 2, and 3 by themselves, and 4 and 5 by themselves. Thus, it is the product of these two cycles,

$$(123)(45) = (45)(123).$$

This product may be written in either order, since the symbols permuted by (123) are left unchanged by (45), which means that successive application of these permutations in either order gives the same result.

Theorem 16. *Any permutation ϕ can be written as a product of cycles, acting on disjoint sets of symbols (more briefly: a product of disjoint†cycles).*

Proof.   Select any symbol, denote it by a_1. Denote $a_1\phi$ by a_2, $a_2\phi$ by $a_3, \cdots , a_{n-1}\phi$ by a_n, until $a_n\phi = a_i$ is some element already named. Since the antecedent of any a_i ($i > 1$) is a_{i-1}, $a_n\phi$ must be a_1. Thus the effect of ϕ on the letters $a_1, \cdots , a_n$ is the cycle $(a_1 a_2 \cdots a_n)$. Moreover, $(a_1 \cdots a_n)$ contains, with any symbol a_i, its antecedent; hence ϕ permutes the remaining symbols among themselves. The result now follows by induction on the number of symbols. In particular, the identity permutation on m letters is represented by m "cycles," each of length one.

Conversely, evidently any product of disjoint cycles represents a permutation. Moreover, one can prove

Theorem 17. *The order of any permutation ϕ is the least common multiple of the lengths of its disjoint cycles.*

Proof.   Write the permutation ϕ as the product $\phi = \gamma_1 \cdots \gamma_r$ of disjoint cycles γ_i. If $i \neq j$, then γ_i and γ_j are disjoint; hence $\gamma_i\gamma_j = \gamma_j\gamma_i$, and the factors γ_i may be rearranged in ϕ and in its powers to give $\phi^n = \gamma_1^n \cdots \gamma_r^n$ for all n. Therefore, $\phi^n = I$ if and only if every γ_i^n is the identity. But by Theorem 15, this means $\phi^n = I$ if and only if n is a common multiple of the lengths of the γ_i, from which the conclusion of Theorem 17 follows immediately.   Q.E.D.

Figure 5

Every finite group is isomorphic with one or more groups of permutations, by Theorem 8 of §6.5. In particular, this is true of finite groups of symmetries of geometrical figures, as we now illustrate by two examples.

Consider the group of symmetries of the rectangle (Figure 5). Under it, the vertices are transformed by the four permutations

$$I = (1)(2)(3)(4), \qquad R = (14)(23), \qquad H = (13)(24), \qquad V = (12)(34).$$

This group is known as the *four group*. According to Theorem 8, it is isomorphic with the group of permutations $\phi_I = (I)(R)(V)(H)$, $\phi_R = (IR)(HV)$, $\phi_H = (IH)(RV)$, $\phi_V = (IV)(RH)$.

---

† Two sets are called *disjoint* when they have no element in common.

The group of symmetries of the square (§6.1) can similarly be represented as a group of permutations of the four vertices. Using Theorem 8, we can also represent it as a group of permutations of the eight symbols which represent the elements of the group. Thus, R corresponds to the permutation effected on right-multiplication of these symbols by "R"; from the column headed "R" in the group table (Table 1), one sees that this permutation is $(IRR'R'')(HD'VD)$. Similarly, H corresponds to $(IH)(RD)(R'V)(R''D')$.

Two cycles of the same length are closely related. For example, if $\gamma = (1234)$ and $\gamma' = (2143)$, then one may compute that $\gamma' = \phi^{-1}\gamma\phi$, where $\phi = (12)(34)$ is the permutation taking each digit of the cycle γ into the corresponding digit in γ'. This is a special case of the following result.

Theorem 18. *Let ϕ and γ be permutations of n letters, where γ is a cyclic permutation $\gamma = (a_1, \cdots, a_m)$, and denote by $\gamma' = (a_1\phi, \cdots, a_m\phi)$ the cycle obtained by replacing each letter a_i in the representation of γ by its image under ϕ. Then $\phi^{-1}\gamma\phi = \gamma'$.*

Proof. The product $\phi^{-1}\gamma\phi$ carries each letter $a_i\phi$ in succession into $a_i\phi\phi^{-1} = a_i$; then to $a_i\gamma = a_{i+1}$, then to $a_i\gamma\phi = a_{i+1}\phi$, and hence has the same effect upon $a_i\phi$ as does γ' (call $a_{m+1} = a_1$). Similarly, one computes that $\phi^{-1}\gamma\phi$ and γ' both carry any letter b not of the form $a_i\phi$ into itself. Hence $\phi^{-1}\gamma\phi = \gamma'$, as asserted.

Corollary. *For any permutations ϕ and ψ, if $\psi = \gamma_1 \cdots \gamma_r$ is written as a product of cycles, we have $\phi^{-1}\psi\phi = \gamma_1' \cdots \gamma_r'$, where the γ_i' are obtained from the γ_i as in Theorem 18.*

Exercises

1. Express as products of disjoint cycles the following permutations:
   (a) $1\phi = 4$, $2\phi = 6$, $3\phi = 5$, $4\phi = 1$, $5\phi = 3$, $6\phi = 2$;
   (b) $1\phi = 5$, $2\phi = 3$, $3\phi = 2$, $4\phi = 6$, $5\phi = 4$, $6\phi = 1$;
   (c) $1\phi = 3$, $2\phi = 5$, $3\phi = 6$, $4\phi = 4$, $5\phi = 1$, $6\phi = 2$.
   Find the order of each of these permutations.
2. Represent the following products as products of disjoint cycles:
   $(1234)(567)(261)(47)$, $(12345)(67)(1357)(163)$, $(14)(123)(45)(14)$.
   Find the order of each product.
3. Find the order of $(abcdef)(ghij)(klm)$ and of $(abcdef)(abcd)(abc)$.
4. Represent the group of the rhombus (equilateral parallelogram) as a group of permutations of its vertices.

5. Describe the right and left cosets of the subgroup of all those permutations of $x_1, \cdots, x_6$ which carry the set $\{x_1, x_2\}$ into itself.
6. Which symmetric groups are Abelian?
7. Let G be the group of all symmetries of the cube leaving one vertex fixed. Represent G as a group of permutations of the vertices (cf. §6.3).
8. (a) Prove that every permutation can be written as a product of (not in general disjoint) cycles of length two ("transpositions").
   ★(b) How does this relate to the proof of the "generalized commutative law" from the law $ab = ba$ (§1.5)?
9. Represent the group of symmetries of the equilateral triangle as a group of permutations of (a) three and (b) six letters.
   ★(c) Do (b) in two essentially different ways.
★10. Prove that the symmetric group of degree n is generated by the cycles $(1, 2, \cdots, n - 1)$ and $(n - 1, n)$.
★11. In what sense is the representation of Theorem 16 unique? Prove your answer.

6.10. Even and Odd Permutations

An important classification of permutations may be found by considering the homogeneous polynomial form

$$P = \prod_{i<j} (x_i - x_j),$$

where i and j run from 1 to n. If $n = 3$, P is

$$(x_1 - x_2)(x_1 - x_3)(x_2 - x_3)$$
$$= x_1{}^2x_2 + x_2{}^2x_3 + x_3{}^2x_1 - x_1{}^2x_3 - x_3{}^2x_2 - x_2{}^2x_1,$$

and P^2 is the discriminant discussed in §5.5. In general, P is a polynomial of degree $n(n - 1)/2$. Clearly, any permutation of the subscripts in P leaves the set of factors of P, and hence P itself, unchanged except as to sign. Moreover, the transposition (x_1x_2) changes $(x_1 - x_2)$ into its negative $(x_2 - x_1)$, interchanges the $(x_1 - x_j)$ and the $(x_2 - x_j)$, $j > 2$, and leaves the other factors unchanged. Hence it does change P to $-P$.

The $n!$ permutations of the subscripts are therefore of two kinds: the *even* permutations leaving P (and $-P$) invariant, and the *odd* permutations interchanging P and $-P$. It follows when we consider the effect of two permutations performed in succession, we have the rules

(12)
$$\text{even} \times \text{even} = \text{odd} \times \text{odd} = \text{even}$$
$$\text{even} \times \text{odd} = \text{odd} \times \text{even} = \text{odd}.$$

It is a corollary of (12) and of Theorem 11 that the even permutations form a subgroup A_n of the symmetric group of degree n. This subgroup is usually called the "alternating group" of degree n.

Moreover, if β is a fixed and ϕ a variable odd permutation, then $\phi\beta^{-1}$ is even, and so $\phi = (\phi\beta^{-1})\beta$ is in the right coset $A_n\beta$. In summary, the odd permutations form a single right coset of A_n. Hence by Lagrange's theorem, the "alternating group" on n symbols contains just $(n!)/2$ elements.

A polynomial $g(x_1, \cdots, x_n)$ in n indeterminates x_i is called "symmetric" if it is invariant under the symmetric group of all permutations of its subscripts. Particular symmetric polynomials are (for $n = 3$)

$$(13) \quad \sigma_1 = x_1 + x_2 + x_3, \quad \sigma_2 = x_1x_2 + x_1x_3 + x_2x_3, \quad \sigma_3 = x_1x_2x_3.$$

They are the coefficients in the expansion

$$(14) \qquad (t - x_1)(t - x_2)(t - x_3) = t^3 - \sigma_1 t^2 + \sigma_2 t - \sigma_3.$$

In general, we call such polynomials *elementary symmetric polynomials* (in n variables); they are

$$(15) \quad \sigma_1 = \sum_i x_i, \quad \sigma_2 = \sum_{i<j} x_i x_j, \quad \sigma_3 = \sum_{i<j<k} x_i x_j x_k, \quad \cdots, \quad \sigma_n = x_1 \cdots x_n.$$

Since $(-1)^k \sigma_k$ is the coefficient of t^{n-k} in the expansion of $p(t) = \Pi_k(t - x_k)$ as a polynomial in t, the expressions σ_i give the coefficients of $p(t)$ as functions of its roots. They derive much of their importance from the so-called "fundamental theorem on symmetric polynomials," which we shall state without proof.[†]

Theorem 19. *Any symmetric polynomial $p(x_1, \cdots, x_n)$ can be expressed as a polynomial in the elementary symmetric polynomials.*

Thus, in the case of two variables x and y,

$$x^2 + y^2 = (x + y)^2 - 2xy = \sigma_1^2 - 2\sigma_2,$$
$$x^3 + y^3 = (x + y)^3 - 3xy(x + y) = \sigma_1(\sigma_1^2 - 3\sigma_2), \text{ and so on.}$$

Even if a polynomial $q(x_1, \cdots, x_n)$ is not symmetric, one can at least ask for the set of all those permutations of the indices which leave the polynomial unchanged. It is clear that this set is a group; it is called the group of the polynomial.

[†] See L. Weisner, *Introduction to the Theory of Equations* (New York: Macmillan, 1938), p. 108. Also see §15.6, Theorem 15, corollary.

Exercises

1. List the odd permutations (a) of three letters, (b) of four letters.
2. For which positive integers n is a cycle of length n even? odd?
3. (a) Show that a product of not necessarily disjoint cycles is odd if and only if it contains an odd number of cycles of even length.
   (b) Are the permutations $(123)(246)(5432)$ and $(12)(345)(67)(891)$ odd or even?
4. (a) Construct sample even and odd permutations of order 14 on 11 letters.
   (b) Prove every permutation of order 10 on 8 letters is odd.
5. Show that a permutation is even if and only if it can be written as a product of an even number of transpositions (Ex. 8, §6.9).
★6. Show that every even permutation can be written as the product of cycles of length three.
7. Find the group of each of the following polynomials:

$$x_1 x_2 + x_3 x_4, \qquad x_2 x_1 + x_3 x_2 + x_2 x_4, \qquad x_1{}^2 x_2 + x_3 x_4{}^2 + x_1{}^2 x_3 + x_2 x_4{}^2.$$

8. Represent each of the following polynomials in terms of the elementary symmetric polynomials:

$$x^2 + y^2 + z^2, \qquad x^2 y + y^2 z + z^2 x + x^2 z + y^2 x + z^2 y.$$

6.11. Homomorphisms

A single-valued transformation from a group G to a group G' may preserve multiplication without being one-one (i.e., without being an isomorphism).

Thus, consider the correspondence between the symmetric group of degree n and the group of ± 1 under multiplication, which carries even permutations into $+1$ and odd ones into -1. By (12), it carries products into products.

Or consider the correspondence $n \mapsto i^n$, where $i = \sqrt{-1}$, between the additive group of the integers and the multiplicative group of the fourth roots of unity. Again the group operation is preserved: $i^{m+n} = i^m i^n$, but the correspondence is many-one.

These and other examples lead to the following concept.

Definition. *A* homomorphism *of a group G to a group G' is a single-valued transformation $x \mapsto x'$ mapping G into G', such that $(xy)' = x'y'$ for all x, y in G.*

Theorem 20. *Under any homomorphism $G \to G'$, the identity e of G goes into the identity of G', and inverses into inverses.*

Proof. Since $e^2 = e$, the image f of e satisfies $f^2 = f = fe'$, where e' is the identity of G'. Hence, by cancellation, $f = e'$, and the identity of G must go into the identity of G'. Likewise, if a goes into a' and a^{-1} into $(a^{-1})'$, then $aa^{-1} = e$ must go into $a'(a^{-1})' = e'$, and so $(a^{-1})'$ is the inverse of a'.

Corollary 1. *Any homomorphic image† of a cyclic group is cyclic.*

For, by Theorem 20, $(a^m)' = (a')^m$ whether m is positive, zero, or negative. Hence if the powers a^m exhaust G, the powers $(a')^m = (a^m)'$ of a' exhaust G'.

Corollary 2. *The set N of all elements of G mapped on the identity e' of G', under a homomorphism of G to G', is a subgroup of G.*

This set N is called the *kernel* of the homomorphism.

Since $e \mapsto e'$, N is nonvoid. Again, by Theorem 20 and hypothesis, $a \mapsto e'$ and $b \mapsto e'$ imply $a^{-1} \mapsto (a')^{-1} = e'^{-1} = e'$ and $ab \mapsto a'b' = e'e' = e'$; hence N is a subgroup.

Direct Products. Any two groups G and H have a *direct product* $G \times H$. The elements of $G \times H$ are the ordered pairs (g, h) with $g \in G, h \in H$; multiplication in $G \times H$ is defined by the formula

(15a) $$(g, h)(g', h') = (gg', hh').$$

Evidently, (e, e) acts as an identity in $G \times H$; (g^{-1}, h^{-1}) is an inverse of (g, h), and multiplication is associative; hence $G \times H$ is a group. Moreover, the function $\alpha(g, h) = g$ defines a homomorphism α from $G \times H$ onto G, and the function $\beta(g, h) = h$ is a homomorphism from $G \times H$ onto H.

It can be proved that *every* Abelian group of finite order is isomorphic to a direct product of cyclic groups of prime-power orders. We content ourselves here with the following, much weaker result.

Theorem 21. *If m and n are relatively prime, then the direct product of cyclic groups of orders m and n is itself a cyclic group, of order mn.*

Proof. Let a and b generate the cyclic groups A and B, of orders m and n, respectively. Then, in $C = A \times B, (a, b)^k = (a^k, b^k)$ is the identity (e, e) if and only if $k \equiv 0 \pmod m$ *and* $k \equiv 0 \pmod n$. By Theorem

---

† Homomorphisms onto are sometimes called epimorphisms, and correspondingly homomorphic images are called epimorphic images.

17 of §1.9, this implies $k \equiv 0 \pmod{mn}$. Hence $(a, b) = c$ is of order mn in C, which contains only mn elements, and is therefore cyclic. Q.E.D.

Exercises

1. In the homomorphism $n \mapsto i^n$, where $i = \sqrt{-1}$ and $n \in \mathbb{Z}$, find the kernel.
2. Show that a cyclic group of order 8 has as homomorphic images (a) a cyclic group of order 4, (b) a cyclic group of order 2.
3. Is the correspondence mapping each x on the complex number $e^{2\pi i x}$ a homomorphism of the additive group of real numbers x? If so, what is its image G', and what is the kernel?
4. If G is a group of permutations of n letters $1, 2, \cdots, n$, in which each permutation ϕ of G carries the subset of letters $1, \cdots, k$ into itself, show that G is homomorphic onto the group G' of the permutations ϕ^* induced on $1, \cdots, k$.
5. In a square let the two diagonals be d and d', the axes h and v. Show that there is a homomorphism $\phi \mapsto \phi^*$ in which each motion ϕ in the group of the square induces a permutation ϕ^* on d, d', h, and v. Exhibit the correspondence $\phi \mapsto \phi^*$ in detail. What is the kernel?
6. If G is homomorphic to G' and G' to G'', prove that G is homomorphic to G''.
7. Which of the following correspondences map the multiplicative group of all nonzero real numbers homomorphically into itself? If the correspondence is a homomorphism, identify the homomorphic image G' and the kernel.
   (a) $x \mapsto |x|$, (b) $x \mapsto 2x$, (c) $x \mapsto x^2$, (d) $x \mapsto 1/x$,
   (e) $x \mapsto -x$, (f) $x \mapsto x^3$, (g) $x \mapsto -1/x$, (h) $x \mapsto \sqrt{x}$.
8. Show that the four group is the direct product of two cyclic groups of order 2.
★9. Show that the multiplicative group of all nonzero complex numbers is the direct product of the group of rotations of the unit circle and the group of all real numbers under addition. (*Hint:* Let $z = re^{i\theta}$.)
10. Prove that for any groups G, H, K, $G \times H$ is isomorphic to $H \times G$ and $G \times (H \times K)$ is isomorphic to $(G \times H) \times K$.

6.12. Automorphisms; Conjugate Elements

Definition. *An isomorphism of a group G with itself is called an* automorphism *of G. Thus an automorphism α of G is a one-one transformation of G onto itself (bijection of G) such that*

$$(16) \qquad (xy)\alpha = (x\alpha)(y\alpha) \qquad \text{for all } x, y \text{ in } G.$$

Theorem 22. *The automorphisms of any group G themselves form a group A.*

Proof. (Cf. Theorem 6.) It is obvious that the identity transformation is an automorphism, and that so is the product of any two automorphisms. Finally, if $x \to x\alpha$ is an automorphism, then by (16)

$$(xy)\alpha^{-1} = [(x\alpha^{-1}\alpha)(y\alpha^{-1}\alpha)]\alpha^{-1} = ([(x\alpha^{-1})(y\alpha^{-1})]\alpha)\alpha^{-1}$$
$$= (x\alpha^{-1})(y\alpha^{-1})$$

so that α^{-1} is an automorphism.   Q.E.D.

A parallel definition and theorem apply to integral domains, and indeed to abstract algebras in general. One can fruitfully regard an "automorphism" of an abstract algebra A as just a symmetry of A.

Definition. *In any group G, $a^{-1}xa$ is called the* conjugate *of x under* "conjugation" *by a.*

In Theorem 18 we have already seen that the conjugate of any cycle in a permutation group is another cycle of the same length. A similar interpretation applies to any group of transformations. Thus if α and ϕ are one–one transformations of a space S onto itself, $\psi = \alpha^{-1}\phi\alpha$ is related to ϕ much as in Theorem 18. Specifically, any point q in S can be written as $q = p\alpha$ for some $p \in S$, and

$$(p\alpha)\psi = p\alpha(\alpha^{-1}\phi\alpha) = (p\alpha\alpha^{-1})\phi\alpha = (p\phi)\alpha.$$

Thus ψ is the transformation $p\alpha \to (p\phi)\alpha$; in other words, the conjugate $\psi = \alpha^{-1}\phi\alpha$ is obtained from ϕ by replacing each point p and its image $r = p\phi$ by $p\alpha$ and $r\alpha$, respectively. For example, in the group of the square, $V = R^{-1}HR$; reflection in the vertical axis is conjugate under R to reflection in the horizontal axis because R carries the horizontal axis into the vertical axis.

Theorem 23. *For any fixed element a of the group G, the conjugation $T_a: x \mapsto a^{-1}xa$ is an automorphism of G.*

Proof. $(a^{-1}xa)(a^{-1}ya) = a^{-1}(xy)a$ for all x, y.

Automorphisms T_a of the form $x \mapsto a^{-1}xa$ are called *inner* automorphisms; all other automorphisms are *outer*.

It may be checked that the group of symmetries of the square has four distinct inner automorphisms; it has four outer automorphisms. On the other hand, the cyclic group of order three has no inner automorphisms except the identity, but has the "outer" automorphism $x \leftrightarrow x^2$.

Theorem 24. *The inner automorphisms of any group G form a sub-group of the group of all automorphisms of G.*

Proof. Since $b^{-1}(a^{-1}xa)b = (ab)^{-1}x(ab)$, the product of the inner automorphisms T_a and T_b is the inner automorphism T_{ab}; similarly, since $(a^{-1})^{-1}(a^{-1}xa)(a^{-1}) = x$, the inverse of the conjugation T_a is $T_{(a^{-1})}$.

Definition (Galois). *A subgroup S of a group G is* normal *(in G) if and only if it is invariant under all inner automorphisms of G (i.e., contains with any element all its conjugates).*

A normal subgroup is sometimes called a "self-conjugate" or an "invariant" subgroup.

Thus, the group of rotations of the square is a normal subgroup of the group of all its symmetries; so is the subgroup $[I, R^2]$. Again, every subgroup of an Abelian group is normal since $a^{-1}xa = a^{-1}ax = x$ for all a, x. The group of translations of the plane is also a normal subgroup of the Euclidean group of all rigid motions of the plane (cf. Chap. 9).

Theorem 25. *The kernel N of any homomorphism* $\theta: G \to H$ *is a normal subgroup of G.*

Proof. It is a subgroup, by Corollary 2 of Theorem 20. Again, if $a \in N$ and $b \in G$, then $\theta(b^{-1}ab) = b'^{-1}\theta(a)b' = b'^{-1}e'b' = e'$ for $b' = \theta(b)$ and $e' = \theta(e)$, since (by Theorem 20) $\theta(b^{-1}) = [\theta(b)]^{-1}$.

In general, let $a^{-1}Sa$ denote the set of all products $a^{-1}sa$ for s in S. The definition then states that S is normal if and only if the set $a^{-1}Sa$ equals S for every a in G.

Theorem 26. *A subgroup S is normal if and only if all its right cosets are left cosets.*

Proof. If S is normal, then $aSa^{-1} = S$ for all s; hence the set Sa of sa ($s \in S$) is the same as the set $(aSa^{-1})a$ of $(asa^{-1})a = as$ ($s \in S$). Thus, $Sa = aS$ for all a. Conversely, if the right coset Sa is a left coset bS, then $a^{-1}Sa = a^{-1}bS$ contains $e = a^{-1}ea$ and so (Lemma 2, §6.8) is $eS = S$.

It is a corollary that every subgroup S with only one other coset is normal; the elements not in S form the right and left coset of S. Hence the alternating group is a normal subgroup of the symmetric group of degree n.

Remark. Consider the correspondence between elements a of a group G and the inner automorphisms T_a which they induce. By the proof of Theorem 24, $T_aT_b = T_{ab}$: it preserves multiplication. Yet, as in the case of the group of symmetries of the square, it is usually not one–one (R^2 and I induce the same inner automorphism); it is a homomorphism. One easily verifies that the kernel of this homomorphism is precisely the center of G.

Exercises

1. How many automorphisms has a cyclic group of order p? of order pq? (p, q distinct primes).
2. List all the automorphisms of the four group. Which are inner?
3. Find all the automorphisms of the cyclic group of order eight.
4. Show that the automorphisms of the cyclic group of order m are the correspondences $a^k \mapsto a^{rk}$, where r is a unit of the ring Z_m.
5. Prove that in any group, the relation "x is conjugate to y" is an equivalence relation.
6. Prove that an element a of a group induces the inner automorphism identity if and only if it is in the center.
★7. (a) Find an automorphism α of the group of the square such that $R\alpha = R$ and $H\alpha = D$.
   (b) Show that α is an outer automorphism. (*Hint*: Represent the group of the square by the generators R and H discussed in § 6.6.)
8. Prove that if G and H are isomorphic groups, the number of different isomorphisms between G and H is the number of automorphisms of G.
9. Enumerate the inner automorphisms, sets of conjugate elements, and normal subgroups of the group of the square.
★10. Let G be any group, and A its group of automorphisms. Show that the couples (α, g) with $\alpha \in A$ and $g \in G$ form a group (the "holomorph" of G) under the multiplication $(\alpha, g)(\alpha', g') = (\alpha\alpha', (g\alpha')g')$.
★11. (a) Show that the holomorph of the cyclic group of order three is the symmetric group of degree three.
   (b) Show that the holomorph of the cyclic group of order four is the group of the square.
12. Prove that if M and N are normal subgroups of a group G, then so is their intersection.
13. Prove that, under the hypotheses of Ex. 12, the set MN of all products xy ($x \in M$, $y \in N$) is a normal subgroup of G.
14. Prove that the inner automorphisms of any group G are a normal subgroup of the group of all automorphisms of G.
★15. (a) Show that for every rational $c \neq 0$, the correspondence $x \mapsto xc$ is an automorphism of the additive group of all rational numbers.
   (b) Show that this group has no other automorphisms.
★16. Let G be a group of order pq (p, q primes). Show that G either is cyclic or contains an element of order p (or q). In the second case, prove G contains either 1 normal or q conjugate subgroups of order p. In the latter case,

show that the $pq - q(p - 1) = q$ elements not of order p form a normal subgroup. Infer that G always has a proper normal subgroup.

★**17.** (a) Show that the defining relations $a^m = b^n = e$, $b^{-1}ab = a^k$ define a group of order mn with a normal subgroup of order m, if $k^n \equiv 1 \pmod{m}$.

(b) Using Ex. 16, find all groups of order 6 and all groups of order 15.

★**18.** Using Ex. 16, find all possible groups of orders (a) 10 and (b) 14.

★**19.** Using the analysis of Ex. 16, show that there are only two nonisomorphic groups of any given prime-square order.

6.13. Quotient Groups

Now we shall show how to construct isomorphic replicas of all the homomorphic images G' of a specified abstract group G.

Indeed, let $x \mapsto x'$ be any homomorphism of G *onto* a group G', and let N be the kernel of this homomorphism. If a and b are any elements of G, we can write $b = at$, so that $b' = a't'$. But by the cancellation law, $a't' = a'$ if and only if $t' = e'$—that is, if and only if $t \in N$. In summary, $b' = a'$ if and only if $b = at$ ($t \in N$).

Lemma 1. *Two elements of G have the same image in G' if and only if they are in the same coset $Nx = xN$ of the kernel N.*

This establishes a one-one correspondence between the elements of G' and the cosets of N in G. Hence the order of G' is the number of cosets (or "index") of N in G.

Lemma 2. *Let x' and y' be elements of G'. Then x'y' may be found as follows. Let Nx and Ny correspond to x' and y', respectively; then x'y' corresponds to the (unique) coset of N containing the set NxNy of all products uv ($u \in Nx$, $v \in Ny$).*

Proof. If $u = ax$, $v = by$ ($a, b \in N$), then

$$(uv)' = a'x'b'y' = e'x'e'y' = x'y'.$$

Thus G' is determined to within isomorphism by G and N: it is isomorphic with the system of cosets of N in G, multiplied by the rule that the "product" $Nx \circ Ny$ of two cosets is the (unique) coset containing all products uv ($u \in Nx$, $v \in Ny$).

We can illustrate the preceding discussion by considering the homomorphism between the group G of the symmetries of the square and the "four group" G': $[e, a, b, c]$ (§ 6.8), under which $[I, R^2] \mapsto e$, $[R, R^3] \mapsto a$, $[H, V] \mapsto b$, $[D, D'] \mapsto c$. (Check from the group table that this is a homomorphism!) The antecedents of e form the normal subgroup $[I, R^2]$, and those of the other elements are the cosets of $[I, R^2]$. Finally, we can derive the sample rule $ab = c$ by computing the products $[RH, RV, R^3H, R^3V]$—which lie in (in fact, form) the coset $[D, D']$ of antecedents of c.

Conversely, let there be given any normal subgroup N of G, not associated a priori with any homomorphism. One can construct from N a homomorphic image G' of G as follows.

The elements of G' are defined as the different cosets Nx of N. The product $[Nx] \circ [Ny]$ of any two cosets Nx and Ny of N is defined as the coset (if any) containing the set $NxNy$ of all products uv ($u \in Nx$, $v \in Ny$). If $u = ax$ and $v = by$ for a, b in N, then $uv = axby = ab'xy$, where $b' = xbx^{-1}$ is also in N because N is normal. Therefore $N(xy)$ is a coset containing $NxNy$; moreover, since distinct cosets are nonoverlapping and the set $NxNy$ is nonvoid, there cannot be two different cosets each containing $NxNy$.

We have thus defined a single-valued binary operation on the elements of G' (alias cosets of G), which may be written as

(17) $$[Nx] \circ [Ny] = N(xy).$$

In words: the product of any two cosets is found by multiplying in G any pair of "representatives" x and y, and forming the coset containing the product xy. The product $[Ne] \circ [Ny] = N(ey) = Ny$, by (17), so the coset $N = Ne$ is a left identity for the system of cosets. Both $([Nx] \circ [Ny]) \circ [Nz]$ and $[Nx] \circ ([Ny] \circ [Nz])$ contain $(xy)z = x(yz)$, so the multiplication of cosets is associative. Finally, the coset $[Nx^{-1}] \circ [Nx]$ contains $x^{-1}x = e$, so must be $Ne = N$; therefore, left-inverses of cosets exist. These results, with Theorem 4, prove the following

Lemma 3. *The cosets of any normal subgroup N of G form a group under multiplication.*

Definition. *The group of cosets of N is called the* quotient-group (*or* factor group) *of G by N and is denoted by†* G/N.

The correspondence $x \mapsto Nx$ is, by (17), a homomorphism of G onto G/N, and the kernel of this homomorphism is N.

† If G is an Abelian group in which the binary operation is denoted by "+," then every subgroup N is normal in G; and the quotient-group is often called the difference group, written $G - N$.

Conversely, we have already seen (following Lemma 2) that for any homomorphism of G onto a group G' in which the kernel is N, the image G' is isomorphic with the quotient-group G/N. We conclude

Theorem 27. *The homomorphic images of a given abstract group G are the quotient-groups G/N by its different normal subgroups, multiplication of cosets of N being defined by* (17).

Remark. The preceding "construction" of quotient-groups from groups and normal subgroups is analogous to the construction of the ring of integers "mod n" from the integral domain of all integers (§§ 1.9–1.10). The cosets of N are the analogues of the residue classes mod n, and the relation $x \equiv y \pmod{n}$ can be paralleled by defining $x \equiv y \pmod{N}$ as the relation $xy^{-1} \in N$—which is equivalent to the assertion that x and y are in the same coset of N (see Ex. 6 of § 6.8).

Exercises

1. List all abstract groups which are homomorphic images of the group of symmetries of a square.
2. Do the same for the group of a regular hexagon.
3. Prove that the center Z of any group G is a normal subgroup of G, and that G/Z is isomorphic with the group of inner automorphisms of G.
4. Prove that in Ex. 6, § 6.8, $x \equiv y \pmod{S}$ implies $ax \equiv ay \pmod{S}$ for all a if and only if S is a normal subgroup.
5. If G is the group of all rational numbers of the form $2^k 3^m 5^n$, with integral exponents k, m, and n, while S is the multiplicative subgroup of all numbers 2^k, describe (a) the cosets of S, (b) G/S.
6. Let $G \to G'$ be a homomorphism. Show that the set of all antecedents of any subgroup S' of G' is a subgroup S of G— and that if S' is normal, then so is S.
★7. If S is a subgroup and N a normal subgroup of a group G, if $S \cap N = e$ and $S \cup N = G$, prove that G/N is isomorphic to S.
★8. If G is a group, elements of the form $x^{-1}y^{-1}xy$ are called *commutators*. Prove that the set C of all products of such commutators forms a normal subgroup of G.
★9. In Ex. 8, prove that G/C is Abelian. Finally, if N is a normal subgroup of G and G/N is Abelian, prove that N contains C.
★10. Two subgroups S and T of a group G are called *conjugate* if $a^{-1}Sa = T$ for some $a \in G$. Prove that the intersection of any subgroup S of G with its conjugates is a normal subgroup of G.
11. (a) Show that if M and N are normal subgroups of G with $M \cap N = 1$, then $ab = ba$ for all $a \in M$, $b \in N$. (*Hint:* Show that $aba^{-1}b^{-1} \in M \cap N$.)
  ★(b) Show that, in (a), if $M \cup N = G$, then $G = M \times N$.

★**12.** Let G be any group, S any subgroup of G. For any $a \in G$, let T_a be the permutation $(Sx) \mapsto (Sxa)$ on the right cosets Sx of S. Prove:
(a) The correspondence $a \mapsto T_a$ is a homomorphism.
(b) The kernel is the normal subgroup of Ex. 10.

13. Prove that the cosets of a nonnormal subgroup do not form a group under the multiplication (17).

★6.14. Equivalence and Congruence Relations

In defining the relation $a \equiv b \pmod n$ between integers, in setting up the rational numbers in terms of a congruence of number pairs $(a, b) \equiv (a', b')$, which was defined to mean that $ab' = a'b$, and elsewhere, we have asserted that any reflexive, symmetric, and transitive relation might be regarded as a kind of equality. We shall now formulate the significance of this assertion.

For convenience, a relation R which has the reflexive, symmetric, and transitive properties,

$$aRa, \qquad aRb \text{ implies } bRa, \qquad aRb \text{ and } bRc \text{ imply } aRc,$$

for all the members a, b, c of a set S, will be called an *equivalence relation* on S. If, as in the case of cosets (§6.13), we are willing to treat suitable *subsets* of S as elements, such as equivalence relation R becomes ordinary equality. Indeed, if a is any element of S, we may denote by $R(a)$ the set of all elements b equivalent to a; $b \in R(a)$ if and only if bRa. These *R-subsets* have various simple properties.

Lemma 1. *aRb implies* $R(a) = R(b)$, *and conversely.*

Proof. Suppose first that aRb, and let c be any element of $R(a)$. Then by definition cRa, hence by the transitive law cRb, which means that $c \in R(b)$. Conversely, since the symmetric law gives bRa, $c \in R(b)$ implies $c \in R(a)$, which means that the two classes $R(a)$ and $R(b)$ have the same members and hence are equal.

Suppose now that $R(a) = R(b)$. By the reflexive law, bRb, so that $b \in R(b)$. Since $R(a) = R(b)$, implies $b \in R(a)$, and so aRb. This completes the proof.

In the particular case when R is the relation of congruence modulo n between integers, the class $R(a)$ determined by an integer a is simply the residue class containing a. Lemma 1 here specializes to the assertion that $a \equiv b \pmod n$ if and only if a and b lie in the same residue class, mod n (cf. §1.10). Other illustrations are given as exercises.

Again, the residue classes mod n divide the whole set $\mathbf{Z}$ of integers into nonoverlapping subclasses, and hence may be said to form a "partition" of $\mathbf{Z}$. In general, a *partition* π of a class S is any collection of subclasses $A, B, C, \cdots$, of S such that each element of S belongs to one and only one of the subclasses (subsets) of the collection. The R-subsets always provide such a partition.

Lemma 2. *Two R-subsets are either identical or have no elements in common, and the collection of all R-subsets is a partition of S.*

Proof. If $R(a)$ and $R(b)$ contain an element c in common, so that cRa and cRb, then by the symmetric and transitive laws, aRb. By Lemma 1 this implies $R(a) = R(b)$. Consequently, if $R(a) \neq R(b)$, the two classes cannot overlap. Finally, every element c of the set S is in the particular R-subset $R(c)$, for, by the reflexive law, cRc, so $c \in R(c)$.

The converse of Lemmas 1 and 2 is immediate. If a set S is divided by a partition π into nonoverlapping subclasses $A, B, C, \cdots$, then a relation aRb may be defined to mean that a and b lie in one and the same subclass of this partition, and this does give an abstract equivalence relation R on S. Moreover, the R-subset $R(a)$ determined by each element a for this relation is exactly that subclass of the partition π which contains a. These conclusions may be summarized as follows:

Theorem 28. *Every equivalence relation R on a set S determines a partition π of S into nonoverlapping R-classes, and, conversely, each partition of S yields an equivalence relation R. There is thus a one-one correspondence $R \leftrightarrow \pi$ between the equivalence relations R on S and the partitions π of S, such that elements a and b of S lie in the same subclass of the partition π if and only if aRb.*

In discussing the requisites for an admissible equality relation (§ 1.11), we also demanded a certain "substitution property" relative to binary operations. In terms of the equivalence relation R and the binary operation $a \circ b = c$ on the set S, this property takes the form

(18) aRa' and bRb' imply $(a \circ b)R(a' \circ b')$.

This condition also has a definite theoretical content.

Indeed, let R be any equivalence relation on S, and let π be the corresponding partition into the R-subsets $A, B, C, \cdots$. Just as with cosets, let us regard the R-subsets as the elements of a new system $\Sigma = S/R$. And just as with quotient-groups (or residue classes mod n), we may try to define a binary operation in Σ from that in S,

(19)
$$A \circ B = C \quad \text{in} \quad \Sigma \quad \text{if and only if}$$

$$a \in A \quad \text{and} \quad b \in B \quad \text{imply} \quad (a \circ b) \in C \quad \text{in } S.$$

Property (18) asserts that if a and a' are both in an R-subset A (i.e., if aRa') and if b and b' are in an R-subset B, then $(a \circ b)$ and $(a' \circ b')$ both lie in the same R-subset. This resulting R-subset C is thus uniquely determined by A and B and is the "product" $A \circ B$ in the sense of (19). In other words, the substitution property (18) is equivalent to the assertion that definition (19) yields a (single-valued) binary operation on R-subsets (i.e., on Σ). This proves

Theorem 29. *Given an equivalence relation R on a set S, any binary operation defined on S and having the substitution property (18) yields a (single-valued) binary operation on the R-subsets of S, as defined by (19).*

For example, if R is the relation of congruence mod n on the set of integers, both addition and multiplication have the substitution property (18), and the theorem yields the addition and multiplication of residue classes in $\mathbf{Z}_n$, as defined in §1.10. More generally, Theorem 29 can be applied to the relation $a - b \in C$, where C is any ideal in any commutative ring, and can even be extended to other algebraic systems with operations which need not be binary. In general, relations satisfying the conditions of Theorem 29 may be called "congruence relations." Similarly, the concepts of isomorphism, automorphism, and homomorphism can be applied to general algebraic systems. Thus if G and H are algebraic with a ternary operation (a, b, c), a homomorphism of G onto H is a map θ of G on H with the property that $(a, b, c)\theta = (a\theta, b\theta, c\theta)$ for all a, b, c in G.

Exercises

1. Which of the following relations R are equivalence relations? In case they are, describe the R-subsets.
   (a) G is a group, S a subgroup, and aRb means $a^{-1}b \in S$.
   (b) G, S as in (a); aRb means $ba^{-1} \in S$.
   (c) $\mathbf{Z}$ is the domain of integers; aRb means that $a - b$ is a prime.
   (d) $\mathbf{Z}$ as in (c); aRb means that $a - b$ is even.
   (e) $\mathbf{Z}$ as in (c); aRb means that $a - b$ is odd.
2. Let G be a group of permutations of the letters $x_1, \cdots, x_n$; let $x_i R x_j$ mean that $x_i \phi = x_j$ for some $\phi \in G$. Is R an equivalence relation? How does G act on each R-subset?

3. Let G consist of the transformations $(x, y) \mapsto (x + a, y)$ of the plane. Let $(x, y)R(x', y')$ mean that $(x, y)\phi = (x', y')$ for some $\phi \in G$. What are the R-subsets in this case?

4. With real numbers a and b, let aRb mean that $a - b$ is an integral multiple of 360.
   (a) Is R an equivalence relation?
   (b) Is it a congruence relation for addition?
   (c) Is it one for multiplication?
   (d) What does this imply regarding addition and multiplication of *angles*?

5. (a) Let C be any ideal in a commutative ring. Show that the relation $(a - b) \in C$ is a congruence relation for addition and multiplication.
   (b) Prove that if R is any congruence relation on a commutative ring, the R-subsets form another commutative ring if addition and multiplication are defined by (19).

6. In Ex. 1(a), show that half the substitution rule (18) holds for any S and that the other half holds if and only if S is normal.

7. Let $\sigma: S^2 \to S$ be a binary operation, and R an equivalence relation on S. Show that if aRa' implies $(a \circ b)R(a' \circ b)$ and $(b \circ a)R(b \circ a')$, then (18) holds.

7

Vectors and
Vector Spaces

7.1. Vectors in a Plane

In physics there arise quantities called *vectors* which are not merely numbers, but which have *direction* as well as magnitude. Thus a parallel displacement in the plane depends for its effect not only on the distance but also on the direction of displacement. It may conveniently be represented by an arrow α of the proper length and direction (Figure 1). The combined effect of two such displacements α and β, executed one after another, is a third "total" displacement γ. If β is applied after α by placing the origin of the arrow β at the terminus of α, then the combined displacement $\gamma = \alpha + \beta$ is the arrow leading from the origin of α to the terminus of β. This is the diagonal of the parallelogram with sides α and β. This rule for finding $\alpha + \beta$ is the so-called *parallelogram law* for the addition of vectors.

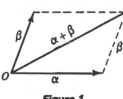

Figure 1

A displacement α may be tripled to give a new displacement $3 \cdot \alpha$, or halved to give a displacement $\frac{1}{2}\alpha$. One may even form a negative multiple such as -2α, representing a displacement twice as large as α in the direction opposite to α. In general, α may be multiplied by any real number c to form a new displacement $c \cdot \alpha$. If c is positive, $c\alpha$ has the direction of α and a magnitude c times as large, while if c is negative, the direction must be reversed. The numbers c are called *scalars* and the product $c\alpha$ a "scalar" product.

Forces acting on a point in a plane, and velocities and accelerations have similar representations by means of vectors—and in all cases the parallelogram law of vector addition, and multiplication by (real) scalars

168

have much the same significance as with displacements. This illustrates the general principle that various physical situations may have the same mathematical representation.

Analytical geometry suggests representing vectors in a plane by pairs of real numbers. We may represent any such vector by an arrow α with origin at $(0, 0)$ and terminus at a suitable point (a_1, a_2), where the coordinates a_1, a_2 are *real* numbers. Then vector sums and scalar products may be computed coordinate by coordinate, using the rules

$$(1) \qquad (a_1, a_2) + (b_1, b_2) = (a_1 + b_1, a_2 + b_2),$$

$$(2) \qquad c(a_1, a_2) = (ca_1, ca_2).$$

From these rules we easily get the various laws of vector algebra,† such as

$$(3) \qquad \alpha + \beta = \beta + \alpha, \qquad \alpha + (\beta + \gamma) = (\alpha + \beta) + \gamma,$$

$$(4) \qquad c(\alpha + \beta) = c\alpha + c\beta, \qquad 1 \cdot \alpha = \alpha,$$

and so on. Many of these (notably the commutative law of vector addition) also correspond to geometrical principles.

Vector operations may be used to express many familiar geometric ideas. For example, the midpoint of the line joining the terminus of the vector $\alpha = (a_1, a_2)$ to that of $\beta = (b_1, b_2)$ is given by the formulas $((a_1 + b_1)/2, (a_2 + b_2)/2)$, hence by the vector sum $\frac{1}{2}(\alpha + \beta)$. The resulting vector is also known as the center of gravity of α and β. A complete list of postulates for vector algebra will be given in §7.3; we shall first describe other examples of vectors.

Exercises

1. Prove the laws (3) and (4) of vector algebra, using the rules (1) and (2).
2. Illustrate the distributive law (4) by a diagram.
3. Show that the vectors in the plane form a group under addition.
4. Show that every vector α in the plane can be represented uniquely as a sum $\alpha = \beta + \gamma$, where β is a vector along the x-axis, γ a vector along the y-axis.

7.2. Generalizations

The example just described can be generalized in two ways. First, the number of *dimensions* (which was two in §7.1) can be arbitrary. The first

---

† We shall systematically use small Greek letters such as α, β, γ, $\cdots$, ξ, η, ζ, $\cdots$ to denote vectors and small Latin letters to denote scalars.

hint of this is seen in the possibility of treating forces and displacements in space in the same way that plane displacements and forces were treated in §7.1. The only difference is that in the case of space, vectors have *three* components (x_1, x_2, x_3) instead of two.

Again, it is shown in the theory of statics that the forces acting on a rigid solid can be resolved into *six* components: three pulling the center of gravity in perpendicular directions and three of torque causing rotation about perpendicular axes. The sum of two forces may again be computed component by component, while multiplication by scalars (real numbers) has the same significance as before.

More generally, for any positive whole number n, the n-tuples $\alpha = (a_1, \cdots, a_n)$ of real numbers form an *n-dimensional* vector space which may be regarded as an *n-dimensional* geometry. Thus, straight lines are the sets of elements of the form $\alpha + t\beta$ (α, β fixed, $\beta \neq 0$; t variable); the center of gravity of $\alpha_1, \cdots, \alpha_m$ is $(1/m)(\alpha_1 + \cdots + \alpha_m)$, and so on (this will be developed in §9.13). To get a complete geometrical theory, one need only introduce *distance* as in § 7.10.

A second line of generalization begins with the observation that, so far as algebraic properties are concerned, the components of vectors and the scalars need not be real numbers, but can be elements of any *field*. Indeed, vectors with *complex* components are constantly used in the theory of electric circuits and in electromagnetism, while we shall in Chap. 14 base the theory of algebraic numbers on the study of vectors with *rational* scalars.

The generalizations described in the last two paragraphs can be combined into a single formulation, valid for any positive integer n (the dimension) and any field F of scalars.

EXAMPLE. The vector space F^n has as elements all n-tuples $\alpha = (a_1, \cdots, a_n)$, $\beta = (b_1, \cdots, b_n), \cdots$, with components a_i and b_i in F. Addition and scalar multiplication in F^n are defined as follows:

(5) $(a_1, \cdots, a_n) + (b_1, \cdots, b_n) = (a_1 + b_1, \cdots, a_n + b_n)$,

(6) $c(a_1, \cdots, a_n) = (ca_1, \cdots, ca_n)$.

Theorem 1. *In the vector space $V = F^n$, vector addition and scalar multiplication have the following properties:*

(7)   *V is an Abelian group under addition;*

(8) $c \cdot (\alpha + \beta) = c \cdot \alpha + c \cdot \beta$, $(c + c') \cdot \alpha = c \cdot \alpha + c' \cdot \alpha$

(9) $(cc') \cdot \alpha = c \cdot (c' \cdot \alpha)$, $1 \cdot \alpha = \alpha$.   *(Distributive laws)*

Proof. We first verify the postulates for a group. Vector addition is associative, since for any vectors α and β as defined above and any $\gamma = (c_1, \cdots, c_n)$, we have:

$$(\alpha + \beta) + \gamma = (a_1 + b_1 + c_1, \cdots, a_n + b_n + c_n) = \alpha + (\beta + \gamma),$$

since $(a_i + b_i) + c_i = a_i + (b_i + c_i)$ for each i by the associative postulate for addition in a field (§ 6.4). The special vector $\mathbf{0} = (0, \cdots, 0)$ acts as identity, while $-\alpha = (-\alpha_1, \cdots, -\alpha_n)$ is *inverse* to α in the sense that $\alpha + (-\alpha) = (-\alpha) + \alpha = \mathbf{0}$. Note that $-\alpha = (-1)\alpha$ is also the product of the vector α by the scalar -1, while $\mathbf{0} = 0 \cdot \alpha$ for any α.

The group is commutative because $a_i + b_i = b_i + a_i$ for each i. Likewise, the definitions (5) and (6) reduce each side of the distributive laws (8) to a corresponding distributive law for fields which holds component by component.

Exercises

1. Let $\alpha = (1, 1, 0)$, $\beta = (-1/2, 0, 2/3)$, $\gamma = (0, 1/4, 2)$.
   Compute: (a) $\alpha + 2\beta + 3\gamma$,   (b) $3(\alpha + \beta) - 2(\beta + \gamma)$.
   (c) What is the center of gravity of α, β, γ?
   (d) Solve $6\beta + 5\xi = \alpha$.
2. Let $\alpha = (1, i, 0)$, $\beta = (0, 1 - i, 2i)$, $\gamma = (1, 2 - i, 1)$.
   Compute: (a) $2\alpha - i\beta$,   (b) $i\alpha + (1 + i)\beta - (i + 3)\gamma$.
   (c) Solve $\alpha - i\xi = \beta$.
3. Divide the line segment $\overrightarrow{\alpha\beta}$ in the ratio $2:1$ in Exs. 1 and 2.
★4. In Ex. 2, can you "divide the line segment $\overrightarrow{\alpha\beta}$ in the ratio $1:2i$"? Explain.
5. Let $\mathbf{Z}_3{}^n$ consist of vectors with n components in the field of integers mod 3.
   (a) How many vectors are then in $\mathbf{Z}_3{}^n$?
   (b) What can you say about $\alpha + \alpha + \alpha$ in $\mathbf{Z}_3{}^n$?
★6. Can you define a "midpoint" between two arbitrary points in $\mathbf{Z}_3{}^n$? A center of gravity for three—for four—arbitrary points? (*Hint*: Try numerical examples.)

7.3. Vector Spaces and Subspaces

We now define the general notion of a vector space; it is essentially just an algebraic system whose elements combine, under vector addition and multiplication by scalars from a suitable field F, so that the rules listed in § 7.2 hold.

Definition. *A vector space V over a field F is a set of elements, called vectors, such that any two vectors α and β of V determine a (unique) vector $\alpha + \beta$ as sum, and that any vector $\alpha \in V$ and any scalar $c \in F$ determine a scalar product $c \cdot \alpha$ in V, with the properties (4) and (7)–(9).*

(Rules (8) and (9) are to hold for all vectors α and β and all scalars c and c'.)

Theorem 1 essentially stated that for any positive integer n and any field, F^n was a vector space. There are also many *infinite*-dimensional vector spaces; they play a fundamental role in modern mathematical analysis.

For example, let S denote the set of all functions $f(x)$ of a real variable x which are single-valued and continuous on the interval $0 \leqq x \leqq 1$. Two such functions $f(x)$ and $g(x)$ have as sum a function $h(x) = f(x) + g(x)$ in S, and the "scalar" product of $f(x)$ by a real constant c is also such a function $cf(x)$. These functions cannot be represented by arrows, but their operations of addition and scalar multiplication have the same formal algebraic properties as our other examples. Vectors in this set S may even be regarded as having one "component" (the value of the function!) at each point x on the line $0 \leqq x \leqq 1$.

Again, consider the functions f whose domain is any set S whatever (say, any plane region), with the field F as codomain, so that f assigns to each $x \in S$ a value $f(x) \in F$. The set of all such functions f forms a vector space over F, if the sum $h = f + g$ and the scalar product $h' = c \cdot f$ are the functions defined for each $x \in S$ by the equations $h(x) = f(x) + g(x)$ and $h'(x) = c \cdot f(x)$.

Conforming with our use of the additive notation for the group operation in a vector space, we shall denote by 0 the identity element of the group; it is the unique "null" or "zero" vector satisfying

$$(10) \qquad \alpha + 0 = 0 + \alpha = \alpha \qquad \text{for all } \alpha.$$

The null vector 0 is not to be confused with the zero scalar 0. However, the two are connected by an identity.

Indeed, the two distributive laws give, for all c and α,

$$c\alpha + 0\alpha = (c + 0)\alpha = c\alpha = c\alpha + 0,$$
$$c\alpha + c0 = c(\alpha + 0) = c \cdot \alpha = c\alpha + 0.$$

Now, cancelling $c\alpha$ on both sides, we get the two laws

$$(11) \qquad 0\alpha = 0 \quad \text{for all } \alpha, \qquad c0 = 0 \quad \text{for all } c.$$

Again, the scalar multiple $(-1)\alpha$ acts as the inverse of any given vector α in the group, for

$$\alpha + (-1)\alpha = 1 \cdot \alpha + (-1)\alpha = (1 + (-1))\alpha = 0\alpha = 0;$$

hence

(12)   the (additive) group *inverse* of any vector α is $(-1)\alpha$.

It follows from (11) and (12) that the cyclic subgroup of the "powers" of any vector α consists of the *multiples* of α by the different integers n.

In ordinary three-dimensional vector space, $\mathbf{R}^3$, the vectors which lie in a fixed plane through the origin form by themselves a "two-dimensional" vector space which is part of the whole space. Similarly, the set S of all vectors lying on a fixed line through the origin is closed under the operations of addition and multiplication by scalars, hence this set is also a "subspace" of $\mathbf{R}^3$.

Definition. *A subspace S of a vector space V is a subset of V which is itself a vector space with respect to the operations of addition and scalar multiplication in V.*

A nonvoid subset S is a subspace if and only if the sum of any two vectors of S lies in S and any product of any vector of S by a scalar lies in S. This statement may be easily checked from the definition. The analogy with earlier definitions of a subfield and subgroup is obvious. Geometrically, a "subspace" is simply a linear subspace (line, plane, etc.) through the origin O.

For example, the vectors of the form $(0, x_2, 0, x_4)$ constitute a subspace of F^4 for any field F. Also, the null vector 0 alone is a subspace of any vector space.

Again, the set of polynomials of degree at most seven is a subspace of the vector space of all polynomials—whether the base field is real or not. Similarly, the set of all continuous functions $f(x)$ defined for $0 \leq x \leq 1$ is a subspace of the linear space of all functions defined on the same domain.

For given vectors $\alpha_1, \cdots, \alpha_m$ in a vector space V, the set of all *linear combinations*

$$c_1\alpha_1 + c_2\alpha_2 + \cdots + c_m\alpha_m \qquad \text{(each } c_i \text{ a scalar)}$$

of the α_i is a subspace. This is because of the identities

(13) $(c_1\alpha_1 + \cdots + c_m\alpha_m) + (c_1'\alpha_1 + \cdots + c_m'\alpha_m)$
$$= (c_1 + c_1')\alpha_1 + \cdots + (c_m + c_m')\alpha_m,$$

(14) $\qquad c'(c_1\alpha_1 + \cdots + c_m\alpha_m) = (c'c_1)\alpha_1 + \cdots + (c'c_m)\alpha_m,$

valid for all vectors α_i and for all scalars c_i, c_i', and c'. This proves

Theorem 2. *The set of all linear combinations of any set of vectors in a space V is a subspace of V.*

This subspace is evidently the smallest subspace containing all the given vectors; hence it is called the subspace *generated* or *spanned* by them. The subspace spanned by a single vector $\alpha_1 \neq \mathbf{0}$ is the set S_1 of all scalar multiples $c\alpha_1$; geometrically, this is simply the line through the origin and α_1. Similarly, the subspace spanned by two noncollinear vectors α_1 and α_2 turns out to be the plane passing through the origin, α_1, and α_2.

Theorem 3. *The intersection $S \cap T$ of any two subspaces of a vector space V is itself a subspace of V.*

Proof. The *intersection* of two given subspaces S and T is defined to be the set $S \cap T$ of all those vectors belonging both to S and to T (cf. Theorem 17 of §6.9, on the intersection of two subgroups). If α and β are two such vectors, their sum $\alpha + \beta$ must be in S (since S is a subspace containing α and β) likewise in T, hence is also in the intersection $S \cap T$. Similarly, any scalar multiple $c \cdot \alpha$ of α is in $S \cap T$. Q.E.D.

Again, any two subspaces S and T of a vector space V determine a set $S + T$ consisting of all sums $\alpha + \beta$ for α in S and β in T. By the commutative, associative, and distributive laws (3) and (4), this set is itself a subspace, called the *linear sum* or *span* of S and T. It clearly contains S and T, and is contained in any other subspace R containing both S and T; hence the concept of linear sum is analogous to that of the join (cf. §6.8) of two subgroups. These properties of $S + T$ may be stated as

(15)
$$S \subset S + T, \qquad T \subset S + T;$$
$$S \subset R \quad \text{and} \quad T \subset R \quad \text{imply} \quad S + T \subset R,$$

where $S \subset R$ means that the subspace S is contained in the subspace R.

Exercises

1. Prove that in any vector space, $c\alpha = 0$ implies $c = 0$ or $\alpha = 0$.

2. In Ex. 1, §7.2, compute $7(2(\alpha - 3\beta) + \frac{1}{3}(3\beta - 6\gamma)) - 2(\alpha - \gamma) + 5\beta + 2\alpha$.

3. In Ex. 2, § 7.2, compute $(1 + 2i)(2\alpha - 3\beta) - 8\alpha - 9i\beta$.

4. Which of the following subsets of $\mathbf{Q}^n$ $(n \geq 2)$ constitute subspaces (here ξ denotes $(x_1, \cdots, x_n)$)?
   (a) all ξ with x_1 an integer,   (b) all ξ with $x_2 = 0$,
   (c) all ξ with either x_1 or x_2 zero,   (d) all ξ such that $3x_1 + 4x_2 = 1$,
   (e) all ξ such that $7x_1 - x_2 = 0$.

5. Which of the following sets of real functions $f(x)$ defined on $0 \leq x \leq 1$ are subspaces of the vector space of all such functions?
   (a) all polynomials of degree four,
   (b) all polynomials of degree $\leq$ four (including $f(x) = 0$),
   (c) all functions f such that $2f(0) = f(1)$,
   (d) all functions such that $0 + f(1) = f(0) + 1$,
   (e) all positive functions,
   (f) all functions satisfying $f(x) = f(1 - x)$ for all x.

6. Which of the sets of functions described in Ex. 3, § 3.3, form vector spaces when D is taken to be a field F?

7. Let S be the subspace of $\mathbf{Q}^3$ consisting of all vectors of the form $(0, x_2, x_3)$, and T the subspace spanned by $(1, 2, 0)$ and $(3, 1, 2)$. Which vectors are in $S \cap T$? In $S + T$?

8. In $\mathbf{Z}_3^3$, how many vectors are spanned by $(1, 2, 1)$ and $(2, 1, 1)$? By $(1, 2, 1)$ and $(2, 1, 2)$?

9. In $\mathbf{Q}^3$ show that the plane $x_3 = 0$ may be spanned by each of the following pairs of vectors: $(1, 0, 0)$ and $(1, 1, 0)$; $(2, 2, 0)$ and $(4, 1, 0)$; $(3, 2, 0)$ and $(-3, 2, 0)$.

10. If S is spanned by ξ_1 and ξ_2, T by η_1, η_2, and η_3, show that $S + T$ is spanned by $\xi_1, \xi_2, \eta_1, \eta_2, \eta_3$. Generalize this result.

11. Construct an addition table for $\mathbf{Z}_2^2$ and list its subspaces.

12. Construct $\mathbf{Z}_2^3$ and tabulate its subspaces.

13. Prove that the set of all solutions $(x_1, \cdots, x_n)$ of a pair of homogeneous linear equations $a_1x_1 + \cdots + a_nx_n = 0$, $b_1x_1 + \cdots + b_nx_n = 0$ is a subspace of F^n, where a_i, b_i, x_i all lie in F.

★14. Prove that the vector space postulate $1 \cdot \alpha = \alpha$ cannot be proved from the other postulates. (*Hint*: Construct in the plane a pseudo-scalar product $c \otimes \alpha$, the projection of $c \cdot \alpha$ on a fixed line.)

★15. Show that the postulate of commutativity for vector addition is redundant. (*Hint*: Expand $(1 + 1)(\alpha + \beta)$ in two ways.)

7.4. Linear Independence and Dimension

The important geometric notion of the dimension of a vector space or subspace remains to be defined abstractly. It will be described as the minimum number of vectors spanning the space (or subspace).

Thus, ordinary space $\mathbf{R}^3$ can be spanned by the three vectors $(1, 0, 0)$, $(0, 1, 0)$, and $(0, 0, 1)$ of unit length lying along the three coordinate axes, but by no set of two vectors (a set of two noncollinear vectors spans a plane through the origin). Hence its dimension is three.

More generally, any F^n is spanned by *n unit vectors*

(16)
$$\begin{aligned}
\varepsilon_1 &= (1, 0, \cdots, 0), \\
\varepsilon_2 &= (0, 1, \cdots, 0), \\
&\;\vdots \\
\varepsilon_n &= (0, 0, \cdots, 1).
\end{aligned}$$

Indeed, any vector of F^n is a linear combination of these, because

(17) $$(x_1, x_2, \cdots, x_n) = x_1\varepsilon_1 + x_2\varepsilon_2 + \cdots + x_n\varepsilon_n.$$

We shall prove in Corollary 2 of Theorem 5 that F^n cannot be spanned by fewer than n vectors. This justifies calling F^n an n-dimensional vector space over the field F.

Not only do $\varepsilon_1, \cdots, \varepsilon_n$ generate the whole of F^n; in addition, $x_1\varepsilon_1 + \cdots + x_n\varepsilon_n = \mathbf{0}$ if and only if $(x_1, \cdots, x_n) = (0, \cdots, 0)$—that is, if and only if $x_1 = \cdots = x_n = 0$. This means that the unit vectors are "linearly independent" in the following sense.

Definition. *The vectors* $\alpha_1, \cdots, \alpha_m$ *are* linearly independent (*over F*) *if and only if, for all scalars* c_i *in F,*

(18) $c_1\alpha_1 + c_2\alpha_2 + \cdots + c_m\alpha_m = \mathbf{0}$

   implies $c_1 = c_2 = \cdots = c_m = 0.$

Vectors which are not linearly independent are called linearly dependent.

It is a trivial consequence of the definition that any subset of a linearly independent set is linearly independent. However, the following relation of dependence to linear combinations is more important:

Theorem 4. *The nonzero vectors* $\alpha_1, \cdots, \alpha_m$ *in a space V are linearly* *dependent if and only if some one of the vectors* α_k *is a linear combination of the preceding ones.*

Proof. In case the vector α_k is a linear combination $\alpha_k = c_1\alpha_1 + \cdots + c_{k-1}\alpha_{k-1}$ of the preceding ones, we have at once a linear relation

$$c_1\alpha_1 + c_2\alpha_2 + \cdots + c_{k-1}\alpha_{k-1} + (-1)\alpha_k = \mathbf{0},$$

with at least one coefficient, (-1), not zero. Hence the vectors are dependent, by (18).

Conversely, suppose that the vectors are linearly dependent, so that $d_1\alpha_1 + d_2\alpha_2 + \cdots + d_m\alpha_m = \mathbf{0}$, and choose the last subscript k for which $d_k \neq 0$. One can then solve for α_k as the linear combination

$$\alpha_k = (-d_k^{-1}d_1)\alpha_1 + \cdots + (-d_k^{-1}d_{k-1})\alpha_{k-1}.$$

This gives α_k as a combination of preceding vectors, except in the case $k = 1$. In this case $d_1\alpha_1 = \mathbf{0}$, with $d_1 \neq 0$, so $\alpha_1 = \mathbf{0}$, contrary to the hypothesis that none of the given vectors equals zero.   Q.E.D.

For instance, the three vectors $\beta_1 = (2, 0, 0)$, $\beta_2 = (1, 3, 0)$, and $\beta_3 = (0, -2, 0)$ do not span the whole of ordinary space $\mathbf{R}^3$ because they all lie in one plane. We can express this linear dependence either by the relation $\beta_1 - 2\beta_2 - 3\beta_3 = \mathbf{0}$ or (solving for β_1) by $\beta_1 = 2\beta_2 + 3\beta_3$. Thus, the set $(\beta_1, \beta_2, \beta_3)$ spans the same subspace as does its proper subset (β_2, β_3). This illustrates

Corollary 1. *A set of vectors is linearly dependent if and only if it contains a proper (i.e., smaller) subset spanning the same subspace.*

Namely, we can delete from the set any one vector which is $\mathbf{0}$ or which is a linear combination of the preceding ones, and show that the remaining vectors generate the same subspace. Now, using induction, we obtain

Corollary 2. *Any finite set of vectors contains a linearly independent subset which spans (generates) the same subspace.*

We can now state the fundamental theorem on linear dependence.

Theorem 5. *Let n vectors span a vector space V containing r linearly independent vectors. Then $n \geqq r$.*

Proof. Let $A_0 = [\alpha_1, \cdots, \alpha_n]$ be a sequence of n vectors spanning V, and let $X = [\xi_1, \cdots, \xi_r]$ be a sequence of r linearly independent vectors of V. Since A_0 spans V, ξ_1 is a linear combination of the α_i, so that the sequence $B_1 = [\xi_1, \alpha_1, \cdots, \alpha_n]$ both spans V and is linearly dependent. By Theorem 4, some vector of B_1 must be dependent on its predecessors. This cannot be ξ_1, since ξ_1 belongs to a set X of independent vectors. Hence some vector α_i is dependent on its predecessors $\xi_1, \alpha_1, \cdots, \alpha_{i-1}$ in B_1. Deleting this term, we obtain, as in Corollary 1, a subsequence $A_1 = [\xi_1, \alpha_1, \cdots, \alpha_{i-1}, \alpha_{i+1}, \cdots, \alpha_n]$ which still spans V.

Now repeat the argument. Construct the sequence $B_2 = [\xi_2, A_1] = [\xi_2, \xi_1, \alpha_1, \cdots, \alpha_{i-1}, \alpha_{i+1}, \cdots, \alpha_n]$. Like B_1, B_2 spans V and is linearly dependent. Hence as before, some vector of B_2 is a linear combination of its predecessors. Because the ξ_i are linearly independent, this vector cannot be ξ_2 or ξ_1, so must be some α_j, with a subscript $j \neq i$ (say, with $j > i$). Deletion of this α_j leaves a new sequence

$$A_2 = [\xi_2, \xi_1, \alpha_1, \cdots, \alpha_{i-1}, \alpha_{i+1}, \cdots, \alpha_{j-1}, \alpha_{j+1}, \cdots, \alpha_n]$$

of n vectors spanning V. This argument can be repeated r times, until the elements of X are exhausted. Each time, an element of A_0 is thrown out. Hence A_0 must have originally contained at least r elements, proving $n \geq r$. Q.E.D.

Theorem 5 has several important consequences. We shall prove these now for convenience, even though the full significance of the concepts of "basis" and "dimension", which they involve, will not become apparent until § 7.8.

Definition. *A* basis *of a vector space is a linearly independent subset which generates (spans) the whole space. A vector space is* finite-dimensional *if and only if it has a finite basis.*

For example, the unit vectors $\varepsilon_1, \cdots, \varepsilon_n$ of (16) are a basis of F^n.

Corollary 1. *All bases of any finite-dimensional vector space V have the same finite number of elements.*

Proof. Since V is finite-dimensional, it has a finite basis $A = [\alpha_1, \cdots, \alpha_n]$; let B be any other basis of V. Since A spans V and B is linearly independent, Theorem 5 shows that B is finite, say with r elements, and that $n \geq r$. On the other hand, B spans V, and A is linearly independent, so $r \geq n$. Hence $n = r$.

The number of elements in any basis of a finite-dimensional vector space V is called the *dimension* of V, and is denoted by $d[V]$. By Theorem 5, we have

Corollary 2. *If a vector space V has dimension n, then* (i) *any $n + 1$*

elements of V are linearly dependent, and (ii) *no set of n − 1 elements can span V.*

Theorem 6. *Any independent set of elements of a finite-dimensional vector space V is part of a basis.*

Proof. Let the independent set be $\xi_1, \cdots, \xi_r$, and let $\alpha_1, \cdots, \alpha_n$ be a basis for V. Form the sequence $C = [\xi_1, \cdots, \xi_r, \alpha_1, \cdots, \alpha_n]$. We can extract (Theorem 4, Corollary 2) an independent subsequence of C which also spans V (hence is a basis for V) by deleting every term which is a linear combination of its predecessors. Since the ξ_i are independent, no ξ_i will be deleted, and so the resulting basis will include every ξ_i.

Corollary. *For n vectors $\alpha_1, \cdots, \alpha_n$ of an n-dimensional vector space to be a basis, it is sufficient that they span V or that they be linearly independent.*

Proof. If $A = \{\alpha_1, \cdots, \alpha_n\}$ spans V, it contains a subset A' which is a basis of V (Theorem 4, Corollary 2); since the dimension of V is n, this subset A' must have n elements (Theorem 5, Corollary 1). Hence $A' = A$, and A is a basis of V. Again, if A is independent, then it is a part of a basis by Theorem 6, and this basis has n elements by Corollary 1 of Theorem 5, and so must be A itself.

Exercises

1. Show that the vectors (a_1, a_2) and (b_1, b_2) in F^2 are linearly dependent if and only if $a_1 b_2 - a_2 b_1 = 0$.
2. Do the vectors $(1, 1, 0)$ and $(0, 1, 1)$ form a basis of $\mathbf{Q}^3$? Why?
3. Prove that if β is not in the subspace S, but is in the subspace spanned by S and α, then α is in the subspace spanned by S and β.
4. Prove that if ξ_1, ξ_2, ξ_3 are independent in $\mathbf{R}^n$, then so are $\xi_1 + \xi_2, \xi_1 + \xi_3, \xi_2 + \xi_3$. Is this true in every F^n?
5. How many elements are in each subspace spanned by four linearly independent elements of $\mathbf{Z}_3{}^7$? Generalize your result.
6. Define a "vector space" over an integral domain D. Which of the postulates and theorems discussed so far fail to hold in this more general case?
★7. Prove: Three vectors with rational coordinates are linearly independent in $\mathbf{Q}^3$ if and only if they are linearly independent in $\mathbf{R}^3$. Generalize this result in two ways.
8. If the vectors $\alpha_1, \cdots, \alpha_m$ are linearly independent, show that the vector β is a linear combination of $\alpha_1, \cdots, \alpha_m$ if and only if the vectors $\alpha_1, \cdots, \alpha_m, \beta$ are linearly dependent.

★9. Show that the real numbers 1, $\sqrt{2}$, and $\sqrt{5}$ are linearly independent over the field of rational numbers.

10. Find four vectors of $\mathbf{C}^3$ which together span a subspace of two dimensions, any two vectors being linearly independent.

11. If $c_1\alpha + c_2\beta + c_3\gamma = 0$, where $c_1c_3 \neq 0$, show that α and β generate the same subspace as do β and γ.

12. If two subspaces S and T of a vector space V have the same dimension, prove that $S \subset T$ implies $S = T$.

★13. (a) How many linearly independent sets of two elements has $\mathbf{Z}_2{}^3$? How many of three elements? of four elements?
   (b) Generalize your formula to $\mathbf{Z}_2{}^n$ and to $\mathbf{Z}_p{}^n$.

★14. How many different k-dimensional subspaces has $\mathbf{Z}_p{}^n$?

7.5. Matrices and Row-equivalence

Problems concerning sets of vectors in F^n with given numerical coordinates can almost always be formulated as problems in simultaneous linear equations. As such, they usually can be solved by the process of *elimination* described in §2.3. We will now begin a systematic study of this process, which centers around the fundamental concept of *matrices* and their *row-equivalence*. We first define the former concept.

Definition. *A rectangular array of elements of a field F, having m rows and n columns, is called an $m \times n$ matrix over F.*

Remark. Evidently, the $m \times n$ matrices A, B, C over any field F form an *mn-dimensional vector space* under the two operations of (i) multiplying all entries by the same scalar c and (ii) adding corresponding components.

We now use the concept of a matrix to determine when two sets of vectors in F^n, $\alpha_1, \cdots, \alpha_m$ and $\beta_1, \cdots, \beta_n$ span the same subspace. Clearly, the vectors $\alpha_1, \cdots, \alpha_m$ define the $m \times n$ matrix

(19)
$$A = \begin{pmatrix} a_{11} & a_{12} & \cdots & a_{1n} \\ a_{21} & a_{22} & \cdots & a_{2n} \\ \vdots & \vdots & & \vdots \\ a_{m1} & a_{m2} & \cdots & a_{mn} \end{pmatrix},$$

whose ith row consists of the components $a_{i1}, \cdots, a_{in}$ of the vector α_i. The matrix (19) may be written compactly as $\| a_{ij} \|$. The *row space* of the matrix A is that subspace of F^n which is spanned by the rows of A, regarded as vectors in F^n. We now ask: when do two $m \times n$ matrices have the same row space? That is, when do their rows span the same subspace of F^n? A partial answer to this question is provided by the concept of *row-equivalence*, which we now define.

We now consider the effect on the matrix A in (19) of the following three types of steps, called *elementary row operations*:

(i)   The interchange of any two rows.
(ii) Multiplication of a row by any nonzero constant c in F.
(iii) The addition of any multiple of one row to any other row.

The $m \times n$ matrix B is called *row-equivalent* to the $m \times n$ matrix A if B can be obtained from A by a finite succession of elementary row operations. Since the effect of each such operation can be undone by another operation of the same type, we have the following

Lemma. *The inverse of any elementary row operation is itself an elementary row operation.*

Hence, if B is row-equivalent to A, then A is row-equivalent to B; that is, the relation of row-equivalence is symmetric. It is clearly also reflexive and transitive, hence it is an equivalence relation.

Theorem 7. *Row-equivalent matrices have the same row space.*

Proof. Denote the successive row vectors of the $m \times n$ matrix A by $\alpha_1, \cdots, \alpha_m$. The row space of A is then the set of all vectors of the form $c_1\alpha_1 + \cdots + c_m\alpha_m$, and the elementary row operations become:

(i)   Interchanging any α_i with any α_j $(i \neq j)$.
(ii) Replacing α_i by $c\alpha_i$ for any scalar $c \neq 0$.
(iii) Replacing α_i by $\alpha_i + d\alpha_j$ for any $j \neq i$ and any scalar d.

It suffices to consider the effect on the row space of a single elementary row operation of each type. Since operations of types (i) and (ii) clearly do not alter the row space, we shall confine our attention to the case of a single elementary operation of type (iii). Take the typical case of the addition of a multiple of the second row to the first row, which replaces the rows $\alpha_1, \cdots, \alpha_m$ of A by the new rows

$$(20) \qquad \beta_1 = \alpha_1 + d\alpha_2, \qquad \beta_2 = \alpha_2, \qquad \cdots, \qquad \beta_m = \alpha_m$$

of the row-equivalent matrix B. Any vector γ in the row space of B has the form $\gamma = \sum c_i\beta_i$, hence on substitution from (20) we have

$$\gamma = c_1(\alpha_1 + d\alpha_2) + c_2\alpha_2 + \cdots + c_m\alpha_m,$$

which shows that γ is in the row space of A. Conversely, by the Lemma, the rows of A can be expressed in terms of the rows of B as

$$\alpha_1 = \beta_1 - d\beta_2, \qquad \alpha_2 = \beta_2, \qquad \cdots, \qquad \alpha_m = \beta_m,$$

so that the same argument shows that the row space of A is contained in the row space of B; thus these row spaces are equal.

This proof gives at once

Corollary 1. *Any sequence of elementary row operations reducing a matrix A to a row-equivalent matrix B yields explicit expressions for the rows of B as linear combinations of the rows of A.*

Simultaneous Linear Equations. We next apply the concept of row-equivalent matrices to reinterpret the process of "Gauss elimination" described in § 2.3. Consider the system of simultaneous linear equations

$$a_{11}x_1 + a_{12}x_2 + \cdots + a_{1n}x_n = a_{1,n+1},$$

$$a_{21}x_1 + a_{22}x_2 + \cdots + a_{2n}x_n = a_{2,n+1},$$

(21)
$$\vdots \qquad \vdots \qquad \qquad \vdots \qquad \vdots$$

$$a_{m1}x_1 + a_{m2}x_2 + \cdots + a_{mn}x_n = a_{m,n+1},$$

where the coefficients a_{ij} are given constants in the field F. We wish to know which *solution vectors* $\xi = (x_1, \cdots, x_n)$, if any, satisfy the given system of equations (21).

It is easy to verify that the set of solution vectors ξ satisfying (21) is invariant under each of the following operations:

(i) the interchange of any two equations.
(ii) multiplication of an equation by any nonzero constant c in F.
(iii) Addition of any multiple of one equation to any other equation.

But as applied to the $m \times (n + 1)$ matrix of constants a_{ij} in (21), these are just the three elementary row operations defined earlier. This proves

Corollary 2. *If A and B are row-equivalent $m \times (n + 1)$ matrices over the same field F, then the system of simultaneous linear equations (21) has the same set of solution vectors $\xi = (x_1, \cdots, x_n)$ as the system*

$$b_{11}x_1 + b_{12}x_2 + \cdots + b_{1n}x_n = b_{1,n+1},$$

$$b_{21}x_1 + b_{22}x_2 + \cdots + b_{2n}x_n = b_{2,n+1},$$

(21′)
$$\vdots \qquad \vdots \qquad \qquad \vdots \qquad \vdots$$

$$b_{m1}x_1 + b_{m2}x_2 + \cdots + b_{mn}x_n = b_{m,n+1}.$$

Exercises

1. If A and B are row-equivalent matrices, prove that the rows of A are linearly independent if and only if those of B are.

2. Show that the meaning of row-equivalence is unchanged if one replaces the operation (iii) by

   (iii') The addition of any row to any other row.

★3. Show that any elementary row operation of type (i) can be effected by a succession of four operations of types (ii) and (iii). (*Hint*: Try 2×2 matrices.)

7.6. Tests for Linear Dependence

We now aim to use elementary row operations to simplify a given $m \times n$ matrix A as much as possible. In any nonzero row of A, the first nonzero entry may be called the "leading" entry of that row. We say that a matrix A is *row-reduced* if:

(a) Every leading entry (of a nonzero *row*) is 1.

(b) Every *column* containing such a leading entry 1 has all its other entries zero.

Sample 4×6 row-reduced matrices are

$$(22) \quad \begin{pmatrix} 0 & 0 & 1 & r_{14} & r_{15} & r_{16} \\ 1 & 0 & 0 & r_{24} & r_{25} & r_{26} \\ 0 & 1 & 0 & r_{34} & r_{35} & r_{36} \\ 0 & 0 & 0 & 0 & 0 & 0 \end{pmatrix}, \quad \begin{pmatrix} 1 & d_{12} & 0 & d_{14} & 0 & d_{16} \\ 0 & 0 & 1 & d_{24} & 0 & d_{26} \\ 0 & 0 & 0 & 0 & 1 & d_{36} \\ 0 & 0 & 0 & 0 & 0 & 0 \end{pmatrix}.$$

Theorem 8. *Any matrix A is row-equivalent to a row-reduced matrix, by elementary row operations of types (ii) and (iii).*

Proof. Suppose that the given matrix A with entries a_{ij} has a nonzero first row with leading entry a_{1t} located in the tth column. Multiply the first row by a_{1t}^{-1}; this leading entry becomes one. Now subtract a_{it} times the first row from the ith row for every $i \neq 1$. This reduces every other entry in column t to zero, so that conditions (a) and (b) are satisfied as regards the first row.

Now let the same construction be applied to the other rows in succession. The application involving row k does not alter the columns containing the leading entries of rows $1, \cdots, k - 1$, because row k already had entry zero in each such column. Hence, after the application involving row k, we have a matrix satisfying conditions (a) and (b) at its first k rows. Theorem 8 now follows by induction on k.

By permutations of rows (i.e., a succession of elementary row operations of type (i)), we can evidently rearrange the rows of a row-reduced matrix R so that

(c) Each zero row of R comes below all nonzero rows of R.

Suppose that there are r nonzero rows and that the leading entry of row i appears in column t_i for $i = 1, \cdots, r$. Since any such column has all its other entries zero, we have $t_i \neq t_j$ whenever $i \neq j$. By a further permutation of rows, we can then arrange R so that

(d) $t_1 < t_2 < \cdots < t_r$ (leading entry of row i in column t_i).

A row-reduced matrix which also satisfies (c) and (d) is called a (row) *reduced echelon matrix* (the leading entries appear "in echelon"). We have proved the

Corollary. *Any matrix is row-equivalent to a reduced echelon matrix.*

For example, the second matrix of (22) is already a reduced echelon matrix; the first matrix of (22) is not, but can be brought to this form by placing the first row after the third row.

Theorem 9. *Let E be a row-reduced matrix with nonzero rows $\gamma_1, \cdots, \gamma_r$, and leading entries 1 in columns $t_1, \cdots, t_r$. Then, for any vector*

$$\beta = y_1 \gamma_1 + \cdots + y_r \gamma_r$$

in the row space of E, the coefficient y_i of γ_i is the entry of β in the column t_i; i.e., the t_i-th entry of β.

Proof. Since all entries of E in column t_i are zero, except that of γ_i, which is one, the t_i-th component of β must be $y_i \cdot 1$.

Corollary 1. *The nonzero rows of a row-reduced matrix are linearly independent.*

For if $\beta = \mathbf{0}$, then every $y_i = 0$ in the preceding theorem.

Corollary 2. *Let the m × n matrix A be row-equivalent to a row-reduced matrix R. Then the nonzero rows of R form a basis of the row space of A.*

Proof. These rows of R are linearly independent, by Corollary 1, and span the row space of R. They are thus a basis of this row space, which by Theorem 7 is identical to the row space of A.

The *rank r* of a matrix A is defined as the dimension of the row space of A. Since this space is spanned by the rows of A, which must contain a linearly independent set of rows spanning the row space, we see that the rank of A can also be described as the maximum number of linearly independent rows of A. By Theorem 7, row-equivalent matrices have the same rank.

In particular, an $n \times n$ (square!) matrix A has rank n if and only if all its rows are linearly independent. One such matrix is the $n \times n$ *identity matrix I_n*, which has entries 1 along the main diagonal (upper left to lower right) and zeros elsewhere.

Corollary 3. *An n × n matrix A has rank n if and only if it is row-equivalent to the n × n identity matrix I_n.*

Proof. A is row-equivalent to a reduced echelon matrix E which also has rank n. This matrix E then has n nonzero rows, hence n leading entries 1 in n different columns and no other nonzero entries in these columns (which include all the columns). Because of the ordering of the rows (condition (d) above), E is then just the identity matrix.   Q.E.D.

In testing vectors for linear independence, or more generally in computing the dimension of a subspace (=rank of a matrix), it is needless to use the *reduced* echelon form. It is sufficient to bring the matrix to any echelon form, such as the form of the following 4 × 7 matrix:

$$E = \begin{pmatrix} 0 & 1 & d_{13} & d_{14} & d_{15} & d_{16} & d_{17} \\ 0 & 0 & 0 & 1 & d_{25} & d_{26} & d_{27} \\ 0 & 0 & 0 & 0 & 1 & d_{36} & d_{37} \\ 0 & 0 & 0 & 0 & 0 & 0 & 0 \end{pmatrix}.$$

Such an echelon matrix may be defined by the condition that the leading entry in each nonzero row is 1, and that in each row after the first the number of zeros preceding this entry 1 is larger than the corresponding number in the preceding row.

Thus, after reduction to echelon form, the rank of a matrix can be found immediately by applying the following theorem.

Theorem 10. *The rank of any matrix A is the number of nonzero rows in any echelon matrix row-equivalent to A.*

The proof will be left as an exercise.

EXAMPLE. Test $\alpha_1 = (1, -1, 1, 3)$, $\alpha_2 = (2, -5, 3, 10)$, and $\alpha_3 = (3, 3, 1, 1)$ for independence. By transformations of type (iii), obtain the new rows $\beta_1 = \alpha_1$, $\beta_2 = \alpha_2 - 2\alpha_1 = (0, -3, 1, 4)$, $\beta_3 = \alpha_3 - 3\alpha_1 = (0, 6, -2, -8)$. Finally, set $\gamma_1 = \beta_1$, $\gamma_2 = -(1/3)\beta_2$, $\gamma_3 = \beta_3 - 6\gamma_2 = \beta_3 + 2\beta_2 = 0$. There results the echelon matrix C with rows γ_1, γ_2, γ_3, sketched below; since C has a row of zeros, the original α_i are linearly

$$C = \begin{pmatrix} 1 & -1 & -1/3 & 3 \\ 0 & 1 & -1/3 & -4/3 \\ 0 & 0 & 0 & 0 \end{pmatrix}$$

dependent. By substitution in the definition of $\gamma_3 = 0$, we have the explicit dependence relation

$$0 = \gamma_3 = \beta_3 + 2\beta_2 = (\alpha_3 - 3\alpha_1) + 2(\alpha_2 - 2\alpha_1) = -7\alpha_1 + 2\alpha_2 + \alpha_3$$

between the α's.

Appendix on Row-equivalence. Reduced echelon matrices provide a convenient test for row-equivalence.

Theorem 11. *There is only one $m \times n$ reduced echelon matrix E with a given row space $S \subset F^n$.*

Proof. Let the reduced echelon matrix E with row space S have the nonzero rows $\gamma_1, \cdots, \gamma_r$, where γ_i has leading entry 1 in column t_i. By condition (d), $t_1 < t_2 < \cdots < t_r$. Let $\beta = y_1\gamma_1 + \cdots + y_r\gamma_r$ be any nonzero vector in the row space of E; by Theorem 9, β has entry y_i in column t_i. If y_s is the first nonzero y_i, then $\beta = y_s\gamma_s + \cdots + y_r\gamma_r$. Because $t_s < \cdots < t_r$, the leading entries of the remaining $\gamma_{s+1}, \cdots, \gamma_r$ lie beyond t_s, so that β has y_s as its leading entry in column t_s. In other words, every nonzero vector β of S has leading entry in one of the columns $t_1, \cdots, t_r$. Each of these columns occurs (as the leading entry of a γ_i); hence the row space S determines the indices $t_1, \cdots, t_r$.

The rows $\gamma_1, \cdots, \gamma_r$ of E have leading entry 1 and entries zero in all but one of the columns $t_1, \cdots, t_r$. If β is any vector of S with leading entry 1 in some column t_i and entries zero in the other columns t_j, then, by Theorem 9, β must be γ_i. Thus the row space and the column indices uniquely determine the rows $\gamma_1, \cdots, \gamma_r$ of E, as was required.   Q.E.D.

Corollary 1. *Every $m \times n$ matrix A is row-equivalent to one and only one reduced echelon matrix.*

This result, which is immediate, may be summarized by saying that the reduced echelon matrices provide a *canonical form* for matrices under row-equivalence: every matrix is row-equivalent to one and only one matrix of the specified canonical form.

Corollary 2. *Two $m \times n$ matrices A and B are row-equivalent if and only if they have the same row space.*

Proof. If A is row-equivalent to B, then A and B have the same row space, by Theorem 7. Conversely, if A and B have the same row space, they are row-equivalent to reduced echelon matrices E and E', respectively. Since E and E' have the same row space, they are identical, by Theorem 11. Hence A is indeed row-equivalent (through $E = E'$) to B.

These results emphasize again the fact that the row-equivalence of matrices is just another language for the study of subspaces of F^n.

Exercises

1. Show the row-equivalence of $\begin{pmatrix} 5 & 2 & 7 \\ -3 & 4 & 1 \\ -1 & -2 & -3 \end{pmatrix}$ and $\begin{pmatrix} 1 & 0 & 1 \\ 0 & 1 & 1 \\ 0 & 0 & 0 \end{pmatrix}$.

2. Reduce each of the following matrices to row-equivalent echelon form:

(a) $\begin{pmatrix} 1 & -1 & 3 \\ 2 & -4 & 1 \\ 0 & 3 & 2 \end{pmatrix}$, (b) $\begin{pmatrix} -5 & 6 & -3 \\ 3 & 1 & 11 \\ 4 & -2 & 8 \end{pmatrix}$, (c) $\begin{pmatrix} 1 & 6 & -2 & 5 \\ 4 & 0 & 4 & -2 \\ 7 & 2 & 0 & 2 \\ -6 & 3 & -3 & 3 \end{pmatrix}$,

(d) $\begin{pmatrix} 2 & -1 & 3 & 2 \\ 0 & 2 & 1 & 4 \\ 4 & -2 & 3 & 9 \\ 2 & -3 & 4 & 5 \end{pmatrix}$, (e) $\begin{pmatrix} i & 1 & -i & 1+i \\ 1 & -i & i & 2-i \\ -1 & 0 & 1 & 0 \\ 2 & i & 2i & 3i \end{pmatrix}$.

3. In Ex. 2, express the rows of each associated echelon matrix as linear combinations of the rows of the original matrix.
4. Test the following sets of vectors for linear dependence:
   (a) $(1, 0, 1), (0, 2, 2), (3, 7, 1)$ in $\mathbf{Q}^3$ and $\mathbf{C}^3$;
   (b) $(0, 0, 0), (1, 0, 0), (0, 1, 1)$ in $\mathbf{R}^3$;
   (c) $(1, i, 1 + i), (i, -1, 2 - i), (0, 0, 3)$ in $\mathbf{C}^3$;
   (d) $(1, 1, 0), (1, 0, 1), (0, 1, 1)$ in $\mathbf{Z}_2^3$ and $\mathbf{Z}_3^2$.
   In every case of linear dependence, extract a linearly independent subset which generates the same subspace.
5. In $\mathbf{Q}^6$, test each of the following sets of vectors for independence, and find a basis for the subspace spanned.
   (a) $(2, 4, 3, -1, -2, 1), (1, 1, 2, 1, 3, 1), (0, -1, 0, 3, 6, 2)$.
   (b) $(2, 1, 3, -1, 4, -1), (-1, 1, -2, 2, -3, 3), (1, 5, 0, 4, -1, 7)$.
6. In Ex. 5, find a basis for the subspace spanned by the two sets of vectors, taken together.
7. Find the ranks and bases for the row spaces of the following matrices:

$$\text{(a) } \begin{pmatrix} 1 & 2 & 3 \\ 2 & 3 & 4 \\ 3 & 4 & 5 \end{pmatrix}, \quad \text{(b) } \begin{pmatrix} 1 & 2 & 1 & 2 \\ 3 & 2. & 3 & 2 \\ -1 & -3 & 0 & 4 \\ 0 & 4 & -1 & -3 \end{pmatrix}, \quad \text{(c) } \begin{pmatrix} 1 & 2 & 4 & 5 & 7 \\ 1 & 2 & 3 & 4 & 5 \\ -1 & -2 & 0 & 2 & 1 \end{pmatrix}.$$

8. List all possible forms for a 2×4 reduced echelon matrix with two nonzero rows. (These yield a cell decomposition of the "Grassmann manifold" whose points are the planes through the origin in 4-space.)
9. Prove: The rank of an $m \times n$ matrix exceeds neither m nor n.
10. If the $m \times (n + k)$ matrix B is formed by adding k new columns to the $m \times n$ matrix A, then rank $(A) \leqq$ rank (B).
11. Prove directly (without appeal to Theorem 8) that any matrix A is row-equivalent to a (not necessarily reduced) echelon matrix.

7.7. Vector Equations; Homogeneous Equations

It is especially advantageous to use elementary row operations on a matrix in place of linear equations (21) when one wishes to solve several vector equations of the form

$$(23) \qquad\qquad \lambda = x_1 \alpha_1 + \cdots + x_m \alpha_m$$

for fixed vectors $\alpha_1, \cdots, \alpha_m$ of F^n and various vectors λ.

For instance, let $\alpha_1, \alpha_2, \alpha_3$ be as in the Example of § 7.6, and let $\lambda = (2, 7, -1, -6)$. Having reduced the matrix A to echelon form, we first solve the equation

$$\lambda = y_1\gamma_1 + y_2\gamma_2 + y_3\gamma_3 = y_1\gamma_1 + y_2\gamma_2.$$

Equating first components, we get $2 = y_1$; equating second components, we then get $7 = -y_1 + y_2$ or $y_2 = 9$. Hence we must have

$$\lambda = 2\gamma_1 + 9\gamma_2 = 2\alpha_1 - 3\beta_2 = 2\alpha_1 - 3(\alpha_2 - 2\alpha_1) = 8\alpha_1 - 3\alpha_2,$$

if λ is a linear combination of the α's at all. Computing the third and fourth components of $8\alpha_1 - 3\alpha_2$, we see that λ is indeed such a linear combination.

Since $\gamma_3 = -7\alpha_1 + 2\alpha_2 + \alpha_3 = \mathbf{0}$, other solutions of (23) in this case are

$$\lambda = (8 - 7y)\alpha_1 + (-3 + 2y)\alpha_2 + y\alpha_3,$$

for arbitrary y. This is actually the most general solution of (23). Had the vector λ been $\lambda' = (2, 7, 1, -6)$, the above procedure would have shown that λ' cannot be expressed as a linear combination of the α's at all.

In fact, when several vectors λ are involved, it is usually best to first transform the $m \times n$ matrix whose rows are $\alpha_1, \cdots, \alpha_m$ to *reduced echelon form* C with nonzero rows $\gamma_1, \cdots, \gamma_r$. Since each elementary row operation on a matrix involves only a finite number of rational operations, and since a given matrix can be transformed to reduced echelon form after a finite number of elementary row operations, this can be done after a finite number of rational operations (i.e., additions, subtractions, multiplications, and divisions).

One can then apply Theorem 9 to get the only possible coefficients which will make $\lambda = y_1\gamma_1 + \cdots + y_r\gamma_r$. If this equation is not satisfied by *all* components of γ, then λ is not in the row space of A, and (23) has no solution. If it is satisfied, then since the rows of C will be known linear combinations $\gamma_i = \sum_{j=1}^{m} e_{ij}\alpha_j$ of the α's, we will obtain a solution for (23) in the form $\lambda = \sum y_i e_{ij}\alpha_j$, whence we have $x_j = y_1 e_{1j} + \cdots + y_r e_{rj}$. This proves the following result.

Theorem 12. *For given vectors $\lambda, \alpha_1, \cdots, \alpha_m$ in F^n, the vector equation $\lambda = x_1\alpha_1 + \cdots + x_m\alpha_m$ can be solved (if a solution exists) by a finite number of rational operations in F.*

Corollary. *Let S and T be subspaces of F^n spanned by vectors $\alpha_1, \cdots, \alpha_m$ and $\beta_1, \cdots, \beta_k$ respectively. Then the relations $S \supset T$, $T \supset S$, and $S = T$ can be tested by a finite number of rational operations.*

For, one can construct from the α's, by elementary operations, a sequence of nonzero vectors $\gamma_1, \cdots, \gamma_r$ forming the rows of a reduced echelon matrix and also spanning S. One then tests as above whether all the β's are linear combinations of the γ's, which is clearly necessary and sufficient for $S \supset T$. By reversing the preceding process, one determines whether or not $T \supset S$. Together these two procedures test for $S = T$; alternatively $S = T$ may be tested by transforming the matrix with rows α and the matrix with rows β to reduced echelon form, for $S = T$ holds if and only if these two reduced forms have the *same* nonzero vectors.

Reduced echelon matrices are also useful for determining the solutions of systems of *homogeneous* linear equations of the form

(24)
$$a_{11}x_1 + a_{12}x_2 + \cdots + a_{1n}x_n = 0,$$
$$\vdots \qquad \vdots \qquad \qquad \vdots$$
$$a_{m1}x_1 + a_{m2}x_2 + \cdots + a_{mn}x_n = 0.$$

Thus let S be the set of all vectors $\xi = (x_1, \cdots, x_n)$ of F^n satisfying (24). It is easy to show that S is a subspace. We shall now show how to determine a basis for this subspace.

First observe, as in §2.3, that elementary row operations on the system (24) transform it into an equivalent system of equations. Specifically, as applied to the $m \times n$ matrix A which has as ith row the coefficients $(a_{i1}, \cdots, a_{in})$ in the ith equation of (24), these operations carry A into another matrix having the same set S of "solution vectors" $\xi = (x_1, \cdots, x_n)$. Now bring A to a reduced echelon form, with leading entries 1 in the columns $t_1, \cdots, t_r$. The corresponding system of equations has r nonzero equations, and the ith equation is the only one containing the unknown x_{t_i}.

To simplify the notation, assume that the leading entries appear in the first r columns (this actually can always be brought about by a suitable permutation, applied to the unknowns x_i and thus to the columns of A). The reduced equations then have the form

(25)
$$\cdot x_1 + c_{1,r+1}x_{r+1} + \cdots + c_{1n}x_n = 0,$$
$$x_2 + c_{2,r+1}x_{r+1} + \cdots + c_{2n}x_n = 0,$$
$$\vdots \qquad \vdots \qquad \qquad \vdots$$
$$x_r + c_{r,r+1}x_{r+1} + \cdots + c_{rn}x_n = 0.$$

In this simplified form, we can clearly obtain all solutions by choosing arbitrary values for $x_{r+1}, \cdots, x_n$, and solving (25) for $x_1, \cdots, x_r$ to give the solution vector

(26)
$$\xi = \left(- \sum_{j=r+1}^{n} c_{1j}x_j, \cdots, - \sum_{j=r+1}^{n} c_{rj}x_j, x_{r+1}, \cdots, x_n\right).$$

In particular, we obtain $n - r$ solutions by setting one of the parameters $x_{r+1}, \cdots, x_n$ equal to 1 and the remaining parameters equal to zero, giving the solutions

$$\xi_{r+1} = (-c_{1,r+1}, \cdots, -c_{r,r+1}, 1, 0, \cdots, 0)$$
$$\vdots \qquad \vdots \qquad \vdots$$
$$\xi_n = (-c_{1n}, \quad \cdots, -c_{rn}, \quad 0, 0, \cdots, 1).$$

These $n - r$ solution vectors are linearly independent (since they are independent even if one neglects entirely the first r coordinates!). Equation (26) states that the general solution ξ is just the linear combination $\xi = x_{r+1}\xi_{r+1} + \cdots + x_n\xi_n$ of these $n - r$ basic solutions. We have thus found a basis for the space S of solution vectors of the given system of equations (24), thereby proving

Theorem 13. *The "solution space" of all solutions* $(x_1, \cdots, x_n)$ *of a system of* r *linearly independent homogeneous linear equations in* n *unknowns has dimension* $n - r$.

Corollary. *The only solution of a system of* n *linearly independent homogeneous linear equations in* n *unknowns* $x_1, \cdots, x_n$ *is*

$$x_1 = x_2 = \cdots = x_n = 0.$$

EXAMPLE. Let S be defined by the equations $x_1 + x_2 = x_3 + x_4$ and $x_1 + x_3 = 2(x_2 + x_4)$. Thus, geometrically, S is the intersection of two three-dimensional "hyperplanes" in four-space. The matrix of these equations reduces as follows:

$$\begin{pmatrix} 1 & 1 & -1 & -1 \\ 1 & -2 & 1 & -2 \end{pmatrix} \rightarrow \begin{pmatrix} 1 & 1 & -1 & -1 \\ 0 & -3 & 2 & -1 \end{pmatrix} \rightarrow \begin{pmatrix} 1 & 4 & -3 & 0 \\ 0 & -3 & 2 & -1 \end{pmatrix}.$$

The final matrix (except for sign and column order) is in reduced echelon form. It yields the equivalent system of equations $x_1 + 4x_2 - 3x_3 = 0$, $-3x_2 + 2x_3 - x_4 = 0$, with the general solution $\xi = (3x_3 - 4x_2, x_2, x_3, -3x_2 + 2x_3)$; a basis for the space of solutions is provided by the cases $x_2 = 0$, $x_3 = 1$, and $x_2 = 1$, $x_3 = 0$, or $(3, 0, 1, 2)$ and $(-4, 1, 0, -3)$.

By duality, one can obtain a basis for the linear equations satisfied by all the vectors of any subspace. Thus, let T be the subspace of F^4 spanned by the vectors $(1, 1, -1, -1)$ and $(1, -2, 1, -2)$. Then the homogeneous linear equation $\Sigma a_i x_i = 0$ is satisfied identically for (x_1, x_2, x_3, x_4) in T if and only if $a_1 + a_2 = a_3 + a_4$ and $a_1 + a_3 = 2(a_2 + a_4)$. A basis for the set of coefficient vectors (a_1, a_2, a_3, a_4) satisfying these equations has been found above, with x's in place of a's.

The linear equations of our example, $x_1 + x_2 - x_3 - x_4 = 0$ and $x_1 - 2x_2 + x_3 - 2x_4 = 0$, are equivalent to the vector equation

$$x_1(1, 1) + x_2(1, -2) + x_3(-1, 1) + x_4(-1, -2) = (0, 0).$$

The solutions are thus all relations of linear dependence between the four vectors $(1, 1)$, $(1, -2)$, $(-1, 1)$, and $(-1, -2)$ of the two-dimensional space F^2. This could also be solved as in §7.5 by reducing to echelon form the 4×2 matrix having these vectors as rows, this matrix being obtained from that displayed above by transposing rows and columns.

Exercises

1. Let $\xi_1 = (1, 1, 1)$, $\xi_2 = (2, 1, 2)$, $\xi_3 = (3, 4, -1)$, $\xi_4 = (4, 6, 7)$. Find numbers c_i not all zero such that $c_1\xi_1 + c_2\xi_2 + c_3\xi_3 + c_4\xi_4 = \mathbf{0}$.
2. Let $\eta_1 = (1 + i, 2i)$, $\eta_2 = (2, -3i)$, $\eta_3 = (2i, 3 + 4i)$. Find *all* complex numbers c_i such that $c_1\eta_1 + c_2\eta_2 + c_3\eta_3 = \mathbf{0}$.
3. Find two vectors which span the subspace of all vectors (x_1, x_2, x_3, x_4) satisfying $x_1 + x_2 = x_3 - x_4 = 0$.
4. Do Ex. 3 for the vectors satisfying

$$3x_1 - 2x_2 + 4x_3 + x_4 = x_1 + x_2 - 3x_3 - 2x_4 = 0.$$

5. Find a basis of the proper number of linearly independent solutions for each of the following four systems of equations:

   (a) $x + y + 3z = 0,$ (b) $x + y + z = 0,$
         $2x + 2y + 6z = 0;$ $y + z + t = 0;$
   (c) $x + 2y - 4z = 0,$ (d) $x + y + z + t = 0,$
         $3x + y - 2z = 0;$ $2x + 3y - z + t = 0,$
                                        $3x + 4y + 2t = 0.$

6. Do Ex. 5 if the equations are taken to be congruences modulo 5.
7. Determine whether each of the following vector equations (over the rational numbers) has a solution, and when this is the case, find one solution.

   (a) $(1, -2) = x_1(1, 1) + x_2(2, 3)$,
   (b) $(1, 1, 1) = x_1(1, -1, 2) + x_2(2, 1, 3) + x_3(1, -1, 0)$,
   (c) $(2, -1, 1) = x_1(2, 0, 3) + x_3(3, 1, 2) + x_3(1, 2, -1)$.

8. In $\mathbf{Q}^4$ let $\alpha_1 = (1, 1, 2, 2)$, $\alpha_2 = (1, 2, 3, 4)$, $\alpha_3 = (0, 1, 3, 2)$, and $\alpha_4 = (-1, 1, -1, 1)$. Express each of the following four vectors in the form $x_1\alpha_1 + x_2\alpha_2 + x_3\alpha_3 + x_4\alpha_4$:
   (a) $(1, 0, 1, 0)$,   (b) $(3, -2, 1, -1)$,   (c) $(0, 1, 0, 0)$,   (d) $(2, -2, 2, -2)$.
9. Show that an $m \times n$ matrix can be put in row-reduced form by at most m^2 elementary operations on its rows.
10. Show that a 4×6 matrix can be put in row-reduced form after at most 56 multiplications, 42 additions and subtractions, and 4 formations of reciprocals. (Do not count computations like $aa^{-1} = 1$, $a - a = 0$, or $0a = 0$.)
★11. State and prove an analogue of Ex. 10 for $n \times n$ matrices.

7.8. Bases and Coordinate Systems

A basis of a space V was defined to be an independent set of vectors spanning V. The real significance of a basis lies in the fact that the vectors of any basis of F^n may be regarded as the unit vectors of the space, under a suitably chosen coordinate system. The proof depends on the following theorem.

Theorem 14. *If $\alpha_1, \cdots, \alpha_n$ form a basis for V, then every vector $\xi \in V$ has a unique expression*

(27) $$\xi = x_1\alpha_1 + x_2\alpha_2 + \cdots + x_n\alpha_n$$

as a linear combination of the α_i.

Proof. Since the α_i form a basis, they span V, hence every vector ξ in V has at least one expression of the form (27). If some $\xi \in V$ has a second such expression $\xi = x_1'\alpha_1 + \cdots + x_n'\alpha_n$, then subtraction from (27) and recombination gives

$$0 = \xi - \xi = (x_1 - x_1')\alpha_1 + \cdots + (x_n - x_n')\alpha_n.$$

Since the α_i are a basis, they are independent, and the preceding equation implies that $(x_1 - x_1') = \cdots = (x_n - x_n') = 0$, so that each $x_i = x_i'$, whence the expression (27) is unique.

We shall call the scalars x_i in (27) the *coordinates* of the vector ξ *relative to the basis* $\alpha_1, \cdots, \alpha_n$. If

$$\eta = y_1\alpha_1 + y_2\alpha_2 + \cdots + y_n\alpha_n$$

is a second vector of V, with coordinates $y_1, \cdots, y_n$, then by the identities of vector algebra

$$(28) \qquad \xi + \eta = (x_1 + y_1)\alpha_1 + (x_2 + y_2)\alpha_2 + \cdots + (x_n + y_n)\alpha_n.$$

In words, the coordinates of a vector sum relative to any basis are found by adding corresponding coordinates of the summands. Similarly, the product of the vector ξ of (27) by a scalar c is

$$(29) \qquad c\xi = c(x_1\alpha_1 + \cdots + x_n\alpha_n) = (cx_1)\alpha_1 + \cdots + (cx_n)\alpha_n,$$

so that each coordinate of $c\xi$ is the product of c and the corresponding coordinate of ξ.

By analogy with the corresponding definitions for integral domains and groups, let us now define an *isomorphism* $C: V \to W$ between two vector spaces V and W, over the same field F, to be a one-one correspondence $\xi \to \xi C$ of V onto W such that

$$(30) \qquad (\xi + \eta)C = \xi C + \eta C \qquad \text{and} \qquad (c\xi)C = c(\xi C)$$

for all vectors ξ, η in V and all scalars c in F. Equations (28) and (29) then show that each basis $\alpha_1, \cdots, \alpha_n$ in a vector space V over F provides an isomorphism of V onto F^n. This isomorphism is the correspondence C_α which assigns to each vector ξ of V the n-tuple of its coordinates relative to α, as in

$$(31) \qquad (x_1\alpha_1 + \cdots + x_n\alpha_n)C_\alpha = (x_1, \cdots, x_n) \in F^n.$$

Since the number n of vectors in a basis is determined by the dimension n, which is an invariant (Theorem 5, Corollary 1), we have proved

Theorem 15. *Any finite-dimensional vector space over a field F is isomorphic to one and only one space F^n.*

We have thus solved the problem of determining (up to isomorphism) all finite-dimensional vector spaces. What is more, we have shown that all bases of the same vector space V are equivalent, in the sense that there is an automorphism of V carrying any basis into any other basis.

A vector space can have many different bases. Thus by Theorem 7, any sequence of vectors of F^n obtained from $\varepsilon_1, \cdots, \varepsilon_n$ by a succession of elementary row operations is a basis for F^n. In particular, $\alpha_1 = (1, 1, 0)$, $\alpha_2 = (0, 1, 1)$, and $\alpha_3 = (1, 0, 1)$ are a basis for F^3 for any field F in which $1 + 1 \neq 0$. Likewise, any three noncoplanar vectors in ordinary three-space define a basis of vectors for "oblique coordinates."

Again, the field **C** of all complex numbers may be considered as a vector space over the field **R** of real numbers, if one ignores all the algebraic operations in **C** except the addition of complex numbers and the ("scalar") multiplication of complex numbers by reals. This space has the dimension 2, for 1 and i form a basis, generating respectively the "subspaces" of real and pure imaginary numbers. The two numbers $1 + i$ and $1 - i$ form another, but less convenient, basis for **C** over **R**.

Or consider the homogeneous linear differential equation $d^2x/dt^2 - 3dx/dt + 2x = 0$. One verifies readily that the sum $x_1(t) + x_2(t)$ of two solutions is itself a solution, and that the product of a solution by any (real) constant is a solution. Therefore the set V of all solutions of this differential equation is a vector space, sometimes called the "solution space" of the differential equation. The easiest way to describe this space is to say that e^t and e^{2t} form a *basis* of solutions, which means precisely that the most general solution can be expressed in the form $x = c_1e^t + c_2e^{2t}$, in one and only one way.

Finally, the domain $F[x]$ of all polynomial forms in an indeterminate x over a field F is a vector space over F, for all the postulates for a vector space are satisfied in $F[x]$. The definition of equality applied to the equation $p(x) = 0$ implies that the powers $1, x, x^2, x^3, \cdots$ are linearly independent over F. Hence $F[x]$ has an infinite *basis* consisting of these powers, for any vector (polynomial form) can be expressed as a linear combination of a finite subset of this basis.

In $\mathbf{R}^3$, a plane S and a line T not in S, both through the origin, span the whole space, and any vector in the space can be expressed uniquely as a sum of a vector in the plane and a vector in the line. More generally, we say that a vector space V is the *direct sum* of two subspaces S and T if every vector ξ of V has one and only one expression

$$(32) \qquad \xi = \sigma + \tau, \ = \ \sigma \in S, \quad \tau \in T$$

as a sum of a vector of S and a vector of T.

Since $(\sigma + \tau) + (\sigma' + \tau') = (\sigma + \sigma') + (\tau + \tau')$, the correspondence $(\sigma, \tau) \to (\sigma + \tau)$ is an isomorphism from the additive group of the vector space V onto the *direct product* (§6.11) of the additive groups of S and T. More generally, F^n is the direct product (as an additive group) of n copies of the additive group of F; in symbols, $F^n = F \times \cdots \times F$ (n factors).

Conversely, if S and T are any two given vector spaces over the same field F, one can define a new vector space $V = S \oplus T$ whose additive group is the direct product of those of S and T, scalar multiplication being defined by the formula $c(\eta, \zeta) = (c\eta, c\zeta)$ for any $c \in F$. In this V, the subsets of $(\eta, 0)$ and $(0, \zeta)$ constitute *subspaces* isomorphic to S and T, respectively; moreover, V is their direct sum in the sense defined above.

One also speaks of $S \oplus T$ as the direct sum of the given vector spaces S and T.

Theorem 16. *If the finite-dimensional vector space V is the direct sum of its subspaces S and T, then the union of any basis of S with any basis of T is a basis of V.*

Proof. Let S and T have the bases $\beta_1, \cdots, \beta_k$ and $\gamma_1, \cdots, \gamma_m$, respectively; we wish to prove that $\beta_1, \cdots, \beta_k, \gamma_1, \cdots, \gamma_m$ is a basis of V. First, these vectors span V, for any ξ in V can be written as $\xi = \eta + \zeta$, where η is a linear combination of the β's and ζ of the γ's. Secondly, these vectors are linearly independent, for if

$$(33) \qquad \mathbf{0} = b_1 \beta_1 + \cdots + b_k \beta_k + c_1 \gamma_1 + \cdots + c_m \gamma_m,$$

then $\mathbf{0}$ is represented as a sum of the vector $\eta_0 = \sum b_i \beta_i$ in S and $\zeta_0 = \sum c_j \gamma_j$ in T. But $\mathbf{0} = \mathbf{0} + \mathbf{0}$ is another representation of $\mathbf{0}$ as a sum of a vector in S and one in T. By assumption, the representation is unique, so that $\mathbf{0} = \eta_0 = \sum b_i \beta_i$ and $\mathbf{0} = \sum c_j \gamma_j$. But the β's and the γ's are separately linearly independent, so that $b_1 = \cdots = b_k = 0$ and $c_1 = \cdots = c_m = 0$. The relation (33) thus holds only when all the scalar coefficients are zero, so that the $\beta_1, \cdots, \beta_k, \gamma_1, \cdots, \gamma_m$ are indeed linearly independent.

This theorem and its proof can readily be extended to the case of a direct sum of a finite number of subspaces.

Corollary. *If the finite-dimensional space V is the direct sum of its subspaces S and T, then*

$$(34) \qquad d[V] = d[S] + d[T].$$

Proof. Since the dimension of a space is the number of vectors in (any) basis, the above proof shows that when $d[S] = k$ and $d[T] = m$, then $d[V] = k + m$. Q.E.D.

When V is the direct sum of S and T, we call S and T *complementary* subspaces of V. We then have

$$(35) \qquad S + T = V, \qquad S \cap T = \mathbf{0}.$$

Indeed, (32) states that V is the linear sum of the subspaces S and T. Secondly, if ξ_1 is any vector common to S and T, then ξ_1 has two representations $\xi_1 = \xi_1 + \mathbf{0}$ and $\xi_1 = \mathbf{0} + \xi_1$ of the form (32); since these two representations must be the same, $\xi_1 = \mathbf{0}$, so that the intersection

$S \cap T$ is zero, as asserted. Conversely, we can prove that under the conditions (35), V is the direct sum of S and T. Thus in this case equation (34) reduces to $d[V] = d[S + T] + d[S \cap T] = d[S] + d[T]$. This latter result holds for any two subspaces.

Theorem 17. *Let S and T be any two finite-dimensional subspaces of a vector space V. Then*

$$(36) \qquad\qquad d[S] + d[T] = d[S \cap T] + d[S + T].$$

Proof. Let $\xi_1, \cdots, \xi_n$ be a basis for $S \cap T$; by Theorem 6, S and T have bases $\xi_1, \cdots, \xi_n, \eta_1, \cdots, \eta_r$ and $\xi_1, \cdots, \xi_n, \zeta_1, \cdots, \zeta_s$ respectively. Clearly, the ξ_i, η_j, and ζ_k together span $S + T$. They are even a basis, since

$$a_1\xi_1 + \cdots + a_n\xi_n + b_1\eta_1 + \cdots + b_r\eta_r + c_1\zeta_1 + \cdots + c_s\zeta_s = 0$$

implies that $\sum b_j\eta_j = -\sum a_i\xi_i - \sum c_k\zeta_k$ is in T, whence $\sum b_j\eta_j$ is in $S \cap T$ and so $\sum b_j\eta_j = \sum d_i\xi_i$ for some scalars d_i. Hence (the ξ_i and η_j being independent) every b_j is 0. Similarly, every $c_k = 0$; substituting, $\sum a_i\xi_i = 0$, and every $a_i = 0$. This shows that the ξ_i, η_j and ζ_k are a basis for $S + T$.

But having proved this, we see that the conclusion of the theorem reduces to the arithmetic rule $(n + r) + (n + s) = n + (n + r + s)$.

Exercises

1. In Ex. 4 of §7.6, which of the indicated sets of vectors are bases for the spaces involved there?
2. In $\mathbf{Q}^4$, find the coordinates of the unit vectors $\varepsilon_1, \varepsilon_2, \varepsilon_3, \varepsilon_4$ relative to the basis

$$\alpha_1 = (1, 1, 0, 0), \quad \alpha_2 = (0, 0, 1, 1), \quad \alpha_3 = (1, 0, 0, 4), \quad \alpha_4 = (0, 0, 0, 2).$$

3. Find the coordinates of $(1, 0, 1)$ relative to the following basis in $\mathbf{C}^3$:

$$(2i, 1, 0), \quad (2, -i, 1), \quad (0, 1 + i, 1 - i).$$

4. In $\mathbf{Q}^4$, find
   (a) a basis which contains the vector $(1, 2, 1, 1)$;
   (b) a basis containing the vectors $(1, 1, 0, 2)$ and $(1, -1, 2, 0)$;
   (c) a basis containing the vectors $(1, 1, 0, 0)$, $(0, 0, 2, 2)$, $(0, 2, 3, 0)$.
5. Show that the numbers $a + b\sqrt{2} + c\sqrt{3} + d\sqrt{6} + e\sqrt{12}$ with rational $a, \cdots, e$ form a commutative ring, and that this ring is a vector space over the rational field $\mathbf{Q}$. Find a basis for this space.

6. In $\mathbf{Q}^4$, two subspaces S and T are spanned respectively by the vectors:

$$S\text{: } (1, -1, 2, -3), \quad (1, 1, 2, 0), \quad (3, -1, 6, -6),$$
$$T\text{: } (0, -2, 0, -3), \quad (1, 0, 1, 0).$$

Find the dimensions of S, of T, or $S \cap T$, and of $S + T$.

★7. Solve Ex. 6 for the most general $\mathbf{Z}_p^{\,4}$.

8. Find the greatest possible dimension of $S + T$ and the least possible dimension of $S \cap T$, where S and T are variable subspaces of fixed dimensions s and t in F^n. Prove your result.

★9. Prove that, for subspaces, $S \cap T = S \cap T'$, $S + T = S + T'$, and $T \subset T'$ imply $T = T'$.

10. If S is a subspace of the finite-dimensional vector space V, show that there exists a subspace T of V such that V is the direct sum of S and T.

★11. V is called the direct sum of its subspaces $S_1, \cdots, S_p$ if each vector ξ of V has a unique expression $\xi = \eta_1 + \cdots + \eta_p$, with $\eta_i \in S_i$. State and prove the analogue of Theorem 16 for such direct sums.

12. Prove that V is the direct sum of S and T if and only if (35) holds.

★13. State and prove the analogue of Ex. 12 for the direct sum of p subspaces.

14. By an "automorphism" of a vector space V is meant an isomorphism of V with itself.

    (a) Show that the correspondence $(x_1, x_2, x_3) \mapsto (x_2, -x_1, x_3)$ is an automorphism of F^3.

    (b) Show that the set of all automorphisms of V is a group of transformations on V.

15. An automorphism of F^2 carries $(1, 0)$ into $(0, 1)$ and $(0, 1)$ into $(-1, -1)$. What is its order? Does your answer depend on the base field?

★16. Establish a one-one correspondence between the automorphisms (cf. Ex. 14) of a finite-dimensional vector space and its ordered bases. How many automorphisms does $\mathbf{Z}_2$ have? What about $\mathbf{Z}_p^{\,n}$?

7.9. Inner Products

Ordinary space is a three-dimensional vector space over the real field; it is $\mathbf{R}^3$. In it one can define *lengths* of vectors and *angles* between vectors (including right angles) by formulas which generalize very nicely not only to $\mathbf{R}^n$, but even to *infinite*-dimensional real vector spaces (see Example 2 of §7.10). This generalization will be the theme of §§7.9–7.11.

To set up the relevant formulas, one needs an additional operation. The most convenient such operation for this purpose is that of forming *inner products*. By the "inner product" of two vectors $\xi = (x_1, \cdots, x_n)$ and $\eta = (y_1, \cdots, y_n)$, with real components, is meant the quantity

$$(37) \qquad\qquad (\xi, \eta) = x_1 y_1 + x_2 y_2 + \cdots + x_n y_n.$$

(Since this is a scalar, physicists often speak of our inner product as a "scalar product" of two vectors.) Inner products have four important properties, which are immediate consequences of the definition (37):

(38) $(\xi + \eta, \zeta) = (\xi, \zeta) + (\eta, \zeta), \qquad (c\xi, \eta) = c(\xi, \eta);$

(39) $(\xi, \eta) = (\eta, \xi), \qquad (\xi, \xi) > 0 \quad \text{unless } \xi = 0.$

The first two laws assert that inner products are *linear* in the left-hand factor; the third is the *symmetric* law, and gives with the first two the linearity of inner products in both factors (*bilinearity*); the fourth is that of *positiveness*.

Thus, the Cartesian formula for the *length* (also called the "absolute value" or the "norm") $|\xi|$ of a vector ξ in the plane $\mathbf{R}^2$ gives the length as the square root of an inner product,

(40) $$|\xi| = (x_1{}^2 + x_2{}^2)^{1/2} = (\xi, \xi)^{1/2}.$$

A similar formula is used for length in three-dimensional space. Again, if α and β are any two vectors, then for the triangle with sides $\alpha, \beta, \gamma = \beta - \alpha$ (Figure 2), the trigonometric law of cosines gives

$$|\beta - \alpha|^2 = |\alpha|^2 + |\beta|^2 - 2|\alpha| \cdot |\beta| \cdot \cos C,$$

$(C = \angle(\alpha, \beta))$. But by (38) and (40),

$$|\beta - \alpha|^2 = (\beta - \alpha, \beta - \alpha) = (\beta, \beta) - 2(\alpha, \beta) + (\alpha, \alpha).$$

Combining and canceling, we get

(41) $$\cos \angle(\alpha, \beta) = (\alpha, \beta)/|\alpha| \cdot |\beta|.$$

In words, the cosine of the angle $\angle(\alpha, \beta)$ between two vectors α and β is the quotient of their inner product by the product of their lengths. It follows that α and β are geometrically *orthogonal* (or "perpendicular") if and only if the inner product (α, β) vanishes.

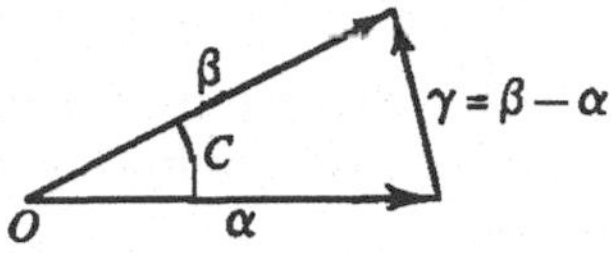

In view of the ease with which the concepts of vector addition and scalar multiplication extend to spaces of an arbitrary dimension over an arbitrary field, it is natural to try to generalize the concepts of length and angle

Figure 2

similarly. When we do this, however, we find that although the *dimension* number can be arbitrary, trouble arises with most fields. Inner products can be defined by (37); but lengths

(42) $$|\xi| = (\xi, \xi)^{1/2} = (x_1{}^2 + x_2{}^2 + \cdots + x_n{}^2)^{1/2}$$

are not definable unless every sum of n squares has a square root. The same applies to distances, while angles cause even more trouble.

For these reasons we shall at present confine our discussion of lengths, angles, and related topics to vector spaces over the *real* field. In §9.12 the corresponding notions for the complex field will be treated.

Exercises

1. In the plane, show by analytic geometry that the square of the distance between $\xi = (x_1, x_2)$ and $\eta = (y_1, y_2)$ is given by $|\xi|^2 + |\eta|^2 - 2(\xi, \eta)$.
2. Use direction cosines in three-dimensional space to show that two vectors ξ and η are orthogonal if and only if $(\xi, \eta) = 0$.
3. If length is defined by the formula (42) for vectors ξ with complex numbers as components, show that there will exist nonzero vectors with zero length.
4. Show that there is a sum of two squares which has no square root in the fields $\mathbf{Z}_3$ and $\mathbf{Q}$.
5. Prove formulas (38) and (39) from the definition (37).
6. Prove formulas analogous to (38) asserting that the inner product is linear in the right-hand factor.
7. Prove that the sum of the squares of the lengths of the diagonals of any parallelogram is the sum of the squares of the lengths of its four sides.
★8. In $\mathbf{R}^3$, define outer products by

$$\xi \times \eta = (x_2 y_3 - x_3 y_2, x_3 y_1 - x_1 y_3, x_1 y_2 - x_2 y_1).$$

   (a) Prove that $(\xi \times \eta, \zeta \times \tau) = (\xi, \zeta)(\eta, \tau) - (\xi, \tau)(\eta, \zeta)$.
   (b) Setting $\xi = \zeta, \eta = \tau$, infer the Schwarz inequality in $\mathbf{R}^3$ as a corollary. (Cf. Theorem 18.)
   (c) Prove that $\xi \times (\eta \times \zeta) = (\xi, \zeta)\eta - (\xi, \eta)\zeta$.

7.10. Euclidean Vector Spaces

Our discussion of geometry without restriction on dimension will be based on the following definition, suggested by the considerations of §7.9.

Definition. *A Euclidean vector space is a vector space E with real scalars, such that to any vectors ξ and η in E corresponds a (real) "inner* product" (ξ, η) which is symmetric, bilinear, and positive in the sense of (38) and (39).

EXAMPLE 1. Any $\mathbf{R}^n$ is an n-dimensional Euclidean vector space if (ξ, η) is defined by equation (37).

EXAMPLE 2. The continuous real functions $\phi(x)$ on the domain $0 \leqq x \leqq 1$ form an *infinite*-dimensional Euclidean vector space, if we make the definition $(\phi, \psi) = \int_0^1 \phi(x)\psi(x)dx$.

The "length" $|\xi|$ of a vector ξ of a Euclidean vector space E may be defined in terms of the inner product as the positive square root $(\xi, \xi)^{1/2}$—the existence of the root being guaranteed by the positiveness condition of (39).

Theorem 18. *In any Euclidean vector space, length has the following properties:*

$$
\begin{array}{lll}
\text{(i)} & |c\xi| = |c| \cdot |\xi|. & \\
\text{(ii)} & |\xi| > 0 \quad \text{unless } \xi = \mathbf{0}. & \\
\text{(iii)} & |(\xi, \eta)| \leqq |\xi| \cdot |\eta| & \textit{(Schwarz inequality)}. \\
\text{(iv)} & |\xi + \eta| \leqq |\xi| + |\eta| & \textit{(triangle inequality)}.
\end{array}
$$

Proof. Since $(c\xi, c\xi) = c^2(\xi, \xi)$, we have (i). Property (ii) is a corollary of the condition of positiveness required in the definition of a Euclidean vector space.

The proof of (iii) is less immediate. If $\xi = \mathbf{0}$ or $\eta = \mathbf{0}$, then (iii) reduces to the trivial inequality $0 \leqq 0$. Otherwise,

$$0 \leqq (a\xi \pm b\eta, a\xi \pm b\eta) = a^2(\xi, \xi) \pm 2ab(\xi, \eta) + b^2(\eta, \eta).$$

Set $a = |\eta|$ and $b = |\xi|$, so that $a^2 = (\eta, \eta)$ and $b^2 = (\xi, \xi)$. Transposing, we then have

$$(43) \qquad \mp 2|\xi| \cdot |\eta| \cdot (\xi, \eta) \leqq 2(\xi, \xi)(\eta, \eta) = 2|\xi|^2 \cdot |\eta|^2.$$

Dividing through by $2|\xi| \cdot |\eta| > 0$, we get (iii).

From (iii) we now get (iv) easily, for

$$|\xi + \eta|^2 = (\xi + \eta, \xi + \eta) = (\xi, \xi) + 2(\xi, \eta) + (\eta, \eta)$$
$$\leqq |\xi|^2 + 2|\xi| \cdot |\eta| + |\eta|^2 = (|\xi| + |\eta|)^2.$$

Now, if we define the *distance* between any two vectors ξ and η of E

as $|\xi - \eta|$, we can show that it has the so-called "metric" properties of ordinary distance, first considered abstractly by Fréchet (1906).

Theorem 19. *Distance has the properties:*

$(M1)$ $|\xi - \xi| = 0$, *while* $|\xi - \eta| > 0$ *if* $\xi \neq \eta$.
$(M2)$ *Distance is symmetric,* $|\xi - \eta| = |\eta - \xi|$.
$(M3)$ $|\xi - \eta| + |\eta - \zeta| \geqq |\xi - \zeta|$.

Proof. First, $|\xi - \xi| = |\mathbf{0}| = |0 \cdot \xi| = 0 \cdot |\xi| = 0$ by (i), while $|\xi - \eta| > 0$ if $\xi - \eta \neq \mathbf{0}$ (or $\xi \neq \eta$) by (ii), proving $M1$. Secondly, $|\xi - \eta| = |(-1)(\eta - \xi)| = |-1| \cdot |\eta - \xi| = |\eta - \xi|$ by (i), proving $M2$. Finally, $M3$ follows from (iv) because

$$|\xi - \eta| + |\eta - \zeta| \geqq |(\xi - \eta) + (\eta - \zeta)| = |\xi - \zeta|.$$

From Schwarz's inequality, we deduce in particular that for any ξ, η not $\mathbf{0}$, we have $-1 \leqq (\xi, \eta)/|\xi| \cdot |\eta| \leqq 1$. Hence $(\xi, \eta)/\|\xi\| \cdot |\eta|$ is the cosine of one and only one angle between $0°$ and $180°$, which we can *define* as the *angle* between the vectors ξ and η (compare the special case (41)). We shall not prove except in the case of *right* angles that the angles so defined have any properties (could you prove that $\angle(\xi, \eta) + \angle(\eta, \zeta) \geqq \angle(\xi, \zeta)$?).

Two vectors ξ and η will be called *orthogonal* (in symbols, $\xi \perp \eta$) whenever $(\xi, \eta) = 0$. This definition, applied to Example 2 above, yields an instance of the important analytical concept of orthogonal functions. It is easy to prove that if $\xi \perp \eta$, then $\eta \perp \xi$ (the orthogonality relation is "symmetric"), and $c\xi \perp c'\eta$ for all c, c'. Also, $\mathbf{0}$ is the only vector orthogonal to itself. Furthermore, whenever $(\eta, \xi_1) = \cdots = (\eta, \xi_m) = 0$, then for any scalars c_i,

$$(\eta, c_1\xi_1 + \cdots + c_m\xi_m) = c_1(\eta, \xi_1) + \cdots + c_m(\eta, \xi_m)$$
$$= c_1 \cdot 0 + \cdots + c_m \cdot 0 = 0,$$

so that η is also orthogonal to every linear combination of the ξ_i. This proves

Theorem 20. *If a vector is orthogonal to* $\xi_1, \cdots, \xi_m$, *then it is orthogonal to every vector in the subspace spanned by* $\xi_1, \cdots, \xi_m$.

Exercises

1. Set $\xi = (1, 2, 3, 4)$, $\eta = (0, 3, -2, 1)$. Compute (ξ, η), $|\xi|$, $|\eta|$, $\angle(\xi, \eta)$.
2. If ξ and η are as in Ex. 1, find a vector of the form $(1, 1, 0, 0) + c_1\xi + c_2\eta$ orthogonal to both ξ and η.

3. (a) Are $\sin 2\pi x$ and $\cos 2\pi x$ "orthogonal" in Example 2 of the text?
   (b) Are $\sin 2m\pi x$ and $\sin 2n\pi x$ orthogonal?
   (c) Find a polynomial of degree two orthogonal to 1 and x.
4. Prove that $|\xi - \eta|^2 + |\xi + \eta|^2 = 2(|\xi|^2 + |\eta|^2)$.
5. Prove that in $\mathbf{R}^3$, there are precisely two vectors of length one perpendicular to two given linearly independent vectors.
6. Prove that there is a vector with rational coordinates in $\mathbf{R}^3$ perpendicular to any two given vectors with rational coordinates.
7. If α and $\beta \neq 0$ are fixed vectors of a Euclidean vector space, find the shortest vector of the form $\gamma = \alpha + t\beta$. Is this orthogonal to β? Draw a figure.
★8. If α is equidistant from β and γ, prove that the midpoint of the segment $\overline{\beta\gamma}$ is the foot of the perpendicular from α to $\overline{\beta\gamma}$.
9. Prove that if $|\xi| = |\alpha|$ in a Euclidean vector space, then $\xi - \alpha \perp \xi + \alpha$. Interpret this geometrically.
★10. (a) Show that the discriminant $B^2 - 4AC$ of the quadratic equation

$$(\xi, \xi)t^2 + 2(\xi, \eta)t + (\eta, \eta) = |t\xi + \eta|^2 = 0$$

is four times $(\xi, \eta)^2 - (\xi, \xi)(\eta, \eta)$.
   (b) Using this fact, prove the Schwarz inequality. (*Hint:* $|t\xi + \eta| = 0$ cannot have two distinct real solutions t unless $\xi = \mathbf{0}$.)
11. Prove $||\xi| - |\eta|| \leq |\xi - \eta|$, in any Euclidean vector space.
12. Show that $\mathbf{R}^3$ becomes a Euclidean vector space if inner products are defined by

$$(\xi, \eta) = (x_1 + x_2)(y_1 + y_2) + x_2 y_2 + (x_2 + 2x_3)(y_2 + 2y_3).$$

7.11. Normal Orthogonal Bases

In Example 1 of §7.10, the "unit vectors" $\varepsilon_1 = (1, 0, \cdots, 0), \cdots,$ $\varepsilon_n = (0, 0, \cdots, 1)$ have unit length and are mutually orthogonal. This is an instance of what is called a "normal orthogonal basis."

Definition. *Vectors $\alpha_1, \cdots, \alpha_n$ are called* normal orthogonal *when* (i) $|\alpha_i| = 1$ *for all* i, (ii) $\alpha_i \perp \alpha_j$ *if* $i \neq j$.

Lemma 1. *Nonzero orthogonal vectors $\alpha_1, \cdots, \alpha_m$ of a Euclidean vector space E are linearly independent.*

Proof. If $x_1\alpha_1 + \cdots + x_m\alpha_m = \mathbf{0}$, then for $k = 1, \cdots, m$

$$0 = (\mathbf{0}, \alpha_k) = x_1(\alpha_1, \alpha_k) + \cdots + x_m(\alpha_m, \alpha_k) = x_k(\alpha_k, \alpha_k),$$

where the last equality comes from the orthogonality assumption. But $\alpha_k \neq \mathbf{0}$ by assumption; hence $(\alpha_k, \alpha_k) > 0$ and $x_k = 0$.   Q.E.D.

Corollary. *Normal orthogonal vectors spanning E are a basis for E (a so-called "normal orthogonal basis").*

We shall now show how to orthogonalize any basis of a Euclidean vector space, using only rational operations. This is called the *Gram–Schmidt orthogonalization process.*

Lemma 2. *From any finite sequence of independent vectors $\gamma_1, \cdots, \gamma_m$ of a finite-dimensional Euclidean vector space E, an orthogonal sequence of nonzero vectors*

$$(44) \qquad \alpha_i = \gamma_i - \sum_{k<i} d_{ik}\gamma_k \qquad (i = 1, \cdots, m)$$

can be constructed, which spans the same subspace of E as the sequence $\gamma_1, \cdots, \gamma_m$.

Proof. By induction on m, we can assume that orthogonal nonzero vectors $\alpha_1, \cdots, \alpha_{m-1}$ have been constructed which span the same subspace S as $\gamma_1, \cdots, \gamma_{m-1}$. We now split γ_m into a part β_m "parallel" to S, and a part α_m perpendicular to S. To do this, set

$$(44') \quad \alpha_m = \gamma_m - \sum_{k<m} c_{mk}\alpha_k, \qquad \text{where} \qquad c_{mk} = (\gamma_m, \alpha_k)/(\alpha_k, \alpha_k).$$

Then for $j = 1, \cdots, m - 1$, we have

$$(\alpha_m, \alpha_j) = (\gamma_m, \alpha_j) - \sum_{k=1}^{m-1} c_{mk}(\alpha_k, \alpha_j) = 0,$$

since $(\alpha_k, \alpha_j) = 0$ if $k \neq j$ by orthogonality, while $c_{mj}(\alpha_j, \alpha_j) = (\gamma_m, \alpha_j)$ by (44'). Substituting in (44), we have by induction on m,

$$\alpha_m = \gamma_m - \sum_{k<m} c_{mk}\alpha_k = \gamma_m - \sum_{k<m} c_{mk}\gamma_k + \sum_{j<k<m} c_{mk}d_{kj}\gamma_j.$$

This proves (44), with

$$d_{mk} = c_{mk} - \sum_{k<j<m} c_{mj}d_{jk}.$$

Since γ_m is not dependent on $\gamma_1, \cdots, \gamma_{m-1}$, it cannot be in S; hence $\alpha_m \neq 0$. Finally, $\gamma_1, \cdots, \gamma_m$ and $\alpha_1, \cdots, \alpha_m$ both span the subspace spanned by S and γ_m. This completes the proof of Lemma 2.

Theorem 21. *Every set* $\gamma_1, \cdots, \gamma_m$ *of normal orthogonal vectors of a finite-dimensional Euclidean vector space E is part of a normal orthogonal basis.*

Proof.   By Theorem 6, the γ_i are part of a basis $\gamma_1, \cdots, \gamma_n$ of E. This basis may be orthogonalized by Lemma 2, and then normalized by setting $\beta_i = \alpha_i / |\alpha_i|$; the process will not change the original vectors $\gamma_1, \cdots, \gamma_m$.

Corollary. *Any finite-dimensional Euclidean vector space E has a normal orthogonal basis.*

The Gram–Schmidt orthogonalization process has other implications. Thus let S be any m-dimensional subspace of a Euclidean vector space E; as above, S has a normal orthogonal basis $\alpha_1, \cdots, \alpha_m$. If γ is any vector not in S, the process represents γ as the sum $\gamma = \alpha + \beta$ of a component β in S and a component α perpendicular to every vector of S. The vector β is called the *orthogonal projection* of γ on S.

We shall conclude by determining all inner products on a given (real) finite-dimensional vector space V. Clearly, if $\alpha_1, \cdots, \alpha_n$ is any basis for V, then for any vectors $\xi = x_1\alpha_1 + \cdots + x_n\alpha_n$ and $\eta = y_1\alpha_1 + \cdots + y_n\alpha_n$, we have by bilinearity

$$(45) \qquad (\xi, \eta) = \left(\sum x_i\alpha_i, \sum y_k\alpha_k \right) = \sum_{i,k} x_i y_k (\alpha_i, \alpha_k).$$

Thus, the inner product of any two vectors is determined by the n^2 real constants $(\alpha_i, \alpha_k) = a_{ik}$ as a certain "bilinear" form $\sum_{i,k} a_{ik}x_i y_k$ in the coordinates x_i and y_k. Because $(\alpha_i, \alpha_k) = (\alpha_k, \alpha_i)$, this form is called "symmetric."

Conversely, any symmetric bilinear form $\sum_{i,k} a_{ik}x_i y_k$ $(a_{ik} = a_{ki})$ in F^n satisfies the first three conditions of (38) and (39). The fourth condition is that the *quadratic* form $\sum a_{ik}x_i x_k$ be "positive definite"—i.e., be positive unless every $x_i = 0$. An algorithm for determining when a square matrix is positive definite will be derived in §9.9.

Relative to a normal orthogonal basis, we have $(\alpha_i, \alpha_k) = 0$ if $i \neq k$, and $(\alpha_i, \alpha_i) = 1$; hence (45) reduces to

$$(46) \qquad (\xi, \eta) = \sum_{i=1}^{n} x_i y_i = x_1 y_1 + \cdots + x_n y_n.$$

This formula enables us to conclude with

Theorem 22. *Relative to any normal orthogonal basis, an "abstract" inner product assumes the "concrete" form* (46).

Thus every finite-dimensional Euclidean vector space is isomorphic to some $\mathbf{R}^n$.

Exercises

1. Find normal orthogonal bases for the subspaces of Euclidean four-space spanned by:
   (a) $(1, 1, 0, 0)$, $(0, 1, 2, 0)$, and $(0, 0, 3, 4)$;
   (b) $(2, 0, 0, 0)$, $(1, 3, 3, 0)$, and $(0, 4, 6, 1)$.
   (*Hint:* First find orthogonal bases, *then* normalize.)
2. Draw a figure to illustrate the orthogonal projection of a vector on a one-dimensional subspace.
3. Find the orthogonal projection of $\beta = (2, 1, 3)$ on the subspace spanned by $\alpha = (1, 0, 1)$.
4. Find the orthogonal projection of $\beta = (0, 0, 0, 3)$ on each of the subspaces of Ex. 1.
5. Let S be any subspace of a Euclidean vector space E. Show that the set $S^\perp$ of all vectors orthogonal to every ξ in S is a subspace satisfying

$$S \cap S^\perp = 0, \qquad S + S^\perp = E, \qquad \text{and} \qquad d[S] + d[S^\perp] = d[E].$$

   (The subspace $S^\perp$ is called the *orthogonal complement* of S.)
6. Find a basis for the orthogonal complement of the subspace spanned by $(2, -1, -2)$ in Euclidean three-space.
7. Find bases for the orthogonal complements of each of the subspaces of Ex. 1.
★8. (a) Exhibit a nontrivial subspace of $\mathbf{Q}^3$ which does not contain any vector of unit length.
   (b) State and prove an analogue of Lemma 2 which is valid for vector spaces with scalars in any ordered field.

7.12. Quotient-spaces

We shall now show that the construction of quotient-groups in §6.13 has an easy extension to vector spaces. Let V be any vector space over a field F, and let S be any subspace of V. Under addition, V is a commutative group, and S is a (necessarily normal) subgroup of V. Hence we can form the additive *quotient-group* V/S.

For example, in Euclidean space $\mathbf{R}^3$, let S consist of the multiples $(0, y, 0)$ of the unit vector $(0, 1, 0)$. Then the coset of any vector $\alpha = (a, b, c)$ will consist of the vectors $(a, b + y, c)$ having the same x-coordinate a and z-coordinate c as α; they are the vectors $(a, \cdot, c)$, where the dot stands for an arbitrary entry. The sum $(a, \cdot, c) + (a', \cdot, c')$ of two such vectors in the quotient-group $\mathbf{R}^3/S$ is clearly $(a + a', \cdot, c + c')$.

In this example, we can also multiply each vector $(a, \cdot, c)$ by any scalar $t \in \mathbf{R}$ to get the new coset $(ta, \cdot, tc)$, and it is evident that the quotient-group $\mathbf{R}^3/S$ is a (real) *vector space* under these operations. We shall now show that a similar construction is possible in general.

Given a vector space V over a field F, we can paraphrase the discussion of §6.13 to obtain a *quotient-space* $V/S = X$. Recall that for any group G and (normal) subgroup N, the elements of the quotient-group G/N are simply the cosets xN of N in G. Hence, given a subspace S of the vector space V, each vector $\alpha \in V$ determines a *coset* of S, defined as the set $\alpha + S$ of all sums $\alpha + \sigma$ for variable $\sigma \in S$. Thus $\alpha = \alpha + \mathbf{0}$ is one of the vectors in this coset; call it a "representative" of the coset. Two cosets $\alpha + S$ and $\beta + S$ are equal (as sets) if and only if $(\alpha - \beta) \in S$; when this holds, α and β represent (are members of) the *same* coset. Geometrically, the different cosets of a subspace S are just its "parallel subspaces" under translation.

Now define the sum of two cosets to be the coset

$$(\alpha + S) + (\beta + S) = (\alpha + \beta) + S;$$

as in Lemma 2 of §6.13, this sum does not depend on the choice of the representatives α and β. Next, define the product of a coset $\alpha + S$ by a scalar c to be the coset

$$c(\alpha + S) = c\alpha + S.$$

Since $(\alpha - \beta) \in S$ implies $(c\alpha - c\beta) \in S$, this product also does not depend on the choice of the representative of the given coset. It is readily verified that these two definitions make the set V/S of all the cosets of S in V into a vector space, called the *quotient-space* of V by S. Moreover, if the function P is defined by $\alpha P = \alpha + S$, then P is an epimorphism of vector spaces with kernel exactly S and range all of V/S. This transformation P is called the *canonical projection* of V onto its quotient-space; we have thus proved

Theorem 23. *Given any subspace S of a vector space V, there exists a quotient-space $X = V/S$ and an epimorphism $P: V \to X$ whose kernel is S and whose range is X.*

Exercises

1. If S is a one-dimensional subspace of the space $\mathbf{R}^3$, show that the cosets of S are the lines parallel to S.

2. For $V = F^3$, F any field, let S be the subspace spanned by $(1, 1, 0)$ and $(1, 1, 1)$.
    (a) Show that two vectors (x, y, z) and (x', y', z') are in the same coset of S if and only if $x + y' = x' + y$.
    (b) For $F = \mathbf{R}$, describe S and its cosets geometrically.
3. Prove that if S is a subspace of $V = F^n$ that is isomorphic to F^m, then V/S is isomorphic to F^{n-m}.
4. Prove in detail that, under the operations displayed in the text, the cosets of any subspace S of a vector space V do form a vector-space.
5. Let $V = \mathbf{R}[x]$ be the space of all real polynomials $f(x)$, and let $\phi \colon f(x) \mapsto \frac{1}{2}[f(x) + f(-x)]$.
    (a) Show that ϕ is a homomorphism of vector spaces.
    (b) Describe its kernel S and the quotient-space V/S.

★7.13. Linear Functions and Dual Spaces

In elementary algebra, a (homogeneous) "linear function" of the coordinates x_i of a variable vector $\xi = (x_1, \cdots, x_n)$ of the finite-dimensional vector space $V = F^n$ is a polynomial function of the special form

$$(47) \qquad f(\xi) = \xi f = c_1 x_1 + \cdots + c_n x_n = x_1 c_1 + \cdots + x_n c_n,$$

where the c_i terms are arbitrary constants in the field F. One easily verifies that any such function f satisfies the identities·

$$(48) \qquad\qquad (\xi + \eta)f = \xi f + \eta f, \qquad (a\xi)f = a(\xi f),$$

for any vectors ξ, η in V and any scalar a in F.

The preceding identities have two advantages over the definition by formula (47): they are *intrinsic* (i.e., they do not depend on the choice of a basis in V), and they apply to infinite-dimensional vector spaces (e.g., to function spaces). We shall therefore define a *linear function f* on any vector space V over any field F as a function from V to F which satisfies the two identities (48).

The first identity, with $\eta = 0$, shows at once that $0f = 0$. The two identities imply the combined identity

$$(49) \qquad (a\xi + b\eta)f = a(\xi f) + b(\eta f), \qquad \xi, \eta \in V; \qquad a, b \in F.$$

Conversely, this one identity yields the first identity of (48), for $a = b = 1$, hence $0f = 0$, and hence the second identity of (48), for $b = 0$. Briefly, a linear function f is one which preserves linear combinations.

The concept of "linear function" just defined is virtually equivalent to that of "coordinate" introduced in §7.8; namely, each x_i in Theorem 14 is a linear function of ξ, as ξ varies over V. The following result is "dual" to Theorem 14, in a sense which will be made precise shortly.

Theorem 24. *If $\beta_1, \cdots, \beta_n$ is a basis of the vector space V over F, and if $c_1, \cdots, c_n$ are n constants in F, then there is one and only one linear function f on V with $\beta_i f = c_i, i = 1, \cdots, n$. This function f is given by the formula*

$$(50) \qquad (x_1\beta_1 + \cdots + x_n\beta_n)f = x_1c_1 + \cdots + x_nc_n.$$

Proof. By induction on n, equation (50) follows directly from (49) for any linear function f with $\beta_i f = c_i, i = 1, \cdots, n$. Conversely, for any basis $\beta_1, \cdots, \beta_n$ of V, each ξ has by Theorem 14 a unique expression $\xi = x_1\beta_1 + \cdots + x_n\beta_n$. For any constants $c_1, \cdots, c_n$ in F, equation (50) therefore defines a single-valued function. This function is linear, because for any ξ and $\eta = y_1\beta_1 + \cdots + y_n\beta_n$,

$$(a\xi + b\eta)f = (\sum(ax_i + by_i)\beta_i)f = \sum(ax_i + by_i)c_i$$
$$= a\sum x_ic_i + b\sum y_ic_i = a(\xi f) + b(\eta f),$$

so that condition (49) is satisfied.

Corollary. *The linear functions on F^n are the functions given by the linear expressions* (47).

Indeed, (47) gives that function f which takes the value c_i at the unit vector ε_i of F^n. Each linear function is thus determined uniquely by the n-tuple $(c_1, \cdots, c_n)$ of coefficients in the formula (47); this suggests that the linear functions themselves form a vector space.

For any vector space V, define the sum $f + g$ of two linear functions f and g to be the function given by the equation

$$(51) \qquad \xi(f + g) = \xi f + \xi g \qquad \text{for all } \xi \in V,$$

and the product fc of the linear function f by a scalar c to be the function given by the equation

$$(52) \qquad \xi(fc) = (\xi f)c \qquad \text{for all } \xi \in V, \quad c \in F.$$

One verifies readily that $f + g$ and fc are again linear functions on V.

Theorem 25. *If V is a vector space over F, the set V^* of all linear functions on V is also a vector space over F, under the operations $f + g$ and fc defined by (51) and (52).*

This space V^* of linear functions on V is called the *dual* or *conjugate* vector space to V; it is fundamental in modern mathematics.

The proof requires only that we verify that the axioms for a vector space hold for the operations $f + g$ and fc. For example, to prove the distributive law $(f + g)c = fc + gc$, observe that for any $\xi \in V$,

$$(53) \quad \xi[(f + g)c] = [\xi(f + g)]c = [\xi f + \xi g]c$$

$$= (\xi f)c + (\xi g)c = \xi(fc) + \xi(gc) = \xi(fc + gc),$$

by the definitions (51) and (52) and the distributive law in V. This equation states that the functions $(f + g)c$ and $fc + gc$ have the same value for any argument ξ, hence are necessarily equal. The proof of the other axioms is similar.

Corollary 1. *If the vector space V has a finite basis $\beta_1, \cdots, \beta_n$, then its dual space V^* has a basis $f_1, \cdots, f_n$, consisting of the n linear functions f_i defined by $(x_1\beta_1 + \cdots + x_n\beta_n)f_i = x_i$, $i = 1, \cdots, n$. The n linear functions f_i are uniquely determined by the formulas*

$$(54) \qquad \beta_i f_j = \begin{cases} 0 & \text{if } i \neq j, \\ 1 & \text{if } i = j, \end{cases} \qquad i, j = 1, \cdots, n.$$

Proof. For n given scalars $c_1, \cdots, c_n$, the linear combination $f = f_1 c_1 + \cdots + f_n c_n$ is a linear function; by (54), its value at any basis vector β_i is

$$\beta_i\left(\sum_j f_j c_j\right) = \sum_j \beta_i f_j c_j = c_i.$$

It follows that the functions $f_1, \cdots, f_n$ are linearly independent in V^*, for if $f = f_1 c_1 + \cdots + f_n c_n = 0$, then $\beta_i f = 0$ for each i, hence $c_1 = c_2 = \cdots = c_n = 0$. It also follows that the n linear functions $f_1, \cdots, f_n$ span V^*: any linear function f is determined, by Theorem 24, by its values $\beta_i f = c_i$, and hence f is equal to the combination $\sum_j f_j c_j$ formed with these values as coefficients.

The basis $f_1, \cdots, f_n$ is called the basis of V^* *dual* to the given basis $\beta_1, \cdots, \beta_n$ of V.

Corollary 2. *The dual V^* of an n-dimensional vector space V has the same dimension n as V.*

The transformation $T\colon V \to V^*$ which maps each vector $\sum x_i \beta_i$ of V into the function $\sum f_i x_i$ of V^* is an isomorphism of V onto V^*; the isomorphism, however, depends upon the choice of the basis in V.

If ξ is a vector in V and f a vector in the dual space V^*, one can also write the value of f at the argument ξ in the symmetric "inner product" notation $\xi f = (\xi, f)$. Equation (49) then becomes

(55) $$(a\xi + b\eta, f) = a(\xi, f) + b(\eta, f),$$

while the definitions (51) and (52) of addition and scalar multiplication become

(56) $$(\xi, fc + gd) = (\xi, f)c + (\xi, g)d.$$

The similarity of these two equations suggests another interpretation. In (ξ, f), hold ξ fixed and let f vary. Then, by (56), ξ determines a linear function of f, and by (55), the vector operations on these functions correspond exactly to the vector operations on the original vectors ξ.

Formally, each ξ in V determines a function F_ξ on the dual space V^*, defined by $F_\xi(f) = (\xi, f)$. Then (56) states that F_ξ is a linear function.

Theorem 26. *Any finite-dimensional vector space V is isomorphic to its second conjugate space $(V^*)^*$, under the correspondence mapping each $\xi \in V$ onto the function F_ξ defined by $F_\xi(f) = \xi f$.*

Proof.   By (55), the correspondence $\tau\colon \xi \to F_\xi$ preserves vector addition and scalar multiplication. We now show that τ is one-one, hence an isomorphism. If $\xi \neq \eta$, then $\zeta = \xi - \eta \neq 0$, and so ζ is a part of a basis of V. Hence, by Theorem 24, there is a linear function f_0 in V^* with $\zeta f_0 = 1 \neq 0$, so that

$$F_\xi(f_0) = F_\eta(f_0) + F_\zeta(f_0) = F_\eta(f_0) + 1 \neq F_\eta(f_0).$$

This proves that τ is one-one, hence an isomorphism of V into $(V^*)^*$. But by Corollary 2 of Theorem 25, V and $(V^*)^*$ have the same dimension, hence τ is onto.   Q.E.D.

This isomorphism $\xi \to F_\xi$, unlike that between V and V^* implied by Corollary 2, is "natural" in that its definition does not depend upon the choice of a basis in V.

With any subspace S of V we associate the set S' consisting of all those linear functions f in V^* such that $(\sigma, f) = 0$ for every σ in S. We call S' the *annihilator* of S. It is clearly a subspace of V^*, for $(\sigma, f) = 0$ and $(\sigma, g) = 0$ imply $(\sigma, fc + gd) = 0$. The correspondence $S \to S'$

between subspaces of V and their annihilators in V^* has the property that

(57) $$S \subset T \quad \text{implies} \quad S' \supset T'$$

(inclusion is reversed). For if $f \in T'$, then $(\sigma, f) = 0$ for every σ in T, and hence for every σ in $S \subset T$. The annihilator of the subspace consisting of **0** alone is the whole dual space V^*, and the annihilator of V is the subspace of V^* consisting of the zero function alone.

Dually, each subspace R of the conjugate space V^* determines as its annihilator the subspace R' of V, consisting of all ξ in V with $(\xi, f) = 0$ for every f in R.

Theorem 27. *If S is a k-dimensional subspace of the n-dimensional vector space V, then the set S' of all linear functions f annihilating S is an $(n - k)$-dimensional subspace of V^*.*

Proof. Choose a basis $\beta_1, \cdots, \beta_k$ of S and extend it, by Theorem 6, to a basis $\beta_1, \cdots, \beta_n$ of V. In the dual basis $f_1, \cdots, f_n$ of V^*, the function $f_1 c_1 + \cdots + f_n c_n$ vanishes in all of S if and only if it vanishes for each $\beta_1, \cdots, \beta_k$; that is, if and only if $c_1 = \cdots = c_k = 0$. This means precisely that the $n - k$ functions $f_{k+1}, \cdots, f_n$ form a basis of the annihilator S' of S.

Theorem 27 is just a reformulation of Theorem 13, about the number of independent solutions of a system of homogeneous linear equations.

The correspondence $S \mapsto S'$ of subspaces to their annihilators leads to the Duality Principle of n-dimensional projective geometry, in which connection the following properties are also basic.

Theorem 28. *The correspondence $S \mapsto S'$ satisfies*

(58) $$(S')' = S, \quad (S + T)' = S' \cap T', \quad (S \cap T)' = S' + T'.$$

Proof. Since $(\xi, f) = 0$ for all ξ in S and all f in S', each ξ in S annihilates every vector $f \in S'$, hence $\xi \in (S')'$, and thus $(S')' \supset S$. But by Theorem 27, the dimension of $(S')'$ is $n - (n - k) = k = d[S]$; therefore $(S')' > S$ is impossible, and $(S')' = S$.

This equation states that the correspondence $S \mapsto S'$ of a subspace to its annihilator when applied twice is the identity correspondence; hence this correspondence has an inverse and is one-one onto. Because it also inverts inclusion by (57), it follows that it carries $S + T$, the smallest subspace containing S and T, into the largest subspace $S' \cap T'$ contained in S' and T', and dually that $(S \cap T)' = S' + T'$.

Corollary 1. *Let $L(V)$ be the set of all subspaces of a finite-dimensional vector space V over a field. There is a one-one correspondence of $L(V)$ onto itself, which inverts inclusion and satisfies (58).*

Proof. Let any *fixed* basis $\beta_1, \cdots, \beta_n$ be chosen in V. For any subspace S of V, let S' be the set of all vectors $\eta = y_1\beta_1 + \cdots + y_n\beta_n$ such that

$$(59) \quad x_1 y_1 + \cdots + x_n y_n = 0 \quad \text{for all } \xi = (x_1\beta_1 + \cdots + x_n\beta_n) \text{ in } S.$$

The arguments leading to Theorem 27 and (58) can be repeated to give the desired result.

Remark 1. In the case of a finite-dimensional Euclidean vector space E, there is a natural isomorphism from E to its dual E^*, which can be defined in terms of the intrinsic inner product (ξ, η). The formula $\xi f_\eta = (\xi, \eta)$ defines for each vector $\eta \in E$ a function f_η on E, which is linear since (ξ, η) is bilinear. The correspondence $\eta \to f_\eta$ can easily be shown to be an isomorphism of E onto E^*.

Remark 2. The isomorphism of V to V^* does not in general hold for an infinite-dimensional space V. For example, let V be the vector space of all sequences $\xi = (x_1, \cdots, x_n, \cdots)$, $x_n \in F$, having only a finite number of nonzero entries, addition and multiplication being performed termwise. Any linear function on V can still be represented in the form $\xi f = \sum x_i c_i$ for an arbitrary infinite list of coefficients $\gamma = (c_1, c_2, \cdots, c_n, \cdots)$. Hence the dual space V^* consists of all such infinite sequences. The spaces V and V^* are not isomorphic; for example, to appeal to more advanced concepts, if F is a countable field, then V is countable but V^* is not.

Exercises

1. Complete the proof of Theorem 25.
2. Let $f_1, \cdots, f_n$ be n linearly independent linear functions on an n-dimensional vector space V, and $c_1, \cdots, c_n$ given constants. Show that there is one and only one vector ξ in V with $\xi f_i = c_i, i = 1, \cdots, n$. Interpret in terms of nonhomogeneous linear equations.
3. (a) Complete the proof in Remark 1.
   (b) Show the connection with Corollary 1 of Theorem 25.
4. In $\mathbf{C}^4$, define $(\xi, \eta) = x_1 y_2 - y_1 x_2 + x_3 y_4 - y_3 x_4$. For each subspace S, define S' as the set of all vectors η with $(\xi, \eta) = 0$ for all $\xi \in S$. Prove (57) and (58), and show that if S is *one*-dimensional, then $S \subset S'$.

8

The Algebra of Matrices

8.1. Linear Transformations and Matrices

There are many ways of mapping a plane into itself *linearly*; that is, so that any linear combination of vectors is carried into the same linear combination of transformed vectors. Symbolically this means that

$$(1) \qquad (c\xi + d\eta)T = c(\xi T) + d(\eta T).$$

Equivalently, it means that T preserves sums and scalar products, in the sense that

$$(2) \qquad (\xi + \eta)T = \xi T + \eta T, \qquad (c\xi)T = c(\xi T).$$

For example, consider the (counterclockwise) rigid rotation R_θ of the plane about the origin through an angle θ. It is clear geometrically that R_θ transforms the diagonal $\xi + \eta$ of the parallelogram with sides ξ and η

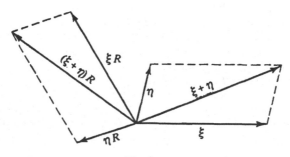

Figure 1

214

into the diagonal $\xi R_\theta + \eta R_\theta$ of the rotated parallelogram with sides ξR_θ and ηR_θ. This is illustrated in Figure 1, where $\theta = 135°$; it shows that $(\xi + \eta)R_\theta = \xi R_\theta + \eta R_\theta$. Also, if c is any real scalar, the multiple $c\xi$ of ξ is rotated into $c(\xi R_\theta)$, so that $(c\xi)R_\theta = c(\xi R_\theta)$. Hence any rigid rotation of the plane is linear; moreover, the same considerations apply to rotations of space about any axis.

Again, consider a simple expansion D_k of the plane away from the origin, under which each point is moved radially to a position k times its original distance from the origin. Thus, symbolically,

(3) $\xi D_k = k\xi$ for all ξ.

This transformation again carries parallelograms into parallelograms, hence vector sums into sums, so that $(\xi + \eta)D_k = \xi D_k + \eta D_k$. Moreover, $(c\xi)D_k = kc\xi = ck\xi = c(\xi D_k)$; hence D_k is linear. Note that if $0 < k < 1$, equation (3) defines a simple contraction toward the origin; if $k = -1$, it defines reflection in the origin (rotation through $180°$), so that these transformations are also linear.

Similar transformations exist in any finite-dimensional vector space F^n. Thus, let T be the transformation of $\mathbf{R}^3$ which carries each vector $\xi = (x_1, x_2, x_3)$ into a vector $\eta = (y_1, y_2, y_3)$ whose coordinates are given by homogeneous linear functions

(4) $y_1 = a_1 x_1 + b_1 x_2 + c_1 x_3,$ $\cdots,$ $y_3 = a_3 x_1 + b_3 x_2 + c_3 x_3$

of x_1, x_2, and x_3. Clearly, if the x_j are all multiplied by the same constant d, then so are the y_i in (4), so that $(d\xi)T = d\eta = d(T\xi)$. Likewise, the transform ζ of the sum $\xi + \xi' = (x_1 + x_1', x_2 + x_2', x_3 + x_3')$ of ξ and the vector $\xi' = (x_1', x_2', x_3')$ may be computed by (4) to have the coordinates

$$z_j = a_j(x_1 + x_1') + b_j(x_2 + x_2') + c_j(x_3 + x_3')$$
$$= (a_j x_1 + b_j x_2 + c_j x_3) + (a_j x_1' + b_j x_2' + c_j x_3'),$$

for $j = 1, 2, 3$. This z_j is just $y_j + y_j'$, where the y_j are given by (4) and the y_j' are corresponding primed expressions; that is, $(\xi + \xi')T = \xi T + \xi'T$.

Conversely, any linear transformation T on $\mathbf{R}^3$ into itself is of the form (4). To see this, denote the transforms of the unit vectors

$$\varepsilon_1 = (1, 0, 0), \qquad \varepsilon_2 = (0, 1, 0), \qquad \varepsilon_3 = (0, 0, 1)$$

by $\alpha = (a_1, a_2, a_3), \qquad \beta = (b_1, b_2, b_3), \qquad \gamma = (c_1, c_2, c_3).$

Then T must carry each $\xi = (x_1, x_2, x_3)$ in $\mathbf{R}^3$ into

$$
\begin{aligned}
\eta = \xi T &= (x_1 \varepsilon_1 + x_2 \varepsilon_2 + x_3 \varepsilon_3) T \\
&= x_1 (\varepsilon_1 T) + x_2 (\varepsilon_2 T) + x_3 (\varepsilon_3 T) = x_1 \alpha + x_2 \beta + x_3 \gamma \\
&= (x_1 a_1 + x_2 b_1 + x_3 c_1, x_1 a_2 + x_2 b_2 + x_3 c_2, x_1 a_3 + x_2 b_3 + x_3 c_3).
\end{aligned}
$$

Hence, if T is linear, it has the form (4).

The preceding construction gives the coefficients of (4) explicitly. Thus consider the counterclockwise rotation R_θ about the origin through an angle θ. The very definition of the sine and cosine functions shows that the unit vector $\varepsilon_1 = (1, 0)$ is rotated into $(\cos \theta, \sin \theta)$, while the unit vector $\varepsilon_2 = (0, 1)$ is rotated into

$$
(\cos (\theta + \pi/2), \sin (\theta + \pi/2)) = (-\sin \theta, \cos \theta).
$$

Thus in (4) we have $a = \cos \theta$, $b = \sin \theta$, $a^* = -\sin \theta$, $b^* = \cos \theta$, so that the equations for R_θ are

(5) $\qquad R_\theta: \quad x' = x \cos \theta - y \sin \theta, \quad y' = x \sin \theta + y \cos \theta.$

Likewise, reflection F_α in a line through the origin making an angle α with the x-axis carries the point whose polar coordinates are (r, θ) into one with polar coordinates $(r, 2\alpha - \theta)$. Hence the effect of F_α is expressed by

(5') $\qquad F_\alpha: \quad x' = x \cos 2\alpha + y \sin 2\alpha, \quad y' = x \sin 2\alpha - y \cos 2\alpha.$

The concept of linearity also applies more generally to transformations between any two vector spaces over the same field.

Definition. *A linear transformation* $T: V \to W$, *of a vector space V to a vector space W over the same field F, is a transformation T of V into W which satisfies* $(c\xi + d\eta)T = c(\xi T) + d(\eta T)$ *for all vectors ξ and η in V and all scalars c and d in F.*

For example, consider the transformation

(6) $\qquad T_1: \quad (x, y) \mapsto (x + y, x - y, 2x) = (x', y', z'),$

defined by the equations $x' = x + y$, $y' = x - y$, $z' = 2x$. This carries the plane vectors $(1, 0)$ and $(0, 1)$ into the orthogonal space vectors $(1, 1, 2)$ and $(1, -1, 0)$, respectively, and transforms the plane linearly into a subset of space.

The finite-dimensional case is most conveniently treated by means of the following principle.

Theorem 1. *If $\beta_1, \cdots, \beta_m$ is any basis of the vector space V, and $\alpha_1, \cdots, \alpha_m$ are any m vectors in W, then there is one and only one linear transformation $T: V \to W$ with $\beta_1 T = \alpha_1, \cdots, \beta_m T = \alpha_m$. This transformation is defined by*

$$(7) \qquad (x_1\beta_1 + \cdots + x_m\beta_m)T = x_1\alpha_1 + \cdots + x_m\alpha_m.$$

For example, let $\beta_1 = (1, 0)$, $\beta_2 = (0, 1)$, $\alpha_1 = (1, 0)$, and $\alpha_2 = (a, 1)$ in the plane. Then Theorem 1 asserts that the horizontal *shear* transformation

$$(8) \qquad S_a: \quad (x, y) \mapsto (x + ay, y)$$

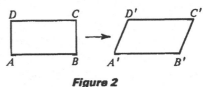

Figure 2

is linear and is the only linear transformation satisfying $\beta_1 S_a = \alpha_1$, $\beta_2 S_a = \alpha_2$. Geometrically, each point is moved parallel to the x-axis through a distance proportional to its altitude above the x-axis, and rectangles with sides parallel to the axes go into parallelograms. (See Figure 2, and picture this with a deck of cards!)

Proof. If T is linear and $\beta_i T = \alpha_i$ $(i = 1, \cdots, m)$, then the definition (1) and induction give the explicit formula (7). Since every vector in V can be expressed uniquely as $x_1\beta_1 + \cdots + x_m\beta_m$, formula (7) defines a single-valued transformation T of V into W; hence there can be no other linear transformation of V into W with $\beta_i T = \alpha_i$. To show that T is linear, let $\eta = \sum y_i\beta_i$ be a second vector of V. Then,

$$(c\xi + d\eta)T = \left[\sum_{i=1}^{m} cx_i\beta_i + \sum_{i=1}^{m} dy_i\beta_i \right]T = \left[\sum_{i=1}^{m} (cx_i + dy_i)\beta_i \right]T$$

$$= \sum_{i=1}^{m} (cx_i + dy_i)\alpha_i = c \sum_{i=1}^{m} x_i\alpha_i + d \sum_{i=1}^{m} y_i\alpha_i$$

$$= c(\xi T) + d(\eta T).$$

Hence T is linear. Q.E.D.

If $V = F^m$ and $W = F^n$, and we let the β_i be the unit vectors $\varepsilon_1 = (1, 0, \cdots, 0), \cdots, \varepsilon_m = (0, 0, \cdots, 1)$ of V_m, we obtain a very important application of Theorem 1. In this case, we can give each α_i its coordinate representation

$$
\begin{aligned}
\varepsilon_1 T = \alpha_1 &= (a_{11}, a_{12}, \cdots, a_{1n}) \\
\varepsilon_2 T = \alpha_2 &= (a_{21}, a_{22}, \cdots, a_{2n}) \\
&\vdots \\
\varepsilon_m T = \alpha_m &= (a_{m1}, a_{m2}, \cdots, a_{mn}).
\end{aligned}
$$

(9)

Theorem 1 states that there is just one linear transformation associated with the formula (9). This transformation is thus determined by the $m \times n$ matrix $A = \|a_{ij}\|$, which has the coordinates $(a_{i1}, \cdots, a_{in})$ as its ith row, and a_{ij} as the entry in its ith row and jth column. We have proved

Theorem 2. *There is a one-one correspondence between the linear transformations $T: F^m \to F^n$ and the $m \times n$ matrices A with entries in the field F. Given T, the corresponding matrix A is the matrix with ith row the row of coordinates of $\varepsilon_i T$; given $A = \|a_{ij}\|$, T is the (unique) linear transformation carrying each unit vector ε_i of F^m into the ith row $(a_{i1}, \cdots, a_{in})$ of A.*

We denote by T_A the linear transformation of F^m into F^n corresponding to A in this fashion. For example, in the plane, the rotation, similitude, and shear of (5), (3), and (8) correspond respectively to the matrices

$$
R_\theta \to \begin{pmatrix} \cos\theta & \sin\theta \\ -\sin\theta & \cos\theta \end{pmatrix}, \quad
D_k \to \begin{pmatrix} k & 0 \\ 0 & k \end{pmatrix}, \quad
S_a \to \begin{pmatrix} 1 & 0 \\ a & 1 \end{pmatrix}.
$$

The general transformation $T = T_A$ of (9) carries any given vector $\xi = (x_1, \cdots, x_m) = x_1\varepsilon_1 + \cdots + x_m\varepsilon_m$ of F^m into the vector

$$
\begin{aligned}
\xi T &= x_1\alpha_1 + \cdots + x_m\alpha_m \\
&= x_1(a_{11}, \cdots, a_{1n}) + \cdots + x_m(a_{m1}, \cdots, a_{mn}) \\
&= (x_1 a_{11} + \cdots + x_m a_{m1}, \cdots, x_1 a_{1n} + \cdots + x_m a_{mn})
\end{aligned}
$$

in $W = F^n$. Hence, if $(y_1, \cdots, y_n)$ are the coordinates of the transformed vector $\eta = \xi T$, T is given in terms of these coordinates by the homogeneous linear equations

$$y_1 = x_1 a_{11} + x_2 a_{21} + \cdots + x_m a_{m1} = \sum_i x_i a_{i1},$$

(10) $$y_2 = x_1 a_{12} + x_2 a_{22} + \cdots + x_m a_{m2} = \sum_i x_i a_{i2},$$

$$\vdots$$

$$y_n = x_1 a_{1n} + x_2 a_{2n} + \cdots + x_m a_{mn} = \sum_i x_i a_{in}.$$

Hence we have the

Corollary. *Any linear transformation T of F^m into F^n can be described by homogeneous linear equations of the form (10). Specifically, each T determines an $m \times n$ matrix $A = \|a_{ij}\|$, so that T carries the vector ξ with coordinates $x_1, \cdots, x_n$ into the vector $\eta = \xi T$ with coordinates $y_1, \cdots, y_n$ given by (10). Conversely, each $m \times n$ matrix A determines, by means of equations (10), a linear transformation $T = T_A : F^m \to F^n$.*

Caution. The rectangular array of the coefficients of (10) is *not* the matrix A appearing in (9); it is the matrix of (9) with its rows and columns interchanged. This $n \times m$ matrix of coefficients of (10), which is obtained from the $m \times n$ matrix A by interchanging rows and columns, is called the *transpose* of A and is denoted by A^T. If $A = \|a_{ij}\|$ has entry a_{ij} in its ith row and jth column, then the *transpose* $B = A^T$ of the matrix A is defined formally by the equations.

(11) $$b_{ij} = a_{ij}^T = a_{ji} \qquad (i = 1, \cdots, n; \quad j = 1, \cdots, m).$$

In this notation, (10) assumes the more familiar form

$$b_{11} x_1 + b_{12} x_2 + \cdots + b_{1m} x_m = y_1$$

$$b_{21} x_1 + b_{22} x_2 + \cdots + b_{2m} x_m = y_2$$

(11′) $$\vdots \qquad \vdots \qquad\qquad \vdots \qquad \vdots$$

$$b_{n1} x_1 + b_{n2} x_2 + \cdots + b_{nm} x_m = y_n.$$

The preceding formulas for linear transformations refer to the spaces F^m and F^n of m-tuples and n-tuples, respectively. More generally, if V and W are any two finite-dimensional vector spaces over F of dimensions m and n, respectively, then any linear transformation $T: V \to W$ can be represented by a matrix A, once we have chosen a basis $\beta_1, \cdots, \beta_m$ in V and a basis $\gamma_1, \cdots, \gamma_n$ in W. For then T is determined by the images $\beta_i T = \sum_j a_{ij} \gamma_j$, and we say that T is represented by the $m \times n$ matrix

$A = \|a_{ij}\|$ of these coefficients, relative to the given bases. This amounts to replacing the spaces V and W by the isomorphic spaces of m- and n-tuples, under the isomorphisms $\sum x_i \beta_i \mapsto (x_1, \cdots, x_m)$; $\sum y_j \gamma_j \mapsto (y_1, \cdots, y_n)$.

Exercises

1. Describe the geometric effect of each of the following linear transformations
   (a) $y' = x$, $\quad x' = y$; (b) $y' = x$, $\quad x' = x$;
   (c) $y' = x$, $\quad x' = 0$; (d) $y' = ky$, $\quad x' = kx + kay$;
   (e) $y' = by$, $\quad x' = cx$.

2. Consider the transformation of the plane into itself which carries every point P into a point P' related to P in the way described below. Determine when the transformation is linear and find its equations.
   (a) P' is two units to the right of P and one unit above (a translation).
   (b) P' is the projection of P on the line of slope $1/2$ through the origin.
   (c) P' lies on the half-line OP joining P to the origin, at a distance from O such that $\overline{OP'} = 4/\overline{OP}$.
   (d) P' is obtained from P by a rotation through $30°$ about the origin, followed by a shear parallel to the y-axis.
   (e) P' is the reflection of P in the line $x = 3$.

3. Find the matrices which represent the symmetries of the equilateral triangle with vertices $(1, 0)$ and $(-1/2, \pm\sqrt{3}/2)$.

4. Describe the geometric effects of the following linear transformations of space:
   (a) $x' = ax$, $\quad y' = by$, $\quad z' = cz$;
   (b) $x' = 0$, $\quad y' = 3y$, $\quad z' = 3z$;
   (c) $x' = x + 2y + 5z$, $\quad y' = y$, $\quad z' = z$;
   (d) $x' = x - y$, $\quad y' = x + y$, $\quad z' = 4z$.

5. What is the matrix of the transformation (6) of the text?

6. Find the matrix which represents the linear transformation described:
   (a) $(1, 1) \mapsto (0, 1)$ $\quad$ and $\quad$ $(-1, 1) \mapsto (3, 2)$;
   (b) $(1, 0) \mapsto (4, 0)$ $\quad$ and $\quad$ $(0, 1) \mapsto (-1, 2)$;
   (c) $(2, 3) \mapsto (1, 0)$ $\quad$ and $\quad$ $(3, 2) \mapsto (1, -1)$;
   (d) $(1, 0, 0) \mapsto (1, 2, 1)$, $\quad$ $(0, 1, 0) \mapsto (3, 1, 1)$, $\quad$ $(0, 0, 1) \mapsto (0, 0, 3)$.

7. By the image of a subspace S of V under a linear transformation T, one means the set $(S)T$ of all vectors ξT for ξ in S. Prove that $(S)T$ is itself a subspace.

8. A linear transformation T takes $(1, 1)$ into $(0, 1, 2)$ and $(-1, 1)$ into $(2, 1, 0)$. What matrix represents T?

8.2. Matrix Addition

The algebra of linear transformations (matrices) involves three operations: addition of two linear transformations (or matrices), multiplication of a linear transformation by a scalar, and multiplication of two linear transformations (matrices). We shall now define the *vector* operations on matrices, namely, the addition of two matrices, and the multiplication of a matrix by a scalar.

The *sum* $A + B$ of two $m \times n$ matrices $A = \|a_{ij}\|$ and $B = \|b_{ij}\|$ is obtained by adding corresponding entries, as

(12) $$\|a_{ij}\| + \|b_{ij}\| = \|a_{ij} + b_{ij}\|.$$

This sum obeys the usual commutative and associative laws because the terms a_{ij} obey them. The $m \times n$ matrix O which has all entries zero acts as a *zero matrix* under this addition, so that

$$O + A = A + O = A \qquad \text{for any } m \times n \text{ matrix } A.$$

The additive inverse may be found by simply multiplying each entry by -1. Under addition, $m \times n$ matrices thus form an Abelian group.

The *scalar product* cA of a matrix A by a scalar c is formed by multiplying each entry by c. One may verify the usual laws for vectors:

(13)
$$1 \cdot A = A, \qquad\qquad c(dA) = (cd)A,$$
$$(c + d)A = cA + dA, \qquad c(A + B) = cA + cB.$$

Theorem 3. *Under addition and scalar multiplication, all $m \times n$ matrices over a field F form a vector space over F.*

Any matrix $\|a_{ij}\|$ may be written as a sum $\sum a_{ij}E_{ij}$, where E_{ij} is the special matrix with entry 1 in the ith row and jth column, entries 0 elsewhere. These matrices E_{ij} are linearly independent, so form a basis for the space of all $m \times n$ matrices. The dimension of this space is therefore mn.

There is a corresponding algebra of linear transformations. One can define the sum $T + U$ of any two linear transformations from a vector space V to a vector space W by

(14) $$\xi(T + U) = \xi T + \xi U \qquad \text{for all } \xi \text{ in } V.$$

Similarly, the scalar product cT is defined by $\xi(cT) = c(\xi T)$. The sum $T + U$ is linear according to definition (1), for

$$(c\xi + d\eta)(T + U) = (c\xi + d\eta)T + (c\xi + d\eta)U$$
$$= c\xi T + c\xi U + d\eta T + d\eta U$$
$$= c\xi(T + U) + d\eta(T + U).$$

The product cT is also linear.

When $V = F^m$ and $W = F^n$, definition (14) implies that $\varepsilon_i(T + U) = \varepsilon_i T + \varepsilon_i U$, whence the matrix C corresponding to $T + U$ in Theorem 2 is the sum of the matrices which correspond to T and U. Since $c(\varepsilon_i T) = c(\varepsilon_i T)$, the operation of scalar multiplication just defined corresponds to that previously defined for $m \times n$ matrices. That is, in the notation introduced following Theorem 2.

(15) $$T_{A+B} = T_A + T_B \quad \text{and} \quad T_{cA} = cT_A.$$

The new definitions have the advantage of being *intrinsic*, in the sense of being independent of the coordinate systems used in V and W (cf. §7.8). They also apply to infinite-dimensional vector spaces.

Finally, it should be observed that a linear transformation of a vector space V into a vector space W is just a *homomorphism* of V into W (both being considered as Abelian groups), which preserves multiplication by scalars as well. For this reason, the vector space of all linear transformations from V into W is often referred to as Hom (V, W).

Exercises

1. For the matrices R_θ, D_k, S_a of §8.1, compute $2R_\theta + D_k$, $2S_a - 3D_k$, and $R_\theta - S_a + 5D_k$.
2. Prove that $(A + B)^T = A^T + B^T$, $(cA)^T = cA^T$.
3. Prove the rules (13).
4. Prove directly, without reference to matrices, that the set of all linear transformations $T: V \to W$ is a vector space under the operations defined in and below (14).

8.3. Matrix Multiplication

The most important combination of two linear transformations T and U is their *product* TU (first apply T, then U, as in §6.2). In this section, we shall consider only the product of two linear transformations T, U of a vector space V into itself. Then TU may be defined as that transformation of V into itself with $\xi(TU) = (\xi T)U$ for every vector ξ.

For instance, if the shear S_a of (8) is followed by a transformation of similitude D_k, which sends (x', y') into $x'' = kx'$, $y'' = ky'$, the combined effect is to take (x, y) into $x'' = kx + kay$, $y'' = ky$. This product $S_a D_k$ is still linear.

Theorem 4. *The product of two linear transformations is linear.*

Proof. By definition, a product TU maps any ξ into $\xi(TU) = (\xi T)U$. By the linearity of T and U, respectively,

(16) $(c\xi + d\eta)TU = [c(\xi T) + d(\eta T)]U = c(\xi TU) + d(\eta TU)$,

which is to say that TU also satisfies the defining condition (1) for a linear transformation.   Q.E.D.

This result implies that the homogeneous linear equations (10) for T and U may be combined to yield homogeneous linear equations for TU. To be specific, let the transformation

(17)
$$\begin{aligned} x' &= xa_{11} + ya_{21}, \\ y' &= xa_{12} + ya_{22}, \end{aligned} \qquad A = \begin{pmatrix} a_{11} & a_{12} \\ a_{21} & a_{22} \end{pmatrix},$$

with matrix A, be followed by a second linear transformation of the plane, mapping (x', y') on (x'', y''), where

(18)
$$\begin{aligned} x'' &= x'b_{11} + y'b_{21}, \\ y'' &= x'b_{12} + y'b_{22}, \end{aligned} \qquad B = \begin{pmatrix} b_{11} & b_{12} \\ b_{21} & b_{22} \end{pmatrix}.$$

The combined transformation, found by substituting (17) in (18), is

(19)
$$\begin{aligned} x'' &= (a_{11}b_{11} + a_{12}b_{21})x + (a_{21}b_{11} + a_{22}b_{21})y, \\ y'' &= (a_{11}b_{12} + a_{12}b_{22})x + (a_{21}b_{12} + a_{22}b_{22})y. \end{aligned}$$

The matrix of coefficients in this product transformation arises from the original matrices A and B by the important rule

(20) $\begin{pmatrix} a_{11} & a_{12} \\ a_{21} & a_{22} \end{pmatrix} \cdot \begin{pmatrix} b_{11} & b_{12} \\ b_{21} & b_{22} \end{pmatrix} = \begin{pmatrix} a_{11}b_{11} + a_{12}b_{21} & a_{11}b_{12} + a_{12}b_{22} \\ a_{21}b_{11} + a_{22}b_{21} & a_{21}b_{12} + a_{22}b_{22} \end{pmatrix}.$

The entry in the *first row* and the *second column* of this result involves only the *first* row of a's and the *second* column of b's, and so on. This multiplication rule is a labor-saving device which spares one the trouble of using variables in substitutions like (19).

Similar formulas hold for $n \times n$ matrices, for Theorems 2 and 4 show that the product of the transformations $T, U: F^n \to F^n$ must yield a suitable product of their matrices. We shall now compute the "matrix product" AB which corresponds to $T_A T_B$, so as to give the rule

(21) $T_A T_B = T_{AB}$.

By Theorem 2, $\varepsilon_i T_A = \sum a_{ij}\varepsilon_j$ and $\varepsilon_j T_B = \sum b_{jk}\varepsilon_k$. Hence

$$\varepsilon_i(T_A T_B) = (\varepsilon_i T_A)T_B = \left(\sum_j a_{ij}\varepsilon_j\right)T_B = \sum_j a_{ij}(\varepsilon_j T_B)$$

$$= \sum_j a_{ij}\left(\sum_k b_{jk}\varepsilon_k\right) = \sum_k c_{ik}\varepsilon_k,$$

where

(22) $\qquad c_{ik} = \sum_j a_{ij}b_{jk} = a_{i1}b_{1k} + a_{i2}b_{2k} + \cdots + a_{in}b_{nk}.$

Hence the *matrix product* $C = AB$ must be defined by (22) in order to make (21) valid; we adopt this definition.

Definition. *The product AB of the $n \times n$ matrix A by the $n \times n$ matrix B is defined to be the $n \times n$ matrix C having for its entry in the ith row and kth column the sum c_{ik} given by (22).*

The product of two matrices may also be described verbally: the entry c_{ik} in the ith row and the kth column of the product AB is found by multiplying the ith row of A by the kth column of B. To "multiply" a row by a column, one multiplies corresponding entries, then adds the results.

It follows immediately from the correspondence (21) between matrix multiplication and transformation multiplication that the multiplication of matrices is associative. In symbols

(23) $\qquad\qquad\qquad A(BC) = (AB)C,$

since these matrices correspond to the transformations $T_A(T_B T_C)$ and $(T_A T_B)T_C$, which are equal by the associative law for the multiplication of transformations (§6.2).

Not only is matrix multiplication associative; it is distributive on matrix sums, for the matrix $(A + B)C$ has entries d_{ik} given by formulas like (22) as

$$d_{ik} = \sum_j (a_{ij} + b_{ij})c_{jk} = \sum_j a_{ij}c_{jk} + \sum_j b_{ij}c_{jk}.$$

This gives d_{ik} as the sum of an entry g_{ik} of AC and an entry h_{ik} of BC and proves the first of the two distributive laws

(24) $\qquad (A + B)C = AC + BC, \qquad A(B + C) = AB + AC.$

For scalar products by d, one may also verify the laws

(25) $\qquad (dA)B = d(AB) \qquad$ and $\qquad A(dB) = d(AB).$

The laws (24) and (25) are summarized by the statement that matrix multiplication is *bilinear,* for the first halves of these laws combine to give $(dA + d^*A^*)B = d(AB) + d^*(A^*B)$. This is exactly the condition that multiplication by B on the right be a linear transformation $X \mapsto XB$ on the vector space of all $n \times n$ matrices X. The other laws of (24) and (25) assert that multiplication by A on the left is also a linear transformation.

Corresponding to the identity transformation T_I of F^n is the $n \times n$ *identity matrix I,* which has entries $e_{ii} = 1$ along the principal diagonal (upper left to lower right) and zeros elsewhere, since $\varepsilon_i T_I = \varepsilon_i$ for all $i = 1, \cdots, n$. Since I represents the identity transformation, it has the property $IA = A = AI$ for every $n \times n$ matrix I.

We may summarize the foregoing as follows:

Theorem 5. *The set of all $n \times n$ matrices over a field F is closed under multiplication, which is associative, has an identity, and is bilinear with respect to vector addition and scalar multiplication.*

However, multiplication is not commutative. Thus

$$\begin{pmatrix} -1 & 0 \\ 0 & 1 \end{pmatrix}\begin{pmatrix} 0 & 1 \\ -1 & 0 \end{pmatrix} = \begin{pmatrix} 0 & -1 \\ -1 & 0 \end{pmatrix}.$$

$$\begin{pmatrix} 0 & 1 \\ -1 & 0 \end{pmatrix}\begin{pmatrix} -1 & 0 \\ 0 & 1 \end{pmatrix} = \begin{pmatrix} 0 & 1 \\ 1 & 0 \end{pmatrix}$$

Hint: What geometric transformations do these matrices induce on the square of §6.1?

Not all nonzero matrices have multiplicative inverses; thus the matrix $\begin{pmatrix} 1 & 0 \\ 3 & 0 \end{pmatrix}$, which represents an oblique projection on the x-axis, does not induce a one-one transformation and is not onto; hence (Theorem 1, §6.2) it has no left-inverse or right-inverse. Similarly, the law of cancellation fails, for there are plenty of divisors of zero, as in

$$\begin{pmatrix} 1 & 0 \\ 0 & 0 \end{pmatrix} \cdot \begin{pmatrix} 0 & 0 \\ 3 & 2 \end{pmatrix} = \begin{pmatrix} 0 & 0 \\ 0 & 0 \end{pmatrix}, \qquad \begin{pmatrix} 2 & 4 \\ -1 & -2 \end{pmatrix} \cdot \begin{pmatrix} 0 & 2 \\ 0 & -1 \end{pmatrix} = \begin{pmatrix} 0 & 0 \\ 0 & 0 \end{pmatrix}.$$

Formulas (15) and (21) assert the following important principle.

Theorem 6. *The algebra of linear transformations of F^n is isomorphic to the algebra of all $n \times n$ matrices over F under the correspondence $T_A \leftrightarrow A$ of Theorem 2.*

This suggests that the formal laws, asserted in Theorem 5 for the algebra of matrices, may in fact be valid for the linear transformations of any vector space whatever. This conjecture is readily verified, and leads directly to certain aspects of the "operational calculus," when applied to suitable vector spaces of infinite dimensions.

EXAMPLE 1.   Let V consist of all functions $f(x)$ of a real variable x, and let J be the transformation or "operator" $[f(x)]J = f(x + 1)$. If I is the identity transformation, the operator $\Delta = J - I$ is known as a "difference operator"; it carries $f(x)$ into $f(x + 1) - f(x)$. Both J and Δ are linear, for $[cf(x) + dg(x)]J = c[f(x)]J + d[g(x)]J$. This definition of linearity applies at once, but observe that we cannot set up the linear homogeneous equations in this infinite space. For fixed $a(x)$ the operation $f(x) \to a(x)f(x)$ is also linear.

EXAMPLE 2.   The derivative operator D applies to the space C^∞ of all functions $f(x)$ which possess derivatives of all orders; it carries $f(x)$ into $f'(x)$. D is linear. Taylor's theorem may be symbolically written as $e^D = J$.

EXAMPLE 3.   For functions $f(x, y)$ of two variables, there are corresponding linear operators J_x, J_y, D_x, D_y, Δ_x, Δ_y. Thus, $[f(x, y)]J_x = f(x + 1, y)$ and $[f(x, y)]D_x = f_x'(x, y)$.

Exercises

1. Compute the products indicated, for the matrices

$$A = \begin{pmatrix} 1 & 1 \\ 0 & 1 \end{pmatrix}, \quad B = \begin{pmatrix} 1 & -1 \\ 0 & 1 \end{pmatrix}, \quad C = \begin{pmatrix} 1 & 3 \\ 2 & 1 \end{pmatrix}, \quad D = \begin{pmatrix} 4 & 0 \\ 2 & 1 \end{pmatrix}.$$

   (a) AB, BA, $A^2 + AB - 2B$;
   (b) $(A + B - I)(A - B + I) - (A + 2B)(B - A)$;
   (c) DB, AC, AD.
   (d) Test the associative law for the products $(AC)D$, $A(CD)$.
2. Use matrix products to compute the equations of the following transformations (notation as in §8.1).
   (a) $D_k S_a$,   (b) $S_a D_k$,   (c) $R_\theta S_a$ for $\theta = 45°$,
   (d) $R_\theta S_a D_k$ for $\theta = 30°$,   (e) $D_k S_a D_k$.

3. When is $S_a D_k = D_k S_a$ (notation as in Ex. 2)?

4. In Ex. 4 of §8.1 denote by T_n the transformation described in part (n). Compute (using matrices) the following products:
   (a) $T_b T_c$,   (b) $T_a T_c$,   (c) $T_b T_a T_b$,   (d) $T_d T_c$,   (e) $T_c T_b T_d$.

5. Prove the laws (25) and the second half of (24).

6. (a) Expand $(A + B)^3$.   (b) Prove that $A^3 A^2 = A^2 A^3$.

7. Prove the associative law for matrix multiplication directly from definition (22).

8. Consider a new "product" $A \times B$ of two matrices, defined by a "row-by-row" multiplication of A by B. Is this product associative?

9. (a) Compute the products BE_1, BE_2, BE_3, $E_2 E_3$, $E_1 E_3$, where

$$B = \begin{pmatrix} 1 & 2 & 1 \\ 1 & 3 & 2 \\ 1 & 4 & 6 \end{pmatrix}, \; E_1 = \begin{pmatrix} 0 & 0 & 1 \\ 0 & 1 & 0 \\ 1 & 0 & 0 \end{pmatrix}, \; E_2 = \begin{pmatrix} 1 & 0 & k \\ 0 & 1 & 0 \\ 0 & 0 & 1 \end{pmatrix}, \; E_3 = \begin{pmatrix} a & 0 & 0 \\ 0 & b & 0 \\ 0 & 0 & c \end{pmatrix}.$$

   (b) If A is any 3×3 matrix, how is AE_3 related to A?
   (c) Describe the effect caused by multiplying any matrix on the right by E_1; by E_2.

10. Without using matrices, prove the laws $R(S + T) = RS + RT$, $(R + S)T = RT + ST$, and $S(cT) = c(ST)$ for any linear transformations R, S, T of V into itself.

★11. Show that if R, S, T are *any* transformations (linear or not) of a vector space, then $R(S + T) = RS + RT$, but that $(R + S)T = RT + ST$ does not hold in general, unless T is linear.

12. Find all matrices which commute with the matrix E_3 of Ex. 9, when a, b, and c are distinct.

★13. Prove that every matrix which commutes with the matrix D of Ex. 1 can be expressed in the form $aI + bD$.

14. If A is any $n \times n$ matrix, prove that the set $C(A)$ of all $n \times n$ matrices which commute with A is closed under addition and multiplication.

★15. Prove that each $n \times n$ matrix A satisfies an equation of the form

$$A^m + c_{m-1} A^{m-1} + \cdots + c_1 A + c_0 I = 0, \qquad m \leqq n^2.$$

★16. (a) Let $A = \|a_{ij}\|$ be an $n \times n$ matrix of real numbers, and let M be the largest of the $|a_{ij}|$. Prove that the entries of A^k are bounded in magnitude by $n^{k-1} M^k$.
   (b) Show that the series $I + A + A^2/2! + A^3/3! + \cdots$ is always convergent. (It may be used to define the exponential function e^A of the matrix A.)

In Exs. 17–21, the notation follows that in Examples 1–3 above.

17. (a) Prove D linear.   (b) Show why $e^D = J$.

18. Prove $D_x D_y = D_y D_x$.

★19. (a) Simplify $xD - Dx$, $x\Delta - \Delta x$, $x\Delta^2 - \Delta^2 x$.
   (b) Simplify $x^i D^j - D^j x^i$, $x^i \Delta^j - \Delta^j x^i$.

★**20.** Define the Laplacian operator ∇^2 by $\nabla^2 = D_x^2 + D_y^2$, and find $x\nabla^2 - \nabla^2 x$, $y(\nabla^2)^2 - (\nabla^2)^2 y$, $\nabla^2(x^2 + y^2) - (x^2 + y^2)\nabla^2$.

★**21.** Expand $\Delta^n = (J - I)^n$ by a "binomial theorem."

8.4. Diagonal, Permutation, and Triangular Matrices

A square matrix $D = \|d_{ij}\|$ is called *diagonal* if and only if $i \neq j$ implies $d_{ij} = 0$; that is, if and only if all nonzero entries of D lie on the principal diagonal (from upper left to lower right). To add or to multiply two diagonal matrices, simply add or multiply corresponding entries along the diagonal (why?). If all the diagonal entries d_{ii} of D are nonzero, the diagonal matrix $E = \|e_{ij}\|$ with $e_{ii} = d_{ii}^{-1}$ is the inverse of D, in the sense that $DE = I = ED$. One may then prove

Theorem 7. *All $n \times n$ diagonal matrices with nonzero diagonal entries in a field F form a commutative group under multiplication.*

A *permutation matrix* P is a square matrix which in each row and in each column has some one entry 1, all other entries zero.

The 3×3 permutation matrices are six in number. They are I and the matrices

$$\begin{pmatrix} 0 & 1 & 0 \\ 1 & 0 & 0 \\ 0 & 0 & 1 \end{pmatrix}, \begin{pmatrix} 1 & 0 & 0 \\ 0 & 0 & 1 \\ 0 & 1 & 0 \end{pmatrix}, \begin{pmatrix} 0 & 0 & 1 \\ 0 & 1 & 0 \\ 1 & 0 & 0 \end{pmatrix}, \begin{pmatrix} 0 & 0 & 1 \\ 1 & 0 & 0 \\ 0 & 1 & 0 \end{pmatrix}, \begin{pmatrix} 0 & 1 & 0 \\ 0 & 0 & 1 \\ 1 & 0 & 0 \end{pmatrix}.$$

Since the rows of a matrix are the transforms of the unit vectors, a matrix P is a permutation matrix if and only if the corresponding linear transformation T_P of V_n permutes the unit vectors $\varepsilon_1, \cdots, \varepsilon_n$. The $n \times n$ permutation matrices therefore correspond one-one with the $n!$ possible permutations of n symbols (§6.9), and this correspondence is an isomorphism.

Theorem 8. *The $n \times n$ permutation matrices form under multiplication a group isomorphic to the symmetric group on n letters.*

There are also other important classes of matrices. A matrix M is *monomial* if each row and column has exactly one nonzero term; any such matrix may be obtained from a permutation matrix by replacing the 1's by any nonzero entries, as for example in

$$(26) \quad M_1 = \begin{pmatrix} 0 & 0 & 5 \\ -2 & 0 & 0 \\ 0 & 3 & 0 \end{pmatrix}, \quad M_2 = \begin{pmatrix} 0 & 7 & 0 \\ 0 & 0 & -3 \\ 4 & 0 & 0 \end{pmatrix}, \quad M_3 = \begin{pmatrix} 0 & 4 \\ -1 & 0 \end{pmatrix}.$$

A square matrix $T = \|t_{ij}\|$ is *triangular* if all the entries below the diagonal are zero; that is, if $t_{ij} = 0$ whenever $i > j$. A matrix S is *strictly triangular* if all the entries on or below the main diagonal are zero. These two patterns may be schematically indicated in the 4×4 case by

$$(27) \quad T = \begin{pmatrix} q & r & s & t \\ 0 & u & v & w \\ 0 & 0 & x & y \\ 0 & 0 & 0 & z \end{pmatrix}, \quad S = \begin{pmatrix} 0 & u & v & w \\ 0 & 0 & x & y \\ 0 & 0 & 0 & z \\ 0 & 0 & 0 & 0 \end{pmatrix},$$

where the letters denote arbitrary entries. Finally, a *scalar matrix* is a matrix which can be written as cI, where I is the identity.

This scheme of prescribing a "pattern" for the nonzero terms of a matrix is not the only method of constructing groups of matrices. Any group of linear transformations may be represented by a corresponding group of matrices. For instance, the group of the square consists of linear transformations. Pick an origin at the center of the square and an x-axis parallel to one side of the square. If the equations giving the motions R, R', H, and D are written out in terms of x and y (see the description in §6.1), they will give transformations with the following matrices,

$$R \to \begin{pmatrix} 0 & -1 \\ 1 & 0 \end{pmatrix}, \quad R' \to \begin{pmatrix} -1 & 0 \\ 0 & -1 \end{pmatrix}, \quad H \to \begin{pmatrix} 1 & 0 \\ 0 & -1 \end{pmatrix}, \quad D \to \begin{pmatrix} 0 & 1 \\ 1 & 0 \end{pmatrix}.$$

The other four elements of the group can be similarly represented. The multiplication table of this group, as given in §6.4, might have been computed by simply multiplying the corresponding matrices here (try it!). In other words, the group of the square is *isomorphic* to a group of eight 2×2 matrices.

The preceding examples show that a given matrix A may have an *inverse* A^{-1}, such that $AA^{-1} = A^{-1}A = I$. Such matrices are called *nonsingular* or *invertible*; they will be studied systematically in §8.6.

Exercises

1. What is the effect of multiplying an $n \times n$ matrix A by a diagonal matrix D on the right?

2. If D is diagonal and all the terms on the diagonal are distinct, what matrices A commute with D (when is $AD = DA$)?

3. Show that a triangular 2×2 matrix with 1's on the main diagonal represents a shear transformation.

4. Exhibit explicitly the isomorphism between the 3×3 permutation matrices and the symmetric group.

5. Let S_i be the one-dimensional subspace of V_n spanned by the ith unit vector ε_i. Prove that a nonsingular matrix D is diagonal if and only if the corresponding linear transformation T_D maps each subspace S_i onto itself.

6. Find a description like that of Ex. 5 for monomial matrices.

7. (a) Prove that a monomial matrix M can be written in one and only one way in the form $M = DP$, where D is nonsingular and diagonal, and P is a permutation matrix. (*Hint:* Use Ex. 5.)

   (b) Write the matrices M_1 and M_2 of the text in the forms DP and PD.

   ★(c) Exhibit a homomorphism mapping the group of monomial matrices onto the group of permutation matrices.

8. Describe the inverse of a monomial matrix M, and find the inverses of M_1 and M_2 in (26).

9. If M is monomial, D diagonal, prove $M^{-1}DM$ diagonal.

10. If P is a permutation matrix and D diagonal, describe explicitly the form of the transform $P^{-1}DP$.

11. How are the rows of PA related to those of A for P as in Ex. 10?

12. A matrix A is called *nilpotent* if some power of A is 0. Prove that any strictly triangular matrix is nilpotent. (*Hint:* Try the 3×3 case.)

13. Represent the group of symmetries of the rectangle as a group of matrices.

14. For the group of symmetries of the square with vertices at $(\pm 1, \pm 1)$, compute the matrices which represent the symmetries H, D, V. Verify that $HD = DV$.

★15. In Ex. 7, show that the formula $M = DP$ defines a group-homomorphism $M \mapsto P$. Find its kernel.

8.5. Rectangular Matrices

So far we have considered only the multiplication of *square* (i.e., $n \times n$) matrices; we now discuss the multiplication of *rectangular* matrices—that is, of $m \times n$ matrices where in general $m \neq n$.

An $m \times n$ matrix $A = \|a_{ij}\|$ and an $n \times r$ matrix $B = \|b_{jk}\|$, with the same n, determine as product $AB = \|c_{ik}\|$ an $m \times r$ matrix C with entries

$$c_{ik} = \sum_j a_{ij}b_{jk},$$

where, in the sum, j runs from 1 to n. This "row-by-column" product cannot be formed unless each row of A is just as long as each column of

B; hence the assumption that the number n of columns of A equals the number n of rows of B. Thus, if $m = 1$, $n = 2$, $r = 3$,

$$(x_1, x_2)\begin{pmatrix} a_{11} & a_{12} & a_{13} \\ a_{21} & a_{22} & a_{23} \end{pmatrix} = (x_1a_{11} + x_2a_{21}, x_1a_{12} + x_2a_{22}, x_1a_{13} + x_2a_{23}).$$

As in our formulas (21)–(22), the matrix product AB corresponds under Theorem 2 to the product $T_A T_B$ of the linear transformations $T_A: F^m \to F^n$, and $T_B: F^n \to F^r$, associated with A and B, respectively. Here, as always, the product of a transformation $T: V \to W$ by a transformation $U: W \to X$ is defined by

$$(28) \qquad \xi(TU) = (\xi T)U \qquad \text{for all } \xi \text{ in } V.$$

The algebraic laws for square matrices hold also for rectangular matrices, provided these matrices have the proper dimensions to make all products involved well defined. For example, the $m \times m$ identity matrix I_m and the $n \times n$ identity I_n satisfy

$$(29) \qquad I_m A = A = A I_n \qquad \text{(if } A \text{ is } m \times n\text{)}.$$

Matrix multiplication is again bilinear, as in (24) and (25). The associative law is

$$(30) \quad A(BC) = (AB)C \qquad (A \text{ is } m \times n; \quad B, \, n \times r; \quad C, \, r \times s).$$

Again, it is best proved by appeal to an interpretation of rectangular matrices as transformations.

As in (11), the *transpose* A^T of an $m \times n$ matrix A is an $n \times m$ matrix A^T with entries $a_{ij}^T = a_{ji}$ $(i = 1, \cdots, n; j = 1, \cdots, m)$. The ith row of this transpose A^T is the ith column of the original A, and vice versa. One may also obtain A^T by reflecting A in its main diagonal. To calculate the transpose C^T of a product $AB = C$, use

$$(31) \qquad c_{ik}^T = c_{ki} = \sum_j a_{kj}b_{ji} = \sum_i b_{ji}a_{kj} = \sum_i b_{ij}^T a_{jk}^T;$$

the result is just the (i, k) element of the product $B^T A^T$. (Note the change in order.) This proves the first of the laws

$$(32) \quad (AB)^T = B^T A^T, \qquad (A + B)^T = A^T + B^T, \qquad (cA)^T = cA^T.$$

The correspondence $A \leftrightarrow A^T$ therefore preserves sums and inverts the order of products, so is sometimes called an anti-automorphism. Since $(A^T)^T = A$, this anti-automorphism is called "involutory."

A systematic use of rectangular matrices has several advantages. For example, a vector ξ in the space F^n of n-tuples over F may be regarded as a $1 \times n$ matrix X with just one row, or "row matrix." This allows us to interpret the equations $y_i = \sum x_i a_{ij}$ of (10) as stating that the row matrix Y is the product of the row matrix X by the matrix A. Thus the linear transformation $T_A: F^m \to F^n$ can be written in the compact form

$$(33) \qquad Y = XA, \qquad X \in F^m, \qquad Y \in F^n.$$

Also, the scalar product cX is just the matrix product of the 1×1 matrix c by the $1 \times n$ (row) matrix X.

Column Vectors. Note that even though Y is a row vector in the equation $XA = Y$, its entries appear in the display (10) in a single column. Hence it is customary to rewrite the matrix equation $XA = Y$ in the *transposed* form $Y^T = A^T X^T$, with Y^T and X^T both column vectors. Changing the notation, there results an equation $BX = Y$ of the form of (11'), with $B = A^T$, and $X = (x_1, \cdots, x_n)^T$ and $Y = (y_1, \cdots, y_n)^T$ both *column* vectors.

In treating bilinear and quadratic forms, row and column vectors are used together. Thus, the inner product $x_1 y_1 + \cdots + x_n y_n$ of two vectors (§7.9) is simply the matrix product of the row matrix X by the column matrix Y^T, so that

$$(34) \qquad (X, Y) = XY^T, \qquad X \text{ and } Y \text{ row matrices.}$$

The row-by-column multiplication of matrices A and B is actually a matrix multiplication of the ith row of A by the kth column of B so that the definition of a product may be written as

$$(35) \qquad AB = \|c_{ik}\| \qquad \text{with } c_{ik} = A_i B^{(k)},$$

where we have employed the notation

$$(36) \qquad A_i = \text{the } i\text{th row of } A, \qquad B^{(k)} = \text{the } k\text{th column of } B.$$

The whole ith row $(c_{i1}, \cdots, c_{in})$ of the product AB uses only the ith row in A and the various columns of B, hence is the matrix product of A_i by all of B. Similarly, the kth column of AB arises only from the kth column of B. In the notation of (36) these rules are

$$(37) \qquad (AB)_i = A_i B, \qquad (AB)^{(k)} = AB^{(k)}.$$

The second rule may be visualized by writing out B as the row of its columns, for then

(38) $A \cdot \| B^{(1)} \quad B^{(2)} \cdots B^{(r)} \| = \| AB^{(1)} \quad AB^{(2)} \cdots AB^{(r)} \|.$

These columns may also be grouped into sets of columns forming larger submatrices. Thus, a 6×5 matrix B could be considered as a 6×2 matrix $D_1 = \| B^{(1)} \quad B^{(2)} \|$ laid side by side with a 6×3 matrix $D_2 = \| B^{(3)} \quad B^{(4)} \quad B^{(5)} \|$ to form the whole 6×5 matrix $B = \| D_1 \quad D_2 \|$. By (38), the rule for multiplication becomes

(39) $A \cdot \| D_1 \quad D_2 \| = \| AD_1 \quad AD_2 \|;$ D_1 and D_2 n-rowed blocks.

If we decompose the $n \times r$ matrix B into n rows $B_1, \cdots, B_n$, and if $Y = (y_1, \cdots, y_n)$ is a row matrix, the product YB is the row matrix

$$\begin{aligned} YB &= (y_1 b_{11} + \cdots + y_n b_{n1}, \cdots, y_1 b_{1r} + \cdots + y_n b_{nr}) \\ &= y_1 (b_{11}, \cdots, b_{1r}) + \cdots + y_n (b_{n1}, \cdots, b_{nr}) \\ &= y_1 B_1 + \cdots + y_n B_n. \end{aligned}$$

The product YB is thus formed by multiplying the row Y by the "column" of rows B_i. For example, the ith row of AB is by definition the product of the row matrix $A_i = (a_{i1}, \cdots, a_{in})$ by B, hence

(40) $(AB)_i = a_{i1} B_1 + \cdots + a_{in} B_n,$ $i = 1, \cdots, m;$

thus each row of AB is a linear combination of the rows of B. These formulas are special instances of a method of multiplying matrices which have been subdivided into "blocks" or submatrices. It is convenient to sketch other instances of this method.

$$\left(\begin{array}{ccc|ccc} a_{11} & \cdots & a_{1s} & a_{1,s+1} & \cdots & a_{1n} \\ \vdots & & \vdots & \vdots & & \vdots \\ a_{m1} & \cdots & a_{ms} & a_{m,s+1} & \cdots & a_{mn} \end{array} \right) \left(\begin{array}{ccc} b_{11} & \cdots & b_{1r} \\ \vdots & & \vdots \\ b_{s1} & \cdots & b_{sr} \\ \hline b_{s+1,1} & \cdots & b_{s+1,r} \\ \vdots & & \vdots \\ b_{n1} & \cdots & b_{nr} \end{array} \right) \begin{array}{l} \Big\} N_1 \\ \\ \Big\} N_2 \end{array}$$

$$\underbrace{\phantom{a_{11} \cdots a_{1s}}}_{M_1} \quad \underbrace{\phantom{a_{1,s+1} \cdots a_{1n}}}_{M_2}$$

Let the n columns of a matrix A consist of the s columns of a submatrix M_1 followed by a submatrix M_2 with the remaining $n - s$

columns. Make a parallel subdivision of the rows of the matrix B, so that B appears as an $s \times r$ matrix N_1 on top of an $(n - s) \times r$ matrix N_2. The product formula for $AB = C$ subdivides into two corresponding sections

(41) $c_{ik} = (a_{i1}b_{1k} + \cdots + a_{is}b_{sk}) + (a_{i,s+1}b_{s+1,k} + \cdots + a_{in}b_{nk}).$

The first parenthesis uses only the ith row from the first block M_1 of A, and only the kth column from the top block N_1 of B. Therefore this first parenthesis is exactly d_{ik}, the entry in the ith row and kth column of the block product M_1N_1. Likewise, the second parenthesis of (41) is the term d_{ik}^* of the product M_2N_2. Therefore $c_{ik} = d_{ik} + d_{ik}^*$, so the whole product AB is the matrix sum, $M_1N_1 + M_2N_2$. Thus,

(42) $$\|M_1 \quad M_2\| \cdot \left\| \begin{matrix} N_1 \\ N_2 \end{matrix} \right\| = M_1N_1 + M_2N_2.$$

This formula is a row-by-column *multiplication of blocks*, just like the row-by-column multiplication of matrix entries. A similar result holds for any subdivision of columns of A, with a corresponding row subdivision of B. When both rows and columns are subdivided, the rule for multiplication is a combination of (42) and rule (39),

(43) $\begin{pmatrix} M_{11} & M_{12} \\ M_{21} & M_{22} \end{pmatrix} \cdot \begin{pmatrix} N_{11} & N_{12} \\ N_{21} & N_{22} \end{pmatrix}$

$$= \begin{pmatrix} M_{11}N_{11} + M_{12}N_{21} & M_{11}N_{12} + M_{12}N_{22} \\ M_{21}N_{11} + M_{22}N_{21} & M_{21}N_{12} + M_{22}N_{22} \end{pmatrix}.$$

This assumes that the subdivisions fit: that the number of columns in M_{11} equals the number of rows in N_{11}. This rule (43) is exactly the rule for the multiplication of 2×2 matrices, as stated in §8.3, (20), except that the entries M_{ij} and N_{ij} are submatrices or "blocks," and not scalars. We conclude that matrix multiplication under any fitting block subdivision proceeds just like ordinary matrix multiplication.

Exercises

1. Let

$$A = \begin{pmatrix} 1 & 0 & 0 \\ 0 & \frac{1}{2} & \frac{1}{2} \end{pmatrix}, \qquad B = \begin{pmatrix} i & 0 & -i \\ 0 & 1 & 1+i \end{pmatrix}, \qquad X = (1, -1), \qquad Y = (i, 0).$$

(a) Find XA, XB, YA, YB.

(b) Find $3A - 4B$, $A + (1 + i)B$, $(X - (1 + i)Y)(iA + 5B)$.
(c) Find BA^T, AB^T, XAB^T, BA^TY^T.

2. Show that if X is any row vector, then XX^T is the inner product of X with itself, while X^TX is the matrix A with $a_{ij} = x_ix_j$.

3. Find AB, BA, AC, and BC, if

$$A = \begin{pmatrix} 2 & 3 & 0 & 0 \\ 5 & 2 & 0 & 0 \\ 0 & 0 & 4 & 0 \\ 0 & 0 & 0 & 2 \end{pmatrix}, \quad B = \begin{pmatrix} 1 & 0 & 0 & 0 \\ 0 & 1 & 0 & 0 \\ 1 & 2 & 1 & 0 \\ 3 & 4 & 0 & 1 \end{pmatrix}, \quad C = \begin{pmatrix} 0 & 1 \\ 1 & 0 \\ 2 & 0 \\ 0 & 2 \end{pmatrix}.$$

4. Let I^* be the $(r + n) \times n$ matrix formed by putting an $n \times n$ identity matrix on top of an $r \times n$ matrix of zeros. What is the effect of multiplying any $n \times (r + n)$ matrix by I^*?

★5. Prove the "block multiplication rule" (43).

8.6. Inverses

Linear transformations of a finite-dimensional vector space are of two kinds: either bijective (both one-one and onto) or neither injective nor surjective (neither one-one nor onto). For instance, the oblique projection $(x, y, z) \mapsto (x, y + z, 0)$ of three-dimensional Euclidean space onto the (x, y)-plane is neither injective nor surjective.

Definition. *A linear transformation T of a vector space V to itself is called* nonsingular *or* invertible *when it is bijective from V onto V. Otherwise T is called* singular.

A nonsingular linear transformation T is a bijection of V onto V which preserves the algebraic operations of sum and scalar product, so is an isomorphism of the vector space V to itself. Hence a nonsingular linear transformation of V may be called an *automorphism* of V.

The most direct way to prove the main facts about singular and nonsingular linear transformations is to apply the theory of linear independence derived in Chap. 7, using a fixed basis $\alpha_1, \cdots, \alpha_n$ for the vector space V on which a given transformation T operates.

Theorem 9. *A linear transformation T of a vector space V with finite basis $\alpha_1, \cdots, \alpha_n$ is nonsingular if and only if the vectors $\alpha_1 T, \cdots, \alpha_n T$ are linearly independent in V. When this is the case, T has a (two-sided) linear inverse T^{-1}, with $TT^{-1} = T_1^{-1}T = I$.*

Proof. First suppose T is nonsingular. If there is a linear relation

$\sum x_i(\alpha_i T) = 0$ between the $\alpha_i T$, then

$$(x_1\alpha_1 + \cdots + x_n\alpha_n)T = x_1(\alpha_1 T) + \cdots + x_n(\alpha_n T) = 0.$$

Since $0T = 0$, and T is one-one, this implies $x_1\alpha_1 + \cdots + x_n\alpha_n = 0$ and hence, by the independence of the α's, $x_1 = \cdots = x_n = 0$. Therefore the $\alpha_i T$ are linearly independent.

Conversely, assume that the vectors $\beta_1 = \alpha_1 T, \cdots, \beta_n = \alpha_n T$ are linearly independent, and recall that a transformation T is one-one onto if and only if it has a two-sided inverse (§6.2). Since V is n-dimensional, the n independent vectors $\beta_1, \cdots, \beta_n$ are a basis of V. By Theorem 1, there is a linear transformation S of V with

$$\beta_1 S = \alpha_1, \qquad \beta_2 S = \alpha_2, \cdots, \qquad \beta_n S = \alpha_n.$$

Thus for each $i = 1, \cdots, n, \beta_i(ST) = \beta_i$. Since the $\beta_1, \cdots, \beta_n$ are a basis, there is by Theorem 1 only one linear transformation R with $\beta_i R = \beta_i$ for every i, and this transformation is the identity. Hence $ST = I$. Similarly, $\alpha_i(TS) = \beta_i S = \alpha_i$, and, since the α's are a basis, $TS = I$. Thus S is the inverse of T, and T is nonsingular.

Thus to test the nonsingularity of T, one may test the linear independence of the images of *any* finite basis of V, as by the methods of §7.6.

Corollary 1. *Let T be a linear transformation of a finite-dimensional vector space V. If T is nonsingular, then* (i) *T has a two-sided linear inverse,* (ii) *$\xi T = 0$ and ξ in V imply $\xi = 0$,* (iii) *T is one-one from V into V,* (iv) *T transforms V onto V. If T is singular, then* (i') *T has neither a left- nor a right-inverse,* (ii') *$\xi T = 0$ for some $\xi \neq 0$,* (iii') *T is not one-one,* (iv') *T transforms V into a proper subspace of V.*

Proof. Condition (i) was proved in Theorem 9. Again, if $\xi T = 0$ for some $\xi \neq 0$, then since $0T = 0$, T would not be one-one, contrary to the definition of "nonsingular." This proves (ii); (iii), and (iv) are parts of the definition. Again, if T is singular, then for any basis $\alpha_1, \cdots, \alpha_n$ of V, the $\alpha_i T$ are linearly dependent by Theorem 9. Therefore

$$0 = x_1\alpha_1 T + \cdots + x_n\alpha_n T = (x_1\alpha_1 + \cdots + x_n\alpha_n)T = \xi T$$

for some $x_1, \cdots, x_n$ not all zero—hence (the α_i being independent) for some $\xi \neq 0$, which proves (ii'). Since $0T = 0$, it follows that T is not one-one, proving (iii'). Again, since the $\alpha_i T$ are linearly dependent and V is n-dimensional, they span a proper subspace of V, by Theorem 5,

Corollary 2, of §7.4, proving (iv'). Finally, by Theorem 1 of §6.2, (iii') and (iv') are equivalent to (i'). Q.E.D.

Note that since the conditions enumerated in Corollary 1 are incompatible in pairs, all eight conditions are "if and only if" (i.e., necessary and sufficient) conditions. Thus if (iv) holds, then (iv') cannot hold, hence T cannot be singular, hence it must be nonsingular.

Corollary 2. *If the product TU of two linear transformations of a finite-dimensional vector space V is the identity, then T and U are both nonsingular, $T = U^{-1}$, $U = T^{-1}$, and $UT = I$.*

Proof. Since $TU = I$, T has a right-inverse, hence is nonsingular by (i') above, and has an inverse T^{-1} by (i). Then $T^{-1} = T^{-1}(TU) = (T^{-1}T)U = U$, as asserted, and the other conclusions follow. Q.E.D.

In view of the multiplicative isomorphism of Theorem 6 between linear transformations of F^n and $n \times n$ matrices over F, the preceding results can be translated into results about matrices. We define an $n \times n$ matrix A to be *nonsingular* if and only if it corresponds under Theorem 2 to a nonsingular linear transformation T_A of F^n; otherwise we shall call A singular. But the transformation T_A is, by Theorem 2, that transformation which takes the unit vectors of F^n into rows of A. Hence the condition of Theorem 9 becomes (cf. the Corollary of Theorem 6, §7.4):

Corollary 3. *An $n \times n$ matrix over a field F is nonsingular if and only if its rows are linearly independent—or, equivalently, if and only if they form a basis for F^n.*

Similarly, conditions (i) and (i') of Corollary 1 translate into the following result.

Corollary 4. *An $n \times n$ matrix A is nonsingular if and only if it has a matrix inverse A^{-1}, such that*

$$(44) \qquad AA^{-1} = A^{-1}A = I \qquad (A, A^{-1}, I \quad all \quad n \times n).$$

If A has an inverse, so does its transpose, for on taking the transpose of either side of (44), one gets by (31) $(A^{-1})^T A^T = A^T (A^{-1})^T = I$, so that

$$(45) \qquad (A^{-1})^T = (A^T)^{-1}.$$

Thus, if A is nonsingular, so is A^T; moreover, the reverse is true similarly. But by Corollary 4, A^T is nonsingular if and only if its rows are linearly independent. These rows are precisely the columns of A; hence we have

Corollary 5. *A square matrix is nonsingular if and only if its columns are linearly independent.*

If Corollary 2 is translated from linear transformations to matrices, by Theorem 6, we obtain

Corollary 6. *Every left-inverse of a square matrix is also a right-inverse.*

If matrices A and B both have inverses, so does their product,

$$(46) \qquad (AB)^{-1} = B^{-1}A^{-1} \qquad \text{(note the order!)},$$

for $(AB)(B^{-1}A^{-1}) = A(BB^{-1})A^{-1} = AIA^{-1} = AA^{-1} = I.$

Inverses of nonsingular matrices may be computed by solving suitable simultaneous linear equations. If we write the coordinates of the basis vectors as

$$(47) \qquad \begin{aligned} I_1 &= (1, 0, 0, \cdots, 0), \\ I_2 &= (0, 1, 0, \cdots, 0), \\ &\vdots \\ I_n &= (0, 0, \cdots, 0, 1), \end{aligned}$$

then in a given matrix $A = \|a_{ij}\|$ each row A_i is given as a linear combination

$$A_i = \sum_j a_{ij} I_j$$

of the basis vectors. One may try to solve these equations for the "unknowns" I_j in terms of the A_i; the result will be linear expressions for the I_j as

$$(48) \qquad I_j = c_{j1}A_1 + \cdots + c_{jn}A_n = \sum_{k=1}^{n} c_{jk}A_k.$$

By (40), this equation states that the matrix $C = \|c_{jk}\|$ satisfies $CA = I$, hence that $C = A^{-1}$. Another construction for A^{-1} appears in §8.8.

EXAMPLE.   To compute the inverse of the matrix

$$\begin{pmatrix} 1 & 2 & -2 \\ -1 & 3 & 0 \\ 0 & -2 & 1 \end{pmatrix},$$

write its rows as $A_1 = I_1 + 2I_2 - 2I_3$, $A_2 = -I_1 + 3I_2$, $A_3 = -2I_2+I_3$. These three simultaneous equations have a solution $I_1 = 3A_1 + 2A_2 + 6A_3$, $I_2 = A_1 + A_2 + 2A_3$, $I_3 = 2A_1 + 2A_2 + 5A_3$. The coefficients c_{jk} in these linear combinations give the inverse matrix, for one may verify that

$$\begin{pmatrix} 3 & 2 & 6 \\ 1 & 1 & 2 \\ 2 & 2 & 5 \end{pmatrix}\begin{pmatrix} 1 & 2 & -2 \\ -1 & 3 & 0 \\ 0 & -2 & 1 \end{pmatrix} = \begin{pmatrix} 1 & 0 & 0 \\ 0 & 1 & 0 \\ 0 & 0 & 1 \end{pmatrix}.$$

Linear transformations from a finite-dimensional vector space V to a second such space W (over the same field) can well be one-one but not onto, or vice versa. The same is true of linear transformations from an infinite-dimensional vector space to itself. For example, the linear transformation $(x_1, x_2, x_3, \cdots) \mapsto (0, x_1, x_2, x_3, \cdots)$ on the space of infinite sequences of real numbers is one-one but not onto, hence has many (linear) right-inverses, but no left-inverse.

However, a two-sided inverse, when it exists, is necessarily linear even if V is a space of infinite dimensions:

Theorem 10. *If the linear transformation $T: V \to W$ is a one-one transformation of V onto W, its inverse is linear.*

Proof. Let ψ denote the unique inverse transformation of W onto V, not assumed to be linear. Take vectors ξ and η in W and scalars c and d. Since ψT is the identity transformation of W, and T is linear

$$(c\xi + d\eta)\psi T = c\xi + d\eta = c(\xi\psi T) + d(\eta\psi T) = [c(\xi\psi) + d(\eta\psi)]T.$$

Apply ψ to both sides; since $T\psi$ is also the identity, one finds that

(49) $$(c\xi + d\eta)\psi = c(\xi\psi) + d(\eta\psi),$$

an equation which asserts that ψ is linear.   Q.E.D.

A one-one linear transformation T of V onto W is an isomorphism of V to W, in the sense of §7.8.

Corollary 1. *An isomorphism T of V onto W carries any set of independent vectors $\alpha_1, \cdots, \alpha_r$ of V into independent vectors in W, and any set $\beta_1, \cdots, \beta_s$ of vectors spanning V into vectors spanning W.*

Proof. If there is a linear relation

$$x_1(\alpha_1 T) + \cdots + x_r(\alpha_r T) = 0$$

between the $\alpha_i T$, we may apply T^{-1} to find $x_1\alpha_1 + \cdots + x_r\alpha_r = 0$, hence $x_1 = \cdots = x_r = 0$; the $\alpha_i T$ are independent. The proof of the second half is similar.

For any transformation $T: V \to W$, the *image* or transform $S' = (S)T$ under T of a subspace S of V is defined as the set of all transforms ξT of vectors in S. This image is always a subspace of W, for each linear combination $c(\xi T) + d(\eta T) = (c\xi + d\eta)T$ of vectors ξT and ηT in S' is again in S'.

Corollary 2. *For an isomorphism $T: V \to W$, the image under T of any finite-dimensional subspace S of V has the same dimension as S. Thus T carries lines into lines, and planes into planes.*

Exercises

1. Find inverses for the matrices A, B, C, D of Ex. 1, §8.3.
2. (a) Prove that $A = \begin{pmatrix} a & b \\ c & d \end{pmatrix}$ is nonsingular if and only if $ad - bc \neq 0$.
   (b) Show that if A is nonsingular, its inverse is $\Delta^{-1}\begin{pmatrix} d & -b \\ -c & a \end{pmatrix}$, where $\Delta = ad - bc$.
3. Find inverses for the linear transformations R_θ, D_k, S_a of §8.1.
4. Find inverses (if any) for the linear transformations of Ex. 4, §8.1.
5. (a) If $\theta = 45°$, compute the matrix of the transformation $R_\theta^{-1}U_bR_\theta$ (see §8.1), where U_b is the transformation $x' = bx, y' = y$.
   (b) Describe geometrically the effect of this transformation.
   (c) Do the same for $R_\theta^{-1}S_aR_\theta$ (with $\theta = 45°$).
6. If A satisfies $A^2 - A + I = 0$, prove that A^{-1} exists and is $I - A$.
7. Find inverses for the matrices E_1, E_2, and E_3 of Ex. 9, §8.3.
8. Find inverses for the matrices A and B of Ex. 3, §8.5. (*Hint:* Use blocks.)
9. (a) Compute a formula for the inverse of a 2×2 triangular matrix.
   (b) The same for a 3×3. (*Hint:* Try a triangular inverse.)
   (c) Prove that every triangular matrix with no zero terms on the diagonal has a triangular inverse.

10. Given A, B, A^{-1}, B^{-1}, and C, find the multiplicative inverse of

(a) $\begin{pmatrix} A & O \\ O & B \end{pmatrix}$,   (b) $\begin{pmatrix} A & C \\ O & B \end{pmatrix}$,   (c) $\begin{pmatrix} A & O \\ C & B \end{pmatrix}$.

11. Prove that all nonsingular $n \times n$ matrices form a group with respect to matrix multiplication.

12. If a product AB of square matrices is nonsingular, prove that both factors A and B are nonsingular.

★**13.** Prove without appeal to linear transformations that a matrix A has a left-inverse if and only if its rows are linearly independent.

14. Prove Corollary 2 of Theorem 10.

15. Exhibit a linear transformation of the space of sequences $(x_1, x_2, x_3, \cdots)$ *onto* itself which is not one-one.

★**16.** If a linear transformation $T: V \to W$ has a right-inverse, prove that it has a right-inverse which is linear (without assuming finite-dimensionality).

8.7. Rank and Nullity

In general (see §6.2), each transformation (function) $T: S \to S_1$ has given sets S and S_1 as *domain* and *codomain*, respectively. The *range* of T is the set of *transforms* (image of the domain under T).

In case T is a *linear* transformation of a vector space V into a second vector space W the image (set of all ξT) cannot be an arbitrary subset of W.

Lemma 1. *The image of a linear transformation $T: V \to W$ is itself a vector space (hence a subspace of W).*

Proof. Since $c(\xi T) = (c\xi)T$ and $\xi T + \eta T = (\xi + \eta)T$, the set of transforms is closed under the vector operations.

Lemma 2. *Let T_A be the linear transformation corresponding to the $m \times n$ matrix A. Then the image of T_A is the row space of A.*

Proof. The transformation $T_A: F^m \to F^n$ carries each vector $X = (x_1, \cdots, x_m)$ of F^m into $Y = XA$ in F^n, so that the image of T_A consists of all n-tuples of the form

$$Y = XA = \left(\sum x_i a_{i1}, \cdots, \sum x_i a_{in} \right) = \sum x_i (a_{i1}, \cdots, a_{in}).$$

These are exactly the different linear combinations of the rows $A_i = (a_{i1}, \cdots, a_{in})$ of A. The range of T_A is thus the set of all linear combinations of the rows of A. But this is the row space of A, as defined in §7.5.   Q.E.D.

The *rank* of a matrix A has been defined (§7.6) as the (linear) dimension of the row space of A; it is therefore the dimension of the range of T_A. More generally, the *rank* of any linear transformation T is defined as the dimension (finite or infinite) of the image of T.

Since the dimension of the subspace spanned by m given vectors is the maximum number of linearly independent vectors in the set, the rank of A also is the maximum number of linearly independent *rows* of A. For this reason, the rank of A as defined above is often called the *row-rank* of A, as distinguished from the *column-rank*, which is the maximum number of linearly independent columns of A.

Dual to the concept of the row space of a matrix or range of a linear transformation is that of its null-space.

Definition. *The* null-space *of a linear transformation T is the set of all vectors ξ such that $\xi T = 0$. The null-space of a matrix A is the set of all row matrices X which satisfy the homogeneous linear equations $XA = 0$.*

Lemma 3. *The null-space of any linear transformation (or matrix) is a subspace of its domain.*

Proof. If $\xi T = 0$ and $\eta T = 0$, then for all c, c',

$$(c\xi + c'\eta)T = c(\xi T) + c'(\eta T) = 0 + 0 = 0.$$

Hence $c\xi + c'\eta$ is in the null-space, which is therefore a subspace. Q.E.D.

The dimension of the null-space of a given matrix A or a linear transformation T is called the *nullity* of A or T. Nullity and rank are connected by a fundamental equation, valid for both matrices and linear transformations. Because of the correspondence between matrices and linear transformations, we need supply the proof only for one case.

Theorem 11. *Rank + nullity = dimension of domain.*

Thus for an $m \times n$ matrix, (row) rank plus (row) nullity equals m.

Proof. If the nullity of T is s, its null-space N has a basis $\alpha_1, \cdots, \alpha_s$ of s elements, which can be extended to a basis $\alpha_1, \cdots, \alpha_s, \beta_1, \cdots, \beta_r$ for the whole domain of T. Since every $\alpha_i T = 0$, the vectors $\beta_j T$ span the image (range) R of T. Moreover, $x_1(\beta_1 T) + \cdots + x_r(\beta_r T) = 0$ implies $x_1\beta_2 + \cdots + x_r\beta_r$ in N, so that $x_1 + \cdots + x_r = 0$. Hence the vectors $\beta_j T$ are independent and form a basis for R. We conclude that the dimension $m = s + r$ of the domain is the sum of the dimensions s of N and r of R, which is all we need.

Theorem 12. *For a linear transformation* $T: F^n \to F^n$ *to be nonsingular, each of the following conditions is necessary and sufficient:*

$$(a)\ rank\ T = n, \qquad (b)\ nullity\ T = 0.$$

Proof. Condition (a) states that T carries F^n onto itself, while condition (b) states that $\xi T = 0$ implies $\xi = 0$ in F^n. Thus Theorem 12 is just a restatement of conditions (iv) and (ii) of Corollary 1 to Theorem 9.

Exercises

1. Find the ranges, null-spaces, ranks, and nullities of the transformations given in Exs. 1(a)–1(d), 4(a), 4(b), §8.1.
2. Construct a transformation of $\mathbf{R}^3$ into itself which will have its range spanned by the vectors $(1, 3, 2)$ and $(3, -1, 1)$.
3. Construct a transformation of $\mathbf{R}^4$ into itself which has a null-space spanned by $(1, 2, 3, 4)$ and $(2, 2, 4, 4)$.
4. Prove that the row-rank of a product AB never exceeds the row-rank of B.
5. If the $n \times n$ matrix A is nonsingular, show that for every $n \times n$ matrix B the matrices AB, B, and BA all have the same rank.
6. Prove that $rank(A + B) \leqq rank(A) + rank(B)$.
7. Given the ranks of A and B, what is the rank of $\begin{pmatrix} A & O \\ O & B \end{pmatrix}$?

8.8. Elementary Matrices

The elementary row operations on a matrix A, introduced in §7.5, may be interpreted as premultiplications of A by suitable factors. For example, two rows in a matrix may be permuted by premultiplying the matrix by a matrix obtained by permuting the same rows of the identity matrix I. Thus

$$\begin{pmatrix} 0 & 1 \\ 1 & 0 \end{pmatrix}\begin{pmatrix} a_1 & a_2 \\ b_1 & b_2 \end{pmatrix} = \begin{pmatrix} 0 \cdot a_1 + 1 \cdot b_1 & 0 \cdot a_2 + 1 \cdot b_2 \\ 1 \cdot a_1 + 0 \cdot b_1 & 1 \cdot a_2 + 0 \cdot b_2 \end{pmatrix} = \begin{pmatrix} b_1 & b_2 \\ a_1 & a_2 \end{pmatrix}.$$

To add the second row to the first row or to multiply the second row by c, simply do the same for an identity factor in front:

$$\begin{pmatrix} 1 & 1 \\ 0 & 1 \end{pmatrix}\begin{pmatrix} a_1 & a_2 \\ b_1 & b_2 \end{pmatrix} = \begin{pmatrix} a_1 + b_1 & a_2 + b_2 \\ b_1 & b_2 \end{pmatrix},$$

$$\begin{pmatrix} 1 & 0 \\ 0 & c \end{pmatrix}\begin{pmatrix} a_1 & a_2 \\ b_1 & b_2 \end{pmatrix} = \begin{pmatrix} a_1 & a_2 \\ cb_1 & cb_2 \end{pmatrix}.$$

Similar results hold for $m \times n$ matrices; the prefactors used to represent the operations are known as *elementary matrices*.

Definition. *An elementary $m \times m$ matrix E is any matrix obtained from the $m \times m$ identity matrix I by one elementary row operation.*

There are thus three types of elementary matrices, samples of which are

$$(50) \quad \begin{pmatrix} 1 & 0 & 0 & 0 \\ 0 & 0 & 0 & 1 \\ 0 & 0 & 1 & 0 \\ 0 & 1 & 0 & 0 \end{pmatrix} \quad \begin{pmatrix} 1 & 0 & 0 & 0 \\ 0 & 1 & 0 & 0 \\ 0 & 0 & 3 & 0 \\ 0 & 0 & 0 & 1 \end{pmatrix} \quad \begin{pmatrix} 1 & 0 & 0 & 0 \\ d & 1 & 0 & 0 \\ 0 & 0 & 1 & 0 \\ 0 & 0 & 0 & 1 \end{pmatrix}.$$
$$\quad H_{24} \qquad\qquad I + 2E_{33} \qquad\qquad I + dE_{21}$$

In general, let I_k denote the kth row of the $m \times m$ identity matrix I. Then the interchange in I of row i with row j gives the elementary permutation matrix $H = \|h_{ij}\|$ whose rows H_k are

$$(51) \qquad H_i = I_j, \qquad H_j = I_i, \qquad H_k = I_k \qquad (k \neq i, j).$$

Similarly, multiplication of row i in I by a nonzero scalar c gives the matrix M whose rows M_k are given by

$$(52) \qquad M_i = cI_i \quad (c \neq 0), \qquad M_k = I_k \quad (k \neq i).$$

If E_{ij} is, as before, the matrix having the single entry 1 in the ith row and the jth column, and all other entries zero, this matrix M can be written as $M = I + (c - 1)E_{ii}$. Finally, the elementary operation of adding d times the ith row to the jth row, when applied to I, gives the elementary matrix $F = I + dE_{ji}$, whose rows F_k are given by

$$(53) \qquad F_j = I_j + dI_i \qquad F_k = I_k \quad (k \neq j).$$

Theorem 13. *Each elementary row operation on an $m \times n$ matrix A amounts to premultiplication by the corresponding elementary $m \times m$ matrix E.*

This may be proved easily by direct computation of the product EA. Consider, for example, the elementary operation of adding the ith row to the jth row of A. The rows F_k of the corresponding elementary matrix F are then given by (53). The rows of any product EA are always found from the rows of the first factor, by formula (37), so

$$(FA)_j = F_jA = (I_i + I_j)A = I_iA + I_jA = (IA)_i + (IA)_j,$$
$$(FA)_k = F_kA = I_kA = (IA)_k \quad (k \neq j).$$

These equations state that the rows of FA are obtained from the rows of $IA = A$ by adding the ith row to the jth row. In other words, the elementary operation in question does carry A into FA, as asserted in Theorem 13.

Corollary 1. *Every elementary matrix E is nonsingular.*

Proof. E is obtained from I by certain operations. The reverse operation corresponds to some elementary matrix E^* and carries E back into I. By Theorem 13 it carries E into E^*E, so $E^*E = I$, E has a left-inverse E^*, and so is nonsingular.

Corollary 2. *If two $m \times n$ matrices A and B are row-equivalent, then $B = PA$, where P is nonsingular.*

For, by Theorem 13, $B = E_nE_{n-1} \cdots E_1A$, where the E_i are elementary, and so nonsingular.

The equivalence between row operations and premultiplication gives to Gauss elimination another useful interpretation, in the usual case that no zeros are produced on the main diagonal, for in this case, not only is the coefficient matrix A reduced to upper triangular form U (which is obvious), but since subtracting multiples of any ith row from *later* rows amounts to premultiplication by a *lower* triangular matrix L_k, we have

$$U = L_sL_{s-1}L_{s-2} \cdots L_1A = LA, \quad s \leqq n(n-1)/2,$$

where $L = L_sL_{s-1} \cdots L_1$ is lower triangular. Therefore $Ax = b$ is equivalent to $Ux = Lb$, where $U = LA$. Hence we can write $A = L^{-1}U$, where L^{-1} is lower and U is upper triangular; this is referred to as the "LU-decomposition" of A.

Matrix inverses can be calculated using elementary matrices. Let A be any nonsingular square matrix. By Corollary 3 of Theorem 9 of §7.6, A can be reduced to I by elementary row operations. Hence, by Theorem 13, for suitable elementary matrices E_i,

$$E_sE_{s-1} \cdots E_1A = I.$$

Multiply each side of this equation by A^{-1} on the right. Then

(54) $$E_sE_{s-1}\cdots E_1I = A^{-1}.$$

The matrix on the left is the result of applying to the identity I the sequence of elementary operations $E_1, \cdots, E_s$. This proves

Theorem 14. *If a square matrix A is reduced to the identity by a sequence of row operations, the same sequence of operations applied to the identity matrix I will give a matrix which is the inverse of A.*

This is an efficient construction for the inverse. Given any A, it will by a finite sequence of rational operations either produce an inverse for A or reduce A to an equivalent singular matrix. In the latter event A has no inverse. For matrices larger than 3×3, this method is more efficient than the devices from determinant theory sometimes used to find A^{-1} (cf. Chap. 10).

Incidentally, any nonsingular matrix P is the inverse $(P^{-1})^{-1}$ of another nonsingular matrix; hence, as in (54), it can be written as a product of elementary matrices. This combines with Corollary 1 of Theorem 13 to yield the following result.

Theorem 15. *A square matrix P is nonsingular if and only if it can be written as a product of elementary matrices,*

(55) $$P = E_sE_{s-1}\cdots E_1.$$

Corollary 1. *Two $m \times n$ matrices A and B are row-equivalent if and only if $B = PA$ for some nonsingular matrix P.*

For B is row-equivalent to A if and only if (by Theorem 13) $B = E_nE_{n-1}\cdots E_1A$, where the E_i are elementary. And by Theorem 15 this amounts to $B = PA$, with P nonsingular.

Theorem 15 has a simple geometrical interpretation in the two-dimensional case. The only 2×2 elementary matrices are

$$H_{12} = \begin{pmatrix} 0 & 1 \\ 1 & 0 \end{pmatrix}, \quad M_1 = \begin{pmatrix} c & 0 \\ 0 & 1 \end{pmatrix}, \quad M_2 = \begin{pmatrix} 1 & 0 \\ 0 & c \end{pmatrix},$$

$$F_{12} = \begin{pmatrix} 1 & 0 \\ d & 1 \end{pmatrix}, \quad F_{21} = \begin{pmatrix} 1 & d \\ 0 & 1 \end{pmatrix}.$$

The corresponding linear transformations are, as in §8.1:

(H_{12}) a reflection of the plane in the 45° line through the origin,
(M_i, for c positive) a compression (or elongation) parallel to the x- or y-axis,
(M_i, for c negative) a compression followed by a reflection in the axis,
(F_{ij}) a shear parallel to one of the axes.

This gives

Corollary 2. *Any nonsingular homogeneous linear transformation of the plane may be represented as a product of shears, one-dimensional compressions (or elongations), and reflections.*

This primarily geometric conclusion has been obtained by algebraic argument on matrices. Analogous results may be stated for a space of three or more dimensions.

The elementary row operations on a matrix involve only manipulations within the given field F. If the elements of a matrix A are rational numbers, while the field is that of all real numbers, the elementary operations can be carried out just as if the field contained only the rational numbers. In either field we get the same echelon form, hence the same number of independent rows.

Theorem 16. *If a matrix A over a field F has its entries all contained in a smaller field F', then the rank of A relative to F is the same as the rank of A relative to the smaller field F'.*

The operations with row-equivalence are exactly those used to solve simultaneous linear equations (§2.3 and §7.5). To state the connection, consider m equations

$$\sum_j a_{ij}x_j = b_i \qquad (i = 1, \cdots, m; \quad j = 1, \cdots, n)$$

in the n unknowns x_j. The coefficients of the unknowns form an $m \times n$ matrix $A = \|a_{ij}\|$, while the constant terms b_i constitute a column vector B^T. The equations may be written in matrix form as $AX^T = B^T$, where X^T is the column vector of unknowns (the transpose of the vector $X = (x_1, \cdots, x_n)$). The column of constants B^T may be adjoined to the given matrix A to form an $m \times (n + 1)$ matrix $\|A, B^T\|$, the so-called *augmented matrix* of the given system of equations. Operations on the *rows* of this matrix correspond to operations carrying the given *equations* into equivalent equations, and so *two systems of equations $AX^T = B^T$ and $A^*X^T = B^{*T}$ have the same solutions X^T if their augmented matrices are row-equivalent.*

Exercises

1. Find the row-equivalent echelon form for each of the matrices displayed in Ex. 9(a) of §8.3.
2. (a) Display the possible 3×3 elementary matrices.
   (b) Draw a diagram to represent each $n \times n$ elementary matrix of the form (51)–(53).
3. Find the inverse of each of the 4×4 elementary matrices $H_{24}, I + 2E_{33}, I + dE_{21}$ displayed in the text.
4. Prove Theorem 13 for 2×2 matrices by direct computation for the five matrices displayed after Theorem 15.
5. Find the inverses of $\begin{pmatrix} 1 & 0 & 3 \\ 2 & 4 & 1 \\ 1 & 3 & 0 \end{pmatrix}$ and $\begin{pmatrix} 0 & 1 & 2 \\ 1 & 0 & 2 \\ 1 & 2 & 0 \end{pmatrix}$.
6. Write each of the following matrices as a product of elementary matrices:
   (a) $\begin{pmatrix} 3 & 6 \\ 2 & 1 \end{pmatrix}$, (b) $\begin{pmatrix} 4 & -2 \\ 3 & -5 \end{pmatrix}$, (c) the first matrix of Ex. 5.
7. Represent the transformation $x' = 2x - 5y, y' = -3x + y$ as a product of shears, compressions, and reflections.
★8. For a three-dimensional space, state and prove an analogue of Corollary 2 to Theorem 15. Using Ex. 3, §7.5, sharpen your result.
9. Prove that any nonsingular 2×2 matrix can be represented as a product of the matrices $\begin{pmatrix} 0 & 1 \\ 1 & 0 \end{pmatrix}$, $\begin{pmatrix} 1 & 1 \\ 0 & 1 \end{pmatrix}$, and $\begin{pmatrix} c & 0 \\ 0 & 1 \end{pmatrix}$, where $c \neq 0$ is any scalar. What does this result mean geometrically?
10. Show that the rank of a product never exceeds the rank of either factor.
11. Prove that a system of linear equations $AX^T = B^T$ has a solution if and only if the rank of A equals the rank of the augmented matrix $\|A, B^T\|$.
12. Let $AX^T = B^T$ be a system of nonhomogeneous linear equations with a particular solution $X^T = X_0{}^T$. Prove that every solution X^T can be written as $X^T = X_0{}^T + Y^T$, where Y^T is a solution of the homogeneous equations $AY^T = 0$, and conversely.
13. Prove: If a system of linear equations with coefficients in a field F has no solutions in F, it has no solutions in any larger field.

8.9. Equivalence and Canonical Form

Operations analogous to elementary row operations can also be applied to columns. An *elementary column operation* on an $m \times n$ matrix A thus means either (i) the interchange of any two columns of A, (ii) the multiplication of any column by any nonzero scalar, or (iii) the addition of any multiple of one column to another column.

The replacement of A by its transpose A^T changes elementary column operations into elementary row operations, and vice versa. In particular, A can be transformed into B by a succession of elementary *column* operations if and only if the transpose A^T can be transformed into B^T by a succession of elementary *row* operations. Applying Corollary 1 of Theorem 15, this means that $B^T = PA^T$, or $B = (B^T)^T = (PA^T)^T = AP^T = AQ$, where $Q = P^T$ is nonsingular. Conversely, $B = AQ$, for a nonsingular Q makes B column-equivalent to A. Hence, application of column operations is equivalent to *post*multiplication by nonsingular factors. The explicit postfactors corresponding to each elementary operation may be found by applying this operation to the identity matrix, much as in Theorem 13.

Column and row operations may be applied jointly. We may define two $m \times n$ matrices A and B to be *equivalent* if and only if A can be changed to B by a succession of elementary row *and* column operations, and we then get the following result.

Theorem 17. *An $m \times n$ matrix A is equivalent to a matrix B if and only if $B = PAQ$ for suitable nonsingular $m \times m$ and $n \times n$ matrices P and Q.*

Using simultaneous row and column operations, we can reduce matrices to a very simple *canonical form* (see §9.5).

Theorem 18. *Any $m \times n$ matrix is equivalent to a diagonal matrix D in which the diagonal entries are either 0 or 1, all the 1's preceding all the 0's on the diagonal.*

Explicitly, if r is the number of nonzero entries in D, where clearly $r \leqq m, r \leqq n$, then $D = D_r$ may be displayed in block form as

$$(56) \quad D_r = \begin{pmatrix} I_r & O_{r,n-r} \\ O_{m-r,r} & O_{m-r,n-r} \end{pmatrix}, \quad I_r \text{ the } r \times r \text{ identity matrix,}$$

where $O_{i,j}$ denotes the $i \times j$ matrix of zeros.

The proof is by induction on the number m of rows of A. If all entries of A are zero, there is nothing to prove. Otherwise, by permuting rows and columns we can bring some nonzero entry c to the a_{11} position. After the first row is multiplied by c^{-1}, the entry a_{11} is 1. All other entries in the first column can be made zero by adding suitable multiples of the first row to each other row, and the same may be done with other elements of the first row. This reduces A to an equivalent matrix of the form

$$(57) \qquad B = \begin{pmatrix} 1 & O \\ O & C \end{pmatrix}, \qquad C \text{ an } (m-1) \times (n-1) \text{ matrix.}$$

Upon applying the induction assumption to C we are done.

Theorem 19. *Equivalent matrices have the same rank.*

Proof. We already know (§7.5, Theorem 7) that row-equivalent matrices have the same row space, and hence the same rank. Hence we need only show that column-equivalent matrices A and $B = AQ$ (Q nonsingular) have the same rank. Again, by Theorem 11, this is true if A and B have the same nullity, which is certainly true if they have the same null-space. But $XA = O$ clearly implies $XB = XAQ = OQ = O$, and conversely, $XB = O$ implies $XA = XAQQ^{-1} = XBQ^{-1} = OQ^{-1} = O$. That is, *column-equivalent matrices have the same null-space.*

Corollary 1. *An $m \times n$ matrix A is equivalent to one and only one diagonal matrix of the form (56); the rank r of A determines the number r of units on the diagonal.*

Corollary 2. *Equivalent matrices have the same column rank.*

Proof. The column rank of A (the maximum number of independent columns of A) equals the (row) rank of its transpose A^T. But the equivalence of A to B entails the equivalence of the transposes A^T and B^T. By the theorem, A^T and B^T have the same rank, so A and B have the same column rank.

In the canonical form (56) the rank is the same as the column rank; since both ranks are unaltered by equivalence, we deduce

Corollary 3. *The (row) rank of a matrix always equals its column rank.*

Corollary 4. *Two $m \times n$ matrices are equivalent if and only if they have the same rank.*

If equivalent, they have the same rank (Theorem 19); if they have the same rank, both are equivalent to the same canonical D; hence to each other.

Corollary 5. *An $n \times n$ matrix A is nonsingular if and only if it is equivalent to the identity matrix I.*

For, by Corollary 4, A is equivalent to I if and only if it has rank n; by Theorem 12, this is true if and only if A is nonsingular.

Exercises

1. Check Corollary 3 of Theorem 19 by computing both the row and the column ranks (a) in Ex. 1 of §7.6, (b) in Exs. 7(a), 7(b) of §7.6.
2. Find an equivalent diagonal matrix for each matrix of Ex. 2, §7.6.
3. Do the same for the matrices of Ex. 7, §7.6.
4. Let T be a linear transformation of an m-dimensional vector space V into an n-space W. Show that by suitably choosing bases in V and in W the equations of T take on the form $y_i = x_i$ $(i = 1, \cdots, r)$, $y_j = 0$ $(j = r + 1, \cdots, n)$.
5. (a) Prove that the transpose of any elementary matrix is elementary.
   (b) Use this to give an independent proof of the fact that the transpose of any nonsingular matrix is nonsingular.
★6. If A and B are $n \times n$ matrices of ranks r and s, prove that the rank of AB is never less than $(r + s) - n$. (*Hint*: Use the canonical form for A.)
★7. (a) Prove Sylvester's law of nullity: the nullity of a product AB never exceeds the sum of the nullities of the factors and is never less than the nullity of A; if A is square, it is also at least the nullity of B.
   (b) Give examples to show that both of these limits can be attained by the nullity of AB.
★8. Let A_k denote the $k \times k$ submatrix of A consisting of the a_{ij} with i, j = 1, $\cdots$, k. Show that if no A_k is singular, $k = 1, \cdots, n$, then $A = LU$, where L is lower and U upper triangular.
9. Prove: An $m \times n$ matrix A has rank at most 1 if and only if it can be represented as a product $A = BC$, where B is $m \times 1$ and C is $1 \times n$.
10. Prove that any matrix of rank r is the sum of r matrices of rank 1.
★11. Let the sequence $E_1, \cdots, E_r$ of elementary row operations, suitably interspersed with the elementary column operations $E_1', \cdots, E_s'$, reduce A to I. Show that $A^{-1} = QP$, where $P = E_r \cdots E_1$ and $Q = E_1' \cdots E_s'$ are obtained from I by the same sequences of elementary operations.
12. Show that if $PAQ = D$, as in Theorem 18, then the system $AX^T = B^T$ of simultaneous linear equations (§8.8) may be solved by solving $DY^T = PB^T$ and then computing $X^T = QY^T$.

★8.10 Bilinear Functions and Tensor Products

Now let V and W be any vector spaces over the same field F. A *bilinear* function $f(\xi, \eta)$ of the *two* variables $\xi \in V$ and $\eta \in W$ is defined as a function with values in F such that

$$(58) \qquad f(a\xi + b\xi', \eta) = af(\xi, \eta) + bf(\xi', \eta), \quad \text{and}$$

$$(58') \qquad f(\xi, c\eta + d\eta') = cf(\xi, \eta) + df(\xi, \eta')$$

for all $\xi, \xi' \in V$ and $\eta, \eta' \in W$. Repeating the argument used to prove Theorem 23 of §7.12, one easily obtains the following result.

Theorem 20. *If V and W have finite bases $\beta_1, \cdots, \beta_m$ and $\gamma_1, \cdots, \gamma_n$, respectively, then every bilinear function $f(\xi, \eta)$ of the variables $\xi = x_1\beta_1 + \cdots + x_m\beta_m \in V$ and $\eta = y_1\gamma_1 + \cdots + y_n\gamma_n \in W$ has the form*

(59) $$f(\xi, \eta) = \sum_{i=1}^{m} \sum_{j=1}^{n} x_i a_{ij} y_j, \qquad a_{ij} = f(\beta_i, \gamma_j).$$

Note that the two equations of (59) describe inverse functions $A \mapsto f$ and $f \mapsto A$ between $m \times n$ matrices $A = \|a_{ij}\|$ over F and bilinear functions $f: F^m \times F^n \to F$, where $F^m \times F^n$ is the Cartesian product of F^m and F^n defined in §1.11. Hence (59) is a *bijection*.

The preceding bijection can be generalized. We can define bilinear functions $h(\xi, \eta)$ of variables ξ and η from vector spaces V and W with values in a third vector space U (U, V, and W all over the same field F). Namely, such a function $h: V \times W \to U$ is *bilinear* when it satisfies (58) and (58').

There are many such functions. For example, the outer product $\xi \times \dot{\eta}$ of two vectors from $\mathbf{R}^3$ is bilinear with $U = V = W = \mathbf{R}^3$. Likewise, if we let $U = V = W = M_n$ be the vector space of all $n \times n$ matrices over F, the "matrix product" function $p(A, B) = AB$ is bilinear from $M_n \times M_n$ to M_n, as stated in Theorems 3 and 5.

The result of Theorem 20 holds in the preceding, more general context, and its proof is similar.

Theorem 21. *Let the vector spaces V and W over F have finite bases $\beta_1, \cdots, \beta_m$ and $\gamma_1, \cdots, \gamma_n$, respectively. Then any mn vectors θ_{ij} in a third vector space U over F determine a bilinear function $h: V \times W \to U$ by the formula*

(60) $$h(\xi, \eta) = \sum_{i=1}^{m} \sum_{j=1}^{n} x_i y_j \theta_{ij}$$

for ξ and η expressed as before. Moreover, any bilinear $h: V \times W \to U$ has this form for $\theta_{ij} = h(\beta_i, \gamma_j)$, so that $H \mapsto h$ is a bijection from the set of $m \times n$ matrices $H = \|\theta_{ij}\|$ with entries in U to the set of bilinear functions $h: V \times W \to U$.

This theorem suggests a way of getting a single standard or "most general" bilinear function $\otimes$ on $V \times W$, with the symbol $\otimes$ usually written between its arguments as $\xi \otimes \eta = \otimes(\xi, \eta)$. The values of this function $\otimes$ lie in a new vector space called $V \otimes W$; indeed, we construct this space so as to have a basis of mn vectors, α_{ij}, for $i = 1, \cdots, m$ and $j = 1, \cdots, n$ which serve as the values $\alpha_{ij} = \beta_i \otimes \gamma_j$ of $\otimes$ on the given basis elements of V and W. This means that the function $\otimes$ can be defined by

$$(61) \quad (x_1\beta_1 + \cdots + x_m\beta_m) \otimes (y_1\gamma_1 + \cdots + y_n\gamma_n) = \sum_{i=1}^{m} \sum_{j=1}^{n} x_i y_j \alpha_{ij},$$

as in (60), with the θ_{ij} replaced by α_{ij}. However, this new space $V \otimes W$ is best described by an intrinsic property not referring to any choice of bases in V and W, as follows:

Theorem 22. *For any given finite-dimensional vector spaces V and W over a field F, there exist a vector space $V \otimes W$ and a bilinear function*

$$\otimes: V \times W \to V \otimes W$$

with the following property: Any bilinear function $h: V \times W \to U$ to any vector space U over F can be expressed in terms of $\otimes: V \times W \to V \otimes W$ as

$$h(\xi, \eta) = (\xi \otimes \eta)T, \quad \xi \in V, \quad \eta \in W,$$

for a unique linear function $T: V \otimes W \to U$.

Proof. We first construct $\otimes$ as above. Then any bilinear h can be expressed, as in (60), in terms of the mn vectors $\theta_{ij} = h(\beta_i, \gamma_j)$. Now the parallel between (60) and (61) leads to the linear transformation $T: V \otimes W \to U$, which is uniquely determined as that transformation which takes each basis vector α_{ij} of $V \otimes W$ to θ_{ij} in U. Then formula (60) becomes

$$h(\xi, \eta) = \sum\sum x_i y_j (\alpha_{ij}T) = \left(\sum\sum x_i y_j \alpha_{ij}\right)T = (\xi \otimes \eta)T$$

as required. On the other hand, if $h(\xi, \eta) = (\xi \otimes \eta)T'$ for some linear $T': V \otimes W \to U$, then

$$\alpha_{ij}T' = (\beta_i \otimes \gamma_j)T' = \theta_{ij}, \quad i = 1, \cdots, m, \quad j = 1, \cdots, n,$$

so T' must be the T used above. Therefore T is unique, as asserted in the theorem.

EXAMPLE. Let $V = F^m$, $W = F^n$, and let the β_1 and γ_j be the standard unit vectors ε_i and $\varepsilon_j{}'$ in these spaces. Then $V \otimes W = F^{mn}$ can be the space of all $m \times n$ matrices $\|a_{ij}\|$, while $\otimes$ maps each $(\xi, \eta) \in V \times W$ into the *rank one* matrix $\|x_i y_j\| = \|a_{ij}\|$. Each bilinear θ: $V \times W \to U$ is then determined by the nm vectors $\theta(\varepsilon_i, \varepsilon_j{}') = h_{ij}$. Then the function θ is clearly the composite $\otimes T$ of $\otimes$ as defined above and the linear function $T: V \otimes W \to U$ defined by the formula $T(\|a_{ij}\|) = \sum a_{ij} h_{ij}$, because for all $\xi \in V$, $\eta \in W$, $(\xi \otimes \eta)T = \sum x_i y_j h_{ij}$.

Universality. This theorem can be represented by a diagram

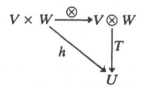

in which the top row is the "standard" bilinear function $\otimes$, and the bottom row is any bilinear function h; the theorem states that there is always exactly one linear transformation T, so that the diagram "commutes" as $\otimes T = h$; that is, so that $h(\xi, \eta) = (\xi \otimes \eta)T$. For this reason, $\otimes$ is called the *universal* bilinear function—any other h can be obtained from it.

In particular, if we constructed any other standard bilinear function $\otimes'$ with the same "universal" property—say, by using different bases for V and W—we would have a diagram

$$V \times W \xrightarrow{\otimes} V \otimes W$$

with $\otimes T = \otimes'$ and $\otimes' T' = \otimes$. This means that $\otimes TT' = \otimes = \otimes I$, with I the identity. In turn, by the theorem, this means that $TT' = I$. Similarly, $T'T = I$, so T is invertible with inverse T' and so is an isomorphism $V \otimes W \cong V \otimes' W$.

The space $V \otimes W$ with this universal property is called the *tensor product* of the spaces V and W; this last result shows that the "universal" property determines this space uniquely *up to an isomorphism*. For example, had we constructed $V \otimes W$ not from the bases $\beta_1, \cdots, \beta_n$ and $\gamma_j, \cdots, \gamma_n$, but from some different bases for V and W, we would have obtained an isomorphic space $V \otimes W$. For that matter, this tensor product space $V \otimes W$ can be constructed in other ways, without using any bases (or using infinitely many basis vectors for infinite-dimensional spaces V and W); it always has the same "universal" property. Our particular construction with its basis $\beta_i \otimes \gamma_j$ shows that its dimension is

$$\dim (V \otimes W) = (\dim V) + (\dim W).$$

Specifically, given one space V and its dual space V^*, we can construct various tensor products:

$$V \otimes V, \qquad V \otimes V \otimes V, \cdots, V \otimes V^*, \qquad V \otimes V^* \otimes V, \cdots$$

These are the spaces of tensors used in differential geometry and relativity theory.

Exercises

1. Show that the mapping $f \mapsto A$ defined by (59) is an isomorphism of vector spaces from the space of all bilinear functions on $V \times W$ to the space of all $m \times n$ matrices over F.
2. Show that the formula $q(x) = a(x)p'(x)$ defines a bilinear function $\phi(a, p) = q$ from the Cartesian product $\mathbf{R}[x] \times \mathbf{R}[x]$ of two copies of the space of all real polynomials to $\mathbf{R}[x]$.
3. Show that the function $p(A, B) = AB$ is bilinear from $V \times W$ to U, where V and W are the spaces of all $m \times r$ and all $r \times n$ matrices over F, respectively. What is U?

In Exs. 4 and 5, let U, V, and W be any vector spaces over a field F.

4. Establish the following *natural* isomorphisms:

$$V \otimes F \cong V, \qquad V \otimes W \cong W \otimes V, \qquad u \otimes (V \otimes W) \cong (U \otimes V) \otimes W.$$

5. Show the set $\mathrm{Hom}\,(V \otimes W, U) = \mathrm{Hom}\,(V, \mathrm{Hom}\,(W, U))$.
★6. Every vector in $V \otimes W$ is a *sum* of terms $\xi \otimes \eta$. Show that there are vectors *not* representable as a single summand $\xi \otimes \eta$. (*Hint:* Take $V = F^2 = W$.)
★7. The Kronecker product $A \otimes B$ of an $m \times m$ matrix A and an $n \times n$ matrix B is the matrix C with entries $c_{pq} = a_{ik}b_{jl}$, where the p and q are the pairs (i, j) and (k, l), suitably ordered. To what linear transformation on $V \otimes W$ does $A \otimes B$ naturally correspond?

★8.11. Quaternions

The algebraic laws valid for square matrices apply to other algebraic systems, such as the quaternions of Hamilton. These quaternions constitute a four-dimensional vector space over the field of real numbers, with a basis of four special vectors denoted by 1, i, j, k. The algebraic operations for quaternions are the usual two vector operations (vector addition and scalar multiplication), plus a new operation of quaternion multiplication.

Definition. *A* quaternion *is a vector* $x = x_0 + x_1i + x_2j + x_3k$, *with real coefficients* x_0, x_1, x_2, x_3. *The product of any two of the quaternions* 1, i, j, k *is defined by the requirement that* 1 *act as an identity and by the table*

(62)
$$i^2 = j^2 = k^2 = -1,$$
$$ij = -ji = k, \quad jk = -kj = i, \quad ki = -ik = j.$$

If c and d are any scalars, while l, m are any two of 1, i, j, k, the product $(cl)(dm)$ is defined as $(cd)(lm)$. These rules, with the distributive law, determine the product of any two quaternions.

Thus, if $x = x_0 + x_1i + x_2j + x_3k$ and $y = y_0 + y_1i + y_2j + y_3k$ are any two quaternions, then their product is

(63)
$$\begin{aligned} xy = {}& x_0y_0 - x_1y_1 - x_2y_2 - x_3y_3 \\ &+ (x_0y_1 + x_1y_0 + x_2y_3 - x_3y_2)i \\ &+ (x_0y_2 + x_2y_0 + x_3y_1 - x_1y_3)j \\ &+ (x_0y_3 + x_3y_0 + x_1y_2 - x_2y_1)k. \end{aligned}$$

Though the multiplication of quaternions is noncommutative, they satisfy every other postulate for a field. Number systems sharing this property are called division rings.

Definition. *A* division ring *is a system R of elements closed under two single-valued binary operations, addition and multiplication, such that*

(i) *under addition, R is a commutative group with an identity* 0;
(ii) *under multiplication, the elements other than* 0 *form a group;*
(iii) *both distributive laws hold:*

$$a(b + c) = ab + ac \quad \text{and} \quad (a + b)c = ac + bc.$$

From these postulates, the rule $a0 = 0a = 0$ and thence the associative law for products with a factor 0 can be deduced easily. It follows that any commutative division ring is a field. We note also that analogues of the results of §§8.1–8.7 are valid over division rings if one is careful about the side on which the scalar factors appear. Thus for the product $c\xi$ of a vector ξ by a scalar c, we write the scalar to the left, but in defining (§8.2) the product of a transformation T by a scalar, we write the scalar on the right $\xi(Tc) = (c\xi)T$, and we likewise multiply a matrix by a scalar on the right. The space of linear transformations T of a left vector space over a division ring R is thus a *right* vector space over R.

Theorem 23. *The quaternions form a division ring.*

The proof of every postulate, except for the existence of multiplicative inverses (which implies the cancellation law, by Theorem 3 of §6.4) and the associative law of multiplication, is trivial. To prove that every nonzero quaternion $x = x_0 + x_1 i + x_2 j + x_3 k$ has an inverse, define the *conjugate* of x as $x^* = x_0 - x_1 i - x_2 j - x_3 k$. It is then easily shown that the *norm* of x, defined by $N(x) = xx^*$, is a real number which satisfies

$$(64) \quad N(x) = xx^* = x^*x = x_0^2 + x_1^2 + x_2^2 + x_3^2 > 0 \quad \text{if } x \neq 0.$$

Hence x has the inverse $x^*/N(x)$.

The proof of the associative law is most easily accomplished using complex numbers. Indeed, it is easily seen from (64) that the quaternions $x = x_0 + x_1 i$ with $x_2 = x_3 = 0$ constitute a subsystem isomorphic with the field of complex numbers. Moreover,

$$(65) \qquad x = (x_0 + x_1 i) + (x_2 + x_3 i)j = z_1 + z_2 j,$$

where z_1 and z_2 behave like ordinary complex numbers. Actually, all the rules of (62) are contained in the expansion (65), the associative and distributive laws, and the rules

$$(66) \qquad z_1 j = j z_1^*, \qquad j^2 = -1,$$

where $z_1^* = x_0 - x_1 i$ is the complex (and quaternion!) conjugate of $z_1 = x_0 + x_1 i$. Indeed, the product of two quaternions in the form (65) is

$$(z_1 + z_2 j)(w_1 + w_2 j) = (z_1 w_1 - z_2 w_2^*) + (z_1 w_2 + z_2 w_1^*)j.$$

Using this formula, we can readily verify the associative law.

Every quaternion x satisfies a quadratic equation $f(t) = 0$ with roots x and x^*, and with real coefficients. This equation is

$$f(t) = (t - x)(t - x^*) = t^2 - (x + x^*)t + xx^* = t^2 - 2x_0 t + N(x).$$

Any quaternion $x = x_0 + x_1 i + x_2 j + x_3 k$ can be decomposed into its real part x_0 and its "pure quaternion" part $x_1 i + x_2 j + x_3 k$. These have various interesting properties (cf. Ex. 2(c), 15); one of the most curious concerns the multiplication of the pure quaternions $\xi = x_1 i + x_2 j + x_3 k$

and $\eta = y_1 i + y_2 j + y_3 k$. By definition,

(67) $$\xi\eta = \xi \times \eta - (\xi, \eta),$$

where $\xi \times \eta = (x_2 y_3 - x_3 y_2)i + (x_3 y_1 - x_1 y_3)j + (x_1 y_2 - x_2 y_1)k$ is the usual *outer product* (or "vector product") of ξ and η, and $(\xi, \eta) = x_1 y_1 + x_2 y_2 + x_3 y_3$ is the "inner product" defined in Chap. 7. Largely because of the identity (67), much of present-day three-dimensional vector analysis was couched in the language of quaternions in the half-century 1850–1900.

It was proved by Eilenberg and Niven in 1944 that any polynomial equation $f(x) = a_0 + a_1 x + \cdots + a_n x^n = 0$, with quaternion coefficients, $a_n \neq 0$, and $n > 0$, has a quaternion solution x.

Exercises

1. Solve $xc = d$ for (a) $c = i$, $d = 1 + j$ and (b) $c = 2 + j$, $d = 3 + k$.
2. (a) Prove that $x^2 = -1$ has an infinity of quaternions x as solutions.
   (b) Show why this does not contradict the Corollary of Theorem 3, §3.2, on the number of roots of a polynomial.
   (c) Show that the real quaternions are those whose squares are positive, while the pure quaternions are those whose squares are negative real numbers. Infer that the set of quaternions satisfying $x^2 < 0$ is closed under addition and subtraction.
   (d) Show that if q is not real, $x^2 = q$ has exactly two quaternion solutions.
3. Let $a = 1 + i + j$, $b = 1 + j + k$.
   (a) Find $a + b$, ab, $a - b$, $ia - 2b$, a^*, aa^*.
   (b) Solve $ax = b$, $xa = b$, $x^2 = b$, $bx + (2j + k) = a$.
4. Derive the multiplication table (66) from (62).
5. (a) Show that the norm $N(x) = xx^*$ of x is $x_0^2 + x_1^2 + x_2^2 + x_3^2$.
   (b) Show that $x^* y^* = (yx)^*$.
6. Show that in the group of nonzero quaternions under multiplication, the "center" consists precisely of the real nonzero quaternions.
7. Prove that the solution of a quaternion equation $xa = b$ is uniquely determined if $a \neq 0$.
8. If a quaternion x satisfies a quadratic equation $x^2 + a_0 x + b_0 = 0$, with real coefficients a_0 and b_0, prove that every quaternion $q^{-1}xq$ satisfies the same quadratic equation (if $q \neq 0$).
9. Prove that the multiplication of quaternions is associative. (*Hint:* Use (65) and (66).)
10. In the algebra of quaternions prove the elements ± 1, $\pm i$, $\pm j$, $\pm k$ form a multiplicative group. (This group, which could be defined directly, is known as the *quaternion group*.)

11. (a) Enumerate the subgroups of the quaternion group (Ex. 10), and show they are all normal.
  (b) Show that the quaternion group is not isomorphic with the group of the square.
12. (a) Prove that the quaternions $x_0 + x_1 i + x_2 j + x_3 k$ with rational coefficients x_i form a division ring.
  (b) Show that this is not the case for quaternions with complex coefficients.
  (*Note:* Do not confuse the scalar $\sqrt{-1} \in \mathbf{C}$ with the quaternion unit i.)
13. In a division ring, show that the commutative law for addition follows from the other postulates. (*Hint:* Expand $(a + b)(1 + 1)$ in two different ways.)
14. How many of the conditions of Theorem 2, §2.1, can you prove in a general division ring, if a/b is interpreted as ab^{-1}?
15. Show that the "outer product" of two vectors is not associative.
16. If the integers a and b are both sums of four squares of integers, show that the product ab is also a sum of four squares. (*Hint:* Use Ex. 5.)
17. Derive all of the rules (62) from $i^2 = j^2 = k^2 = ijk = -1$.
18. Does $(AB)^T = B^T A^T$ hold for matrices with quaternion entries?

9

Linear Groups

9.1. Change of Basis

The coordinates of a vector ξ in a space V depend upon the choice of a basis in V (see §7.8); hence any change in the basis will cause a change in the coordinates of ξ. For example, in the real plane $\mathbf{R}^2$, the vector $\beta = 4\varepsilon_1 + 2\varepsilon_2$ has by definition the coordinates $(4, 2)$ relative to the basis of unit vectors $\varepsilon_1, \varepsilon_2$. The vectors

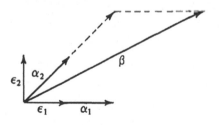

Figure 1

(1) $\quad \alpha_1 = 2\varepsilon_1, \qquad \alpha_2 = \varepsilon_1 + \varepsilon_2$

also form a basis; in terms of this basis, β is expressed as $\beta = \alpha_1 + 2\alpha_2$. The coefficients 1 and 2 are the coordinates of β, relative to this new basis (i.e., relative to the oblique coordinate system shown in Figure 1).

More generally, the coordinates $x_1{}^*, x_2{}^*$ of any vector ξ relative to the "new" basis α_1, α_2 may be found from the "old" coordinates x_1, x_2 of ξ as follows. By definition (§7.8), these coordinates are the coefficients in the expressions

$$\xi = x_1\varepsilon_1 + x_2\varepsilon_2, \qquad \xi = x_1{}^*\alpha_1 + x_2{}^*\alpha_2$$

of ξ in terms of the two bases. Solving the vector equations (1) for ε_1 and

ε_2, we find

$$\varepsilon_1 = \tfrac{1}{2}\alpha_1, \qquad \varepsilon_2 = \alpha_2 - \tfrac{1}{2}\alpha_1.$$

Substituting in the first expression for ξ the values of ε_1 and ε_2, we find

$$\xi = x_1(\tfrac{1}{2}\alpha_1) + x_2(\alpha_2 - \tfrac{1}{2}\alpha_1) = \tfrac{1}{2}(x_1 - x_2)\alpha_1 + x_2\alpha_2.$$

Hence the new coordinates of ξ are given by the linear homogeneous equations

(2) $$x_1^* = (x_1 - x_2)/2, \qquad x_2^* = x_2.$$

Conversely, the old coordinates can be expressed in terms of the new, as $x_1 = 2x_1^* + x_2^*, x_2 = x_2^*$.

Similar relations hold in n dimensions. If $\alpha_1, \cdots, \alpha_n$ is a given basis with its vectors arranged in a definite order, and $\alpha_1^*, \cdots, \alpha_n^*$ a new (ordered) basis, then each vector α_i^* of the new basis can be expressed as a linear combination of vectors of the old basis, in the form

(3) $$\alpha_i^* = p_{i1}\alpha_1 + \cdots + p_{in}\alpha_n = \sum_{j=1}^{n} p_{ij}\alpha_j, \qquad i = 1, \cdots, n.$$

Formally, the expression (3) can be written as the matrix equation $\alpha^* = \alpha P^T$, where P^T is the transpose of P.

The matrix $P = \|p_{ij}\|$ of the coefficients in these expressions has as its ith row the row of old coordinates $(p_{i1}, \cdots, p_{in})$ of the vector α_i^*. Since the vectors α_i^* form a basis, the rows of P are linearly independent, and hence the matrix P is nonsingular (§8.6, Theorem 9). Conversely, if $P = \|p_{ij}\|$ is any nonsingular matrix, and $\alpha_1, \cdots, \alpha_n$ any basis of V, the vectors α_i^* determined as in (3) by P are linearly independent, hence form a new basis of V. This proves

Theorem 1. *If $\alpha_1, \cdots, \alpha_n$ is a basis of the vector space V, then for each nonsingular matrix $P = \|p_{ij}\|$, the n vectors $\alpha_i^* = \sum p_{ij}\alpha_j$, with $i = 1, \cdots, n$, constitute a new basis of V, and every basis of V can be obtained in this way from exactly one nonsingular $n \times n$ matrix P.*

One may also express the old basis in terms of the new basis by equations $\alpha_k = \sum q_{ki}\alpha_i^*$ with a coefficient matrix $Q = \|q_{ki}\|$. Upon substituting the values of α_i^* in terms of the α's, we obtain

$$\alpha_k = \sum_i q_{ki}\left(\sum_j p_{ij}\alpha_j\right) = \sum_j \left(\sum_i q_{ki}p_{ij}\right)\alpha_j.$$

But there is only one such expression for the vectors α_k in terms of themselves; namely, $\alpha_k = \alpha_k$. Hence the coefficient $\sum q_{ki}p_{ij}$ of α_j here must be 0 or 1, according as $k \neq j$ or $k = j$. These coefficients are exactly the (k, j) entries in the matrix product QP; hence $QP = I$, and $Q = P^{-1}$ is the inverse of P.

The parallel result for change of coordinates is as follows:

Theorem 2. *If the basis $\alpha_1, \cdots, \alpha_n$ of the vector space V is changed to a new basis $\alpha_1{}^*, \cdots, \alpha_n{}^*$ expressed in the form $\alpha_i{}^* = \sum_j p_{ij}\alpha_j$, then the coordinates x_i of any vector ξ relative to the old basis α_i determine the new coordinates $x_i{}^*$ of ξ relative to the $\alpha_i{}^*$ by the linear homogeneous equations*

$$(4) \qquad x_j = x_1{}^*p_{1j} + \cdots + x_n{}^*p_{nj} = \sum_{i=1}^{n} x_i{}^*p_{ij}$$

Proof. By definition (§7.8), the coordinates $x_i{}^*$ of ξ relative to the basis $\alpha_i{}^*$ are the coefficients in the expression $\xi = \sum x_i{}^*\alpha_i{}^*$ of ξ as a linear combination of the $\alpha_i{}^*$. Substitution of the formula (3) for $\alpha_i{}^*$ yields

$$\xi = \sum_i x_i{}^* \left(\sum_j p_{ij}\alpha_j \right) = \sum_j \left(\sum_i x_i{}^*p_{ij} \right)\alpha_j.$$

The coefficient of each α_j here is the old coordinate x_j of ξ; hence the equations (4).

The equations (4) may be written in matrix form as $X = X^*P$, where $X = (x_1, \cdots, x_n)$ is the row matrix of old coordinates and $X^* = (x_1{}^*, \cdots, x_n{}^*)$ is the row matrix of new coordinates. Since the α_i and $\alpha_i{}^*$ are bases, P is nonsingular, and one may solve for X^* in terms of X as $X^* = XP^{-1}$.

If one compares this matrix equation with the matrix formulation $\alpha^* = \alpha P^T$ of (3) already mentioned, one gets the interesting relation

$$(5) \qquad \text{bases: } \alpha^* = \alpha P^T, \qquad \text{coordinates: } X^* = XP^{-1}.$$

The matrix P^{-1} of the second equation is the *transposed inverse* of the matrix P^T of the first. (This situation is sometimes summarized by the statement that the change of coordinates is contragredient to the corresponding change of basis.)

Exercises

1. Let T carry the usual unit vectors ε_i (in V_2 or V_3) into the vectors α_i·specified below. Find the corresponding equations for the new coordinates in terms of the old coordinates, and for the old coordinates in terms of the new coordinates. In cases (a) and (b) draw a figure.
   (a) $\alpha_1 = (1, 1), \alpha_2 = (1, -1)$;   (b) $\alpha_1 = (2, 3), \alpha_2 = (-2, -1)$,
   (c) $\alpha_1 = (1, 1, 0), \alpha_2 = (1, 0, 1), \alpha_3 = (0, 1, 1)$;
   (d) $\alpha_1 = (i, 1, i), \alpha_2 = (0, 1, i), \alpha_3 = (0, i, 1)$, where $i^2 = -1$.

2. If a new basis $\alpha_j{}^*$ is given indirectly by equations of the form $\alpha_i = \sum\limits_j q_{ij}\alpha_j{}^*$,

   work out the equations for the corresponding change of coordinates.

3. Give the equations for the transformation of coordinates due to rotation of axes in the plane through an angle θ.

9.2. Similar Matrices and Eigenvectors

A linear transformation $T: V \to V$ of a vector space V may be represented by various matrices, depending on the choice of a basis (coordinate system) in V. Thus, in the plane the transformation defined by $\varepsilon_1 \mapsto 3\varepsilon_1$, $\varepsilon_2 \mapsto -\varepsilon_1 + 2\varepsilon_2$ is represented, in the usual coordinate system of $\mathbf{R}^2$, by the matrix A whose rows are the coordinates of the transforms of ε_1 and ε_2, as displayed below:

$$A = \begin{pmatrix} 3 & 0 \\ -1 & 2 \end{pmatrix}, \qquad D = \begin{pmatrix} 3 & 0 \\ 0 & 2 \end{pmatrix}.$$

But relative to the new basis $\alpha_1 = 2\varepsilon_1$, $\alpha_2 = \varepsilon_1 + \varepsilon_2$ discussed in §9.1 the transformation is $\alpha_1 \mapsto 3\alpha_1$ and $\alpha_2 \mapsto 2\alpha_2$; hence it is represented by the simpler diagonal matrix D displayed above. We shall say that two such matrices A and D are similar.

To generalize this result, let us recall how a matrix represents a transformation. Take any (ordered) basis $\alpha_1, \cdots, \alpha_n$ of a vector space V and any linear transformation $T: V \mapsto V$. Then the images under T of the basis vectors α_i may be written by formula (9) of §8.1 as

(6) $$\alpha_i T = \sum_j a_{ij}\alpha_j, \qquad A = \|a_{ij}\|.$$

Hence T is represented, relative to the basis $\alpha = \{\alpha_1, \cdots, \alpha_n\}$, by the $n \times n$ matrix A. This relation can also be expressed in terms of coordinates. Let $\xi = \sum x_i\alpha_i$ be a vector of V with the n-tuple $X = (x_1, \cdots, x_n)$

of coordinates relative to the basis α. Then the image $\eta = \xi T$ is

$$\xi T = \left(\sum x_i \alpha_i\right) T = \sum x_i (\alpha_i T) = \sum_i \sum_j x_i a_{ij} \alpha_j = \sum_j \left(\sum_i x_i a_{ij}\right) \alpha_j.$$

The coordinates y_i of η are then just the coefficients of α_j, so that

$$y_j = \sum_i x_i a_{ij},$$

and the coordinate vector Y of η is just the matrix product $Y = XA$. Briefly,

(7) $Y = XA,$ where $X = \alpha$-coordinates of ξ,

$Y = \alpha$-coordinates of $\eta = \xi T.$

Either of the equivalent statements (6) or (7) means that T is *represented, relative to the basis α, by the matrix A.*

Now let $\alpha_1^*, \cdots, \alpha_n^*$ be a second basis. Then, by Theorem 1, the new basis is expressed in terms of the old one by a nonsingular $n \times n$ matrix P as in (3); and the new coordinates of ξ and $\eta = \xi T$ are given in terms of the old coordinates, by Theorem 2, as $X^* = XP^{-1}$ and $Y^* = YP^{-1}$. Then by (7)

$$Y^* = YP^{-1} = XAP^{-1} = X^*(PAP^{-1}).$$

Hence, by (7) again, the matrix B representing T in the new coordinate system has the form PAP^{-1}. The equivalence relation $B = PAP^{-1}$ is formally like that of conjugate elements in a group (§6.12); it is of fundamental importance, and called the relation of similarity.

Definition. *Two $n \times n$ matrices A and B with entries in a field F are similar (over F) if and only if there is a nonsingular $n \times n$ matrix P over F with $B = PAP^{-1}$.*

Our discussion above proves

Theorem 3. *Two $n \times n$ matrices A and B over a field F represent the same linear transformation $T: V \to V$ on an n-dimensional vector space V over F, relative to (usually) different coordinate systems, if and only if the matrices A and B are similar.*

More explicitly, we may restate this as

Theorem 3′. *Let the linear transformation $T: V \to V$ be represented by a matrix A relative to a basis $\alpha_1, \cdots, \alpha_n$ of V, let $P = \|p_{ij}\|$ be a nonsingular matrix, and $\alpha_i{}^* = \sum_j p_{ij} \alpha_j$ the corresponding new basis of V. Then T is represented relative to the new basis by the matrix PAP^{-1}.*

The algebra of matrices applies especially smoothly to diagonal matrices: to add or multiply any two diagonal matrices, one simply adds (or multiplies) corresponding diagonal entries. For this and other reasons, it is important to know which matrices are similar to diagonal matrices—and also which pairs of diagonal matrices are similar to each other. The answer to these questions involves the notions of characteristic vector and characteristic root—also called *eigenvector* and *eigenvalue*.

Definition. *An* eigenvector *of a linear transformation $T: V \to V$ is a nonzero vector $\xi \in V$ such that $\xi T = c\xi$ for some scalar c. An* eigenvalue *of T is a scalar c such that $\xi T = c\xi$ for some vector ξ not 0. An eigenvector or eigenvalue of a square matrix A is, correspondingly, a vector $X = (x_1, \cdots, x_n)$ such that $XA = cX$. The set of all eigenvalues of T (or T_A) is called its* spectrum.

Thus each eigenvector ξ of T determines an eigenvalue c, and each eigenvalue belongs to at least one eigenvector. Since similar matrices correspond to the same linear transformation under different choice of bases, similar matrices have the same eigenvalues. Explicitly, the n-tuple $X \neq 0$ is an eigenvector of the $n \times n$ matrix A if $XA = cX$ for some scalar c. If the matrix $B = PAP^{-1}$ is similar to A, then $(XP^{-1})B = XP^{-1}PAP^{-1} = c(XP^{-1})$, so that XP^{-1} is an eigenvector of B belonging to the same eigenvalue c. Note also that any nonzero scalar multiple of an eigenvector is an eigenvector.

The connection between eigenvectors and diagonal matrices is provided by

Theorem 4. *An $n \times n$ matrix A is similar to a diagonal matrix D if and only if the eigenvectors of A span F^n; when this is the case, the eigenvalues of A are the diagonal entries in D.*

In particular, this means that the eigenvalues of a diagonal matrix are the entries on the diagonal.

Proof. Suppose first that A is similar to a diagonal matrix D with diagonal entries $d_1, \cdots, d_n$. The unit vectors $\varepsilon_1 = (1, 0, \cdots, 0), \cdots,$

$\varepsilon_n = (0, \cdots, 0, 1)$ are then characteristic vectors for D, since $\varepsilon_1 D = d_1\varepsilon_1, \cdots, \varepsilon_n D = d_n\varepsilon_n$. Also, the diagonal entries $d_1, \cdots, d_n$ are the corresponding eigenvalues of D and hence of A. They are the only eigenvalues, for let $X = (x_1, \cdots, x_n) \neq 0$ be any characteristic vector of D, so that $XD = cX$ for a suitable eigenvalue c. Now $XD = (d_1 x_1, \cdots, d_n x_n)$, so that $d_i x_i = cx_i$ for all i. Since some $x_i \neq 0$, this proves that $d_i = c$ for this i, and the eigenvalue c is indeed some d_i.

Conversely, suppose there are enough eigenvectors of the matrix A to span the whole space F^n on which T_A operates. Then (§7.4, Theorem 4, Corollary 2) we can extract a subset of eigenvectors $\beta_1, \cdots, \beta_n$ which is a basis for F^n. Since each β_i is an eigenvector, $\beta_1 T_A = c_1\beta_1, \cdots, \beta_n T_A = c_n\beta_n$ for eigenvalues $c_1, \cdots, c_n$. Hence, relative to the basis $\beta_1, \cdots, \beta_n$, T_A is represented as in (6) by the diagonal matrix D with diagonal entries $c_1, \cdots, c_n$, and A is similar to this matrix D.

Corollary. *If P is a matrix whose rows are n linearly independent eigenvectors of the $n \times n$ matrix A, then P is nonsingular and PAP^{-1} is diagonal.*

Proof. We are given n linearly independent n-tuples $X_1, \cdots, X_n$ which are eigenvectors of A, so that $X_i A = c_i X_i$ for characteristic roots $c_1, \cdots, c_n$. The matrix P with rows $X_1, \cdots, X_n$ is nonsingular because the rows are linearly independent. By the block multiplication rule

$$(8) \quad \begin{pmatrix} X_1 \\ \vdots \\ X_n \end{pmatrix} A = \begin{pmatrix} c_1 X_1 \\ \vdots \\ c_n X_n \end{pmatrix} = \begin{pmatrix} c_1 & & 0 \\ & \ddots & \\ 0 & & c_n \end{pmatrix} \begin{pmatrix} X_1 \\ \vdots \\ X_n \end{pmatrix}.$$

This asserts that $PA = DP$, and hence that $PAP^{-1} = D$, where D is the diagonal matrix with diagonal entries $c_1, \cdots, c_n$. The matrix P is, in fact, exactly the matrix required for the change of basis involved in the direct proof of Theorem 4. Q.E.D.

On the other hand, there are matrices which are not similar to any diagonal matrix (cf. Ex. 5 below).

To explicitly construct a diagonal matrix (if it exists!) similar to a given matrix, one thus searches for eigenvalues and eigenvectors. This search is greatly facilitated by the following consideration.

If a scalar λ is an eigenvalue of the $n \times n$ matrix A, and if I is the $n \times n$ identity matrix, then $XA = \lambda X = \lambda XI$ and consequently $X(A - \lambda I) = O$ for some nonzero n-tuple X. The n homogeneous linear equations with matrix $A - \lambda I$ thus have a nontrivial solution; hence by Theorem 9 of §8.6, we have

Theorem 5. *The scalar λ is an eigenvalue of the matrix A if and only if the matrix $A - \lambda I$ is singular.*

For example, the 2×2 matrix

$$(9) \qquad A - \lambda I = \begin{pmatrix} a_{11} - \lambda & a_{12} \\ a_{21} & a_{22} - \lambda \end{pmatrix}$$

is readily seen to be singular if and only if

$$(10) \qquad \lambda^2 - (a_{11} + a_{22})\lambda + a_{11}a_{22} - a_{12}a_{21} = 0.$$

(This merely states that the determinant of $A - \lambda I$ is zero.) Hence we find all eigenvalues by solving this equation. Moreover, for each root λ there is at least one eigenvector, found by solving

$$x_1 a_{11} + x_2 a_{21} = \lambda x_1$$
$$x_1 a_{12} + x_2 a_{22} = \lambda x_2.$$

EXAMPLE. Find a diagonal matrix similar to the matrix $\begin{pmatrix} -3 & 4 \\ 2 & -1 \end{pmatrix}$. The polynomial (10) is $\lambda^2 + 4\lambda - 5$. The roots of this are 1 and -5; hence the eigenvectors satisfy one or the other of the systems of homogeneous equations

$$\begin{array}{cc} -3x + 2y = x & -3x + 2y = -5x \\ \text{or} & \\ 4x - y = y, & 4x - y = -5y. \end{array}$$

Solving these, we get eigenvectors $(1, 2)$ and $(1, -1)$. Using these as a new basis, the transformation takes on a diagonal form. The new diagonal matrix may be written, according to Theorem 3', as a product

$$\begin{pmatrix} 1 & 2 \\ 1 & -1 \end{pmatrix} \begin{pmatrix} -3 & 4 \\ 2 & -1 \end{pmatrix} \begin{pmatrix} 1 & 2 \\ 1 & -1 \end{pmatrix}^{-1} = \begin{pmatrix} 1 & 0 \\ 0 & -5 \end{pmatrix}.$$

Exercises

1. Show that the equations $2x' = (1 + b)x + (1 - b)y$, $2y' = (1 - b)x + (1 + b)y$ represent a compression on the $45°$ line through the origin. Compute the eigenvalues and the eigenvectors of the transformation and interpret them geometrically.

2. Compute the eigenvalues and the eigenvectors of the following matrices over the complex field:

(a) $\begin{pmatrix} 2 & 4 \\ 5 & 3 \end{pmatrix}$, (b) $\begin{pmatrix} 3 & 2 \\ -2 & 3 \end{pmatrix}$, (c) $\begin{pmatrix} 1 & 2 \\ 2 & -2 \end{pmatrix}$, (d) $\begin{pmatrix} -1 & 2i \\ -2i & 2 \end{pmatrix}$.

3. For each matrix A given in Ex. 2 find, when possible, a nonsingular matrix P for which PAP^{-1} is diagonal.

4. (a) Find the complex eigenvalues of the matrix representing a rotation of the plane through an angle θ.

(b) Prove that the matrix representing a rotation of the plane through an angle θ ($0 < \theta < \pi$) is not similar to any real diagonal matrix.

5. Prove that no matrix $\begin{pmatrix} 1 & c \\ 0 & 1 \end{pmatrix}$ is similar to a real or complex diagonal matrix if $c \neq 0$. Interpret the result geometrically.

6. Show that the slopes γ of the eigenvectors of a 2×2 matrix A satisfy the quadratic equation: $a_{21}\gamma^2 + (a_{11} - a_{22})\gamma - a_{12} = 0$.

7. Prove that the set of all eigenvectors belonging to a fixed eigenvalue of a given matrix constitutes a subspace when $\mathbf{0}$ is included among the eigenvectors.

8. Prove that any 2×2 real symmetric matrix not a scalar matrix has two distinct real eigenvalues.

9. (a) Show that two $m \times n$ matrices A and B are equivalent if and only if they represent the same linear transformation $T: V \to W$ of an m-dimensional vector space V into an n-dimensional vector space W, relative to different bases in V and in W.

(b) Interpret Theorem 18, §8.9, in the light of this remark.

★10. Let both A and B be similar to diagonal matrices. Prove that $AB = BA$ if and only if A and B have a common basis of eigenvectors (Frobenius).

★11. (a) Show that if $A = \begin{pmatrix} a & b \\ c & d \end{pmatrix}$ is similar to an orthogonal matrix, then $ad - bc = \pm 1$. (For definition of orthogonal, see §9.4.)

(b) Show that if $ad - bc = 1$, then A is similar to an orthogonal matrix if and only if $A = \pm I$ or $-2 < a + d < 2$.

(c) Show that if $ad - bc = -1$, then A is similar to an orthogonal matrix if and only if $a + d = 0$.

9.3. The Full Linear and Affine Groups

All nonsingular linear transformations of an n-dimensional vector space F^n form a group because the products and inverses of such transformations are again linear and nonsingular (§8.6, Theorem 9). This group is called the *full linear group* $L_n = L_n(F)$. In the one-one correspondence of linear transformations to matrices, products correspond to products, so the full linear group $L_n(F)$ is isomorphic to the group of all nonsingular $n \times n$ matrices with entries in the field F.

The translations form another important group. A translation of the plane moves all the points of the plane the same distance in a specified direction. The distance and direction may be represented by a vector κ of the appropriate magnitude and direction; the translation then carries the end-point of each vector ξ into the end-point of $\xi + \kappa$. A *translation* in any space F^n is a transformation $\xi \mapsto \xi + \kappa$ for κ fixed. Relative to any coordinate system, the coordinates y_i of the translated vector are $y_1 = x_1 + k_1, \cdots, y_n = x_n + k_n$, where the k_i are coordinates of κ. The product of a translation $\xi \mapsto \eta = \xi + \kappa$ by $\eta \mapsto \zeta = \eta + \lambda$ is found by substitution to be the translation $\xi \mapsto \zeta = \xi + (\kappa + \lambda)$. It corresponds exactly to the sum of the vectors κ and λ. Similarly, the inverse of a translation $\xi \mapsto \xi + \kappa$ is $\eta \mapsto \eta - \kappa$. Thus we have proved the following special case of Cayley's theorem (§6.5, Theorem 8):

Theorem 6. *All the translations $\xi \mapsto \xi + \kappa$ of F^n form an Abelian group isomorphic to the additive group of the vectors κ of F^n.*

A linear transformation T followed by a translation yields

$$(11) \qquad \xi \mapsto \eta = \xi T + \kappa \qquad (T \text{ linear}, \kappa \text{ a fixed vector}).$$

An *affine* transformation H of F^n is any transformation of this form. The affine transformations include the linear transformations (with $\kappa = 0$) and the translations (with $T = I$). If one affine transformation (11) is followed by a second, $\eta \mapsto \eta U + \lambda$, the product is

$$(12) \qquad \xi \mapsto (\xi T + \kappa)U + \lambda = \xi(TU) + (\kappa U + \lambda).$$

The result is again affine because $\kappa U + \lambda$ is a fixed vector of F^n. Every translation is one-one and onto, hence it has an inverse; hence the affine transformation (11) will be one-one and onto if and only if its linear part T is one-one. Its inverse will consequently be the affine transformation $\eta \mapsto \xi = \eta T^{-1} - \kappa T^{-1}$, found by solving (11) for ξ. This proves

Theorem 7. *The set of all nonsingular affine transformations of F^n constitutes a group, the affine group $A_n(F)$. It contains as subgroups the full linear group and the group of translations.*

What are the equations of an affine transformation relative to a basis? The linear part T yields a matrix $A = \|a_{ij}\|$; the translation vector has as coordinates a row $K = (k_1, \cdots, k_n)$. The affine transformation thus carries a vector with coordinates $X = (x_1, \cdots, x_n)$ into a vector with coordinates,

$$(13) \quad Y = XA + K, \quad y_j = \sum_{i=1}^{n} x_i a_{ij} + k_j \quad (j = 1, \cdots, n).$$

A transformation is affine if and only if it is expressed, relative to some basis, by nonhomogeneous linear equations such as these.

The product of the transformation (13) by $Z = YB + L$ is

$$(14) \quad Z = X(AB) + KB + L \quad (K, L \text{ row matrices});$$

the formula is parallel to (12). The same multiplication rule holds for a matrix of order $n + 1$ constructed from the transformation (13) by bordering the matrix A to the right by a column of zeros, below by the row K, below and to the right of the single entry 1:

$$(15) \quad \{Y = XA + K\} \leftrightarrow \begin{pmatrix} A & O \\ K & 1 \end{pmatrix} \quad \begin{matrix} (O \text{ is } n \times 1) \\ (K \text{ is } 1 \times n) \end{matrix}.$$

The rule for block multiplication (§8.5, (43)) gives

$$(16) \quad \begin{pmatrix} A & O \\ K & 1 \end{pmatrix}\begin{pmatrix} B & O \\ L & 1 \end{pmatrix} = \begin{pmatrix} AB + O \cdot L & A \cdot O + O \cdot 1 \\ KB + 1 \cdot L & K \cdot O + 1 \cdot 1 \end{pmatrix}$$

$$= \begin{pmatrix} AB & O \\ KB + L & 1 \end{pmatrix};$$

the result is precisely the bordered matrix belonging to the product transformation (14). This proves

Theorem 8. *The group of all nonsingular affine transformations of an n-dimensional space is isomorphic to the group of all those nonsingular $(n + 1) \times (n + 1)$ matrices in which the last column is $(0, \cdots, 0, 1)$. The isomorphism is explicitly given by the correspondence (15).*

Each affine transformation $\xi H = \xi T + \kappa$ determines a unique linear transformation T, and the product of two affine transformations determines as in (12) the product of the corresponding linear parts. This correspondence $H \mapsto T$ maps the group of nonsingular affine transformations H onto the full linear group, and is a homomorphism in the sense of group theory (§6.11). In any homomorphism the objects mapped on the identity form a normal subgroup; in this case the affine transformations H with $T = I$ are exactly the translations. This proves

Theorem 9. *The group of translations is a normal subgroup of the affine group.*

Equation (13) was interpreted above as a transformation of points (vectors), which carried each point $X = (x_1, \cdots, x_n)$ into a new point Y having coordinates $(y_1, \cdots, y_n)$ in the same coordinate system. We could equally well have interpreted equation (13) as a change of *coordinates*. We call the first interpretation an *alibi* (the point is moved elsewhere) and the second an *alias* (the point is renamed).

Thus, in the plane, the equations

$$y_1 = x_1 + 2, \qquad y_2 = x_2 - 1$$

can be interpreted (alibi) as a point transformation which translates the whole plane two units east and one unit south, or (alias) as a change of coordinates, in which the original coordinate network is replaced by a parallel network, with new origin two units west and one unit north of the given origin.

A similar double interpretation applies to all groups of substitutions.

Exercises

1. (a) Represent each of the following affine transformations by a matrix

$$H_1: \qquad x' = 3x + 6y + 2, \qquad y' = 3y - 4;$$
$$H_2: \qquad x' = x + y + 3, \qquad y' = x - y + 5.$$

   (b) Compute the products $H_1 H_2$, $H_2 H_1$.

   (c) Find the inverses of H_1 and H_2.

2. Prove that the set of all affine transformations $x' = ax + by + e$, $y' = cx + dy + f$, with $ad - bc = 1$, is a normal subgroup of the affine group $A_2(F)$.

★3. Given the circle $x^2 + y^2 = 1$, prove that every nonsingular affine transformation of the plane carries this circle into an ellipse or a circle.

4. Which of the following sets of $n \times n$ matrices over a field are subgroups of the full linear group?

   (a) All scalar matrices cI.   (b) All diagonal matrices.

   (c) All nonsingular diagonal matrices. (d) All permutation matrices.

   (e) All monomial matrices.   (f) All triangular matrices.

   (g) All strictly triangular matrices.

(h) All matrices with zeros in the second row.

(i) All matrices in which at least one row consists of zeros.

5. Exhibit a group of matrices isomorphic with the group of all translations of F^n.

6. (a) If Z_2 is the field of integers modulo 2, list all matrices in $L_2(Z_2)$.
   (b) Construct a multiplication table for this group $L_2(Z_2)$.

★7. What is the order of the full linear group $L_2(Z_p)$ when Z_p is the field of integers mod p?

8. Let G be the group of all matrices $A = \begin{pmatrix} a & b \\ 0 & d \end{pmatrix}$ with $ad \neq 0$. Show that the correspondence $A \to a$ is a homomorphism.

9. Map the group of nonsingular 3×3 triangular matrices homomorphically on the nonsingular 2×2 triangular matrices. (*Hint:* Proceed as in Ex. 8, but use *blocks*.)

10. If two fields F and K are isomorphic, prove that the groups $L_n(F)$ and $L_n(K)$ are isomorphic.

11. If $n < m$, prove that $L_n(F)$ is isomorphic to a subgroup of $L_m(F)$.

12. (a) Prove that the center of the linear group $L_n(F)$ consists of scalar matrices cI ($c \neq 0$). (*Hint:* They must commute with every $I + E_{ij}$.)
    (b) Prove that the identity is the only affine transformation that commutes with every affine transformation.

★13. If $L_n(F)$ is the full linear group, show that two affine transformations H_1 and H_2 fall into the same right coset of $L_n(F)$ if and only if $OH_1 = OH_2$ (O is the origin!).

14. Prove that the quotient group $A_n(F)/T_n(F)$ is isomorphic to $L_n(F)$, where A_n denotes the affine group, T_n the group of translations.

15. (a) Show that all one-one transformations $y = (ax + b)/(cx + d)$ with $ad \neq bc$ form a group (called the *linear fractional group*).
    (b) Prove that this group is isomorphic to the quotient group of the full linear group modulo the subgroup of nonzero scalar matrices.
    ★(c) Extend the results to matrices larger than 2×2.

16. (a) Show that the set of all nonsingular matrices of the form $\begin{pmatrix} A & 0 \\ 0 & B \end{pmatrix}$, with A $r \times r$ and B $s \times s$, is a group isomorphic with the direct product $L_r(F) \times L_s(F)$.
    (b) What is the geometric character of the linear transformations of $\mathbf{R}^3$ determined by such a matrix if $r = 2$, $s = 1$?

9.4. The Orthogonal and Euclidean Groups

In Euclidean geometry, length plays an essential role. Hence we seek those linear transformations of Euclidean vector spaces which preserve the lengths $|\xi|$ of all vectors ξ.

Definition. *A linear transformation T of a Euclidean vector space is* orthogonal *if it preserves the length of every vector ξ, so that $|\xi T| = |\xi|$.*

We now determine all orthogonal transformations $Y = XA$ of the Euclidean plane. The transforms

$$(17) \qquad (1,0)\begin{pmatrix} a_1 & a_2 \\ b_1 & b_2 \end{pmatrix} = (a_1, a_2), \qquad (0,1)\begin{pmatrix} a_1 & a_2 \\ b_1 & b_2 \end{pmatrix} = (b_1, b_2)$$

of the unit vectors $(1,0)$ and $(0,1)$ have the length 1, since A is orthogonal. According to the Pythagorean formula for length, this means

$$(18) \qquad\qquad a_1^{\,2} + a_2^{\,2} = 1, \qquad b_1^{\,2} + b_2^{\,2} = 1.$$

In addition, the vector $(1,1)$ has a transform $(a_1 + b_1, a_2 + b_2)$ of length $\sqrt{2}$, so $(a_1 + b_1)^2 + (a_2 + b_2)^2 = 2$. Expanding, and subtracting (18), we find

$$(18') \qquad\qquad a_1 b_1 + a_2 b_2 = 0.$$

By (18) there is an angle θ with $\cos\theta = a_1$, $\sin\theta = a_2$. Then by (18'), $\tan\theta = a_2/a_1 = -b_1/b_2$, whence by (18) $b_2 = \pm\cos\theta$, $b_1 = \pm\sin\theta$. The two choices of sign give exactly the two matrices

$$(19) \qquad\qquad \begin{pmatrix} \cos\theta & \sin\theta \\ -\sin\theta & \cos\theta \end{pmatrix} \qquad \begin{pmatrix} \cos\theta & \sin\theta \\ \sin\theta & -\cos\theta \end{pmatrix}$$

By §8.1, formulas (5) and (5'), these represent rotation through an angle θ and reflection in a line making an angle $\alpha = \theta/2$ with the x-axis, respectively. Hence *every orthogonal transformation of the plane is a rotation or a reflection.*

Geometrically, the inverse of the first orthogonal transformation (19) is obtained by replacing θ by $-\theta$; hence it is the *transpose* of the original. This fact (unlike the trigonometric formulas) generalizes to $n \times n$ orthogonal matrices.

Theorem 10. *An orthogonal transformation T has, for every pair of vectors ξ, η, the properties*
  (i) *T preserves distance, or $|\xi - \eta| = |\xi T - \eta T|$.*
  (ii) *T preserves inner products, or $(\xi, \eta) = (\xi T, \eta T)$.*
  (iii) *T preserves orthogonality, or $\xi \perp \eta$ implies $\xi T \perp \eta T$.*
  (iv) *T preserves magnitude of angles, or $\cos \angle(\xi, \eta) = \cos \angle(\xi T, \eta T)$.*

Proof. Since T is linear, the definition gives (i). Since $\xi \perp \eta$ means $(\xi, \eta) = 0$ and since angle is also definable in terms of inner products (§7.9, (41)), properties (iii) and (iv) will follow immediately from (ii). As for (ii), the "bilinearity" of the inner product proves that $(\xi + \eta, \xi + \eta) = (\xi, \xi) + 2(\xi, \eta) + (\eta, \eta)$. This equation may be solved for (ξ, η) in terms of the "lengths" such as $|\xi| = (\xi, \xi)^{1/2}$, in the form

$$(20) \qquad 2(\xi, \eta) = |\xi + \eta|^2 - |\xi|^2 - |\eta|^2.$$

The orthogonal transformation T leaves invariant the lengths on the right, hence also the inner product on the left of this equation. This proves (ii). Q.E.D.

Conversely, a transformation T known to preserve all inner products must preserve length and hence be orthogonal, for length is defined in terms of an inner product.

Next we ask, which matrices correspond to orthogonal linear transformations? The question is easily answered, at least relative to normal orthogonal bases.

Theorem 11. *Relative to any normal orthogonal basis, a real $n \times n$ matrix A represents an orthogonal linear transformation if and only if each row of A has length one, and any two rows, regarded as vectors, are orthogonal.*

Proof. Any orthogonal transformation T must, by Theorem 10, carry the given basis $\varepsilon_1, \cdots, \varepsilon_n$ into a basis $\alpha_1 = \varepsilon_1 T, \cdots, \alpha_n = \varepsilon_n T$ which is normal and orthogonal. Conversely, if T has this property, then for any vector $\xi = x_1\varepsilon_1 + \cdots + x_n\varepsilon_n$ having the transform $\xi T = x_1\alpha_1 + \cdots + x_n\alpha_n$, we know by Theorem 22 of §7.11 that the length is given by the usual formula as

$$|\xi| = (x_1{}^2 + \cdots + x_n{}^2)^{1/2} = |\xi T|,$$

whence T is orthogonal. The proof is completed by the remark (cf. §8.1) that the ith row of A represents the coordinates of $\alpha_i = \varepsilon_i T_A$ relative to the original basis $\varepsilon_1, \cdots, \varepsilon_n$.

In coordinate form, the conditions on A stated in the theorem are equivalent to the equations

$$(21) \qquad \sum_{k=1}^{n} a_{ik}a_{ik} = 1 \quad \text{for all } i, \qquad \sum_{k=1}^{n} a_{ik}a_{jk} = 0 \quad \text{if } i \neq j.$$

The conclusions (21) are exactly those already found explicitly in (18) and (18') for a two-rowed matrix. If we write A_i for the ith row of the matrix A and A_i^T for its transpose, the inner product of A_i by A_j is the matrix product $A_i A_j^T$ (see (34), §8.5), so the conditions (21) may be written as

(21')
$$A_i A_i^T = 1, \qquad A_i A_j^T = 0 \qquad \text{if } i \neq j.$$

In the row-by-column product AA^T of A by its transpose, the equations (21') state that the ith row times the jth column is $A_i A_j^T = \delta_{ij}$, where δ_{ij} is the element in the ith row and the jth column of the identity matrix $I = \| \delta_{ij} \|$, with diagonal entries $\delta_{ii} = 1$ and all nondiagonal entries zero. (The symbol δ_{ij} is called the *Kronecker delta*.) We have proved

Theorem 12. *A real $n \times n$ matrix represents an orthogonal transformation if and only if $AA^T = I$.*

The equation $AA^T = I$ has meaning over any field, so that the concept of an orthogonal matrix can be defined in general.

Definition. *A square matrix A over any field is called* orthogonal *if and only if $AA^T = I$.*

This means that the transpose A^T of an orthogonal matrix A is a right-inverse of A; hence by Theorem 9 of §8.6, every orthogonal matrix A is nonsingular, with $A^{-1} = A^T$. Therefore $A^TA = I$. This equation may be written as $A^T(A^T)^T = I$, whence A^T is orthogonal: the transpose of any orthogonal matrix A is also orthogonal. From this it also follows that a matrix A is orthogonal if and only if each *column A* has length one, and any two *columns* are orthogonal,

(22)
$$\sum_{k=1}^{n} a_{ki} a_{ki} = 1 \quad \text{for all } i, \qquad \sum_{k=1}^{n} a_{ki} a_{kj} = 0 \quad \text{if } i \neq j.$$

All $n \times n$ orthogonal matrices form a group. This is clear, since the inverse $A^{-1} = A^T$ of an orthogonal matrix is orthogonal and the product of the two orthogonal matrices A and B is orthogonal: $(AB)^T = B^TA^T = B^{-1}A^{-1} = (AB)^{-1}$. This subgroup of the full linear group $L_n(F)$ is called the *orthogonal group* $O_n(F)$; it is isomorphic to the group of all orthogonal transformations of the given Euclidean space if $F = \mathbf{R}$.

By a *rigid motion* of a Euclidean vector space E is meant a nonsingular transformation U of E which preserves distance, i.e., which satisfies $|\xi U - \eta U| = |\xi - \eta|$ for all vectors ξ, η. Any translation of E preserves vector differences $\xi - \eta$, hence their lengths, and so is a rigid motion. Therefore if an affine transformation $\xi \mapsto \xi T + \kappa$ is rigid, so is $\xi \mapsto$

$(\eta - \kappa) = \xi T$; conversely, if T is rigid, so is $\xi \mapsto \eta = \xi T + \kappa$. But by Theorem 10, a linear transformation is rigid if and only if it is orthogonal. We conclude, *an affine transformation* (11) *is a rigid motion if and only if T is orthogonal.* It follows as in the proof of Theorem 7, since the orthogonal transformations form a group, that the totality of rigid affine transformations constitutes a subgroup of the affine group, called the *Euclidean group.* It is the basis of Euclidean geometry.†

Various other geometrical groups exist. A familiar one is the group of similarity transformations T, consisting of those linear transformations T which alter all lengths by a numerical factor $c_T > 0$, so that $|\xi T| = c_T |\xi|$. One may prove that they do in fact form a group which contains the orthogonal group as subgroup. The "extended" similarity group consists of all affine transformations $\xi \mapsto \xi T + \kappa$ in which T is a similarity transformation.

Exercises

1. Test the following matrices for orthogonality. If a matrix is orthogonal, find its inverse:

    (a) $\begin{pmatrix} 1/2 & \sqrt{3}/2 \\ -\sqrt{3}/2 & 1/2 \end{pmatrix}$,   (b) $\begin{pmatrix} 1/2 & \sqrt{3}/2 \\ \sqrt{3}/2 & 1/2 \end{pmatrix}$,   (c) $\begin{pmatrix} .6 & .8 \\ .8 & -.6 \end{pmatrix}$.

2. Find an orthogonal matrix whose first row is a scalar multiple of $(5, 12, 0)$.

3. If the columns of an orthogonal matrix are permuted, prove that the result is still orthogonal.

4. If A and B are orthogonal, prove that $\begin{pmatrix} A & 0 \\ 0 & B \end{pmatrix}$ and $\begin{pmatrix} 0 & A \\ B & 0 \end{pmatrix}$ are also.

5. Multiply the following two matrices, and test the resulting product for orthogonality:

$$\begin{pmatrix} \cos\phi & \sin\phi & 0 \\ -\sin\phi & \cos\phi & 0 \\ 0 & 0 & 1 \end{pmatrix}, \quad \begin{pmatrix} 1 & 0 & 0 \\ 0 & \cos\theta & \sin\theta \\ 0 & -\sin\theta & \cos\theta \end{pmatrix}.$$

6. Show that the Euclidean group is isomorphic to a group of matrices.

7. Prove that all translations form a normal subgroup of the Euclidean group.

8. As an alternative proof of (ii) in Theorem 10, show from first principles that $4(\xi, \eta) = |\xi + \eta|^2 - |\xi - \eta|^2$.

9. Show that an affine transformation H commutes with every translation if and only if H is itself a translation.

10. Prove that any similarity transformation S can be written in the form $S = cT$ as a product of a positive scalar and an orthogonal transformation T in one and only one way.

† It is a fact that any rigid motion is necessarily affine; hence the Euclidean group is actually the group of all rigid motions.

11. Give necessary and sufficient conditions that a matrix A represent a similarity transformation relative to a normal orthogonal basis (cf. Theorems 11 and 12).

12. (a) Prove that all similarity transformations form a group S_n.
    (b) Prove that O_n is a normal subgroup of S_n.
    (c) Prove that the quotient group S_n/O_n is isomorphic to the multiplicative group of all positive real numbers.

13. How many 3×3 orthogonal matrices are there with coefficients in $\mathbf{Z}_2$? in $\mathbf{Z}_3$?

14. (a) Show that the correspondence $A \mapsto \theta(A) = (A^{-1})^T$ is an automorphism of the full linear group $L_n(F)$.
    (b) Show that $\theta^2(A) = A$ for all A.
    (c) For which matrices does $\theta(A) = A$?

9.5. Invariants and Canonical Forms

The full linear, the affine, the orthogonal, and the Euclidean groups form examples of linear groups. Another is the unitary group (§9.12). In the following sections, we shall see how far one can go in "simplifying" polynomials, quadratic forms, and various geometrical figures, by applying suitable transformations from these groups. These simplifications will be analogous to the simplification made in reducing a general matrix to row-equivalent reduced echelon form, whose rank was proved to be an invariant under the transformations considered. The notions of a simplified "canonical form" and "invariant" can be formulated in great generality, as follows.†

Let G be a group of transformations (§6.2) on any set or "space" S. Call two elements x and y of S *equivalent* under G (in symbols, xE_Gy) if and only if there is some transformation T of G which carries x into y. Then T^{-1} carries y back into x, so yE_Gx, and the relation of equivalence is symmetric. Similarly, using the other group properties, one proves that equivalence under any G is also a reflexive and transitive relation (an equivalence relation). A subset C of S is called a set of *canonical forms* under G if each $x \in S$ is equivalent under G to one and only one element c in C; this element c is then *the* canonical form of x. A function $F(x)$ defined for all elements x of S and with values in some convenient other set, say a set of numbers, is an *invariant* under G if $F(xT) = F(x)$ for every point x in S and every transformation T in G; in other words, F must have the same value at all equivalent elements. A collection of invariants $F_1, \cdots, F_n$ is a *complete set of invariants* under G if $F_1(x) = F_1(y), \cdots, F_n(x) = F_n(y)$ imply that x is equivalent to y.

† The reader is advised to return again to this discussion when he has finished the chapter.

For example, let the space S be the set M_n of all $n \times n$ matrices over some field. W already have at hand three different equivalence relations to such matrices; these are listed below, together with three new cases which will be discussed in subsequent sections (§9.8, §9.10, and §9.12, respectively).

A row-equivalent to B	$B = PA,$	P nonsingular,
A equivalent to B	$B = PAQ,$	P, Q nonsingular,
A similar to B	$B = PAP^{-1},$	P nonsingular,
A congruent to B	$B = PAP^{T},$	P nonsingular,
A orthogonally equivalent to B	$B = PAP^{-1}$	P orthogonal,
A unitary equivalent to B	$B = PAP^{-1},$	P unitary.

The first line is to be read "A is row-equivalent to B if and only if there exists P such that $B = PA$, with P nonsingular," and similarly for the other lines.

Each of these equivalence relations is the equivalence relation E_G determined by a suitable group G acting on M_n, and arises naturally from one of the various interpretations of a matrix.

The first relation, that of row equivalence, arises from the study of a fixed subspace of F^n represented as the row space of a matrix A; in this case, the full linear group of matrices P acts on A by $A \mapsto PA$, and the reduced echelon form is a canonical form under this group. The rank of A is a (numerical) invariant under this group, but it does not give a complete system of invariants, since two matrices A and B with the same rank need not be row-equivalent.

The second relation of equivalence (in the technical sense $B = PAQ$, not to be confused with the general notion of an equivalence relation) arises when we are studying the various matrix representations of a linear transformation of one vector space into a second such space (cf. §9.2, Ex. 9). Here, by Theorem 18 of §8.9, the rank is a complete system of invariants under the group $A \mapsto PAQ$. The set of all diagonal matrices with entries 1 and 0 along the diagonal, the 1's preceding the 0's, is a set of canonical forms. Note that we might equally well have chosen a *different* set of canonical forms—say the same sort of diagonal matrices, but with the 0's preceding the 1's on the diagonal.

The relation of similarity arises when we study the various matrix representations of a linear transformation of a vector space into itself; in this case, the full linear group acts on A by $A \mapsto PAP^{-1}$. Under similarity the rank of the matrix A is an invariant, since two similar matrices are certainly equivalent, and rank is even invariant under equivalence. The set of all eigenvalues of a matrix is also an invariant under similarity, by §9.2, but is not a complete system of invariants. The formulation of a complete set of canonical forms under similarity is one of the major problems of matrix theory; for the field of complex numbers, it gives rise to the Jordan canonical form of a matrix (see §10.10).

The relation of congruence $(B = PAP^T)$, as will appear subsequently, arises from the representation of a quadratic form by a (symmetric) matrix.

As still another example of equivalence under a group, consider the simplification of a quadratic polynomial $f(x) = ax^2 + bx + c$, with $a \neq 0$, by the group of all translations $y = x + k$. Substituting $x = y - k$, we find the result of translating $f(x)$ to be

$$g(y) = a(y - k)^2 + b(y - k) + c = ay^2 + (b - 2ak)y + ak^2 - bk + c.$$

In particular, we obtain the familiar "completion of the square"—the new polynomial will have no linear terms if and only if $k = b/2a$, and in this case the polynomial is

(23) $$g(y) = ay^2 - d/(4a), \quad \text{where } d = b^2 - 4ac.$$

Thus $f(x)$ is equivalent under the group of translations to one and only one polynomial of the form $ay^2 + h$, so that the quadratic polynomials without linear terms are canonical forms under this group. On the other hand, any transformed polynomial has the same leading coefficient a and the same discriminant $d = (b - 2ak)^2 - 4a(ak^2 - bk + c)$ as the original polynomial $f(x)$. Hence the first coefficient and the discriminant of $f(x)$ are invariants under the group. They constitute a complete set of invariants because the canonical form can be expressed in terms of them as shown in (23).

To give a last example, recall that the full linear group $L_n(F)$ is a group of transformations on the vector space F^n. Each transformation of this group carries a subspace S of F^n into another subspace. By Corollary 2 of Theorem 10, §8.6, the dimension of any subspace S is an invariant under the full linear group. This one invariant is actually a complete set of invariants for subspaces of F^n under the full linear group (see Ex. 5 below).

Exercises

1. Find canonical forms for all *monic* quadratic polynomials $x^2 + bx + c$ under the group of translations.

2. Find canonical forms for all quadratic polynomials $ax^2 + bx + c$ with $a \neq 0$ under the affine group $y = hx + k$, $h \neq 0$.

3. In Ex. 2, show that $d/a = b^2/a - 4c$ is an affine invariant.
4. Show that over any field in which $1 + 1 \neq 0$, any quartic polynomial is equivalent under translation to a polynomial in which the cubic term is absent.
5. Let V be an n-dimensional vector space. Show that a complete system of invariants for ordered pairs of subspaces (S_1, S_2) of V under the full linear group is given by the dimensions of S_1, of S_2, and of their intersection.
6. Consider the set of homogeneous quadratic functions ax^2 with rational a under the group $x \mapsto rx$, with $r \neq 0$ and rational. Show that the set of integral coefficients a which are products of distinct primes ("square free") provide canonical forms for this set.
7. If $f(x)$ is any polynomial in one variable, prove that the degree of f and the number of real roots are both invariant under the affine group.
8. For a polynomial in n variables, show that the coefficient of the term of highest degree is invariant under the group of translations.
9. Show that a real cubic polynomial is equivalent under the affine group to one and only one polynomial of the form $x^3 + ax + b$.
★10. Find canonical forms for the quadratic functions $x^2 + bx + c$ under the group of translations, in case b and c are elements from the field $\mathbf{Z}_2$ of integers modulo 2.

9.6. Linear and Bilinear Forms

A *linear form* in n variables over a field F is a polynomial of the form

$$(24) \qquad f(x_1, \cdots, x_n) = b_1 x_1 + \cdots + b_n x_n + c,$$

with coefficients $b_1, \cdots, b_n$ and c in F; to exclude the trivial case, we assume that some coefficient b_j is not zero. The form is *homogeneous* if $c = 0$. Any form (24) may be regarded as a function $f(X)$ of the vector $X = (x_1, \cdots, x_n)$ of F^n. Distinct forms determine distinct functions, for the function $f(X)$ determines the coefficients of the form by the formulas $f(0, \cdots, 0) = c, f(1, 0, \cdots, 0) = b_1 + c, \cdots, f(0), \cdots, 0, 1) = b_n + c$.

To any linear form we may apply a nonsingular affine transformation

$$(25) \qquad x_i = \sum_j a_{ij} y_j + k_i, \qquad \|a_{ij}\| \text{ nonsingular},$$

to yield upon substitution in (24) the new linear form

$$(26) \qquad g(y_1, \cdots, y_n) = \sum_j \left(\sum_i b_i a_{ij} \right) y_j + \left(\sum_i b_i k_i + c \right).$$

We say that f and g are *equivalent* forms under the affine group, if there exists such a nonsingular affine transformation carrying f into g.

A canonical form may be obtained readily. First, since some $b_j \neq 0$, the translation $x_j = y_j - c/b_j$ and $x_i = y_i$ for $i \neq j$ will remove the constant term. The permutation $z_1 = y_j$, $z_j = y_1$, and $z_i = y_i$ for $i \neq 1$ or j will then give a new form like (24) with $b_1 \neq 0$ and $c = 0$. If this form is written with the variables x_j, the new affine transformation with the equations

$$y_1 = b_1 x_1 + \cdots + b_n x_n, \qquad y_2 = x_2, \qquad \cdots, \qquad y_n = x_n$$

is nonsingular, and carries any f with $c = 0$ to the equivalent function $g(y_1, \cdots, y_n) = y_1$. Therefore *all nonzero linear forms are equivalent under the affine group.*

Consider now the equivalence of real linear forms under the Euclidean group (i.e., with $A = \|a_{ij}\|$ in (25) an orthogonal matrix). Call $d = (b_1^2 + \cdots + b_n^2)^{1/2}$ the *norm* of the form (24). As before, we can remove the constant c by a translation. By the choice of d, $(b_1/d, \cdots, b_n/d)$ is a vector of unit length. Hence there is an orthogonal matrix $\|h_{ij}\|$ with this vector as its first row. The transformation $y_i = \sum h_{ij} x_j$ is then in the Euclidean group; since $dy_1 = b_1 x_1 + \cdots + b_n x_n$, it carries the form f, with $c = 0$, into the form $g = dy_1$.

This form dy_1 is a canonical form for linear forms under the Euclidean group. To show this, we need only prove that the norm d is invariant under the Euclidean group. Now the norm d of f is just the length of the coefficient vector $\beta = (b_1, \cdots, b_n)$, and (26) shows that in the transformed form the coefficient vector is the transform βA of the original coefficient vector by the orthogonal matrix $\|a_{ij}\|$; hence the norm is indeed invariant. We have proved

Theorem 13. *Under the Euclidean group, every linear form (24) is equivalent to one and only one of the canonical forms dy, with positive d, where $d = (b_1^2 + \cdots + b_n^2)^{1/2}$ is an invariant under this group.*

A (homogeneous) *bilinear form* in two sets of variables $x_1, \cdots, x_m$ and $y_1, \cdots, y_n$ is a polynomial of the form

$$(27) \qquad b(x_1, \cdots, x_m, y_1, \cdots, y_n) = \sum_{i=1}^{m} \sum_{j=1}^{n} x_i a_{ij} y_j;$$

it is determined by the matrix of coefficients $A = \|a_{ij}\|$. In terms of the vectors $X = (x_1, \cdots, x_m)$ and $Y = (y_1, \cdots, y_n)$ the bilinear form may be written as the matrix product

$$(28) \qquad b(X, Y) = XAY^T.$$

As a function of X and Y, it is linear in each argument separately.

More generally, let V and W be any vector spaces having finite dimensions m and n, respectively, over the same field F, and let $B(\xi, \eta)$ be any function, with values in F, defined for arguments $\xi \in V$ and $\eta \in W$, which is bilinear in the sense that for a_1 and $a_2 \in F$,

$$(29) \quad B(a_1\xi_1 + a_2\xi_2, \eta) = a_1B(\xi_1, \eta) + a_2B(\xi_2, \eta),$$
$$\xi_1, \xi_2 \in V, \quad \eta \in W;$$
$$(29') \quad B(\xi, a_1\eta_1 + a_2\eta_2) = B(\xi, \eta_1)a_1 + B(\xi, \eta_2)a_2,$$
$$\xi \in V, \quad \eta_1, \eta_2 \in W.$$

Choose a basis $\alpha_1, \cdots, \alpha_m$ in V and a basis $\beta_1, \cdots, \beta_n$ in W and let the scalars a_{ij} be defined as $a_{ij} = B(\alpha_i, \beta_j)$. Then for any vectors ξ and η in V and W, expressed in terms of the respective bases, we have

$$B(\xi, \eta) = B(x_1\alpha_1 + \cdots + x_m\alpha_m, y_1\beta_1 + \cdots + y_n\beta_n)$$

and hence, by (29) and (29'),

$$B(\xi, \eta) = \sum_{i,j} x_iB(\alpha_i, \beta_j)y_j = \sum_{i,j} x_ia_{ij}y_j.$$

In other words, any bilinear function B on V and W has a unique expression, relative to given bases, as a bilinear form (27). Equivalently, in the notation of §8.5, a bilinear form is just the product XBY of a row m-vector X, an $m \times n$ matrix B, and a column n-vector Y.

A change of basis in both spaces corresponds to nonsingular transformations $X = X^*P$ and $Y = Y^*Q$ of each set of variables. These transformations replace (28) by a new bilinear form $X^*(PAQ^T)Y^{*T}$, with a new matrix PAQ^T. Since any nonsingular matrix may be written as the transpose Q^T of a nonsingular matrix, we see that two bilinear forms are equivalent (under changes of bases) if and only if their matrices are equivalent. Hence, by Theorem 18 of §8.9 on the equivalence of matrices, *any bilinear form is equivalent to one and only one of the canonical forms*

$$x_1y_1 + \cdots + x_ry_r.$$

The integer r, which is the rank of the matrix of the form, is a (complete set of) invariants.

Exercises

1. Find a canonical form for homogeneous real linear functions under the similarity group.

2. Treat the same question (a) under the diagonal group of transformations $y_1 = d_1x_1, \cdots, y_n = d_nx_n$, (b) under the monomial group of all transformations $Y = XM$ with a monomial matrix M.

3. Prove that any bilinear form of rank r is expressible as

$$\sum_i (b_{i1}x_1 + \cdots + b_{in}x_n)(c_{i1}y_1 + \cdots + c_{in}y_n), \quad \text{where} \quad i = 1, \cdots, r;$$

that is, as the sum of r products of linear forms.

4. Find new variables x^*, y^*, z^* and u^*, v^*, w^* which will reduce to canonical form the bilinear function

$$xu + xv + xw + yu + yv + yw + zu + zv + zw.$$

9.7. Quadratic Forms

The next four sections are devoted to the study of canonical forms for quadratic functions under various groups of transformations. The simplest problems of this type arise in connection with central conics in the plane (ellipses or hyperbolas with "tilted" axes). Such conics have equations $Ax^2 + Bxy + Cy^2 = 1$, in which the left-hand side is a "quadratic form" Such quadratic forms (homogeneous quadratic expressions in the variables) arise in many other instances: in the equations for quadric surfaces in space, in the projective equations for conics in homogeneous coordinates, in the formula $|X|^2 = (x_1^2 + x_2^2 + \cdots + x_n^2)$ for the square of the length of a vector, in the formula $(m/2)(u^2 + v^2 + w^2)$ for the kinetic energy of a moving body in space with three velocity components u, v, and w, in differential geometry, in the formula for the length of arc ds in spherical coordinates of space, $ds^2 = dr^2 + r^2d\phi^2 + r^2\sin\phi d\theta^2$.

Such quadratic forms can be expressed by matrices. To obtain a matrix from a quadratic form such as $5x^2 + 6xy + 2y^2$, first adjust the form so that the coefficients of xy and yx are equal, as $5x^2 + 3xy + 3yx + 2y^2$. The result can be written as a matrix product,

$$(x, y)\begin{pmatrix} 5 & 3 \\ 3 & 2 \end{pmatrix}\begin{pmatrix} x \\ y \end{pmatrix} = (x, y)\begin{pmatrix} 5x + 3y \\ 3x + 2y \end{pmatrix} = 5x^2 + 6xy + 2y^2.$$

The 2×2 matrix of coefficients which arises here is symmetric, in the sense that it is equal to its transpose.

In general, a square matrix A is called *symmetric* if it is equal to its own transpose, $A^T = A$; in other words, $\|a_{ij}\|$ is symmetric if and only if $a_{ij} = a_{ji}$ for all i and j. Similarly, a matrix C is *skew-symmetric* if $C^T = -C$. To split a matrix B into symmetric and skew-symmetric parts, write

(30) $\qquad B = (B + B^T)/2 + (B - B^T)/2 = S + K,$

where $S = (B + B^T)/2$, $K = (B - B^T)/2$. By the laws for the transpose, $(B \pm B^T)^T = B^T \pm B^{TT} = B^T \pm B$, so S is symmetric and K skew. No other decomposition $B = S_1 + K_1$ with S_1 symmetric and K_1 skew is possible, since any such decomposition would give $B^T = S_1^T + K_1^T = S_1 - K_1$, $B + B^T = 2S_1$, $B - B^T = 2K_1$, and $S_1 = S$, $K_1 = K$. Formulas (30) apply over any field in which $2 = 1 + 1 \neq 0$, but are meaningless for matrices over the field $\mathbf{Z}_2$, where $1 + 1 = 0$. In conclusion, any matrix can be expressed uniquely as the sum of a symmetric matrix and a skew matrix, provided $1 + 1 \neq 0$.

A *homogeneous quadratic form* in n variables $x_1, \cdots, x_n$ is by definition a polynomial

$$\sum_i \sum_j x_i b_{ij} x_j$$

in which each term is of degree two. This form may be written as a matrix product XBX^T. If the matrix B of coefficients is skew-symmetric, $b_{ij} = -b_{ji}$ and the form equals zero. In general, write the matrix B as $B = S + K$, according to (30); the form then becomes

$$XBX^T = X(S + K)X^T = XSX^T + XKX^T = XSX^T, \qquad K \text{ skew.}$$

Hence if $1 + 1 \neq 0$, any quadratic form may be expressed uniquely, with S denoted by A, as

(31) $\qquad \displaystyle\sum_{i=1}^n \sum_{j=1}^n x_i a_{ij} x_j = XAX^T, \qquad A = \|a_{ij}\| \text{ symmetric.}$

If a vector ξ has coordinates $X = (x_1, \cdots, x_n)$, each quadratic form determines a quadratic function $Q(\xi) = XAX^T$ of the vector ξ. A change of basis in the space gives new coordinates X^* related to the old coordinates by an equation $X = X^*P$, with P nonsingular. In terms of the new coordinates of ξ, the quadratic function becomes

$$Q(\xi) = XAX^T = (X^*P)A(X^*P)^T = X^*(PAP^T)X^{*T};$$

this is another quadratic form with a new matrix PAP^T. The new matrix, like A, is symmetric; $(PAP^T)^T = P^{TT}A^TP^T = PAP^T$.

Theorem 14. *A change of coordinates replaces a quadratic form with matrix A by a quadratic form with matrix PAP^T, where P is nonsingular.*

Symmetric matrices A and B are sometimes called *congruent* when (as in this case) $B = PAP^T$ for some nonsingular P.

Reinterpreted, Theorem 14 asserts that the problem of reducing a homogeneous quadratic form to canonical form, under the full linear group of nonsingular linear homogeneous substitutions on its variables, is equivalent to the problem of finding a canonical form for symmetric matrices A under the group $A \mapsto PAP^T$.

Exercises

1. Prove that A^TA and AA^T are always symmetric.
2. Prove: If A is skew-symmetric, then A^2 is symmetric.
3. Represent each matrix of Ex. 1, §8.3, in the form $S + K$.
4. Find the symmetric matrix associated with each of the following quadratic forms:
   (a) $2x^2 + 3xy + 6y^2$,   (b) $8xy + 4y^2$,
   (c) $x^2 + 2xy + 4xz + 3y^2 + yz + 7z^2$,   (d) $4xy$,
   (e) $x^2 + 4xy + 4y^2 + 2xz + z^2 + 4yz$.
5. Prove: (a) If S is symmetric and A orthogonal, then $A^{-1}SA$ is symmetric, (b) If K is skew-symmetric and A orthogonal, then $A^{-1}KA$ is skew-symmetric.
6. Describe the symmetry of the matrix $AB - BA$ in the following cases:
   (a) A and B both symmetric,   (b) A and B both skew-symmetric,
   (c) A symmetric and B skew-symmetric.
7. Prove: If A and B are symmetric, then AB is symmetric if and only if $AB = BA$.
8. (a) Prove: Over the field $\mathbf{Z}_2$ (integers mod 2) every skew-symmetric matrix is symmetric.
   (b) Exhibit a matrix over $\mathbf{Z}_2$ which is not a sum $S + K$ (cf. (30)).
9. Let D be a diagonal matrix with no repeated entries. Show that $AD = DA$ if and only if A is also diagonal.
10. If $Q(\xi)$ is a quadratic function, prove that

$$Q(\alpha + \beta + \gamma) - Q(\alpha + \beta) - Q(\beta + \gamma) - Q(\gamma + \alpha)$$
$$+ Q(\alpha) + Q(\beta) + Q(\gamma) = 0.$$

11. A bilinear form $B(\xi, \eta)$, with ξ, η both in V, is symmetric when $B(\xi, \eta) = B(\eta, \xi)$. Prove that if B is a symmetric bilinear form, then $Q(\xi) = B(\xi, \xi)$ is a quadratic form, with $2B(\xi, \eta) = Q(\xi + \eta) - Q(\xi) - Q(\eta)$.
12. Show that a real $n \times n$ matrix A is symmetric if and only if the associated linear transformation $T = T_A$ of Euclidean n-space satisfies $(\xi T, \eta) = (\xi, \eta T)$ for any two vectors ξ and η.
★13. Show that if the real matrix S is skew-symmetric and $I + S$ is nonsingular, then $(I - S)(I + S)^{-1}$ is orthogonal.

9.8. Quadratic Forms Under the Full Linear Group

The familiar process of "completing the square" may be used as a device for simplifying a quadratic form by linear transformations. For two variables, the procedure gives

$$ax^2 + 2bxy + cy^2 = a[x^2 + 2(b/a)xy + (b^2/a^2)y^2] + [c - (b^2/a)]y^2$$
$$= a[x + (b/a)y]^2 + [c - (b^2/a)]y^2.$$

The term in brackets suggests the new variables $x' = x + (b/a)y$, $y' = y$. Under this linear change of variables, the form becomes $ax'^2 + [c - (b^2/a)]y'^2$; the cross term has been eliminated.

This argument requires $a \neq 0$. If $a = 0$, but $c \neq 0$, a similar transformation works. Finally, if $a = c = 0$, the original form is $2bxy$, and the corresponding equation $2bxy = 1$ represents an equilateral hyperbola. In this case, the transformation $x = x' + y'$, $y = x' - y'$ will reduce the form to

$$2b(x' + y')(x' - y') = 2b(x'^2 - y'^2);$$

the result again contains only square terms. (*Hint:* How is the transformation used here related to a rotation of the axes of the hyperbola?)

An analogous preparatory device may be applied to forms in more than two variables.

Lemma. *By a nonsingular linear transformation, any quadratic form* $\sum x_i a_{ij} x_j$ *not identically zero can be reduced to a form with leading coefficient* $a_{11} \neq 0$, *provided only that* $1 + 1 \neq 0$.

Proof. By hypothesis, at least one coefficient $a_{ij} \neq 0$. If there is a diagonal term $a_{ii} \neq 0$, one can get a new coefficient $a_{11}' \neq 0$ by interchanging the variables x_1 and x_i (this is a nonsingular transformation because its matrix is a permutation matrix). In the remaining case, all diagonal terms a_{ii} are zero, but there are indices $i \neq j$, with $a_{ij} \neq 0$. By permuting the variables, we can make $a_{12} \neq 0$; by the symmetry of the matrix $a_{12} = a_{21}$. The given quadratic form is then $a_{12}x_1x_2 + a_{21}x_2x_1 = 2a_{12}x_1x_2$, plus terms involving other variables. Just as in the case of the equilateral hyperbola, this may be reduced to a form $2a_{12}(y_1^2 - y_2^2)$, with a leading coefficient $2a_{12} \neq 0$, by a transformation

$$x_1 = y_1 - y_2, \qquad x_2 = y_1 + y_2, \qquad x_3 = y_3, \cdots, x_n = y_n.$$

This transformation is nonsingular, for by elimination one easily shows that it has an inverse

$$y_1 = (x_1 + x_2)/2, \qquad y_2 = (x_2 - x_1)/2, \qquad y_3 = x_3, \cdots, y_n = x_n.$$

Query: Where does this argument use the hypothesis $1 + 1 \neq 0$?

Now for the completion of the square in any quadratic form! By the Lemma, we make $a_{11} \neq 0$, so the form can be written as $a_{11}(\sum x_i b_{ij} x_j)$, where $b_{ij} = a_{ij}/a_{11}$ and $b_{11} = 1$. Because of the symmetry of the matrix, the terms which actually involve x_1 are then

$$x_1^2 + 2 \sum_{j=2}^{n} b_{1j} x_1 x_j = \left(x_1 + \sum_{j=2}^{n} b_{1j} x_j \right)^2 - \left(\sum_{j=2}^{n} b_{1j} x_j \right)^2.$$

The formation of this "perfect square" suggests the transformation

$$y_1 = x_1 + \sum_{j=2}^{n} b_{1j} x_j, \qquad y_2 = x_2, \cdots, y_n = x_n;$$

then y_1 will appear only as y_1^2. The original form is now $a_{11}y_1^2 + \sum y_j c_{jk} y_k$, where the indices j and k run from 2 to n. This residual part in $y_2, \cdots, y_n$ is a quadratic form in $n - 1$ variables; to this form the same process applies. The process may be repeated (an induction argument!) till the new coefficients in one of the residual quadratic forms are all zero. Hence we have

Theorem 15. *By nonsingular linear transformations of the variables, a quadratic form over any field with $1 + 1 \neq 0$ can be reduced to a diagonal quadratic form,*

$$(32) \qquad d_1 y_1^2 + d_2 y_2^2 + \cdots + d_r y_r^2, \qquad each\ d_i \neq 0.$$

The number r of nonzero diagonal terms is an invariant.

This number r is called the *rank* of the given form XAX^T. Its invariance is immediate, for r is the rank of the diagonal matrix D of the reduced form (32). This rank must equal the rank of the matrix A of the original quadratic form, for by Theorem 14 our transformations reduce A to $D = PAP^T$, and we already know (§8.9, Theorem 19) that rank is invariant under the more general transformation $A \mapsto PAQ$.

A quadratic form XAX^T in n variables is called nonsingular if its rank is n, since this means that the matrix A is nonsingular.

In the diagonal form (32) the rank r is an invariant, but the coefficients are not, since different methods of reducing the form may well yield different sets of coefficients $d_1, \cdots, d_r$. We shall now get a complete set of invariants for the special case of the real field.

Exercises

1. Over the field of rational numbers reduce each of the quadratic forms of Ex. 4, §9.7, to diagonal form.
2. Reduce $2x^2 + xy + 3y^2$ to diagonal form over the field of integers modulo 5.
3. Over the field $\mathbf{Z}_5$, prove that every quadratic form may be reduced by linear transformations to a form $\sum d_i y_i^2$, with each coefficient $d_i = 0$, 1, or 2.
4. Over the field of rational numbers show that the quadratic form $x_1^2 + x_2^2$ can be transformed into both of the distinct diagonal forms $9y_1^2 + 4y_2^2$ and $2z_1^2 + 8z_2^2$.
5. Find a P such that PAP^T is diagonal if

   (a) $A = \begin{pmatrix} 3 & 1 \\ 1 & 0 \end{pmatrix}$,   (b) $A = \begin{pmatrix} 1 & 2 \\ 2 & 4 \end{pmatrix}$,   (c) $A = \begin{pmatrix} 0 & 1 & 0 \\ 1 & 0 & 2 \\ 0 & 2 & 0 \end{pmatrix}$.

6. Find all linear transformations which carry the real quadratic form $x_1^2 + \cdots + x_n^2$ into $y_1^2 + \cdots + y_n^2$.
7. Show rigorously that the quadratic form xy is not equivalent to a diagonal form under the group $L_2(\mathbf{Z}_2)$.

9.9. Real Quadratic Forms Under the Full Linear Group

Conic sections and quadric surfaces are described in analytic geometry by real quadratic polynomial functions. Over the real field, each term of the diagonal form (32) can be simplified further by making the substitution $y_i' = (\pm d_i)^{1/2} y_i$, so that the term $d_i y_i^2$ becomes $\pm y_i'^2$. Carrying out these substitutions simultaneously on all the variables will reduce the quadratic form to $\sum \pm y_i^2$. In this sum the variables may be permuted so that the positive squares come first. This proves

Theorem 16. *Any quadratic function over the field of real numbers can be reduced by nonsingular linear transformation of the variables to a form*

$$(33) \qquad z_1^2 + \cdots + z_p^2 - z_{p+1}^2 - \cdots - z_r^2.$$

Theorem 17. *The number p of positive squares which appear in the reduced form* (33) *is an invariant of the given function Q, in the sense that p depends only on the function and not on the method used to reduce it* (*Sylvester's law of inertia*).

Proof. Suppose that there is another reduced form

(34) $$y_1^2 + \cdots + y_q^2 - y_{q+1}^2 - \cdots - y_r^2$$

with q positive terms. Since both are obtained from the same Q by nonsingular transformations, there is a nonsingular transformation carrying (33) into (34). We may regard the equations of this transformation as a change of coordinates ("alias"); then (33) and (34) represent the same quadratic function $Q(\xi)$ of a fixed vector ξ with coordinates z_i relative to one basis, y_j relative to another.

Suppose $q < p$. Then $Q(\xi) \geqq 0$ whenever $z_{p+1} = \cdots = z_r = 0$ in (33). The ξ's satisfying these $r - p$ equations form an $n - (r - p)$ dimensional subspace S_1 (in this subspace there are $n - (r - p)$ coordinates $z_1, \cdots, z_p, z_{r+1}, \cdots, z_n$). Similarly, (34) makes $Q(\xi) < 0$ for each $\xi \neq 0$ with coordinates $y_1 = \cdots = y_q = y_{r+1} = \cdots = y_n = 0$. These conditions determine an $(r - q)$-dimensional subspace S_2. The sum of the dimensions of these subspaces S_1 and S_2 is

$$n - (r - p) + (r - q) = n + (p - q) > n.$$

Therefore S_1 and S_2 have a nonzero vector ξ in common, for according to Theorem 17 of §7.8, the dimension of the intersection $S_1 \cap S_2$ is positive. For this common vector ξ, $Q(\xi) \geqq 0$ by (33) and $Q(\xi) < 0$ by (34), a manifest contradiction. The assumption $q > p$ would lead to a similar contradiction, so $q = p$, completing the proof.

This result shows that any real quadratic form can be reduced by linear transformations to one and only one form of the type (33). The expressions $\sum \pm z_i^2$ of this type are therefore *canonical* for quadratic forms under the full linear group. This canonical form itself is uniquely determined by the so-called *signature* $\{+, \cdots, +, -, \cdots, -\}$ which is a set of p positive and $r - p$ negative signs, r being the *rank* of the form. This set of signs is determined by r and by $s = p - (r - p) = 2p - r$ (s is the number of positive signs diminished by the number of negative signs). Sometimes this integer s is called the signature. Together, r and s form a complete system of numerical invariants, since two forms are equivalent if and only if they reduce to the same canonical form (33).

Theorem 18. *Two real quadratic forms are equivalent under the full linear group if and only if they have the same rank and the same signature.*

A real quadratic form $Q = XAX^T$ in n variables is called *positive definite* when $X \neq 0$ implies $Q > 0$; a real symmetric matrix A is called positive definite under the same conditions. If we consider the canonical reduced form (33), it is evident that this is the case if and only if the reduced form is $z_1^2 + \cdots + z_n^2$. This is because a sum of n squares is positive unless all terms are individually zero, and because for $X = \varepsilon_n$, the nth unit vector, $XAX^T \leqq 0$ in (33) unless $p = n$. That is, we have proved

Theorem 19. *A real quadratic form is positive definite if and only if its canonical form is* $z_1^2 + \cdots + z_n^2$.

By Theorem 14, this means that $A = PIP^T$, which gives the following further result.

Theorem 20. *A real symmetric matrix A is positive definite if and only if there exists a real nonsingular matrix P such that $A = PP^T$.*

A quadratic form XAX^T defines in n-dimensional real space a locus, consisting of all points X satisfying $XAX^T = 1$. The canonical form (33) means that a suitable nonsingular linear transformation will reduce this locus to one with an equation

$$z_1^2 + \cdots + z_p^2 - z_{p+1}^2 - \cdots - z_r^2 = 1.$$

For example, in the plane, the reduced equations of rank 2 are

$$x^2 + y^2 = 1, \qquad x^2 - y^2 = 1, \qquad -x^2 - y^2 = 1.$$

They represent, respectively, a circle, an equilateral hyperbola, or no locus. The only form of rank 0 is $0 = 1$; those of rank 1 are $x^2 = 1$ (which represents the two lines $x = \pm 1$) or $-x^2 = 1$ (no locus). In §8.8 it was proved (Theorem 15, Corollary 2) that any nonsingular linear transformation of the plane can be represented as a product of shears, compressions, and reflections. Hence any "central conic" with an equation $ax^2 + bxy + cy^2 = 1$ can be reduced to one of the forms we have listed by a succession of shears, compressions, and reflections. Geometrically, this result is reasonable: an ellipse could be compressed along one axis to make a circle; but, clearly, no sequence of linear transformations could reduce a circle $x^2 + y^2 = 1$ to an equilateral hyperbola $x^2 - y^2 = 1$. This is the geometric significance of the invariance of the signature in this case.

The signature is useful in studying the maxima and minima of functions of two variables. Let $z = f(x, y)$ be a smooth function whose first partial derivates f_x and f_y both vanish at $x = x_0$, $y = y_0$, so that there are no first-degree terms in the Taylor's series expansion of z in powers of $h = (x - x_0)$ and $k = (y - y_0)$. This expansion is

$$f(x_0 + h, y_0 + k) = f(x_0, y_0) + (1/2)[ah^2 + 2bhk + ck^2] + \cdots,$$

the coefficients being the partial derivatives

$$a = f_{xx}(x_0, y_0), \qquad b = f_{xy}(x_0, y_0), \qquad c = f_{yy}(x_0, y_0).$$

For small values of h and k the controlling term is the one in brackets; it is a quadratic form in h and k with real coefficients. If this form has rank 2, it can be expressed in terms of transformed variables h' and k', as $\pm h'^2 \pm k'^2$. If both signs are plus, nearby values of $f(x_0 + h, y_0 + k)$ must always exceed $f(x_0, y_0)$, and z has a relative *minimum*. If both signs are minus, z has a *maximum*. If one sign is plus and one sign is minus, the quadratic form may take on both positive and negative values, so x_0, y_0 is neither a maximum nor a minimum, but a *saddle-point* (like a saddle or a pass between two mountain peaks, where motion in one direction increases the altitude z, in another decreases z). Maxima, minima, and saddle-points of f are therefore distinguished by the signature of the quadratic form. Similar results hold for *critical points* of functions of three or more variables.

Exercises

1. Prove that the real quadratic function $ax^2 + bxy + cy^2$ is positive definite if and only if $a > 0$ and $4ac - b^2 > 0$.
2. Show that a positive definite symmetric matrix has all positive entries on its main diagonal.
3. Reduce the following real quadratic forms to the canonical form of Theorem 16. Find the rank and signature of each form.
   (a) $9x_1^2 + 12x_1x_2 + 79x_2^2$, (b) $2x_1^2 - 12x_1x_2 + 18x_2^2$,
   (c) $-2x_1^2 - 4x_1x_2 + 22x_2^2 + 12x_2x_3 + 6x_3x_1 - x_3^2$.
4. Describe the geometrical loci corresponding to the various possible canonical forms for real quadratic forms in three dimensions.
5. Prove: A homogeneous quadratic form with complex coefficients is always equivalent under the full complex linear group to a sum of squares $z_1^2 + \cdots + z_r^2$.
6. Prove that two quadratic forms in n variables with complex coefficients are equivalent under the full linear group if and only if they have the same rank.
7. Prove that the bilinear function XAY^T is an "inner product" if and only if A is symmetric and positive definite.
8. A quadratic form is called *positive semidefinite* if its rank equals its signature. State and prove an analogue of Theorem 19 for such forms.
9. Do the same for Theorem 20.
10. (a) List all the types of nonsingular quadratic forms in four variables.
    (b) Describe geometrically at least two of the corresponding loci in $\mathbf{R}^4$.

9.10. Quadratic Forms Under the Orthogonal Group

How far can a real quadratic form be simplified by transformations restricted to be orthogonal? An orthogonal transformation $Y = XP$ changes XAX^T into $Y(P^{-1}AP^{-1T})Y^T$, *since P is orthogonal*, the new matrix may be written† $P^{-1}AP^{-1T} = P^{-1}AP$.

In the plane an orthogonal transformation (rotation or reflection) of an ellipse will never yield a circle; at most one can hope to rotate the axes of the ellipse into standard position. The major axis might be characterized as the longest diameter. To reformulate this maximum property, consider any real quadratic function $Q(\xi) = ax^2 + cy^2$ with $a \le c$ and no xy term. Then $Q(\xi) \le cx^2 + cy^2 = c(x^2 + y^2)$; this means that the maximum value assumed by Q for all points on the unit circle $x^2 + y^2 = 1$ is c, and this maximum is taken on at the point $y = 1$, $x = 0$. Conversely, the latter statement insures the absence of an xy term in Q.

Lemma. *If a real quadratic function $Q = ax^2 + 2bxy + cy^2$ has among all points on the unit circle $x^2 + y^2 = 1$ a maximum value at $x = 0$, $y = 1$, then $b = 0$.*

Proof. Consider Q as a (two-valued) function of one variable x, where y is given implicitly in terms of x by $x^2 + y^2 = 1$. Differentiating, we get $2x + 2y(dy/dx) = 0$, so the derivative $y' = dy/dx$ is $y' = -x/y$. The derivative of Q is

$$Q' = (ax^2 + 2bxy + cy^2)' = 2ax + 2by + 2bxy' + 2cyy'.$$

Putting in the value of y' and setting $y = 1$, $x = 0$, one finds $Q' = 2b$. But at the maximum $y = 1$, $x = 0$, this derivative must be zero, hence $2b = 0$.   Q.E.D.

Now return to quadratic forms in n variables. In n-space the unit hypersphere $\sum x_i^2 = 1$ is a closed and bounded set S; its points are all vectors of length one. On this hypersphere the values taken on by a real quadratic form $Q(\xi) = \sum_{i,j} x_i a_{ij} x_j$ have an upper bound $\sum_{i,j} |a_{ij}|$. Therefore, since $Q(\xi)$ is a continuous function of ξ, $Q(\xi)$ has a maximum‡ λ_1 on S. In other words, among all vectors ξ of unit length there is one, ξ_0, at which $Q(\xi)$ takes as its maximum value λ_1. Since ξ_0 has length 1, we may

---

† Two symmetric matrices A and $P^{-1}AP$, with P orthogonal, are sometimes called *orthogonally congruent.*

‡ Here, as in calculus, we assume the fact that a function continuous on a bounded closed set has a maximum value on this set.

choose $\alpha_1 = \xi_0$ as the first vector of a new normal orthogonal basis $\alpha_1, \cdots, \alpha_n$ (Theorem 21, §7.11). In terms of the new coordinates $y_1, \cdots, y_n$ of ξ relative to this basis, the quadratic form is now expressed as $Q(\xi) = \sum y_i b_{ij} y_j$ with a new matrix of coefficients b_{ij}. The maximum value λ_1 of Q is given by the vector α_1 with coordinates $(1, 0, \cdots, 0)$; so by substitution the maximum value λ_1 is b_{11}. This maximum will remain the maximum if we further restrict the variables so that all but two, y_1 and y_i, are zero. Therefore, $y_1 = 1$, $y_i = 0$ is the maximum of the form $b_{11}y_1^2 + 2b_{1i}y_1y_i + b_{ii}y_i^2$, subject to the condition $y_1^2 + y_i^2 = 1$. The Lemma (with x replaced by y_i) then asserts that the cross product coefficient b_{1i} is zero. This argument applies to each $i = 2, \cdots, n$. Therefore Q, in these coordinates y_i, loses all cross product terms involving y_1 and becomes

$$(35) \qquad Q(\xi) = \lambda_1 y_1^2 + \sum_{i=2}^{n} \sum_{j=2}^{n} y_i b_{ij} y_j, \qquad B = \|b_{ij}\| = B^T.$$

The first coefficient λ_1 is *not* a vector, but a scalar (the maximum of $Q(\xi)$ on the sphere $|\xi| = 1$).

The difference $Q^*(\xi) = Q(\xi) - \lambda_1 y_1^2$ in (35) is a quadratic form in $n - 1$ variables $y_2, \cdots, y_n$. These variables are coordinates in the space S_{n-1} spanned by the $n - 1$ new basis vectors $\alpha_2, \cdots, \alpha_n$. In this space (which is the orthogonal complement of the first basis vector ξ_0), we may reapply the same device of choosing a new normal orthogonal basis which makes $Q^*(\xi)$ a maximum for $|\xi| = 1$; this splits another diagonal term off the form. One finally finds a basis of *principal axes* for which

$$(36) \quad Q(\xi) = \lambda_1 z_1^2 + \lambda_2 z_2^2 + \cdots + \lambda_n z_n^2, \qquad \lambda_1 \geqq \lambda_2 \geqq \cdots \geqq \lambda_n.$$

Here $z_1, \cdots, z_n$ are the coordinates of ξ relative to a basis $\alpha_1, \beta_2, \gamma_3, \cdots$ which has been chosen step by step by successive maximum requirements. The first vector α_1 gives $Q(\xi)$ its maximum value λ_1, subject only to the restriction $|\xi| = 1$. The second basis vector β_2 was chosen as a vector in the space orthogonal to α_1; that is, $\eta = \beta_2$ makes $Q(\eta)$ a maximum λ_2 among all vectors η for which $|\eta| = 1$, $(\eta, \alpha_1) = 0$. The third basis vector yields a maximum for $Q(\zeta)$ among all vectors $|\zeta| = 1$ orthogonal to α_1 and β_2, and so on. These successive maximum problems may be visualized (in inverted form) on an ellipsoid with three different axes $a > b > c > 0$. The shortest principal axis c is the minimum diameter; the next principal axis b is the minimum diameter among all those perpendicular to the shortest axis, etc.

The coefficients λ_i of (36) may be thus characterized as the solutions of certain maximum problems which depend only on Q, and not on a particular coordinate system. An ambiguity in the reduction process could arise only if the first maximum (or some later maximum) were given by two or more distinct vectors ξ_0 and η_0 of length 1. Even in this case the λ_i can still be proved unique (§10.4).

This proves the following Principal Axis Theorem.

Theorem 21. *Any real quadratic form in n variables assumes the diagonal form (36), relative to a suitable normal orthogonal basis.*

This new basis $\alpha_1^*, \cdots, \alpha_n^*$ can, by Theorem 1, be expressed in terms of the original basis $\varepsilon_1 = (1, 0, \cdots, 0), \cdots, \varepsilon_n = (0, \cdots, 0, 1)$ as $\alpha_i^* = \sum_j p_{ij} \varepsilon_j$; furthermore, since the vectors $\alpha_1^*, \cdots, \alpha_n^*$ are normal and orthogonal, the matrix $p = \|p_{ij}\|$ of coefficients is an orthogonal matrix. As in Theorem 2, the old coordinates $x_1, \cdots, x_n$ are then expressed in terms of the new coordinates $x_1^*, \cdots, x_n^*$ as $x_j = \sum_i x_i^* p_{ij}$; in other words, we have made an orthogonal transformation of the variables in the quadratic form. The "alias" result of Theorem 21 thus may be rewritten in "alibi" form as

Corollary 1. *Any real homogeneous quadratic function of n variables can be reduced to the diagonal form (36) by an orthogonal point-transformation.*

Either of these two results is known as the "Principal Axis Theorem." If the quadratic form is replaced by its symmetric matrix, the theorem asserts

Corollary 2. *For any real symmetric matrix A there is a real orthogonal matrix P such that $PAP^T = PAP^{-1}$ is diagonal.*

In other words, we have shown that any real symmetric matrix is *similar* to a diagonal matrix. Comparing with Theorem 4, we see that the λ_i in the canonical form (36) are just the eigenvalues of A.

In the plane the canonical forms of equations $Q(\xi) = 1$ are simply $\lambda_1 x^2 + \lambda_2 y^2 = 1$; they include the usual standard equations for an ellipse ($\lambda_1 \geqq \lambda_2 > 0$) or hyperbola ($\lambda_1 > 0 > \lambda_2$); the coefficients determine the lengths of the axes. In three-space a similar remark applies to the three coefficients $\lambda_1, \lambda_2, \lambda_3$. If all are positive, the locus $Q = 1$ is an ellipsoid; if one is negative, a hyperboloid of one sheet; if two are negative, a hyperboloid of two sheets; if all three are negative, no locus. (Note again the role of the signature and of the rank.)

Comparing with the Corollary of Theorem 4, we see that (for A symmetric) the principal axes of the quadratic function XAX^T are precisely the eigenvectors of the linear transformation $X \mapsto XA$. There follows

Corollary 3. *For A real symmetric, the linear transformation $X \mapsto XA$ has a basis of orthogonal eigenvectors with real eigenvalues.*

Corollary 4. *Every nonsingular real matrix A can be expressed as a product $A \doteq SR$, where S is a symmetric positive definite matrix and R is orthogonal.*

Proof. We already know (Theorem 20) that AA^T is symmetric and positive definite. By the present theorem, there is an orthogonal matrix P with $P^{-1}AA^TP$ diagonal and positive definite. The diagonal entries are thus positive; by extracting their square roots, we obtain a positive definite diagonal matrix T with $T^2 = P^{-1}AA^TP$ and hence a positive definite symmetric matrix $S = PTP^{-1}$ with $S^2 = AA^T$. The corollary will be proved if we show that $R = S^{-1}A$ is orthogonal, for then $A = SR$, as desired.   But $RR^T = S^{-1}AA^T(S^{-1})^T = S^{-1}S^2(S^{-1})^T = S(S^{-1})^T = SS^{-1} = I$, since $(S^{-1})^T = S^{-1}$ for symmetric S.

Corollary 5. *Let A be any real symmetric matrix, and B any positive definite (real) symmetric matrix. Then there exists a real nonsingular matrix P such that PAP^{-1} and PBP^{-1} are simultaneously diagonal.*

We leave the proof as an exercise; to find a basis of vectors ξ_j such that $A\xi_j = \lambda_j B\xi_j$ is called the *generalized* eigenvector problem; its solution plays a basic role in vibration theory.

Exercises

1. Consider the real quadratic form $ax^2 + 2bxy + cy^2$.
   (a) Show that $a + c$ and $b^2 - ac$ are invariant under orthogonal transformations.
   (b) If $\cot 2\alpha = (a - c)/2b$, show that the form is diagonalized by the orthogonal substitution $x = x'\cos\alpha - y'\sin\alpha$, $y = x'\sin\alpha + y'\cos\alpha$.
2. Prove that every real skew-symmetric matrix A has the form $A = P^{-1}BP$, where P is orthogonal and B^2 diagonal.
3. Reduce the following quadratic forms to diagonal forms by orthogonal transformations, following the method given:
   (a) $5x^2 - 6xy + 5y^2$,   (b) $2x^2 + 4\sqrt{3}xy - 2y^2$.

4. To the quadratic form $9x_1{}^2 - 9x_2{}^2 + 18x_3{}^2$ apply the orthogonal transformation:

$$3x_1 = 2y_1 - y_2 + 2y_3, \quad 3x_2 = -y_1 + 2y_2 + 2y_3, \quad 3x_3 = 2y_1 + 2y_2 - y_3.$$

For the resulting form Q in y_1, y_2, and y_3 show directly that the vector $(2/3, 2/3, -1/3)$ yields the maximum value 18 for Q when $y_1{}^2 + y_2{}^2 + y_3{}^2 = 1$. Check by the calculus.

5. Consider the quadratic form $ax^2 + 2bxy + cy^2$ on the unit circle $x = \cos \theta$, $y = \sin \theta$. Show that its extreme values are (cf. Ex. 1):

$$(a + c) \pm \sqrt{(a + c)^2 - 4\Delta}/2, \quad \Delta = ac - b^2.$$

6. Show that there is no orthogonal matrix with rational entries which reduces xy to diagonal form.

7. A Lorentz transformation is defined to be a linear transformation leaving $x_1{}^2 + x_2{}^2 + x_3{}^2 - x_4{}^2$ invariant. Show that a matrix P defines a Lorentz transformation if and only if $P^{-1} = SP^TS = SP^TS^{-1}$, where S is the special diagonal matrix with diagonal entries, 1, 1, 1, −1.

8. (a) If $A = SR$, with S symmetric and R orthogonal, prove that $S^2 = AA^T$.
   ★(b) Show that there is only one positive definite symmetric matrix S which satisfies $S^2 = AA^T$. (*Hint:* Any eigenvector for S^2 must be one for S.)

★9. Prove Corollary 5 of Theorem 21. (*Hint:* Consider XAX^T as a quadratic function in the Euclidean vector space with inner product XBX^T and write $B = PP^T$ by Theorem 20.)

9.11. Quadrics Under the Affine and Euclidean Groups

Consider next an arbitrary *nonhomogeneous* quadratic function of a vector ξ with coordinates $x_1, \cdots, x_n$,

$$(37) \quad f(\xi) = \sum_i \sum_j x_i a_{ij} x_j + \sum_k b_k x_k + c \quad (i, j, k = 1, \cdots, n).$$

This may be written $f(\xi) = XAX^T + BX^T + c$, where $A = \|a_{ij}\|$ is a symmetric matrix and $B = (b_1, \cdots, b_n)$ a row matrix. In the simple case of a function $f = ax^2 + bx + c$ of one variable, observe that a translation $x = y + k$ leaves invariant the quadratic coefficient a, for

$$(38) \quad \begin{aligned} f &= a(y + k)^2 + b(y + k) + c \\ &= ay^2 + (2ak + b)y + ak^2 + bk + c. \end{aligned}$$

A similar computation works for n variables; a translation $X \mapsto Y = X - K$ (K a row matrix) gives

$$f(\xi) = (Y + K)A(Y + K)^T + B(Y + K)^T + c$$
$$= YAY^T + KAY^T + YAK^T + KAK^T + BY^T + BK^T + c.$$

The product YAK^T (row matrix × matrix × column matrix) is a scalar, hence equals its transpose $KA^TY^T = KAY^T$; all told,

$$(39) \qquad f(\xi) = YAY^T + (2KA + B)Y^T + KAK^T + BK^T + c,$$

an exact analogue of the formula (38). This proves the

Lemma. *A translation leaves unaltered the matrix A of the homogeneous quadratic part of a quadratic function* $f(\xi)$.

On the other hand, a homogeneous linear transformation $X = YP$ changes $f(\xi)$ to $Y(PAP^T)Y^T + (BP^T)Y^T + c$; in this quadratic function the new matrix of quadratic terms is PAP^T, just as in the case of transformation of a homogeneous form alone.

Now to reduce the real function $f(\xi)$ by a rigid motion with equations $X = YP + K$, P orthogonal! By the remarks above, the orthogonal transformation by P alone may be used to simplify the matrix A of the quadratic terms, exactly as for a homogeneous quadratic form. As in §9.10, one finds (with new coefficients b'_i)

$$f(\xi) = \lambda_1 z_1^2 + \cdots + \lambda_n z_n^2 + b'_i z_1 + \cdots + b'_n z_n + c.$$

The b'_j associated with nonzero λ_j can now be eliminated by the simple device of "completing the square," using a translation $y_j = z_j + b'_j/2\lambda_j$. Now, permuting the variables so that the nonzero λ's come first, we get

$$f(\xi) = \lambda_1 y_1^2 + \cdots + \lambda_r y_r^2 + b'_{r+1} z_{r+1} + \cdots + b'_n z_n + c'.$$

If the linear part of this function is not just the constant c', it may be changed by a suitable translation and orthogonal transformation, as in Theorem 13, to the form dy_{r+1}. This transformation need not affect the first r variables. The result is one of the forms

$$(40) \qquad\qquad f(\xi) = \lambda_1 y_1^2 + \cdots + \lambda_r y_r^2 + dy_{r+1},$$

$$(41) \qquad\qquad f(\xi) = \lambda_1 y_1^2 + \cdots + \lambda_r y_r^2 + c',$$

where $\lambda_1 \geqq \lambda_2 \geqq \cdots \geqq \lambda_r$, no $\lambda_i = 0$, $d > 0$.

Theorem 22. *Under the Euclidean group of all rigid motions, any real quadratic form* (37) *is equivalent to one of the forms* (40) *or* (41).

These reduced forms are actually canonical under the Euclidean group, but the proof is much more difficult. In outline, it goes as follows.

The λ_i are (see § 9.10) the eigenvalues of the matrix A of (37); the uniqueness of these (including multiplicity) will be proved in § 10.4. In particular, the number r of squares in (40) or (41) is an invariant; note that r is also the rank of A, unaltered under $A \mapsto PAP^T$. The invariants d and c' are most simply characterized intuitively using the calculus. Consider the locus where the vector

$$\operatorname{grad} f = \left(\frac{\partial f}{\partial y_1}, \cdots, \frac{\partial f}{\partial y_n} \right)$$

is the zero vector. In (41) it is the subspace $y_1 = \cdots = y_r = 0$, and c' is the constant value of $f(\xi)$ on this (invariant) locus. In (40), the locus is empty, since $\partial f/\partial y_{r+1} = d \neq 0$; but d can be characterized as the minimum of $|\operatorname{grad} f|$; this minimum can also be proved invariant under the Euclidean group.

For affine transformations $X = YP + K$, with P nonsingular, a similar treatment applies. In reducing the quadratic part to diagonal form, the coefficients now can all be made ± 1, as in § 9.9. The linear part is then treated as in §9.6.

Theorem 23. *By an affine transformation (or by an affine change of coordinates) any real quadratic function in n variables may be reduced to one of the forms*

$$(42) \qquad y_1^2 + \cdots + y_p^2 - y_{p+1}^2 - \cdots - y_r^2 + c \qquad (r \leqq n),$$

$$(43) \qquad y_1^2 + \cdots + y_p^2 - y_{p+1}^2 - \cdots - y_r^2 + y_{r+1} \qquad (r < n).$$

Since the quadratic terms are unaffected by translation, the rank r and the number p of positive terms must be invariants by the law of inertia (Theorem 17).

From a geometrical point of view, each quadratic function $f(\xi) = XAX^T + BX^T + c$ defines a *figure* or *locus*, which consists of all those vectors ξ which satisfy the equation $f(\xi) = 0$. In two-dimensional space, the figure found from such a quadratic equation is simply an ordinary conic section; in three-space, it is a quadric surface; and in general it may be called a *hyperquadric* (or a quadric hypersurface). An affine transformation $Y = XP + K$ applied to the equation of this surface amounts simply to applying the same transformation to the points of the figure, and the new figure is said to be *equivalent* to the old one under the given affine transformation.

Clearly, the results found above for the classification of quadratic functions under equivalence will yield a similar classification of the corresponding figures. Observe first, however, that an equation $f(\xi) = 0$ and a scalar multiple $cf(\xi) = 0$ of the same equation give identical loci. This may be used to simplify the canonical forms such as $y_1^2 - y_2^2 + c = 0$ found above. When $c \neq 0$, this equation gives the same locus as does $(c^{-1})y_1^2 - (c^{-1})y_2^2 + 1 = 0$; when $c > 0$, this may be reduced by an affine transformation $y_1 = \sqrt{cz_1}$, $y_2 = \sqrt{cz_2}$ to the form $z_1^2 - z_2^2 + 1 = 0$, while for $c < 0$ the transformation $y_i = \sqrt{-cz_i}$ gives a similar result $z_2^2 - z_1^2 + 1 = 0$. In general, this device can always be applied to change the constant c which appears in (43) to 1 or 0. Therefore, in an n-dimensional vector space over the field of real numbers, any hyperquadric is equivalent under the affine group to a locus given by an equation of one of the following forms:

$$(44) \qquad y_1^2 + \cdots + y_p^2 - y_{p+1}^2 - \cdots - y_r^2 + 1 = 0,$$

$$(45) \qquad y_1^2 + \cdots + y_p^2 - y_{p+1}^2 - \cdots - y_r^2 + y_{r+1} = 0,$$

$$(46) \qquad y_1^2 + \cdots + y_p^2 - y_{p+1}^2 - \cdots - y_r^2 = 0,$$

where $0 \leq p \leq r \leq n$, with $r < n$ in the case of (45).

In (44) distinct forms represent affinely inequivalent loci, but in (45) the transformation $y_{r+1} \mapsto -y_{r+1}$ interchanges p and $r - p$, which are thus equivalent.

For example, in the plane the possible types of loci with $r > 0$ are:

$r = 2$		$r = 1$	
$x^2 + y^2 + 1 = 0$	no locus	$\pm x^2 + y = 0$	parabola
$x^2 - y^2 + 1 = 0$	hyperbola	$x^2 + 1 = 0$	no locus
$-x^2 - y^2 + 1 = 0$	circle	$-x^2 + 1 = 0$	two parallel lines
$\pm(x^2 + y^2) = 0$	one point	$x^2 = 0$	one line.
$x^2 - y^2 = 0$	two intersecting lines		

Observe in particular that the *different* canonical functions $x^2 + y^2 + 1$ and $x^2 + 1$ give the *same* locus (namely, the figure consisting of no points at all). So do the canonical functions $x^2 + y^2$ and $-x^2 - y^2$, and so do $x^2 + y$ and $-x^2 + y$.

Exercises

1. Classify under the Euclidean group the forms
   (a) $4xz + 4y^2 + 8y + 8$,
   (b) $9x^2 - 4xy + 6y^2 + 3z^2 + 2\sqrt{5}x + 4\sqrt{5}y + 12z + 16$.
2. Classify under the affine group the forms
   (a) $x^2 + 4y^2 + 9z^2 + 4xy + 6xz + 12yz + 8x + 16y + 24z + 15$,
   (b) $x^2 - 6xy + 10y^2 + 2xz - 20z^2 - 10yz - 40z - 17$,
   (c) $x^2 + 4z^2 + 4xz + 4x + 4z - 6y + 6$,
   (d) $-2x^2 - 3y^2 - 7z^2 + 2xy - 8yz - 6xz - 4x - 6y - 14z - 6$.
3. In a quadratic function $XAX^T + BX^T + c$ with a nonsingular matrix A, prove that the linear terms may be removed by a translation.
4. (a) Show that a nontrivial real quadric $XAX^T = 1$ is a surface of revolution if and only if A has a double eigenvalue.
   (b) Describe the quadric $xy + yz + zx = 3$.
5. Generalize the affine classification of quadratic functions given in Theorem 23 to functions with coefficients in any field in which $1 + 1 \neq 0$.
6. (a) List the possible affine types of quadric surfaces in three-space.
   (b) Give a brief geometric description of each type.
7. Classify (a) ellipses, (b) parabolas, and (c) hyperbolas under the extended similarity group (§ 9.4, end). Find complete sets of numerical invariants in each case.
★8. Classify quadric hypersurfaces in n-dimensional Euclidean space under the group of rigid motions (use Theorem 22).
9. Find a hexagon of maximum area inscribed in the ellipse $x^2 + 3y^2 = 3$.

★9.12. Unitary and Hermitian Matrices

For the complex numbers the orthogonal transformations of real quadratic forms are replaced by "unitary" transformations of certain "hermitian" forms. A single complex number $c = a + ib$ is defined as a pair of real numbers (a, b) or a vector with components (a, b) in two-dimensional real space $\mathbf{R}^2$. The norm or absolute value $|c|$ of the complex number is just the length of the real vector

(47) $\qquad |c|^2 = |a + ib|^2 = a^2 + b^2 = (a + ib)(a - ib) = cc^*,$

where c^* denotes the complex conjugate $a - ib$. On the same grounds, a vector γ with n complex components $(c_1, \cdots, c_n)$, each of the form $c_j = a_j + ib_j$, may be considered as a vector with $2n$ components $(a_1, b_1, \cdots, a_n, b_n)$ in a real space of twice the dimensions. The length of this real vector is given by the square root of

(48)
$$|(c_1, \cdots, c_n)|^2 = (a_1^2 + b_1^2) + \cdots + (a_n^2 + b_n^2)$$
$$= \sum_{j=1}^{n} (a_j + ib_j)(a_j - ib_j)$$
$$= c_1 c_1^* + \cdots + c_n c_n^*.$$

Since each product $c_j c_j^* = a_j^2 + b_j^2 \geqq 0$, this expression has the crucial property of *positive definiteness*: the real sum $\sum_{j=1}^{n} c_j c_j^*$ is positive unless all $c_j = 0$. In this respect (48) resembles the usual Pythagorean formula for the length of a *real* vector. We adopt (48) as the definition of the length of the complex row vector $K = (c_1, \cdots, c_n)$. The formula $\sum c_j c_j^*$ may be written in matrix notation as KK^{*T}, where K^* is the vector obtained by forming the conjugate of each component of K.

Definition. *In the complex vector space* $\mathbf{C}^n$, *let* ξ *and* η *be vectors with coordinates* $X = (x_1, \cdots, x_n)$ *and* $Y = (y_1, \cdots, y_n)$, *and introduce an inner product*

(49)
$$(\xi, \eta) = x_1 y_1^* + \cdots + x_n y_n^* = XY^{*T}.$$

The length of ξ *is then* $|\xi| = (\xi, \xi)^{1/2}$.

Much as in the case of the ordinary inner product, one may then prove the basic properties

Linearity: $(c\xi + d\eta, \zeta) = c(\xi, \zeta) + d(\eta, \zeta).$

Skew-symmetry: $(\xi, \eta) = (\eta, \xi)^*.$

Positiveness: If $\xi \neq \mathbf{0}$, (ξ, ξ) is real and $(\xi, \xi) > 0$.

The skew-symmetry clearly implies a *skew*-linearity in the second factor:

$$(\xi, c\eta + d\zeta) = (c\eta + d\zeta, \xi)^* = c^*(\eta, \xi)^* + d^*(\zeta, \xi)^*$$
$$= c^*(\xi, \eta) + d^*(\xi, \zeta),$$

so that

(50)
$$(\xi, c\eta + d\zeta) = c^*(\xi, \eta) + d^*(\xi, \zeta).$$

If desired, one may adopt the properties of linearity, skew-symmetry, and positiveness as postulates for an inner product (ξ, η) in an abstract vector space over the complex field; the space is then called a *unitary space* (compare the Euclidean vector spaces of §7.10).

Two vectors ξ and η are *orthogonal* ($\xi \perp \eta$) if $(\xi, \eta) = 0$. By the skew-symmetry, $\xi \perp \eta$ implies $\eta \perp \xi$. A set of n vectors $\alpha_1, \cdots, \alpha_n$ in the (n-dimensional) space is a *normal unitary basis* of the space if each vector has length one and if any two are orthogonal:

(51) $|\alpha_1| = \cdots = |\alpha_n| = 1, \qquad (\alpha_i, \alpha_j) = 0 \qquad (i \neq j).$

Such a set is necessarily a basis in the ordinary sense. The original basis vectors $\varepsilon_1 = (1, 0, \cdots, 0), \cdots, \varepsilon_n = (0, \cdots, 0, 1)$ do form such a basis. By the methods of §7.11, one may construct other such bases and prove

Theorem 24. *Any set of $m < n$ mutually orthogonal vectors of length one of a unitary space forms part of a normal unitary basis of the space.*

In particular, if $\alpha_1, \cdots, \alpha_m$ are orthogonal nonzero vectors, and $c_i = (\xi, \alpha_i)/(\alpha_i, \alpha_i)$, then $\alpha_{m+1} = \xi - c_1\alpha_1 - \cdots - c_m\alpha_m$ is orthogonal to $\alpha_1, \cdots, \alpha_m$, for any ξ.

An $n \times n$ matrix $U = \|u_{ij}\|$ of complex numbers is called *unitary* if $UU^{*T} = I$, where U^* denotes the matrix found by taking the conjugate of each entry of U. This is clearly equivalent to the condition that $\sum_k u_{ik}u_{jk}^* = \delta_{ij}$, where δ_{ij} is the Kronecker delta (§9.4); in other words, that each row of U has length one, and any two rows of U are orthogonal. This means that the linear transformation of $\mathbf{C}^n$ defined by U carries $\varepsilon_1, \cdots, \varepsilon_n$ into a normal unitary basis. It is also equivalent, by Theorem 9, Corollary 6, of §8.6, to the condition $U^{*T}U = I$, which states that each column of U has length one, and that any two columns are orthogonal.

An arbitrary linear transformation $X \mapsto XA$ of $\mathbf{C}^n$ carries the inner product XY^{*T} into $XAA^{*T}Y^{*T}$. This new product is again equal to $XY^{*T} = XIY^{*T}$ for all vectors X and Y if and only if $AA^{*T} = I$, i.e., if and only if A is unitary. Thus a matrix A is unitary if and only if the corresponding linear transformation T_A preserves complex inner products XY^{*T}. A similar argument shows that A is unitary if and only if T_A preserves lengths $(XX^{*T})^{1/2}$. Geometrically, a linear transformation T of a unitary space is said to be *unitary* if T preserves lengths, $|\xi T| = |\xi|$, and hence inner products. The set of all unitary transformations of n-space is thus a group, isomorphic to the group of all $n \times n$ unitary matrices.

Quadratic forms are now replaced by "hermitian" forms, of which the simplest example is the formula $\sum x_i x_i^*$ for the length. In general, a *hermitian form* is an expression with complex coefficients h_{ij}

(52) $$\sum_{i,j=1}^n x_i h_{ij} x_j^* = XHX^{*T}, \qquad H = \|h_{ij}\|,$$

in which the coefficient matrix H has the property $H^{*T} = H$. A matrix H of this sort is called *hermitian*; in the special case when the terms h_{ij} are real, the hermitian matrix is symmetric. The form (52) may be considered as a function $h(\xi) = XHX^{*T}$ of the vector ξ with coordinates $x_1, \cdots, x_n$ relative to some basis. The value XHX^{*T} of this function is always a *real* number. To prove this, it suffices to show that this number is equal to its conjugate (or, equally well, to its conjugate transpose). But since H is hermitian,

$$(XHX^{*T})^{*T} = (X^*H^*X^{**T})^T = X^{TT}H^{*T}X^{*T} = XHX^{*T},$$

as asserted.

A unitary transformation $Y = XU$, $X = YU^{-1} = YU^{*T}$, applied to a hermitian form, yields

$$XHX^{*T} = (YU^{-1})H(YU^{*T})^{*T} = YU^{-1}H(UY^{*T}) = Y(U^{-1}HU)Y^{*T}.$$

The coefficient matrix $U^{-1}HU$ is still hermitian,

$$(U^{-1}HU)^{*T} = U^{*T}H^{*T}(U^{-1})^{*T} = U^{-1}HU, \quad \text{since } U^{-1} = U^{*T}.$$

Exactly the same effect on the form may be had by changing to a new normal unitary coordinate system, for such a change will give new coordinates Y for ξ related to the old coordinates by an equation $Y = XU$ with a *unitary* matrix U.

Using this interpretation of the substitution, one may transform any hermitian form to principal axes. The new axes are chosen by successive maximum properties exactly as in the discussion of the principal axes of a quadratic form under orthogonal transformations. The first axis α_1 is chosen as a vector of length one which makes $h(\xi)$ a maximum among all ξ with $|\xi| = 1$; one may then find a normal unitary basis involving α_1 by Theorem 24. Relative to this basis the cross product terms $x_1 x_j^*$ for $j \neq 1$ again drop out. Since the values of the form are all real, the successive maxima λ_i are real numbers. This process proves the following Principal Axis Theorem.

Theorem 25. *Any hermitian form XHX^{*T} can be reduced to real diagonal form,*

$$(53) \qquad YHY^{*T} = \lambda_1 y_1 y_1^* + \lambda_2 y_2 y_2^* + \cdots + \lambda_n y_n y_n^*,$$

by a unitary transformation $Y = XU$.

This theorem may be translated into an assertion about the matrix H of the given form, as follows:

Theorem 26. *For each hermitian matrix H there exists a unitary matrix U such that $U^{-1}HU = U^{*T}HU$ is a real diagonal matrix.*

The methods of Chap. 10 will again prove the diagonal coefficients λ_i of (53) unique.

Exercises

1. Which of the following matrices are unitary or hermitian?

$$\begin{pmatrix} (1+i)/2 & (1-i)/2 \\ (1-i)/2 & (1+i)/2 \end{pmatrix}, \quad \begin{pmatrix} 3 & 1-i \\ 1+i & \sqrt{2} \end{pmatrix}, \quad \begin{pmatrix} 1 & i \\ i & 1 \end{pmatrix}.$$

2. Find a normal unitary basis for the subspace of vectors orthogonal to $(1/2, i/2, (1+i)/2)$.

3. Prove that $\|h_{ij}\|$ is hermitian if and only if $h_{ij}^* = h_{ji}$ for all i and j.

4. Show that if ω is a primitive nth root of unity, then $n^{-1/2}\|\omega^{ij}\|$ is unitary, for $i, j = 1, \cdots, n$.

5. Show that the complex matrix $\begin{pmatrix} \cosh\theta & i\sinh\theta \\ -i\sinh\theta & \cosh\theta \end{pmatrix}$ is unitary for any real θ. Compute its eigenvalues and elgenvectors.

6. Show that all $n \times n$ unitary matrices form a group (the unitary group) which is isomorphic with a subgroup of the group of all $2n \times 2n$ real orthogonal matrices.

7. Prove the linearity, skew-symmetry, and positiveness properties of the hermitian inner product (ξ, η).

8. Give a detailed proof of Theorem 24 on normal unitary bases.

9. Show that a monomial matrix is unitary if and only if all its nonzero entries have absolute value one.

10. Prove a lemma like that of § 9.10 for a hermitian form in two variables with a maximum at $x = 0$, $y = 1$. (*Hint*: Split each variable into its real and imaginary parts.)

★11. Give a detailed proof of the principal axis theorem for hermitian forms.

12. Reduce the form $xy^* + x^*y$ to diagonal form by a unitary transformation of x and y. (*Hint*: Consider the corresponding real quadratic form.)

13. Reduce $zz^* - 2ww^* + 2i(zw^* - wz^*)$ to diagonal form under the unitary group.

14. Show that any real skew-symmetrix matrix A has a basis of complex eigenvectors with pure imaginary characteristic values. (*Hint*: Show iA is hermitian.)

15. Show that the spectrum of any unitary matrix lies on the unit circle in the complex plane.
16. Show that a complex matrix C is positive definite and hermitian if and only if $C = PP^{*T}$ for some nonsingular P.
17. Show that a hermitian matrix is positive definite if and only if all its eigenvalues are positive.

★9.13. Affine Geometry

Affine geometry is the study of properties of figures invariant under the affine group, just as Euclidean geometry treats properties invariant under the Euclidean group. The affine group, acting on a finite-dimensional vector space V, consists as in (11) of the transformations H of V which carry a point (vector) ξ of V into the point

$$(54) \qquad \xi H = \eta = \xi T + \kappa;$$

here κ is a fixed vector, and T a fixed nonsingular linear transformation of V. We assume that V is a vector space over a field F in which $1 + 1 \neq 0$ (e.g., F is not the field $\mathbf{Z}_2$).

In affine geometry, just as in Euclidean geometry, any two points α and β are equivalent, for the translation $\xi \mapsto \xi + (\beta - \alpha)$ carries α into β. This distinguishes affine geometry from the vector geometry of V (under the full linear group), where the origin O plays a special role as the **0** of V. When considering properties preserved under the affine group, one usually refers to vector spaces as *affine* spaces.

In plane analytic geometry, the line joining the two points (x_1, y_1) and (x_2, y_2) has the equation

$$y - y_1 = \frac{y_2 - y_1}{x_2 - x_1}(x - x_1), \qquad x_2 \neq x_1.$$

Introduce the parameter $t = (x - x_1)/(x_2 - x_1)$; then one obtains $y = y_1 + t(y_2 - y_1)$ and $x - x_1 = t(x_2 - x_1)$; in other words, the line has the parametric equations

$$(55) \qquad x = (1 - t)x_1 + tx_2, \qquad y = (1 - t)y_1 + ty_2,$$

which may be written in vector form as $(x, y) = \xi = (1 - t)\xi_1 + t\xi_2$. Geometrically, the point (x, y) of (55) is the point dividing the line segment from (x_1, y_1) to (x_2, y_2) in the ratio $t : (1 - t)$. For $t = \frac{1}{2}$, this point is the *midpoint*.

In any affine space, the point dividing the "segment" from α to β in the ratio $t:(1-t)$ is defined to be the point

$$(56) \qquad\qquad \gamma = (1-t)\alpha + t\beta,$$

and the (affine) *line* $\overline{\alpha\beta}$ joining α to β, for $\alpha \neq \beta$, is defined to be the set of all such points for t in F.

Theorem 27. *Any nonsingular affine transformation carries lines into lines.*

Proof. By substituting (54) in (56), we have

$$\begin{aligned}
\gamma H &= \gamma T + \kappa = (1-t)\alpha T + t\beta T + \kappa \\
&= (1-t)(\alpha T + \kappa) + t(\beta T + \kappa) = (1-t)(\alpha H) + t(\beta H).
\end{aligned}$$

Hence H carries the affine line $\overline{\alpha\beta}$ through α and β into the affine line through αH and βH. Q.E.D.

If $\gamma = (1-t)\alpha + t\beta$ and $\delta = (1-u)\alpha + u\beta$ are any two distinct points of $\overline{\alpha\beta}$, then, since

$$(1-v)\gamma + v\delta = (1 - t + vt - vu)\alpha + (t - vt + vu)\beta,$$

$\overline{\alpha\beta}$ contains every point of $\overline{\gamma\delta}$. The converse may be proved similarly, whence $\overline{\alpha\beta} = \overline{\gamma\delta}$. That is, *a straight line is determined by any two of its points.*

An ordinary plane is sometimes characterized by the property of *flatness*: it contains with any two points the entire straight line through these points. We may use this property to define an *affine subspace* of V as any subset M of V with the property that when α and β are in M, then the entire line $\overline{\alpha\beta}$ lies in M. Clearly, an affine transformation maps affine subspaces onto affine subspaces. Furthermore, the affine subspaces of V are exactly the subspaces obtained by translating vector subspaces of V, in the following sense.

Theorem 28. *If M is any affine subspace of V, then there is a linear subspace S of V and a vector κ such that M consists of all points $\xi + \kappa$ for ξ in S. Conversely, any S and κ determine in this way an affine subspace $M = S + \kappa$.*

Proof. Let κ be any point in M, and define S to be the set of all vectors $\alpha - \kappa$ for α in M; in other words, S is obtained by translating M by $-\kappa$. Clearly, M has the required form in terms of S and κ; it remains only to prove that S is a vector subspace. Since straight lines translate into straight lines, the hypotheses on M insure a like property for S: the line joining any two vectors of S lies in S. For any α in S, the line joining O (in S) to α lies in S, which therefore contains all scalar multiples $c\alpha$. If S contains α and β, it contains 2β and 2α and all the line $\xi = 2\alpha + t(2\beta - 2\alpha)$ joining them. (Draw a figure!) In particular, for $t = 1/2$, it contains $\xi = 2\alpha + (\beta - \alpha) = \beta + \alpha$, the sum of the given vectors. Thus, we have demonstrated that S is closed under sum and scalar product, hence is a vector subspace, as desired.   Q.E.D.

The case $F = \mathbf{Z}_2$ is a genuine exception: the triple of vectors $(0, 0)$, $(1, 0)$, $(0, 1)$ is a "flat" which contains with any two points α and β, all $(1 - t)\alpha + t\beta$; yet this triple is not an affine subspace.

The converse assertion is readily established; it asserts in other words that an affine subspace is just a coset of a vector subspace in the additive group of vectors. In particular, an affine line is a coset (under translation) of a one-dimensional vector subspace.

The preceding results involve another concept of affine geometry: that of parallelism.

Definition. *Two subsets S and S^* of an affine space V are called parallel if and only if there exists a translation $L: \xi \mapsto \xi + \lambda$ of V which maps S onto S^*.*

Theorem 29. *Any affine transformation of V carries parallel sets into parallel sets.*

Proof. Let S and $S^* = S + \lambda$ be the given parallel sets; let U and U^* be their transforms under $H: \xi \mapsto \xi T + \kappa$. The theorem asserts that U^* is the set of all $\xi + \mu$ for variable $\xi \in U$ and some fixed translation vector μ. By definition, U^* is the set of $(\sigma + \lambda)T + \kappa = (\sigma T + \kappa) + \lambda T$ for $\sigma \in S$. And U is the set of all $\xi = \sigma T + \kappa$ for $s \in S$. Setting $\mu = \lambda T$, the conclusion is now obvious.   Q.E.D.

Equivalence under the affine group over the real field $\mathbf{R}$ has a number of interesting elementary geometrical applications. Under the affine group any two triangles are equivalent. To prove this, it suffices to show that any triangle $\alpha\beta\gamma$ is equivalent to the particular *equilateral* triangle with vertices at $O = (0, 0)$, $\beta_0 = (2, 0)$, and $\gamma_0 = (1, \sqrt{3})$ (see Figure 2). By a translation, the vertex α may be moved to the origin O; the other vertices then take up positions β' and γ'. Since these vectors β' and γ' are linearly independent, there then exists a linear transformation $x\beta' + y\gamma' \mapsto x\beta_0 + y\gamma_0$ carrying β' into β_0, γ' into γ_0. The product of the translation by this linear transformation will carry $\alpha\beta\gamma$ into $O\beta_0\gamma_0$, as desired; hence the two triangles are equivalent.

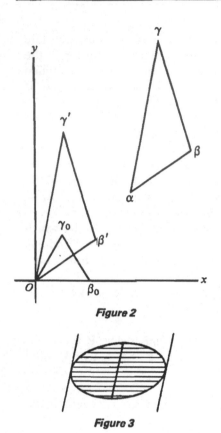

Figure 2

Figure 3

Thus, every triangle is equivalent to an equilateral triangle. But in the latter, the three medians must by symmetry meet in one point (the center of gravity). An affine transformation, however, carries midpoints into midpoints and hence medians into medians. This proves the elementary theorem that the medians of *any* triangle meet in a point. Again, one may prove very easily that the point of intersection divides the medians of an *equilateral* triangle in the ratio $1:2$; hence the same property holds for any triangle.

Moreover, any ellipse is affine equivalent to a circle. But any *diameter* through the center of a circle has parallel tangents at opposite extremities; furthermore, the *conjugate diameter* which is parallel to these tangents bisects all chords parallel to the given diameter. It follows that the same two properties hold for any ellipse, for an affine transformation leaves parallel lines parallel and carries tangents into tangents (but observe that conjugate diameters in an ellipse need not be orthogonal, Figure 3).

Appendix. Centroids and Barycentric Coordinates. The point (56) dividing a line segment in a given ratio is a special case of the notion of a centroid. Given $m + 1$ points $\alpha_0, \cdots, \alpha_m$ in V and $m + 1$ elements $x_0, \cdots, x_m$ in F such that $x_0 + \cdots + x_m = 1$, the *centroid* of the points $\alpha_0, \cdots, \alpha_m$ with the weights $x_0, \cdots, x_m$ is defined to be the point

$$(57) \qquad \xi = x_0\alpha_0 + \cdots + x_m\alpha_m, \qquad x_0 + \cdots + x_m = 1.$$

(More generally, whenever $w = w_0 + \cdots + w_m \neq 0$, the "centroid" of the points $\alpha_0, \cdots, \alpha_m$ with weights $w_0, \cdots, w_m$ is defined by (57), where $x_i = w_i/w$.)

If H is any affine transformation (54), then

$$\xi H = (x_0\alpha_0 + \cdots + x_m\alpha_m)T + \kappa$$
$$= x_0(\alpha_0 T) + \cdots + x_m(\alpha_m T) + (\sum x_i)\kappa$$
$$= x_0(\alpha_0 H) + \cdots + x_m(\alpha_m H).$$

In other words, *an affine transformation carries centroids to centroids with the same weights.*

Theorem 30. *An affine subspace M contains all centroids of its points.*

The proof is by induction on the number $m + 1$ of points in (57). If $m = 0$, the result is immediate, and if $m = 1$, a centroid of α_0 and α_1 is just a point on the line through α_0 and α_1, hence lies in M by definition. Assume $m > 1$, and consider ξ as in (57). Then some coefficient x_i, say x_m, is not equal to 1. Set $t = x_0 + \cdots + x_{m-1}$; then $x_m = 1 - t$, $t \neq 0$, and the point $\beta = (x_0/t)\alpha_0 + \cdots + (x_{m-1}/t)\alpha_{m-1}$ is a centroid of $\alpha_0, \cdots, \alpha_{m-1}$ and lies in M by the induction assumption. Furthermore, $\xi = t\beta + (1 - t)\alpha_m$ is on the line joining $\beta \in M$ to $\alpha_m \in M$, hence ξ is in M, as asserted.

Centroids may be used to describe the subspace M spanned by a given set of points $\alpha_0, \cdots, \alpha_m$, as follows.

Theorem 31. *The set of all centroids (57) of $m + 1$ points $\alpha_0, \cdots, \alpha_m$ of V is an affine subspace M. This subspace M contains each α_i and is contained in any affine subspace N containing all of $\alpha_0, \cdots, \alpha_m$.*

Proof. Let the ξ of (57) and

$$(57') \qquad \eta = y_0\alpha_0 + \cdots + y_m\alpha_m, \qquad y_0 + \cdots + y_m = 1$$

be any two centroids. Then

$$(1 - t)\xi + t\eta = [(1 - t)x_0 + ty_0]\alpha_0 + \cdots + [(1 - t)x_m + ty_m]\alpha_m$$

is also a centroid of $\alpha_0, \cdots, \alpha_m$, since the sum of the coefficients $(1 - t)x_i + ty_i$ is 1. Hence M is indeed an affine subspace. That it contains each α_i is clear. On the other hand, any affine subspace N containing all the α_i must, by Theorem 30, contain all of M. Q.E.D.

The $m + 1$ points $\alpha_0, \cdots, \alpha_m$ are called *affinely independent* if the m vectors $\alpha_1 - \alpha_0, \cdots, \alpha_m - \alpha_0$ are linearly independent. For an affine transformation H, one has $(\alpha_i - \alpha_0)T = \alpha_i H - \alpha_0 H$; hence a nonsingular affine transformation carries affinely independent points into affinely independent points. In this definition of affine independence, the initial point α_0 plays a special role. The following result will show that affine independence does not depend on the choice of an initial point.

Theorem 32. *The $m + 1$ points $\alpha_0, \cdots, \alpha_m$ are affinely independent if and only if every point ξ in the affine subspace M spanned by $\alpha_0, \cdots, \alpha_m$ has a unique representation as a centroid (57) of the α_i.*

Proof. Suppose that the points α_i are independent, but that some point ξ in M has two representations $\xi = \sum x_i \alpha_i$, $\xi = \sum x_i' \alpha_i$ as a centroid, both with $\sum x_i = 1 = \sum x_i'$. Then

$$x_0' - x_0 = (x_1 - x_1') + \cdots + (x_m - x_m'),$$

and the zero vector $\mathbf{0} = O$ has a representation

$$\mathbf{0} = \sum_{i=0}^{m} (x_i - x_i')\alpha_i = \sum_{i=1}^{m} (x_i - x_i')\alpha_i - (x_0' - x_0)\alpha_0$$

$$= \sum_{i=1}^{m} (x_i - x_i')(\alpha_i - \alpha_0).$$

Since the vectors $\alpha_i - \alpha_0$ are linearly independent, we conclude that $x_i = x_i'$, for $i = 1, \cdots, m$. Since $x_0 = 1 - (x_1 + \cdots + x_m)$, we also have $x_0 = x_0'$. The representation of ξ as a centroid is thus unique.

Secondly, suppose the points $\alpha_0, \cdots, \alpha_m$ to be affinely dependent. There is then a linear relation $\sum c_i(\alpha_i - \alpha_0) = 0$ with some coefficient, say c_1, not zero. By division we can assume $c_1 = 1$. Then

$$\alpha_1 = -c_2\alpha_2 - \cdots - c_m\alpha_m + (c_2 + \cdots + c_m + 1)\alpha_0,$$

a representation of α_1 in which the sum of the coefficients is 1. But α_1 has a second such representation as $\alpha_1 = 1 \cdot \alpha_1$; hence the representation as a centroid is not unique.   Q.E.D.

When the points $\alpha_0, \cdots, \alpha_m$ are affinely independent, the scalars $x_0, \cdots, x_m$ appearing in the representation (57) of points in the space spanned by $\alpha_0, \cdots, \alpha_m$ are called the *barycentric coordinates* of ξ relative to $\alpha_0, \cdots, \alpha_m$. Note that any m of these coordinates determines the remaining coordinate, in virtue of $x_0 + \cdots + x_m = 1$.

Exercises

1. For each of the following pairs of points, find the parametric equations of the line joining the two points and represent the line in the form $S_1 + \lambda$ (i.e., find the space S_1).

    (a) $(2, 1)$ and $(5, 0)$,   (b) $(1, 3, 2)$ and $(-1, 7, 5)$,

    (c) $(1, 2, 3, 4)$ and $(4, 3, 2, 1)$.

2. Represent the line through $(1, 3)$ and $(4, 2)$ in the form $S + \lambda$, with four different choices of λ. Draw a figure.

3. Prove: Through three vectors α, β, γ not on a line there passes one and only one two-dimensional affine subspace (a plane!). Prove that the vectors in this plane have the form $\xi = \alpha + s(\beta - \alpha) + t(\gamma - \alpha)$ for variables s and t.

4. Find the parametric equations (in the form of Ex. 3) for the plane through each of the following triples of points:

    (a) $(1, 3, 2), (4, 1, -1), (2, 0, 0)$,   (b) $(1, 1, 0), (1, 0, 1), (0, 1, 1)$,

    (c) $(2, -1, 3), (1, 1, 1), (3, 0, 4)$.

5. In each part of Ex. 4, find a basis for the parallel plane through the origin.

★6. Prove that Theorem 28 is valid over every field *except* $\mathbf{Z}_2$.

7. Prove, assuming only the relevant definitions, that any affine transformation carries midpoints into midpoints.

8. Show that every parallelogram is affine equivalent to a square.

9. Give an affine proof that the diagonals of a parallelogram always bisect each other.

10. (a) Find an affine transformation of $\mathbf{R}^2$ which will take the triangle with vertices $(0, 0)$, $(0, 1)$, and $(1, 0)$ into the equilateral triangle with vertices $(1, 0)$, $(-1, 0)$, $(0, \sqrt{3})$.

    (b) The same problem, if the first triangle has vertices $(1, 1)$, $(1, 2)$, and $(3, 3)$.

11. Prove by affine methods that in a trapezoid the two diagonals and the line joining the midpoints of the parallel sides go through a point.

12. Prove that any parallelepiped is affine equivalent to a cube.

13. Prove that the four diagonals of any parallelepiped have a common midpoint (it is the center of gravity).

14. (a) Show that over any field F any two triangles are equivalent under the affine group.

   ★(b) Show that if $1 + 1 \neq 0$ and $1 + 1 + 1 \neq 0$ in F, then the medians of any triangle meet in a point.

15. Show that a one-one transformation T of a vector space V is affine if and only if $\gamma = (1 - t)\alpha + t\beta$ always implies $\gamma T = (1 - t)\alpha T + t(\beta T)$.

16. If an affine subspace M is spanned by $m + 1$ affinely independent points $\alpha_0, \cdots, \alpha_m$, prove that M is parallel to an m-dimensional vector subspace.

17. By definition, a hyperplane in F^n is an affine subspace of dimension $n - 1$.

    (a) Prove that the set of all vectors ξ whose coordinates satisfy a linear equation $a_1 x_1 + \cdots + a_n x_n = c$ is a hyperplane, provided the coefficients a_i are not all zero.

    (b) Conversely, prove that every hyperplane has such an equation.

    (c) Find the equation of the hyperplane through $(1, 0, 1, 0)$, $(0, 1, 0, 1)$, $(0, 1, 1, 0)$, $(1, 0, 0, 1)$.

18. Let $a_0, \cdots, \alpha_n$ be $n + 1$ affinely independent points of an n-dimensional vector space V, and let $\beta_0, \cdots, \beta_n$ be any $n + 1$ points in V. Prove that there is one and only one affine transformation of V carrying each α_i into β_i.

19. Prove: If an affine subspace M is spanned by $m + 1$ affinely independent points $\alpha_0, \cdots, \alpha_m$ and by $r + 1$ affinely independent points $\beta_0, \cdots, \beta_n$, then $m = r$.

★9.14. Projective Geometry

In the real affine plane, any two points lie on a unique line, and any two *nonparallel* lines "intersect" in a unique point. We shall now construct a real *projective* plane, in which

(i) Any two distinct points lie on a unique line.
(ii) Any two distinct lines intersect in a unique point.

The incidence properties (i) and (ii) are clearly dual to each other, in the sense that the interchange of the words "point" and "line," plus a minor change in terminology, changes property (i) into property (ii) and vice versa.

One way to construct the *real projective plane* $P_2 = P_2(\mathbf{R})$ is as follows. Take a three-dimensional vector space V_3 over the field $\mathbf{R}$ of real numbers, and call a one-dimensional vector (not affine) subspace S of V_3 a "point" of P_2, and a two-dimensional subspace L of V_3 a "line" of P_2. Furthermore, say that the point S lies *on* the line L if and only if the subspace S is contained in the subspace L.

We prove that the "points" and "lines" of $P_2(\mathbf{R})$ satisfy (i) and (ii), as follows. If the points S_1 and S_2 are the one-dimensional subspaces spanned by the vectors α_1 and α_2, then $S_1 \neq S_2$ if and only if α_1 and α_2 are linearly independent. The unique line L in which both S_1 and S_2 lie is, then, the two-dimensional vector subspace spanned by α_1 and α_2; this proves (i). Secondly, if the lines (two-dimensional subspaces) L_1 and L_2 are distinct, the subspace $L_1 + L_2$, which is their linear sum, must have a higher dimension and is then the whole three-dimensional space V_3. Therefore, by Theorem 17, §7.8,

$$\dim (L_1 \cap L_2) = \dim L_1 + \dim L_2 - \dim (L_1 + L_2) = 2 + 2 - 3 = 1,$$

so that the one-dimensional subspace $L_1 \cap L_2$ is the unique point lying on both L_1 and L_2. This proves (ii).

To obtain suitable projective coordinates in $P_2 = P_2(\mathbf{R})$, take V_3 to be the space $\mathbf{R}^3$ of triples (x_1, x_2, x_3) of real numbers. Then each nonzero triple (x_1, x_2, x_3) determines a point S of P_2; the triples (x_1, x_2, x_3) and (cx_1, cx_2, cx_3) determine the same point S if $c \neq 0$. We call these triples, with the identification

$$(x_1, x_2, x_3) = (cx_1, cx_2, cx_3), \qquad c \neq 0,$$

homogeneous coordinates of the point S. Since any two-dimensional subspace L of V_3 may be described as the set of vector solutions of a single homogeneous linear equation, a line L of P_2 is the locus of points whose homogeneous coordinates satisfy an equation

$$(58) \qquad a_1x_1 + a_2x_2 + a_3x_3 = 0, \qquad (a_1, a_2, a_3) \neq (0, 0, 0).$$

We may call (a_1, a_2, a_3) *homogeneous coordinates* of the line L; clearly, the coordinates (a_1, a_2, a_3) and (ca_1, ca_2, ca_3), for $c \neq 0$, determine the same line.

The real projective plane has a very simple geometrical representation. Any homogeneous coordinates (x_1, x_2, x_3) of a point S can be normalized, by multiplication with $(x_1{}^2 + x_2{}^2 + x_3{}^2)^{-1/2}$, so that the new coordinates (y_1, y_2, y_3) satisfy $y_1{}^2 + y_2{}^2 + y_3{}^2 = 1$ and lie on the unit sphere, and two antipodal points (y_1, y_2, y_3) and $(-y_1, -y_2, -y_3)$ on this sphere determine the same point of P_2. In other words, the points of P_2 may be obtained by identifying diametrically opposite points on the unit sphere. Since any two-dimensional vector subspace L of V_3 cuts the unit sphere in a great circle, we may say that a line of P_2 consists of the pairs of antipodal points on a great circle of the unit sphere. It is thus again clear that two projective lines (two great circles) intersect in one projective point (one pair of antipodal points on the sphere).

A "projective plane" $P_2(F)$ can be defined in just the same way over any field F. In any case, it is clear that each one-dimensional vector subspace (cx_1, cx_2, cx_3), with $x_3 \neq 0$, intersects the affine plane $x_3 = 1$ in exactly one point $(x_1/x_3, x_2/x_3, 1)$; the ratios $(x_1/x_3, x_2/x_3)$ are called the *nonhomogeneous* coordinates of the projective point (cx_1, cx_2, cx_3). But the locus $x_3 = 0$ is a projective line, called the "line at infinity." It may be verified that each line

$$L : a_1x_1 + a_2x_2 + a_3x_3 = 0$$

of the projective plane P_2 is either the line at infinity (if $a_1 = a_2 = 0$) or a line $a_1(x_1/x_3) + a_2(x_2/x_3) + a_3 = 0$ of the affine plane, plus one point $(a_2, -a_1, 0)$ on the line at infinity.

An n-dimensional projective space P can be constructed over any field F. The essential step is to start with a vector space $V = F^{n+1}$ of one greater dimension. Then $P = P_n(F)$ is described as follows: a *point* of P is a one-dimensional subspace S of V; an *m-dimensional subspace* of P is the set of all points S or P lying in some $(m + 1)$-dimensional vector subspace L of V. Clearly, each such subspace is itself isomorphic to the m-dimensional projective space P_m determined in the same fashion by the $(m + 1)$-dimensional vector space L. If V is represented (say by coordinates relative to a given basis) as the space of $(n + 1)$-tuples of elements of F, then each point S of P_n can be given $n + 1$ homogeneous coordinates $(x_1, \cdots, x_{n+1})$, and the coordinates $(cx_1, \cdots, cx_{n+1})$, with $c \neq 0$, determine the same point.

A hyperplane (a subspace of dimension $n - 1$) in $P = P_n(F)$ is again the locus given by a single homogeneous equation

$$(59) \quad a_1 x_1 + \cdots + a_{n+1} x_{n+1} = 0, \qquad (a_1, \cdots, a_{n+1}) \neq (0, \cdots, 0).$$

The numbers $(a_1, \cdots, a_{n+1})$ may be regarded as the homogeneous coordinates of the hyperplane; the relations between the projective space P and the *dual* projective space whose points are the hyperplanes of P are exactly the same as the relation between the vector space V and the dual space V^*. By Theorem 13 in §7.7 about the dimension of the set of solutions of homogeneous linear equations, it follows that a set of r linearly independent equations such as (59) determines a projective subspace of dimension $n - r$.

Let $T: V \to V$ be a nonsingular linear transformation. We know (§8.6, Theorem 10, Corollary 2) that T carries each one-dimensional subspace S of V into a one-dimensional subspace S^* of V. Hence T induces a transformation $S \mapsto S^* = ST^*$ of the points of the projective space P, and this transformation T^* carries projective subspaces into projective subspaces, with preservation of the dimension. We call T^* a *projective transformation* of P. If T_1 and T_2 are two such linear transformations of V, the product $T_1 T_2$ induces a transformation $(T_1 T_2)^*$ on P which is the product of the induced transformations, $T_1^* T_2^*$. Hence the set of all projective transformations constitutes a group, the *n-dimensional projective group*; and the correspondence $T \mapsto T^*$ is a homomorphism of the full linear group in $(n + 1)$ dimensions onto the projective group in n dimensions over the field F.

Relative to a given system of coordinates in V, the linear transformation T is determined by a nonsingular $(n + 1) \times (n + 1)$ matrix $\|a_{ij}\|$. The transformation T^* then carries the point with homogeneous coordinates $(x_1, \cdots, x_{n+1})$ into the point with homogeneous coordinates $y_1, \cdots, y_{n+1}$ given by

$$(60) \qquad y_j = x_1 a_{1j} + \cdots + x_{n+1} a_{n+1,j} \qquad (j = 1, \cdots, n + 1).$$

Theorem 33. *The $(n + 1) \times (n + 1)$ matrix A determines the identi-*

cal projective transformation T^* of P_n if and only if A is a scalar multiple cI of the identity matrix I, with $c \neq 0$.

Proof. If $A = cI$ in (60), then $y_j = cx_j$: the homogeneous coordinates $(x_1, \cdots, x_{n+1})$ and $(cx_1, \cdots, cx_{n+1})$ determine the same point of P, and T^* is indeed the identity. Conversely, suppose that T^* is the identity. Then T must carry each of the $n + 1$ unit vectors ε_i into some scalar multiple $c_i\varepsilon_i$, hence A must be the diagonal matrix with diagonal entries $c_1, \cdots, c_{n+1}$. But T must also carry the vector $(1, 1, \cdots, 1)$ into some scalar multiple of itself, while A carries this vector into $(c_1, \cdots, c_{n+1})$. This is a scalar multiple of $(1, \cdots, 1)$ if and only if all the c_i are equal. Therefore A is indeed a scalar multiple of I.

Corollary. *The projective group in n dimensions over the field F is isomorphic to the quotient-group of the full linear group in $n + 1$ dimensions by the subgroup of nonzero scalar multiples of the identity.*

Proof. The map $T \mapsto T^*$ is a homomorphism of the full linear group into the projective group; Theorem 33 asserts that the kernel of this homomorphism is precisely the set of scalar multiples of the identity transformation. Hence the result follows by Theorem 28 of §7.13.

It also follows that two matrices A and A_1 determine the same projective transformation if and only if $A_1 = cA$ for some scalar c.

For the one-dimensional projective line, a projective transformation has the form

$$(61) \qquad y_1 = ax_1 + bx_2, \qquad y_2 = cx_1 + dx_2, \qquad ad \neq bc.$$

In terms of the nonhomogeneous coordinates $z = x_1/x_2$ and $w = y_1/y_2$, this transformation may be written as a linear fractional substitution

$$(62) \qquad\qquad w = (az + b)/(cz + d),$$

obtained by dividing the first equation of (61) by the second. Formula (62) is to be interpreted as follows: if $c = 0$, then (62) carries the point $z = \infty$ into the point $w = \infty$; if $c \neq 0$, then (62) carries the point $z = \infty$ into the point a/c, and the point $z = -d/c$ into the point $w = \infty$. The correctness of these symbolic interpretations may be verified by reverting to homogeneous coordinates and using (61). A similar representation

$$(62') \qquad w_i = \frac{z_1 a_{1i} + \cdots + z_n a_{ni} + a_{n+1,i}}{z_1 b_1 + \cdots + z_n b_n + b_{n+1}} \qquad (b_j = a_{j,n+1}; \quad i = 1, \cdots, n)$$

of projective transformations by linear fractional substitutions is possible in n dimensions.

We have already seen that projective transformations of $P_n(F)$ carry lines into lines. Conversely, it is a classical result that any one-one transformation of a *real* projective space $P_n(\mathbf{R})$, which carries lines into lines, is projective if $n \geq 2$ (see Ex. 6).

A homogeneous quadratic form in three variables determines a locus

$$(63) \qquad \sum_{i,j} x_i b_{ij} x_j = 0 \qquad (i, j = 1, 2, 3)$$

in the projective plane, for if the coordinates (x_1, x_2, x_3) satisfy this equation, then any scalar multiple (cx_1, cx_2, cx_3) also satisfies the equation. This locus is called a *projective conic*; the (projective) *rank* of the conic is the rank of the matrix B of coefficients. If the line at infinity is deleted, the projective conic (63) becomes an ordinary conic. In the real projective plane, any nondegenerate conic (i.e., ellipse, hyperbola, or parabola) is equivalent by § 9.9 to one having one of the four equations

$$(64) \qquad x_1{}^2 + x_2{}^2 + x_3{}^2 = 0, \qquad x_1{}^2 + x_2{}^2 - x_3{}^2 = 0,$$

$$(64') \qquad -x_1{}^2 - x_2{}^2 - x_3{}^2 = 0, \qquad x_1{}^2 - x_2{}^2 - x_3{}^2 = 0.$$

A change of sign in the whole left-hand side of such an equation does not alter the locus; hence the conics given by (64') are essentially those given by (64). For the first conic of (64), the locus is empty. Hence we conclude that *any two nondegenerate conics are projectively equivalent in the real projective plane.*

Exercises

1. In the projective three-space over a field F, prove:
   (a) Any two distinct points lie on one and only one line.
   (b) Any three points not on a line lie on one and only one plane.
2. Generalize Ex. 1 to projective n-space.
3. List all points and lines, and the points on each line in the projective plane over the field $\mathbf{Z}_2$.
4. In the projective plane over a finite field with n elements, show that there are $n^2 + n + 1$ points, $n^2 + n + 1$ lines, and $n + 1$ points on each line.
5. The cross-ratio of four distinct numbers z_1, z_2, z_3, z_4 is defined as the ratio $(z_3 - z_1)(z_4 - z_2)/(z_3 - z_2)(z_4 - z_1)$ (with appropriate conventions when one of the z_i is ∞). Prove that the cross-ratio is invariant under any linear fractional transformation (62).

6. Show that the transformation $(z_1, z_2, z_3) \mapsto (z_1^*, z_2^*, z_3^*)$ carries lines into lines, in the complex projective plane, but is not projective. (Asterisks denote complex conjugates.)

7. What does the projective conic $x_1{}^2 = 2x_2 x_3$ represent in the affine plane if the "line at infinity" $x_3 = 0$ is deleted?

8. (a) Show that every nondegenerate real quadric surface is projectively equivalent to a sphere or to a hyperboloid of one sheet.
   (b) To which of the above is an elliptic paraboloid projectively equivalent? a hyperbolic paraboloid?
   (c) Show that a sphere is not projectively equivalent to a hyperboloid of one sheet.

9. Show that, given any two triples of distinct points z_1, z_2, z_3 and w_1, w_2, w_3 in the projective line, there exists a projective transformation (62) which carries each z_i into the corresponding w_i.

10. Let p_1, p_2, p_3, p_4 and q_1, q_2, q_3, q_4 be any two quadruples of points in the projective plane. Show that there exists a projective transformation (62') which carries each p_i into the corresponding q_i.

10

Determinants and Canonical Forms

10.1. Definition and Elementary Properties of Determinants

Over any field each square matrix A has a determinant; though the determinant can be used in the elementary study of the rank of a matrix and in the solution of simultaneous linear equations, its most essential application in matrix theory is to the definition of the characteristic polynomial of a matrix. In this chapter we shall define determinants, examine their geometric properties, and show the relation of the characteristic polynomial of a matrix A to its characteristic roots (eigenvalues). These concepts will then be applied to the study of canonical forms for matrices under similarity.

The formulas for the solution of simultaneous linear equations lead naturally to determinants. Two linear equations $a_1x + b_1y = k_1$, $a_2x + b_2y = k_2$ have the unique solution

$$x = (k_1b_2 - k_2b_1)/(a_1b_2 - a_2b_1), \qquad y = (a_1k_2 - a_2k_1)/(a_1b_2 - a_2b_1),$$

provided $a_1b_2 - a_2b_1 \neq 0$. The polynomials which appear here in numerator and denominator are known as *determinants*,

$$(1) \qquad \begin{vmatrix} a_1 & b_1 \\ a_2 & b_2 \end{vmatrix} = a_1b_2 - a_2b_1, \qquad \begin{vmatrix} k_1 & b_1 \\ k_2 & b_2 \end{vmatrix} - k_1b_2 = k_2b_1.$$

Similarly, one may compute the simultaneous solutions of three linear equations $\sum a_{ij}x_j = k_i$. The denominator of each solution x_j turns out to be

(2)
$$\begin{vmatrix} a_{11} & a_{12} & a_{13} \\ a_{21} & a_{22} & a_{23} \\ a_{31} & a_{32} & a_{33} \end{vmatrix} = \begin{array}{l} a_{11}a_{22}a_{33} + a_{12}a_{23}a_{31} + a_{13}a_{21}a_{32} \\ -a_{11}a_{23}a_{32} - a_{12}a_{21}a_{33} - a_{13}a_{22}a_{31}. \end{array}$$

On the right are six products. Each involves one factor a_{1i} from the first row, one from the second row, and one from the third. Each column is also represented in every product, so that a term of (2) has the form $a_{1\_}a_{2\_}a_{3\_}$, with the blanks filled by some permutation of the column indices 1, 2, 3. Of the six possible permutations, the three even permutations I, (123), (132) appear in products with a prefix +, while the odd permutations are associated with a minus sign. Experience has shown that the solutions of n equations in n unknowns are expressed by analogous formulas.

Definition. *The determinant $|A|$ of an $n \times n$ matrix $A = \|a_{ij}\|$ is the following polynomial in the entries† $a_{ij} = a(i,j)$:*

(3)
$$\det (A) = |A| = \sum_{\phi} \operatorname{sgn} \phi \left[\prod_{i=1}^{n} a_{i,i\phi} \right]$$
$$= \sum_{\phi} (\operatorname{sgn} \phi) a(1, 1\phi) a(2, 2\phi) \cdots a(n, n\phi).$$

Summation is over the $n!$ different permutations ϕ of the integers $1, \cdots, n$. The factor sgn ϕ prefixing each product $\Pi a_{i,i\phi}$ is +1 or −1, according as ϕ is an even or an odd permutation.

Thus, the determinant $|A| = \|a_{ij}\|$ is a sum of $n!$ terms $\pm a_{1\_}a_{2\_} \cdots a_{n\_}$, where the blanks are filled in by a variable permutation ϕ of the digits $1, \cdots, n$. Writing a_{ij} as $a(i,j)$, and letting $i\phi$ be the image of i under ϕ, the general term can be written $\pm a(1, 1\phi) a(2, 2\phi) \cdots a(n, n\phi)$, where the sign $\pm$ is called sgn ϕ (for *signum* ϕ). Each term has exactly one factor from each row, and exactly one factor from each column.

Each row appears once and only once in each term of $|A|$, which means that $|A|$ is a linear homogeneous function of the entries $a_{i1}, \cdots, a_{in}$ in the ith row of A. Collecting the coefficients of each such a_{ij}, we get an expression

(4)
$$|A| = A_{i1}a_{i1} + A_{i2}a_{i2} + \cdots + A_{in}a_{in},$$

where the coefficient A_{ij} of a_{ij} is called the *cofactor* of a_{ij}; it is a

† The entries are elements of a field F or, more generally, of a commutative ring.

polynomial in the entries of the remaining rows of A. This cofactor can also be described as the partial derivative $A_{ij} = \partial |A|/\partial a_{ij}$. Since each term of $|A|$ involves each row and each column only once, the cofactor A_{ij} can involve neither the ith row nor the jth column. It contains only entries from the "minor" or submatrix M_{ij}, which is the matrix obtained from A by crossing out the ith row and the jth column.

Rows and columns enter symmetrically in $|A|$:

Theorem 1. *If A^T is the transpose of A, $|A^T| = |A|$.*

Proof. The entry $a_{ij}{}^T = a_{ji}$ of A^T is found by inverting subscripts. A sample term of $|A|$ with $j = i\phi$ and $i = j\phi^{-1}$ is

$$(\text{sgn } \phi) \prod_i a(i, i\phi) = (\text{sgn } \phi) \prod_j a(j\phi^{-1}, j) = (\text{sgn } \phi) \prod_j a^T(j, j\phi^{-1}).$$

This result is a sample term of $|A^T|$, for every permutation is the inverse ϕ^{-1} of some permutation ϕ. Even the signs $(\text{sgn } \phi) = (sgn\ \phi^{-1})$ agree, for ϕ is even (i.e., in the alternating group) if and only if its inverse ϕ^{-1} is also even (§ 6.10). Hence $|A| = |A^T|$. Q.E.D.

What is the effect of elementary row operations on a determinant?

RULE 1. To multiply the ith row of A by a scalar $c \neq 0$, multiply the determinant $|A|$ by c, for in the linear homogeneous expression (4), an extra factor c in each term $a_{i1}, \cdots, a_{in}$ from the ith row simply gives an extra factor c in $|A|$.

RULE 2. To permute two rows of A, change the sign of $|A|$. By symmetry (Theorem 1), we may prove instead that the interchange of two columns changes the sign. This interchange is represented by an odd permutation ϕ_0 of the column indices; thus it replaces A by $B = \|b_{ij}\|$ where $b(i, j) = a(i, j\phi_0)$. Then,

$$|B| = \sum_\phi (\text{sgn } \phi) \prod_i b(i, i\phi) = \sum_\phi (\text{sgn } \phi) \prod_i a(i, i\phi\phi_0).$$

Since the permutations form a group, the products $\phi\phi_0$ (with ϕ_0 fixed) include all permutations, so that $|B|$ above has all the terms of $|A|$. Only the signs of the terms are changed, for ϕ_0 is odd, so $\phi\phi_0$ is even when ϕ is odd, and vice versa: sgn $\phi\phi_0 = -$sgn ϕ. This gives the rule.

Lemma 1. *If A has two rows alike, $|A| = 0$.*

Proof. By Theorem 1, it suffices to prove that $|A| = 0$ if A has two like columns. Let ψ be the transposition which interchanges the two like columns. Then the summands $(\text{sgn } \phi)\Pi a(i, i\phi)$ in (3) occur in pairs $\{\phi, \psi\phi\}$, consisting of the cosets of the two-element subgroup generated by ψ. Since ψ is odd, $\text{sgn } \phi = -\text{sgn } \psi\phi$; since the columns are alike, $\Pi a(i, i\phi) = \Pi a(i, i\psi\phi)$. Hence the paired summands are equal in magnitude and opposite in sign, and their sum is zero. Q.E.D.

For the consideration of adjoints (§10.2), it is convenient to express this lemma by an equation. In A, replace row i by row k. Then two rows become alike, and the determinant is zero. But this determinant may be found by replacing row i by row k in the linear homogeneous expression (4), so that

$$(5) \qquad 0 = A_{i1}a_{k1} + A_{i2}a_{k2} + \cdots + A_{in}a_{kn} \qquad (i \neq k).$$

RULE 3. The addition of a constant c times row k to row i leaves $|A|$ unchanged. This operation replaces each a_{ij} by $a_{ij} + ca_{kj}$; by the linear homogeneous expression (4), the new determinant is

$$\sum_j A_{ij}(a_{ij} + ca_{kj}) = \sum_j A_{ij}a_{ij} + c\sum_j A_{ij}a_{kj} = |A| + 0,$$

by (4) and (5). The determinant is indeed unchanged.

These rules may be summarized in terms of elementary matrices. Any elementary row operation carries the identity I into an elementary matrix E, and A into its product EA. The determinant $|I| = 1$ is thereby changed to $|E| = c$, -1, or 1 (Rule 1, 2, or 3), while $|A|$ goes to $|EA| = c|A|$, $(-1)|A|$, or $|A|$ as the case may be. This proves $|EA| = |E||A|$; by symmetry (Theorem 1) the same applies to postfactors E. This establishes

Theorem 2. *If E is an elementary matrix,*

$$|EA| = |E||A| = |AE|$$

Another rule is that for explicitly getting the cofactors from the submatrices M_{ij} discussed above.

RULE 4. $A_{ij} = (-1)^{i+j}|M_{ij}|$; in words, each cofactor A_{ij} is found from the determinant of the corresponding submatrix by prefixing the sign $(-1)^{i+j}$. This is the sign to be found in the (i, j) position on a $\pm$ checkerboard which starts with a plus in the upper left-hand corner. First, consider the proof of this rule for $i = j = 1$. The definition (3) shows at once that the terms involving a_{11} are exactly the terms belonging to permutations ϕ with $1\phi = 1$. An even (odd) permutation of this type is actually an even (odd) permutation of the remaining digits $2, \cdots, n$, so the terms with a_{11} removed are exactly the terms in the expansion of $|M_{11}|$. Any other cofactor A_{ij} may now be reduced to this special case by moving the a_{ij} term to the upper left position by means of $i - 1$ successive interchanges of adjacent rows and $j - 1$ interchanges of adjacent columns. These operations do not alter $|M_{ij}|$ because the relative position of the rows and columns in M_{ij} is unaffected, but they do change the sign of $|A|$, and hence the sign of the cofactor of a_{ij}, $i + j - 1 - 1$ times. This reduction proves the rule.

An especially useful case is that in which all the first row is zero except for the first term. The expansion (4) then need involve only the first cofactor $|M_{11}| = A_{11}$, so

$$(6) \qquad \begin{vmatrix} c & O \\ K & B \end{vmatrix} = c|B|,$$

where O is $1 \times (n - 1)$, K is $(n - 1) \times 1$, and B is $(n - 1) \times (n - 1)$. By this rule and induction, one obtains the following result.

Lemma 2. *The determinant of a triangular matrix is the product of its diagonal entries.*

The preceding rules provide a system for computing a determinant $|A|$. Reduce A by elementary operations to a triangular form T, and record t, the number of interchanges of rows (or columns) used, and $c_1, \cdots, c_s$, the various scalars used to multiply rows (or columns) of A. By Theorem 2, $|A| = (-1)^t (c_1 \cdots c_s)^{-1} |T|$. The computation is completed by setting $|T| = t_{11} \cdots t_{nn}$, using Lemma 2.

Exercises

1. Prove Lemma 2 directly from the definition of a determinant.

2. Compute the determinant of the matrices of Ex. 2, § 7.6.

3. (a) If $\quad A = \begin{pmatrix} 1 & -1 & 0 \\ -1 & 0 & 1 \\ 2 & 1 & 1 \end{pmatrix}$, compute $|A|$ both by the minors of the first row and by the minors of the first column, and compare the results.

(b) Compute $|A|$ on the assumption that the entries of A are integers modulo 2.

4. Write out the positive terms in the expansion of a general 4×4 determinant.

5. If n is odd and $1 + 1 \neq 0$, show that an $n \times n$ skew-symmetric matrix A has determinant 0.

6. (a) Deduce the following expansion of the "Vandermonde" determinant:

$$\begin{vmatrix} 1 & x_1 & x_1^2 \\ 1 & x_2 & x_2^2 \\ 1 & x_3 & x_3^2 \end{vmatrix} = (x_2 - x_1)(x_3 - x_1)(x_3 - x_2).$$

  (b) Generalize the result to the 4×4 case.

  (c) Generalize to the $n \times n$ case, by proving that if $a_{ij} = x_i^{j-1}$, then
$$|A| = \prod_{i>j} (x_i - x_j).$$

7. Show that for any 4×4 skew-symmetric matrix A,

$$|A| = (a_{12}a_{34} - a_{13}a_{24} + a_{14}a_{23})^2.$$

8. (a) Show that the determinant of any permutation matrix is ± 1.

  (b) Show that the determinant of a monomial matrix is the product of the nonzero entries, times ± 1.

9. A real $n \times n$ matrix is called *diagonally dominant* if $\sum_{j \neq i} |a_{ij}| < a_{ii}$ for
$i = 1, \cdots, n$. Show that if A is diagonally dominant, then $|A| > 0$.

10. In the plane show that the line joining the point (a_1, a_2) to the point (b_1, b_2) has the equation

$$\begin{vmatrix} x_1 & x_2 & 1 \\ a_1 & a_2 & 1 \\ b_1 & b_2 & 1 \end{vmatrix} = 0.$$

★11. (a) If each entry a_{ij} in a matrix A is a function of x, show that

$$\frac{d|A|}{dx} = \sum_{j,k=1}^{n} \frac{da_{jk}}{dx} A_{jk}.$$

  (b) Use this to verify that $A_{ij} = \dfrac{\partial |A|}{\partial a_{ij}}$.

★12. If A and C are square matrices, prove that $\begin{vmatrix} A & B \\ 0 & C \end{vmatrix} = |A||C|$.

★13. If Ω is the $n \times n$ matrix $\|\omega^{ij}\|$, where ω is a primitive complex nth root of unity, show that $|\Omega| = n^{n/2}$, provided $n \equiv 1 \pmod 4$.

10.2. Products of Determinants

Under elementary row and column operations, any square matrix A is equivalent to a diagonal matrix D (Theorem 18, § 8.9), so A can be obtained from D by pre- and postmultiplication by elementary matrices E_i and $E^{(i)}$, as in Theorem 13 of § 8.8,

$$(7) \qquad\qquad A = E_s \cdots E_1 D E^{(1)} \cdots E^{(t)}.$$

The rules $|EA| = |E| \cdot |A|$ and $|AE| = |A| \cdot |E|$ of Theorem 2 show that in the determinant of the product (7) the factors $|E|$ may be taken out one at a time to give

$$(8) \qquad |A| = |E_s| \cdots |E_1| |D| |E^{(1)}| \cdots |E^{(t)}|.$$

Since each $|E_i| \neq 0$, the whole determinant $|A| \neq 0$ if and only if $|D| \neq 0$. The canonical form D has exactly r entries 1 along the diagonal, where r is the rank of A, while the determinant $|D|$ is the product of its n diagonal entries. Hence $|D| \neq 0$ if and only if $r = n$; that is, if and only if A is nonsingular. Therefore (8) proves

Theorem 3. *A square matrix A is nonsingular if and only if $|A| \neq 0$.*

Computing Determinants. Formula (8) also provides an efficient algorithm for computing $n \times n$ determinants numerically. One proceeds as in Gaussian elimination, forming the product of the diagonal entries which are replaced by 1 as one proceeds; since the determinants of the other elementary matrices used are 1, this suffices. Thus

$$\begin{vmatrix} 2 & 3 & 4 & 1 \\ 4 & -1 & 2 & 3 \\ -6 & 5 & 2 & 6 \\ 8 & 5 & 7 & -2 \end{vmatrix} = 2 \begin{vmatrix} 1 & 3/2 & 2 & 1/2 \\ 0 & -7 & -6 & 1 \\ 0 & 14 & 14 & 9 \\ 0 & -7 & -9 & -6 \end{vmatrix} = -14 \begin{vmatrix} 1 & 3/2 & 2 & 1/2 \\ 0 & 1 & 6/7 & -1/7 \\ 0 & 0 & 2 & 11 \\ 0 & 0 & -3 & -7 \end{vmatrix}$$

whence the determinant is $(-14)(19) = -266$.

A nonsingular matrix A is a product $A = E_t \cdots E_1$ of elementary matrices. If $B = E_s^* \cdots E_1^*$ is a second such matrix, the product AB has a determinant which may be computed, as in (8), as

$$|AB| = |E_t \cdots E_1 E_s^* \cdots E_1^*|$$
$$= |E_t| \cdots |E_1| \cdot |E_s^*| \cdots |E_1^*| = |A| \cdot |B|.$$

Theorem 4. *The determinant of a matrix product is the product of the determinants*: $|AB| = |A| \cdot |B|$.

Proof. The computation above proves this rule only when A and B are both nonsingular. But if A or B is singular, so is AB, and both sides of $|AB| = |A| \cdot |B|$ are zero. Q.E.D.

The inverse of a matrix A with a determinant $|A| \neq 0$ exists and may be found explicitly by using cofactors of A. The original equations (4) and

(5) involving the cofactors may be written as

$$(9) \quad a_{k1}A_{i1} + \cdots + a_{kn}A_{in} = \delta_{ki}|A|, \quad \text{where } \delta_{ki} = \begin{cases} 1 & \text{if } i = k, \\ 0 & \text{if } i \neq k. \end{cases}$$

This number δ_{ki} is exactly the (k, i) entry of the identity matrix $I = \|\delta_{ki}\|$. The equation (9) is much like a matrix product; if the subscripts of the cofactors A_{ij} are interchanged, the left side of (9) gives the (k, i) entry of the product of $A = \|a_{kj}\|$ by the transposed matrix of cofactors. On the right of (9) is the (k, i) entry of the identity multiplied by a scalar $|A|$, so

$$(10) \quad A\|A_{ij}\|^T = |A|I.$$

The matrix $\|A_{ij}\|^T$ which appears in this equation is the *transposed* matrix of the cofactors of elements of A, and is known as the *adjoint* of A. In case $|A| = 1$, the equation (10) states that the adjoint is the inverse of A; in general, if $|A| \neq 0$, (10) proves

Theorem 5. *If $|A| \neq 0$, the inverse of A is $A^{-1} = |A|^{-1}\|A_{ij}\|^T$.*

Cramer's rule for solving n linear equations in n unknowns is a consequence of this formula for the inverse. A given system of equations has the form

$$\sum_j a_{ij}x_j = b_i,$$

where i and j range from 1 to n. In matrix notation the equation is $AX = B$ (X and $B = (b_1, \cdots, b_n)^T$ column n-vectors). If A is nonsingular, this equation premultiplied by A^{-1} gives the unique vector solution $X = (x_1, \cdots, x_n)^T = A^{-1}B$. This solution may be expanded if we observe that the (i, j) entry in the inverse A^{-1} is just $A_{ji}/|A|$. This proves

Theorem 6 (Cramer's Rule). *If n linear equations*

$$\sum_j a_{ij}x_j = b_i$$

in n unknowns have a nonsingular matrix $A = \|a_{ij}\|$ of coefficients, there is a unique solution

$$(11) \quad x_j = (A_{1j}b_1 + \cdots + A_{nj}b_n)/|A|, \quad j = 1, \cdots, n,$$

where A_{ij} is the cofactor of a_{ij} in the coefficient matrix A.

The numerator of this formula may itself be written as a determinant, for it is the expansion by cofactors of the jth column of a determinant obtained from A by replacing the jth column by the column of constants b_j. Observe, however, that large sets of simultaneous equations may usually be solved more efficiently by reducing the matrix (or augmented matrix) to a row-equivalent "echelon" form, as in §7.7.

Cramer's rule evidently applies to any field—and so in particular to all equations discussed in §2.3 (cf. Ex. 9 below). It is especially convenient for solving simultaneous linear equations in 2 and 3 unknowns.

Appendix. Determinants and Rank. A *submatrix* (or "minor") of a rectangular matrix A is any matrix obtained from A by crossing out certain rows and certain columns of A (this is to include the case when *no* rows or *no* columns are omitted). A "determinant rank" d for any rectangular matrix $A \neq O$ may be defined as the number of rows in the biggest square minor of A with a nonvanishing determinant; in other words, d has the properties: (i) A has at least one $d \times d$ minor M, with $|M| \neq 0$; (ii) if $h > d$, every $h \times h$ square minor N of A has $|N| = 0$. It can be shown that *the rank of any matrix equals its determinant rank.*

Exercises

1. Write out the adjoint of a 2×2 matrix $A = \begin{pmatrix} a & b \\ c & d \end{pmatrix}$ and the product of A by its adjoint.
2. (a) Compute the adjoint of the matrix of Ex. 2(a), §7.6, and verify in this case the rule for the product of a matrix by its adjoint.
   (b) Do the same for the matrix of Ex. 2(b), §7.6.
3. By the adjoint method, find the inverses of the 4×4 elementary matrices H_{24}, $I + 2E_{33}$, and $I + dE_{21}$ of §8.8 (50).
4. Find the inverses of Ex. 5, §8.8, by the adjoint method.
5. If A is nonsingular, prove that $|A^{-1}| = |A|^{-1}$.
6. Prove that the product of a singular matrix by its adjoint is the zero matrix.
7. Prove that the adjoint of any orthogonal matrix is its transpose.
8. Write out Cramer's Rule for three equation in 3 unknowns.
9. Solve the simultaneous congruences of Ex. 1, §2.3, by Cramer's Rule.
10. (a) Show that the pair of homogeneous linear equations

$$a_1 x + b_1 y + c_1 z = 0, \qquad a_2 x + b_2 y + c_2 z = 0$$

has a simultaneous solution

$$x = \begin{vmatrix} b_1 & c_1 \\ b_2 & c_2 \end{vmatrix}, \qquad y = \begin{vmatrix} c_1 & a_1 \\ c_2 & a_2 \end{vmatrix}, \qquad z = \begin{vmatrix} a_1 & b_1 \\ a_2 & b_2 \end{vmatrix}.$$

   (b) When is this solution a basis for the whole set of solutions?
   (c) Derive similar formulas for three equations in 4 unknowns.

11. Prove that the determinant of an orthogonal matrix is ± 1.

12. Show that the determinant of the adjoint of a matrix A is $|A|^{n-1}$.

13. Prove that the adjoint of the adjoint of A is $|A|^{n-2}A$.

14. Show directly from the definition of determinant rank that an elementary row operation does not alter the determinant rank.

★15. (a) If A and B are 3×3 matrices, show that the determinant of any 2×2 submatrix of AB is the sum of a number of terms, each of which is a product of the determinant of a 2×2 submatrix of A by that of a 2×2 submatrix of B.

(b) Generalize this result and use it to prove that rank $(AB) \leq$ rank A.

★16. If an $n \times n$ matrix A has rank r, prove that the rank s of the adjoint of A is determined as follows: If $r = n$, then $s = n$; if $r = n - 1$, then $s = 1$; if $r < n - 1$, then $s = 0$.

★17. Prove that the rank of any matrix equals its determinant rank.

10.3. Determinants as Volumes

Determinants of *real* $n \times n$ matrices can be interpreted geometrically as volumes in n-dimensional Euclidean space. The connection is suggested by the formula for the area of a parallelogram.

Each real 2×2 matrix A with rows α_1 and α_2 may be represented as a parallelogram with vertices at

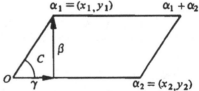

Figure 1

$$O, \alpha_1, \alpha_2, \alpha_1 + \alpha_2;$$

and conversely, each such parallelogram determines a matrix (cf. Figure 1). The area of the parallelogram is

(12) $$\text{base} \times \text{altitude} = |\alpha_1| \cdot |\alpha_2| \cdot |\sin C|,$$

where C denotes the angle between the given vectors α_1 and α_2. By the cosine formula (41) of §7.9, the *square* of the area is

$$(\alpha_1, \alpha_1)(\alpha_2, \alpha_2)(1 - \cos^2 C) = (\alpha_1, \alpha_1)(\alpha_2, \alpha_2) - (\alpha_1, \alpha_2)(\alpha_2, \alpha_1).$$

The result looks very much like the determinant of a 2×2 matrix; it is in fact the determinant of $\|(\alpha_i, \alpha_j)\| = AA^T$.

A similar formula holds for parallelograms in Euclidean space of any dimension—and can even be extended to m-dimensional analogues of parallelograms in n-dimensional Euclidean space. These analogues are called *parallelepipeds*.

To establish the generalization, let A be any $m \times n$ matrix, with rows $\alpha_1, \cdots, \alpha_m$. These rows represent vectors issuing from the origin in n-dimensional Euclidean space E_n. The parallelepiped Π in E_n spanned by the m vectors α_i consists of all vectors of the form

$$t_1\alpha_1 + \cdots + t_m\alpha_m \qquad (0 \leqq t_i \leqq 1; \quad i = 1, \cdots, m).$$

(Picture this in case $m = n = 3$; you will get something affinely equivalent to a cube!) This construction establishes a correspondence between real $m \times n$ matrices and m-dimensional parallelepipeds in n-dimensional space; the α_i are called *edges* of the parallelepiped Π.

The m-dimensional *volume* (including as special cases length if $m = 1$ and area if $m = 2$) $V(\Pi)$ of this figure can be defined by induction on m. Let the parallelepiped with the edges $\alpha_2, \cdots, \alpha_m$ be called the *base* of Π. The *altitude* is the component of α_1 orthogonal to $\alpha_2, \cdots, \alpha_m$; it is to be found from the remaining edge α_1 by writing α_1 as the sum of a component γ in the space S_{m-1} spanned by $\alpha_2, \cdots, \alpha_m$ and a component β orthogonal to S_{m-1} (see Figure 1; this is always possible by §7.11).

$$(13) \qquad \alpha_1 = \beta + \gamma, \qquad \beta \perp S_{m-1}, \qquad \gamma \text{ in } S_{m-1}.$$

The volume of Π is defined as the *product* of the $(m - 1)$-dimensional volume of the *base* by the length $|\beta|$ of the *altitude*.

Theorem 7. *The square of the volume of the parallelepiped with edges $\alpha_1, \cdots, \alpha_m$ is the determinant $|AA^T|$, where A is the matrix with the coordinates of α_i in the ith row.*†

Note. Since a permutation of the rows of A replaces A by PA, where P is an $m \times m$ permutation matrix with $|P| = |P^T| = \pm 1$, and

$$|(PA)(PA)^T| = |P| \cdot |AA^T| \cdot |P^T| = |AA^T|,$$

the "volume" of Π is independent of which $m - 1$ vectors are said to span its "base."

Proof. Since A is an $m \times n$ matrix, the product AA^T is an $m \times m$ square matrix. We now argue by induction on m. If $m = 1$, the matrix A is a row, and the "inner product" $AA^T = (\alpha_1, \alpha_1)$ is the square of the

---

† Throughout §10.3, the coordinates of a vector are taken relative to a fixed normal orthogonal basis. Theorem 7 degenerates to the equation $0 = 0$ if $m > n$.

length, as desired. Suppose that the theorem is true for matrices of $m - 1$ rows, and consider the case of m rows. As in (13), the first row A_1 may be written $A_1 = B_1 + C_1$, where the "altitude" B_1 is orthogonal to each of the rows $A_2, \cdots, A_m$ $(B_1 A_i^T = 0)$, while $C = c_2 A_2 + \cdots + c_m A_m$ is a linear combination of them. Subtract successively c_i times the ith row from the first row of A. This changes A into a new matrix A^* with first row B_1; furthermore, the elementary row operations involved each pre-multiply A by an elementary matrix of determinant 1, hence $A^* = PA$, where $|P| = 1$, and $|A^*A^{*T}| = |PAA^TP^T| = |P||AA^T||P^T| = |AA^T|$. But if D is the block composed of the $m - 1$ rows $A_2, \cdots, A_m$ of A^*, then

$$A^*A^{*T} = \begin{pmatrix} B_1 \\ D \end{pmatrix}(B_1^T D^T) = \begin{pmatrix} B_1 B_1^T & B_1 D^T \\ D B_1^T & D D^T \end{pmatrix} = \begin{pmatrix} B_1 B_1^T & O \\ O & D D^T \end{pmatrix}$$

where $B_1 D^T = O$ because $B_1 A_i^T = 0$ for each row A_i of D. By (6), the determinant is

$$|AA^T| = |A^*A^{*T}| = (B_1 B_1^T) \cdot |DD^T|.$$

Here D is the matrix whose rows $A_2, \cdots, A_m$ span the base of Π, so $|DD^T|$ is the square of the volume of the base, by induction on m. Furthermore, the scalar $B_1 B_1^T$ is the square of the length of the altitude, so we have the desired base × altitude formula for AA^T. Q.E.D.

In the special case when the number of rows is n, evidently $|AA^T| = |A| \cdot |A^T| = |A|^2$, and we have proved†

Theorem 8. *Let A be any real $n \times n$ matrix with rows $\alpha_1, \cdots, \alpha_n$. The determinant of A is (except possibly for sign) the volume of the parallelepiped in E_n having the vectors $\alpha_1, \cdots, \alpha_n$ as edges.*

The absolute value of a determinant is unaltered by any permutations of the rows, so this theorem shows also that our definition of the volume of a parallelepiped is independent of the arrangement of the edges in a sequence. This argument applies also to the formula of Theorem 7, when $m < n$. When $m = n$, the determinant $|A|$ is often called the "signed" volume of the parallelepiped with edges $\alpha_1, \cdots, \alpha_n$; its sign is reversed by any odd permutation.

Theorem 9. *A linear transformation $Y = XP$ of an n-dimensional Euclidean vector space multiplies the volumes of all n-dimensional parallelepipeds by the factor $\pm|P|$.*

†The line of argument in this proof was originally suggested to us by Professor J. S. Frame.

Proof. Consider a parallelepiped which has n edges with respective coordinates $A_1, \cdots, A_n$. The row vectors $A_1, \cdots, A_n$ are transformed into $A_1P, \cdots, A_nP$: the matrix with these new rows is simply the matrix product AP, where A has the rows $A_1, \cdots, A_n$. The new signed volume is then $|AP| = |A||P|$, where $|A|$ is the old volume.

From this it follows that the transformation $Y = XP$ preserves signed volumes if and only if its matrix satisfies $|P| = +1$. The set of all matrices (or of all transformations) with this property is known as the *unimodular group*. Sometimes this group is enlarged to include all P with $|P| = \pm 1$ (i.e., all transformations which preserve the absolute magnitude of volumes).

The volume of any region f in n-dimensional Euclidean space may be defined loosely as follows: circumscribe f by a finite set of parallelepipeds $\Pi_1, \cdots, \Pi_s$ of given shape and orientation, take the sum $\sum V(\Pi_i)$, and define the volume of f to be the greatest lower bound (Chap. 4) of all these sums for different such sets of parallelepipeds. (This is commonly done in the integral calculus, the parallelepipeds being cubes with sides parallel to the coordinate axes.)

By Theorem 9, a linear transformation with matrix P changes the volume of any parallelepiped in the ratio $1:|P|$; hence it changes the volume of f in the same ratio. Since translations leave volumes unaltered, we obtain the following result.

Corollary. *An affine transformation $Y = XP + K$ alters all volumes by the factor $|P|$ (or rather, its absolute value).*

Exercises

1. (a) Compute the area of the parallelogram with vertices $(0,0), (3,0), (1,4)$, and $(4,4)$ in the plane.
   (b) Do the same for the parallelepiped in space with the adjacent vertices $(0,2,0), (2,0,0), (1,1,5)$, and $(0,0,0)$.
2. Show that the medians of any triangle divide it into six parts of equal area. (*Hint:* Reduce to the case of an equilateral triangle, by an affine transformation.)
3. Prove that the diagonals of any parallelogram divide it into four parts of equal area.
4. (a) If P is the intersection of the diagonals of a parallelogram, prove that any line through P bisects the area of the parallelogram.
   (b) Extend this result to three dimensions.
5. Describe three planes which divide a tetrahedron into six parts of equal volume.

6. Using trigonometry, prove directly that the area A of the parallelogram spanned by the vectors $\xi = (x_1, x_2)$ and $\eta = (y_1, y_2)$ satisfies

$$A^2 = \begin{vmatrix} x_1 & x_2 \\ y_1 & y_2 \end{vmatrix}^2.$$

7. (a) If m vectors $\alpha_1, \cdots, \alpha_m$ in E_n are linearly dependent, prove that the parallelepiped which they span has m-dimensional volume zero.
   (b) State and prove the converse of this result.
8. In the group of orthogonal matrices, show that the matrices with $|A| = +1$ (the "proper" orthogonal matrices) form a normal subgroup of index 2.
9. (a) Show that the correspondence $A \mapsto |A|$ maps the full linear group homomorphically onto the multiplicative group of nonzero scalars.
   (b) Infer that the unimodular group is a normal subgroup of the full linear group.
   (c) Is the extended unimodular group (all P with $|P| = \pm 1$) a normal subgroup of the full linear group?
10. (a) Prove that if A is any matrix with rows α_i, then AA^T is the matrix of inner products $\|(\alpha_i, \alpha_j)\|$.
    (b) Using (a), prove that if the α_i are orthogonal, then

$$|AA^T| = (|\alpha_1| \cdots |\alpha_m|)^2.$$

11. (a) If A is a real $m \times n$ matrix, use the proof of Theorem 7 to show that $|AA^T| \geq 0$. Show that the case $m = 2$ of this result is the Schwarz inequality of §7.10, Theorem 18.
    (b) Show that the area of a triangle with vertices $(0, 0, 0)$, (x_1, y_1, z_1), and (x_2, y_2, z_2) is $(1/2)|AA^T|^{1/2}$, where $A = \begin{pmatrix} x_1 & y_1 & z_1 \\ x_2 & y_2 & z_2 \end{pmatrix}$.
   ★(c) The volume of the tetrahedron with three unit edges along the x-, y-, and z-axes is $1/6$. Prove the volume of a tetrahedron with vertices $\alpha_1, \alpha_2, \alpha_3, \alpha_4$ is $(1/6)|BB^T|^{1/2}$, where B is the $3 \times n$ matrix with rows $(\alpha_2 - \alpha_1)$, $(\alpha_3 - \alpha_1)$, $(\alpha_4 - \alpha_1)$.
   ★(d) Generalize to "tetrahedra" of higher dimensions.
★12. Let $-K \leq a_{ij} \leq K$ for $i, j = 1, \cdots, n$.
    (a) Show that if $\alpha_i = (a_{i1}, \cdots, a_{in})$, then $|\alpha_i| \leq K\sqrt{n}$.
    (b) Infer $|A| \leq |\alpha_1| \cdot |\alpha_2| \cdots |\alpha_n| \leq K^n n^{n/2}$ (Hadamard's determinant theorem).

10.4. The Characteristic Polynomial

We have already seen (§9.2, Theorem 5) that λ is a characteristic root (eigenvalue) of the $n \times n$ matrix A if and only if the matrix $A - \lambda I$ is singular. By Theorem 3, this is the case if and only if $|A - \lambda I| = 0$, which proves the following lemma.

Lemma. *The characteristic roots (eigenvalues) of a matrix A are the scalars λ such that $|A - \lambda I| = 0$.*

This lemma provides a straightforward means for reducing a matrix to diagonal form, when such a reduction is possible.

EXAMPLE. Let A be the real symmetric matrix

$$A = \begin{pmatrix} 1 & 3 & 0 \\ 3 & -2 & -1 \\ 0 & -1 & 1 \end{pmatrix}.$$

Then, expanding $|A - \lambda I|$ by minors of the first row,

$$|A - \lambda I| = \begin{vmatrix} 1 - \lambda & 3 & 0 \\ 3 & -2 - \lambda & -1 \\ 0 & -1 & 1 - \lambda \end{vmatrix} = -\lambda^3 + 13\lambda - 12.$$

Factoring, we have $|A - \lambda I| = -(\lambda - 1)(\lambda + 4)(\lambda - 3)$, so that the characteristic roots of A are $1, 3, -4$. (In general, to find the characteristic roots of a 3×3 matrix, one must solve a cubic equation, as in §4.4 or §5.5.) For each characteristic root there is a characteristic vector of the transformation T_A.
Since

$$(x, y, z)T_A = (x + 3y, 3x - 2y - z, -y + z),$$

a vector $\xi = (x, y, z)$ is characteristic, with characteristic root $\lambda = 1$, if and only if $x + 3y = x$, $3x - 2y - z = y$, $-y + z = z$; i.e., if and only if $y = 0$ and $z = 3x$, giving $\xi = (x, 0, 3x)$. Similarly, it is characteristic for $\lambda = 3$ if and only if $x + 3y = 3x$, $3x - 2y - z = 3y$, $-y + z = 3z$; this is the case only for scalar multiples of $(3, 2, -1)$. The characteristic vectors for $\lambda = -4$ are likewise the scalar multiples of $(-3, 5, 1)$. The three characteristic vectors

$$(1, 0, 3), \qquad (3, 2, -1), \qquad (-3, 5, 1)$$

are mutually orthogonal, hence linearly independent. The matrix P with these vectors as rows is nonsingular. Relative to the new basis formed by these three vectors, the transformation T_A is nonsingular with matrix PAP^{-1} (cf. §9.2, Theorems 3' and 4). We may also normalize this basis, to obtain the normal orthogonal basis of characteristic vectors

$$\alpha_1 = \frac{1}{\sqrt{10}}(1, 0, 3), \quad \alpha_2 = \frac{1}{\sqrt{14}}(3, 2, -1), \quad \alpha_3 = \frac{1}{\sqrt{35}}(-3, 5, 1).$$

The matrix Q with these rows is orthogonal, and $QAQ^{-1} = QAQ^T$ is the diagonal matrix with diagonal entries $1, 3, -4$.

The 3×3 symmetric matrix A displayed above is the matrix of the quadratic form $x^2 + 6xy - 2y^2 - 2yz + z^2$. The preceding analysis shows that this quadratic form, relative to the normal orthogonal basis ("principal axes") $\alpha_1, \alpha_2, \alpha_3$, assumes the diagonal form $x^2 + 3y^2 - 4z^2$.

In general, let A be any $n \times n$ matrix. Since a determinant is a polynomial, linear in the entries of each row, the determinant $|A - \lambda I|$ is a polynomial of degree n in the indeterminate λ, of the form

$$(14) \qquad |A - \lambda I| = (-1)^n \lambda^n + b_{n-1} \lambda^{n-1} + \cdots + b_1 \lambda + b_0.$$

We shall define the *characteristic polynomial* of A as the polynomial $c_A(\lambda) = |A - \lambda I|$, and the *characteristic equation* of A as the equation $|A - \lambda I| = 0$. We can now restate the lemma above as follows:

Theorem 10. *The characteristic roots (eigenvalues) of a matrix A are the roots of the characteristic equation of A.*

Since a complex polynomial has at least one root, we infer the following

Corollary. *Over the complex field, a linear transformation has at least one (nonzero) characteristic vector.*

Theorem 11. *Similar matrices have the same characteristic polynomial.* ·

Proof. Let the matrices be A and $B = P^{-1}AP$. Since $|P^{-1}| = |P|^{-1}$ and $|P|$ are scalars, they commute, and so the rule for multiplying determinants gives

$$|P^{-1}AP - \lambda I| = |P^{-1}AP - \lambda P^{-1}IP| = |P^{-1}(A - \lambda I)P|$$
$$= |P^{-1}| \cdot |A - \lambda I| \cdot |P| = |A - \lambda I|.$$

It is a corollary that the successive coefficients $b_0 = |A|, b_1, \cdots,$

$$b_{n-2} = (-1)^n \sum_{i<j} (a_{ii}a_{jj} - a_{ij}a_{ji}),$$

$$b_{n-1} = (-1)^n (a_{11} + \cdots + a_{nn})$$

of $|A - \lambda I|$ are *invariants* of the matrix A, under the group $A \mapsto P^{-1}AP$. Suitable polynomials in the b_i give other useful invariants. One such invariant is

$$\sum_{i,j=1}^n a_{ij}a_{ji} = \sum_{i=1}^n a_{ii}^2 + 2 \sum_{i<j} a_{ij}a_{ji} = b_{n-1}^2 + (-1)^{n-1}2b_{n-2}.$$

In the case of symmetric matrices, this invariant is simply $\sum a_{ij}^2$.

Since $|A^T - \lambda I| = |(A - \lambda I)^T| = |A - \lambda I|$, by Theorem 1, we also have the

Corollary. *A matrix A and its transpose A^T have the same characteristic polynomial, hence the same characteristic roots.*

Theorem 12. *The characteristic polynomial of a triangular matrix T with diagonal entries $d_1, \cdots, d_n$ is*

$$|T - \lambda I| = (d_1 - \lambda)(d_2 - \lambda) \cdots (d_n - \lambda).$$

The proof follows from Lemma 2 of §10.1, since $T - \lambda I$ is itself a triangular matrix. It is a corollary that the set of diagonal entries (with multiplicity) consists of the *roots* (with multiplicity) of the characteristic polynomial. Hence the set of diagonal entries and the number of occurrences of each diagonal entry are the same for any two similar diagonal matrices. This can be stated as follows:

Corollary. *Two diagonal matrices are similar if and only if they differ only in the order of their diagonal terms.*

The properties of similarity throw a new light on the orthogonal transformation of a real quadratic form (§9.10). If a quadratic from XAX^T with matrix A has been reduced by an orthogonal transformation $Z = XP$ to a diagonal form $\lambda_1 z_1^2 + \cdots + \lambda_n z_n^2$, the diagonal matrix D of this new form is $D = PAP^T$. Since P is orthogonal, $P^T = P^{-1}$ and $D = PAP^{-1}$; hence the new matrix D and the original matrix A are similar. The eigenvalues $\lambda_1, \cdots, \lambda_n$ of D are therefore the same as those of the given matrix A. This gives the following sharpened form of Theorem 21 of §9.10.

Theorem 13. *Any real quadratic form XAX^T may be reduced by an orthogonal transformation to a diagonal form $\lambda_1 z_1^2 + \cdots + \lambda_n z_n^2$, in which the coefficients λ_i are the roots of the characteristic equation $|A - \lambda I| = (\lambda_1 - \lambda) \cdots (\lambda_n - \lambda)$ of A.*

But the characteristic equation, and hence its roots, is uniquely determined by A. This proves the essential uniqueness of the diagonal form—and gives a direct way to compute the coefficients. Knowing the coefficients, one can also compute the principal axes as the associated eigenvectors in the way indicated above.

Since we know that any real symmetric matrix is orthogonally equivalent to a real diagonal matrix, we get the

Corollary. *All eigenvalues of a real symmetric matrix are real.*

Remark. If A is symmetric, then eigenvectors X_1 and X_2 having distinct characteristic values $\lambda_1 \neq \lambda_2$ are necessarily orthogonal, for the bilinear expression $X_1 A X_2^T$ may be computed in two ways as

$$(X_1 A)X_2^T = \lambda_1(X_1 X_2^T), \qquad X_1(AX_2^T) = X_1(X_2 A)^T = \lambda_2(X_1 X_2^T).$$

Since $\lambda_1 \neq \lambda_2$, $X_1 X_2^T$ must be zero, and X_1 is therefore orthogonal to X_2.

Hence if the $n \times n$ symmetric matrix A has n distinct eigenvalues $\lambda_1, \cdots, \lambda_n$, any n associated eigenvectors $X_1, \cdots, X_n$ will be orthogonal, and the unit vectors $X_1/|X_1|, \cdots, X_n/|X_n|$ will form the rows of an orthogonal matrix P such that $PAP^T = PAP^{-1}$ will be diagonal.

Exercises

1. Let D be a diagonal matrix with diagonal entries 3, 1, and -1, while P is a traingular matrix with rows $(1, 2, -3)$, $(0, -1, 4)$, $(0, 0, 1)$. Compute the characteristic equation of $P^{-1}DP$ and compare with that of D.

2. Compute the eigenvalues and eigenvectors of the matrices:

$$\text{(a)} \begin{pmatrix} -1 & 2 & 2 \\ 2 & 2 & 2 \\ -3 & -6 & -6 \end{pmatrix}, \quad \text{(b)} \begin{pmatrix} 3 & 2 & 2 \\ 1 & 4 & 1 \\ -2 & -4 & -1 \end{pmatrix}, \quad \text{(c)} \begin{pmatrix} 4 & 9 & 0 \\ 0 & -2 & 8 \\ 0 & 0 & 7 \end{pmatrix}.$$

3. Find the lengths of the principal axes of the quadric $xy + yz + zx + x + y + z = 1$.

4. Write down a diagonal quadratic form equivalent under orthogonal transformation to the expressions given below.

(a) $-2x^2 - 11y^2 - 5z^2 + 4xy + 16yz + 20xz$. (*Hint:* Show that all integral eigenvalues are multiples of 9.)

(b) $3x^2 - y^2 - 3z^2 - t^2 - 4xz - 10yt$.

5. Exhibit an orthogonal transformation which reduces each form of Ex. 4 to its diagonal equivalent.

6. Find a necessary and sufficient condition that the eigenvalues of a 2×2 matrix be equal.

7. Find all 2×2 matrices with eigenvalues $+1$ and -1.

8. Show that if A and B are square matrices, the characteristic polynomial of the matrix $\begin{pmatrix} A & 0 \\ C & B \end{pmatrix}$ is the product of those for A and B.

9. Prove that in (14), $b_{n-1} = \pm(a_{11} + \cdots + a_{nn})$. (The invariant $a_{11} + \cdots + a_{nn}$ is called the *trace* of A.)

10. Prove that in (14), $b_{n-2} = (-1)^n \sum (a_{ii}a_{jj} - a_{ij}a_{ji})$.

11. Prove the formula $\sum a_{ij}^2 = b_{n-1}^2 + (-1)^{n-1}2b_{n-2}$ for symmetric matrices.

12. Prove directly from the definition that all eigenvalues of a real symmetric A are real. (*Hint:* For X an eigenvector, show $XAX^{*T} = \lambda XX^{*T} = \lambda^* XX^{T*}$, where X^* denotes the complex conjugate of X.)

13. (a) Prove that all eigenvalues of a hermitian matrix are real.
    (b) Prove that the eigenvectors span the space of all vectors.

★14. Show that every unitary matrix U has an eigenvector ξ with $\xi U = d\xi$, where $|d| = 1$.

15. (a) Show that if a matrix A has r linearly independent eigenvectors with eigenvalue λ_j, then $c_A(\lambda)$ is a multiple of $(\lambda - \lambda_j)^r$.
    (b) Construct, for any r, an $r \times r$ matrix A with $c_A(\lambda) = (\lambda - \lambda_1)^r$, but having no two linearly independent eigenvectors with eigenvalue λ_1.

★16. Prove the principal axis theorem for a real symmetric matrix A by the following analysis of the linear transformation $X \mapsto XA$.
    (a) The matrix A has an eigenvector α_1 of length 1.
    (b) If α_1 is chosen as the first vector in a new normal orthogonal basis, the new matrix for the given transformation has zeros in the first column and the first row, except for the first entry.
    (c) The argument is continued by induction.

★17. Prove that the volume of the ellipsoid $\sum a_{ij}x_ix_j \leq 1$ is $(4\pi/3) \cdot |A|^{-1/2}$, where $A = \|a_{ij}\|$. (*Hint:* Transform to principal axes, and use Theorem 9.)

10.5. The Minimal Polynomial

The construction of canonical forms for a matrix under similarity depends upon the study of the polynomial equations satisfied by the matrix or by the corresponding transformation. Specifically, let V be an n-dimensional vector space over a field F, and $T: V \to V$ a linear transformation of V. The various powers T^m of T are then also linear transformations of V. Since transformations can also be added or multiplied by scalars, we can consider for each polynomial form $f(x) = a_0 + a_1 x + \cdots + a_k x^k$ with coefficients a_i in F the corresponding polynomial

$$(15) \qquad\qquad f(T) = a_0 I + a_1 T + \cdots + a_k T^k$$

in T. It represents a linear transformation $f(T): V \to V$, and in particular, the constant polynomial $f(x) = 1$ yields the identity transformation $I: V \to V$. since powers of T are permutable ($T^m T^q = T^q T^m = T^{m+q}$), the polynomials $f(T)$ are added and multiplied like the polynomials $f(x)$.

Similarly, each $n \times n$ matrix A with entries in F yields polynomials

$$(16) \qquad\qquad f(A) = a_0 I + a_1 A + \cdots + a_k A^k$$

in A; they are again $n \times n$ matrices with entries in F. Since there are exactly n^2 linearly independent $n \times n$ matrices over F, the $n^2 + 1$ matrices $I, A, \cdots, A^{n^2}$ are certainly linearly dependent, and the dependence relation provides a nonzero polynomial $f(x)$ of degree at most n^2 with $f(A) = O$. Because of the isomorphism $A \mapsto T_A$ between $n \times n$ matrices and linear transformations of V_n, there will also exist for each linear transformation T of an n-dimensional vector space V a nonzero polynomial $f(x)$ with $f(T) = O$.

Theorem 14. *For each linear transformation T of a finite-dimensional vector space V over F, the polynomials $f(x)$ over F such that $f(T) = O$ are the multiples of a unique monic polynomial $m(x)$.*

Proof. Consider the set M of all polynomials $f(x)$ over F such that $f(T) = O$. We have just seen that M contains a nonzero polynomial. Moreover, M is closed under addition, subtraction, and multiplication by any polynomial $g(x)$: it is an *ideal* of the ring $F[x]$. Hence, by Theorem 11 of §3.8, M consists of the multiples of the monic polynomial $m(x)$ of least degree with $m(T) = O$.

We call $m(x)$ the *minimal polynomial* of T. It is the monic polynomial characterized by the properties

$$(17) \qquad m(T) = O; \qquad f(T) = O \quad \text{implies} \quad m(x) \,|\, f(x),$$

where the symbol $m(x) \,|\, f(x)$ means that $m(x)$ divides $f(x)$ in the polynomial ring $F[x]$, as in Chap. 3. The *minimal polynomial* of an $n \times n$ matrix A is described similarly; it is identical with the minimal polynomial of the corresponding transformation T_A of F^n. Since similar matrices are different representations of the same linear transformation, we have

Corollary. *Similar matrices over a field F have the same minimal polynomial over F.*

As an illustration, consider a *nilpotent* transformation (or matrix); that is, a linear transformation T with $T^m = O$ for some m. Since T then satisfies $T^m = O$, its minimal polynomial is x^h for some integer h; indeed, h is the least positive integer with $T^h = O$.

As a special case, suppose that $h = n$. Since $T^{h-1} = T^{n-1} \neq O$, there is a vector α with $\alpha T^{n-1} \neq O$. We assert then that the n vectors $\alpha, \alpha T, \alpha T^2, \cdots, \alpha T^{n-1}$ are linearly independent. If not, there would be a linear dependence relation $O = a_0\alpha + a_1\alpha T + \cdots + a_{n-1}\alpha T^{n-1}$ with coefficients a_i not all zero. If a_j is the first nonzero coefficient, we apply T^{n-j-1} to the equation to get $O = OT^{n-j-1} = a_j\alpha T^j T^{n-j-1} = a_j\alpha T^{n-1}$; but α was chosen so that $\alpha T^{n-1} \neq O$; hence $a_j = 0$, a contradiction.

When these independent vectors $\alpha, \alpha T, \cdots, \alpha T^{h-1}$ are chosen as a basis, T carries each vector of the basis into the next and the last vector to zero, and hence is represented by the $n \times n$ matrix

$$\begin{pmatrix} 0 & 1 & 0 & \cdots & 0 \\ 0 & 0 & 1 & \cdots & 0 \\ \cdot & \cdot & \cdot & & \cdot \\ 0 & 0 & 0 & \cdots & 1 \\ 0 & 0 & 0 & \cdots & 0 \end{pmatrix},$$

in which the only nonzero entries are 1's along the diagonal just above the principal diagonal. This matrix, which is clearly nilpotent, is known as the "companion matrix" of the polynomial x^n.

More generally, to each monic polynomial

$$g(x) = c_0 + c_1x + \cdots + c_{n-1}x^{n-1} + x^n$$

of degree n we can construct an $n \times n$ matrix with minimal polynomial $g(x)$. This matrix, called the *companion matrix* of $g(x)$ is, for $n = 4$,

$$(18) \qquad C_g = \begin{pmatrix} 0 & 1 & 0 & 0 \\ 0 & 0 & 1 & 0 \\ 0 & 0 & 0 & 1 \\ -c_0 & -c_1 & -c_2 & -c_3 \end{pmatrix};$$

for any n, C_g has entries zero except for entries 1 in the diagonal just above the main diagonal, and entries $-c_0, \cdots, -c_{n-1}$ in the last row.

Theorem 15. *For each monic polynomial $g(x)$, the companion matrix C_g has minimal polynomial $g(x)$ and characteristic polynomial $(-1)^n g(\lambda)$.*

Proof. Let T be the linear transformation of F^n represented by the companion matrix C_g of (18). Since the rows of the matrix are the coordinates of the transforms of the unit vectors $\varepsilon_1, \cdots, \varepsilon_n$ of F^n, we have

$$\varepsilon_1 T = \varepsilon_2, \quad \cdots, \quad \varepsilon_{n-1} T = \varepsilon_n, \quad \varepsilon_n T = -c_0 \varepsilon_1 - \cdots - c_{n-1} \varepsilon_n.$$

In other words, the vectors $\varepsilon_1, \varepsilon_1 T, \cdots, \varepsilon_1 T^{n-1}$ are a basis of F^n, so that any vector ξ can be written uniquely as

$$(19) \qquad \xi = a_0 \varepsilon_1 + a_1 \varepsilon_1 T + \cdots + a_{n-1} \varepsilon_1 T^{n-1} = \varepsilon_1 f(T),$$

where $f(x) = a_0 + a_1 x + \cdots + a_{n-1} x^{n-1}$ is a polynomial of degree at most $n - 1$. Furthermore, $\varepsilon_1 T^n = -c_0 \varepsilon_1 - \cdots - c_{n-1} \varepsilon_1 T^{n-1}$, so that $\varepsilon_1 g(T) = 0$. Therefore, for any vector ξ,

$$\xi g(T) = \varepsilon_1 f(T) g(T) = \varepsilon_1 g(T) f(T) = 0,$$

which asserts that T satisfies the monic polynomial equation $g(T) = 0$. For any $f(x) \neq 0$ of smaller degree, $\varepsilon_1 f(T) = \xi \neq 0$ by (19), hence $f(T) \neq 0$. Thus $g(x)$ is indeed the minimal polynomial of C_g.

The characteristic polynomial of C_g is found by expanding the determinant $|C_g - \lambda I|$ by minors of the last row. Since the minor of $-c_k$ is triangular with k diagonal entries $-\lambda$ and the others 1, $|C_g - \lambda I|$ is exactly $(-1)^n g(\lambda)$; the sign $(-1)^n$ occurs because the characteristic polynomial of any $n \times n$ matrix has $(-1)^n \lambda^n$ as its leading term.

Exercises

1. (a) Show that any 2×2 matrix which satisfies $X^2 = O$ is similar to $\begin{pmatrix} 0 & 1 \\ 0 & 0 \end{pmatrix}$ or is $\begin{pmatrix} 0 & 0 \\ 0 & 0 \end{pmatrix}$.

   (b) Prove a corresponding result for 3×3 matrices.
2. Show that every real 2×2 matrix whose determinant is negative is similar to a diagonal matrix. Interpret geometrically.

3. (a) For any nonsingular $n \times n$ matrix P, prove that the correspondence $A \mapsto PAP^{-1}$ is an automorphism of the algebra of all $n \times n$ matrices A.

  (b) Deduce from (a) a direct proof that similar matrices have the same minimal polynomial.

4. Show that the characteristic polynomial of any diagonal matrix is a multiple of its minimal polynomial. When are they the same?

★**5.** (a) Show that every real 2×2 orthogonal matrix, whose determinant is negative, is a rigid reflection. (*Hint:* See Ex. 2 or §9.4.)

  (b) Show that every 2×2 orthogonal matrix, whose determinant is positive, is a rigid rotation.

★**6.** (a) Show that any real 3×3 matrix A has a real eigenvector.

  (b) Show that any orthogonal 3×3 matrix is similar, under an orthogonal change of basis, to a matrix of the form $\begin{pmatrix} \pm 1 & 0 \\ 0 & B \end{pmatrix}$, where B is an orthogonal 2×2 matrix.

  (c) Using Ex. 5, show that if A is an orthogonal 3×3 matrix and $|A| > 0$, then A has an eigenvalue $+1$ and is a rigid rotation. (This is Euler's theorem.)

★**7.** Show that if λ is an eigenvalue of A, and $q(\lambda)$ is any polynomial, then $q(\lambda)$ is an eigenvalue of $q(A)$.

★**8.** (a) Show that the eigenvalues of the matrix C displayed to the right are $\pm 1, \pm i$, the complex fourth roots of unity.

  (b) What are the complex eigenvectors of C?

  (c) To what complex diagonal matrix is C similar?

$$\begin{pmatrix} 0 & 1 & 0 & 0 \\ 0 & 0 & 1 & 0 \\ 0 & 0 & 0 & 1 \\ 1 & 0 & 0 & 0 \end{pmatrix}$$

★**9.** An $n \times n$ matrix A is called a *circulant* matrix when $a_{i,j} = a_{i+1,j+1}$ for all i, j—subscripts being all taken modulo n. Show that the eigenvalues $\lambda_1, \cdots, \lambda_n$ of any circulant matrix are

$$\lambda_p = a_{11} + a_{12}\omega^p + \cdots + a_{1n}\omega^{(n-1)p},$$

where ω is a primitive nth root of unity. (*Hint:* Use Exs. 7 and 8.)

10.6. Cayley–Hamilton Theorem

We shall now show that every square matrix A satisfies its characteristic equation—that is, the minimal polynomial of A divides the characteristic polynomial of A.

This is eaily proved using the concept of a matric polynomial or λ-*matrix*. By this is meant a matrix like $A - \lambda I$ whose entries are polynomials in a symbol λ. Collecting terms involving like powers of λ, one can write any nonzero λ-matrix $B(\lambda)$ in the form

$$B(\lambda) = B_0 + \lambda B_1 + \cdots + \lambda^r B_r,$$

where the B_r are matrices of constants and $B_r \neq O$. (Equality means equality in every coefficient of λ in each entry.)

Lemma. *If $C = B(\lambda)(A - \lambda I)$ is a matrix of constants, then $C = O$.*

Proof. Expanding $B(\lambda)(A - \lambda I)$, we get

$$-\lambda^{r+1}B_r + \sum_{k=1}^{r} \lambda^k(B_k A - B_{k-1}) + B_0 A,$$

where $B_r \neq O$ unless $B(\lambda) = O$. The conclusion is now obvious.

Theorem 16 (Cayley–Hamilton). *Every square matrix satisfies its characteristic equation.*

This means that if each power λ^i in the characteristic polynomial $f(\lambda) = |A - \lambda I|$ of (14) is replaced by the same power A^i of the matrix (*and if λ^0 is replaced by $A^0 = I$*), the result is zero:

$$(20) \qquad b_0 I + b_1 A + \cdots + b_{n-1} A^{n-1} + (-1)^n A^n = O.$$

Proof. In the matrix $A - \lambda I$ the entries are linear polynomials in λ, so that its nonzero minors are determinants which are also polynomials in λ of degree $n - 1$ or less. Each entry in the adjoint C of $A - \lambda I$ is such a minor, so that this adjoint may be written as a sum of n matrices, each involving terms in a fixed power $\lambda^0, \lambda^1, \cdots, \lambda^{n-1}$ of λ. In other words, the adjoint $C = C(\lambda)$ is a λ-matrix $C = C(\lambda) = \sum \lambda^i C_i$. According to (10), the product of $A - \lambda I$ by its adjoint is

$$(21) \qquad C(\lambda)(A - \lambda I) = |A - \lambda I| \cdot I = f(\lambda) \cdot I,$$

where $f(\lambda)$ is the characteristic polynomial.

Now observe that the familiar factorization ($i \geq 1$)

$$A^i - \lambda^i I = (A^{i+1} + \lambda A^{i-2} + \cdots + \lambda^{i-1} I)(A - \lambda I)$$

will give, in terms of the coefficients b_i of the characteristic polynomial (14),

$$f(A) - f(\lambda) \cdot I = \sum_{i=0}^{n} b_i A^i - \sum_{i=0}^{n} b_i \lambda^i I = \sum_{i=1}^{n} b_i (A^i - \lambda^i I)$$

$$= \sum_{i=1}^{n} b_i (A^{i-1} + \lambda A^{i-2} + \cdots + \lambda^{i-1} I)(A - \lambda I),$$

where $f(A)$ is obtained from the characteristic polynomial $f(\lambda)$ by substituting A for λ. That is,

(22) $$f(A) - f(\lambda) \cdot I = -G(\lambda)(A - \lambda I),$$

where $G(\lambda)$ is a new λ-matrix. If we add (22) to (21), we get

$$[C(\lambda) - G(\lambda)](A - \lambda I) = f(A),$$

where $f(A)$ is a matrix of constants. By the lemma, this result implies $f(A) = O$.

Exercises

1. Prove by direct substitution that every 2×2 matrix satisfies its characteristic equation.
2. Show that if A is nonsingular and has the characteristic polynomial (14), then the adjoint of A is given by

$$-[b_1 I + b_2 A + \cdots + b_{n-1} A^{n-2} + (-1)^n A^{n-1}].$$

3. In the notation of Ex. 2, show that the characteristic polynomial of A^{-1} is

$$(-1)^n \left[\lambda^n + \frac{b_1}{|A|} \lambda^{n-1} + \cdots + \frac{(-1)^n}{|A|} \right].$$

4. (a) Prove the Cayley–Hamilton theorem for strictly triangular matrices, by direct computation.
   ★(b) Same problem for triangular matrices.
5. (a) Show by explicit computation that the 4×4 companion matrix C_g of (18) satisfies its characteristic equation.
   (b) Do the same for the companion matrix of a polynomial of degree n.

10.7. Invariant Subspaces and Reducibility

If a linear transformation T satisfies a polynomial equation which can be factored, the matrix representing T can often be correspondingly simplified. Suppose, for example, that T satisfies $T^2 = I$ (is of period two); we assume that $1 + 1 \neq 0$ in the base field F, so that the factors of $(T - I)(T + I) = O$ will be relatively prime. The eigenvectors for T include all nonzero vectors $\eta = \xi(T + I)$ in the range of $T + I$, since

$$(\xi(T + I))T = \xi(T^2 + T) = \xi(T + I);$$

they belong to the eigenvalue $+1$. All nonzero vectors in the range of $T - I$ are also characteristic, with eigenvalue -1, for

$$(\xi(T - I))T = \xi(T^2 - T) = \xi(I - T) = -(\xi(T - I)).$$

But since $1 + 1 \neq 0$, any vector ξ can be written as a sum

$$\xi = (1/2)[\xi(T + I) - \xi(T - I)];$$

hence the eigenvectors with eigenvalues ± 1 span the whose space. Therefore, by Theorem 4 of §9.2, T can be represented by a diagonal matrix with diagonal entries ± 1 on the diagonal.

Specifically, if the entries are all $+1$, T is the identity, and the minimal polynomial of T is $x - 1$; if the entries are all -1, the minimal polynomial is $x + 1$; if both $+1$ and -1 occur, the minimal polynomial of T is $x^2 - 1$. This analysis is a special case of

Theorem 17. *If the minimal polynomial $m(x)$ of a linear transformation $T: V \to V$ can be factored over the base field F of V as $m(x) = f(x)g(x)$, with $f(x)$ and $g(x)$ monic and relatively prime, then any vector in V has a unique expression as a sum*

(23) $$\xi = \eta + \zeta, \qquad \eta f(T) = 0, \qquad \zeta g(T) = 0.$$

Proof. Since f and g are relatively prime, the Euclidean algorithm provides polynomials $h(x)$ and $k(x)$ with coefficients in F so that

(24) $$1 = h(x)f(x) + k(x)g(x).$$

Substitution of T for x yields $I = h(T)f(T) + k(T)g(T)$. Thus, for any vector ξ,

$$\xi = \xi I = \eta + \zeta, \qquad \eta = \xi k(T)g(T), \qquad \zeta = \xi h(T)f(T).$$

As $\eta f(T) = \xi k(T)g(T)f(T) = \xi k(T)m(T) = 0$, and similarly $\zeta g(T) = 0$, this is the required decomposition.

The decomposition (23) is unique, for if $\xi = \eta_1 + \zeta_1 = \eta_2 + \zeta_2$ are two decompositions, then $\alpha = \eta_1 - \eta_2 = \zeta_2 - \zeta_1$ is a vector such that $\alpha f(T) = 0$ and also $\alpha g(T) = 0$; hence by (24),

$$\alpha I = \alpha h(T)f(T) + \alpha k(T)g(T) = 0, \qquad \text{and} \qquad \eta_1 = \eta_2, \quad \zeta_1 = \zeta_2.$$

Theorem 17 can be restated in another way. The subspace S_1 consists of all vectors η with $\eta f(T) = 0$, and S_2 of all ζ with $\zeta g(T) = 0$. That is,

S_1 is the null-space of $f(T)$, and S_2 the null-space of $g(T)$. Moreover V is the direct sum of the subspaces S_1 and S_2, in the sense of §8.8. Each of these subspaces is mapped into itself by T; thus it is an "invariant" subspace in the sense of the following general definition.

A subspace S of a vector space V is said to be *invariant* under a linear transformation $T: V \to V$, if $\xi \in S$ implies $\xi T \in S$. In this event, the correspondence $\xi \mapsto \xi T$ is called the transformation *induced* on S by T.

Evidently, if S is invariant under T, and $h(x)$ is any polynomial, then S is invariant under $h(T)$.

In Theorem 17, $\eta f(T) = 0$ for every $\eta \in S_1$; hence, if T_1 is the linear transformation of S_1 induced by T, the minimal polynomial of T_1 is a divisor $f_1(x)$ of $f(x)$. Similarly the minimal polynomial of T_2 on S_2 is a divisor $g_2(x)$ of $g(x)$. Therefore, for any vector ξ, represented as in (23),

$$(25) \qquad \xi f_1(T)g_2(T) = [\eta f_1(T)]g_2(T) + [\zeta g_2(T)]f_1(T) = 0 + 0 = 0.$$

Hence the product $f_1(x)g_2(x)$ is divisible by the minimal polynomial $m(x) = f(x)g(x)$. Since f and g are relatively prime, this proves that $f(x)$ divides $f_1(x)$, $g(x)$ divides $g_2(x)$. But $f_1(x)$ also divides $f(x)$, so that $f_1 = f$, and likewise $g_2 = g$. We thus get the following result.

Theorem 17'. *If S_1 and S_2 are the null-spaces of $f(T)$ and $g(T)$, respectively, in Theorem 17, then V is the direct sum of S_1 and S_2, and the transformations T_1 and T_2 induced by T on S_1 and S_2 have the minimal polynomials $f(x)$ and $g(x)$, respectively.*

Invariant subspaces arise in many ways. Thus if $f(x)$ is any polynomial, then the range of the transformation $f(T): V \to V$—that is, the set of all vectors $\xi f(T)$ with $\xi \in V$—is invariant under T, for $\xi f(T)T = (\xi T)f(T)$ is in this range. A special class of invariant subspaces are the *cyclic* subspaces generated by one vector, which we shall now define.

Given $T: V \to V$ and a vector α in V, clearly any subspace of V which contains α and is invariant under T must contain all transforms $\alpha f(T)$ of α by polynomials in T. But the set Z_α of all such transforms is an invariant subspace which contains α; we call it the *T-cyclic subspace* generated by α.

Consider now the sequence $\alpha = \alpha I, \alpha T, \alpha T^2, \cdots$ of transforms of α under successive powers of T. Clearly, there is a first one αT^d which is linearly dependent on its predecessors. We will then have

$$(26) \qquad \alpha T^d + c_{d-1}\alpha T^{d-1} + \cdots + c_0\alpha I = \alpha m_\alpha(T) = 0,$$

where $\alpha, \alpha T, \cdots, \alpha T^{d-1}$ are linearly independent. Thus $m_\alpha(x) =$

$x^d + c_{d-1}x^{d-1} + \cdots + c_0$ is the *minimal polynomial* for the transformation T_α induced by T on the T-cyclic subspace Z_α; the polynomial $m_\alpha(x)$ is called the *T-order* of α. Note that T carries each vector of the basis $\alpha, \alpha T, \cdots, \alpha T^{d-1}$ for Z_α into its immediate successor, except for αT^{d-1}, which is carried into

$$(27) \qquad \alpha T^d = -c_0 \alpha - c_1 \alpha T - \cdots - c_{d-1}\alpha T^{d-1}.$$

Relative to the basis $\alpha, \alpha T, \cdots, \alpha T^{d-1}$ of Z_α, T_α is thus represented by the matrix whose rows are the coordinates $(0, 1, 0, \cdots, 0)$, $(0, 0, 1, \cdots, 0), \cdots, (0, \cdots, 0, 1)$, $(-c_0, \cdots, -c_{d-1})$ of the transforms of the basis vectors. This matrix is exactly the companion matrix of the polynomial $m_\alpha(x)$, so that we have proved

Theorem 18. *The transformation induced by T on a T-cyclic subspace Z_α with T-order $m_\alpha(x)$ can be represented by the companion matrix of $m_\alpha(x)$.*

Conversely, the companion matrix C_f of a monic polynomial f of degree n represents a transformation $T = T_{C_f}: F^n \to F^n$ which carries each unit vector ε_i of F^n into the next one and the last unit vector ε_n into $\varepsilon_1 T^n$; hence, as in (19), the whole space F^n is a T-cyclic subspace generated by ε_1, with the T-order $f(x)$.

Theorem 19. *If $T: V \to V$ has the minimal polynomial $m(x)$, then the T-order of every vector α in V is a divisor of $m(x)$.*

Proof. Since $m(T) = O$, $\alpha m(T) = 0$; therefore, by (26), $m(x)$ is a multiple of the T-order $m_\alpha(x)$ of α.

Corollary. *Two vectors α and β in V span the same T-cyclic subspace $Z_\alpha = Z_\beta$ if and only if $\beta = \alpha g(T)$, where the polynomial $g(x)$ is relatively prime to the T-order $m_\alpha(x)$ of α.*

The proof is left as an exercise (Ex. 8).

Exercises

1. (a) Show that any real 2×2 matrix satisfying $A^2 = -I$ is similar to the matrix $\begin{pmatrix} 0 & 1 \\ -1 & 0 \end{pmatrix}$.

    (b) Show that no real 3×3 matrix satisfies $A^2 = -I$.

    (c) What can be said of real 4×4 matrices A satisfying $A^2 = -I$?

2. (a) Show that the range and null-space of any "idempotent" linear trans-formation T satisfying $T^2 = T$ are complementary subspaces (cf. §7.8, (35)).

   (b) Show that any two idempotent matrices which have the same rank are similar. (*Hint:* Use the result of (a).)

3. (a) Classify all 3×3 complex matrices which satisfy $A^3 = I$.

   (b) Do the same for real 3×3 matrices.

4. Every plane shear satisfies $A^2 + I = A + A$. Find a canonical form for 2×2 matrices satisfying this equation. (*Hint:* Form $A - I$.)

★5. Under what conditions on the field of scalars is a matrix with minimal polynomial $x^2 + x - 2$ similar to a diagonal 2×2 matrix?

6. (a) In Theorem 17, show that the range of $g(T)$ is identical with the null-space of $f(T)$.

   (b) Show that if $f(T)g(T) = O$, where $f(x)$ and $g(x)$ are relatively prime polynomials, then the conclusion of Theorem 17 holds, even if $f(x)g(x)$ is not the minimal polynomial of T.

7. Prove: The T-order of a vector α is the monic polynomial $f(x)$ of least degree such that $\alpha f(T) = \mathbf{0}$.

8. Prove the Corollary of Theorem 19.

9. Prove: Given $T: V \to V$ and vectors α and β in V with the relatively prime T-orders $f(x)$ and $g(x)$, then $\alpha + \beta$ has the T-order $f(x)g(x)$.

★10. Prove: Every invariant subspace of a T-cyclic space is itself T-cyclic. (*Hint:* Consider the corresponding property of cyclic groups.)

11. Prove: If $f(T) = O$, while $f(x)$ and $g(x)$ are relatively prime, then T and $g(T)$ have the same cyclic subspaces.

10.8. First Decomposition Theorem

The construction used in proving Theorems 17 and 17′ can be used to decompose a general linear transformation into "primary" components, whose minimal polynomials are powers of irreducible polynomials. In this decomposition, the concept of a direct sum of k subspaces plays a central role.

Definition. *A vector space V is said to be the* direct sum *of its subspaces $S_1, \cdots, S_k$ (in symbols, $V = S_1 \oplus \cdots \oplus S_k$) when every vector ξ in V has a unique representation*

$$(28) \qquad \xi = \eta_1 + \cdots + \eta_k \qquad (\eta_i \in S_i; \quad i = 1, \cdots, k).$$

Exactly as in §7.8, Theorem 16, one can prove

Theorem 20. *If V has subspaces $S_1, \cdots, S_k$, where each S_i is of dimension n_i and has a basis $\alpha_{i1}, \cdots, \alpha_{in_i}$, then V is the direct sum of $S_1, \cdots, S_k$ if and only if*

$$(29) \qquad \alpha_{11}, \cdots, \alpha_{1n_1}; \alpha_{21}, \cdots, \alpha_{2n_2}; \cdots; \alpha_{k1}, \cdots, \alpha_{kn_k}$$

is a basis of V.

It follows that the dimension of V is the sum $n_1 + \cdots + n_k$ of the dimensions of the direct summands S_i.

Corollary. *If V is spanned by the subspaces $S_1, \cdots, S_k$, and*

$$d[V] = d[S_1] + \cdots + d[S_k],$$

then V is the direct sum of $S_1, \cdots, S_k$.

A linear transformation $T: V \to V$ (or a matrix representing T) is said to be *fully reducible* if the space V can be represented as a direct sum of proper invariant subspaces.

Theorem 21. *If V is the direct sum of invariant subspaces $S_1, \cdots, S_k$ on each of which the transformation induced by a given transformation $T: V \to V$ is represented by a matrix B_i, then T can be represented on V by the matrix*

$$(30) \qquad B = \begin{pmatrix} B_1 & 0 & \cdots & 0 \\ 0 & B_2 & \cdots & 0 \\ \vdots & \vdots & & \vdots \\ 0 & 0 & \cdots & B_k \end{pmatrix}$$

This matrix B, consisting of blocks $B_1, \cdots, B_k$ arranged along the diagonal, with zeros elsewhere, is called the *direct sum* of the matrices $B_1, \cdots, B_k$. Observe that any polynomial $f(B)$ in B is the direct sum of $f(B_1), \cdots, f(B_k)$.

Proof. Choose a basis $\alpha_{i1} \cdots \alpha_{in_i}$ for each invariant subspace S_i, so that B_i is the matrix representing the transformation T on S_i relative to this basis. Then these basis vectors combine to yield a basis (29) for the whole space. Furthermore, T carries the basis vectors $\alpha_{i1}, \cdots, \alpha_{in_i}$ into vectors of the ith subspace, and hence T is represented, relative to the basis (29), by the indicated direct sum matrix (30). Q.E.D.

Now consider the factorization of the minimal polynomial $m(x)$ of T as a product of powers of distinct monic polynomials $p_i(x)$ irreducible over the base field, in the form

$$(31) \qquad m(x) = p_1(x)^{c_1} \cdots p_k(x)^{e_k}, \qquad e_i > 0.$$

Since distinct $p_i(x)^{e_i}$ are relatively prime, repeated use of Theorem 17′ will yield

Theorem 22. *If the minimal polynomial of the linear transformation $T: V \to V$ has the factorization (31) into monic factors $p_i(x)$ irreducible over the base field F, then V is the direct sum of invariant subspaces $S_1, \cdots, S_k$ where S_i is the null-space of $p_i(T)^{e_i}$. The transformation T_i induced by T on S_i has the minimal polynomial $p_i(x)^{e_i}$.*

This is our first decomposition theorem; the subspaces S_i are called the "primary components" of V under T. They are uniquely determined by T because the decomposition (31) is unique.

An important special case is the

Corollary. *A matrix A with entries in F is similar over F to a diagonal matrix if and only if the minimal polynomial $m(x)$ of A is a product of distinct linear factors over F.*

Proof. Let $T = T_A: F^n \to F^n$ be the transformation corresponding to A. If

$$(32) \quad m(x) = (x - \lambda_1) \cdots (x - \lambda_k), \qquad \lambda_1, \cdots, \lambda_k \text{ distinct scalars,}$$

the theorem shows that V is the direct sum of spaces S_i, where S_i consists of all vectors η_i with $\eta_i T = \lambda_i \eta_i$; that is, of all eigenvectors belonging to the eigenvalue λ_i. Any basis of S_i must consist of such eigenvectors, so that the matrix representing T on S_i is $\lambda_i I$. Combining these bases as in (29), we have T represented by a diagonal matrix with the entries $\lambda_1, \cdots, \lambda_k$ on the diagonal.

Conversely, if D is any diagonal matrix whose distinct diagonal entries are $c_1, \cdots, c_k$, then the transformation represented by the product $f(D) = (D - c_1 I) \cdots (D - c_k I)$ carries each basis vector into $\mathbf{0}$, and consequently $f(D) = O$. The minimal polynomial of D—and of any other matrix similar to D—is a factor of the product $(x - c_1) \cdots (x - c_k)$, and hence is a product of distinct linear factors.

Exercises

1. Prove Theorem 20.
2. In Theorem 22, set $q_i(x) = m(x)/p_i(x)^{e_i}$, and prove that the subspace S_i there is the range of $q_i(T)$.
★3. Give a direct proof of Theorem 22, not using Theorem 17′.
4. If the $n \times n$ matrix A is similar to a diagonal matrix D, prove that the number of times an entry λ_i occurs on the diagonal of D is equal to the dimension of the set of eigenvectors belonging to the eigenvalue λ_i.
5. Prove: the minimal polynomial of the direct sum of two matrices B_1 and B_2 is the least common multiple of the minimal polynomials of B_1 and B_2.
6. Show that the minimal polynomial of a matrix A can be factored into linear factors if and only if the characteristic polynomial of A can be so factored.
★7. Let A be a complex matrix whose minimal polynomial $m(x) = (x - \lambda_1)^{e_1} \cdots (x - \lambda_r)^{e_r}$ equals its characteristic polynomial. Show that A is similar to a direct sum of r triangular $e_i \times e_i$ matrices B_i, of the form sketched below:

$$B_i = \begin{pmatrix} \lambda_i & 1 & & & 0 \\ & \lambda_i & 1 & & \\ & & \ddots & \ddots & \\ & & & \lambda_i & 1 \\ 0 & & & & \lambda_i \end{pmatrix}.$$

★8. Prove that if $m(x)$ is the minimal polynomial for T, there exists a vector α with T-order exactly $m(x)$. (*Hint:* Use Ex. 9, §10.7, considering first the case $m(x) = p(x)^e$, where $p(x)$ is irreducible.)

10.9. Second Decomposition Theorem

We shall show below that each "primary" component S_i of a linear transformation $T: V \to V$ is itself a direct sum of T-cyclic subspaces. In proving this, we shall use the concept of the *quotient-space* V/Z of a vector space V by a subspace S. We recall (§7.12) that the elements of this quotient-space $V' = V/S$ are the cosets $\xi + S$ of S, and that the projection $P: V \to V/S = V'$ given by $\xi P = \xi + S$ is a linear transformation. In particular, for given $T: V \to V$, if the subspace S is invariant under T, then in the formula

(33) $$(\xi + S)T' = \xi T + S,$$

$\xi T + S$ does not depend upon the choice of the representative ξ of $\xi' = \xi + S$, for if another representative $\eta = \xi + \zeta$ were chosen, then

$$\eta T + S = \xi T + \zeta T + S = \xi T + S,$$

since $\zeta \in S$ implies $\zeta T \in S$. Hence the linear transformation $T': V' \to V'$ defined by (33) is single-valued; it is easily verified that T' is also linear. We call T' the transformation of $V/S = V'$ *induced* by T. Moreover, for any polynomial $f(T)$ in T, (33) gives with the formulas of §7.12,

(34) $(\xi + S)f(T') = \xi f(T) + S.$

In particular, $f(T) = O$ implies $f(T') = O'$ in V', so that the T'-order of ξ' in V' divides the T-order of ξ in V.

We are now ready to prove the second decomposition theorem.

Theorem 23. *If the linear transformation $T: V \to V$ has a minimal polynomial $m(x) = p(x)^e$ which is a power of a monic polynomial $p(x)$ irreducible over the field F of scalars of V, then V is the direct sum*

(35) $V = Z_1 \oplus \cdots \oplus Z_r,$

of T-cyclic subspaces Z_i with the respective T-orders

(36) $p(x)^{e_1}, p(x)^{e_2}, \cdots, p(x)^{e_r}, \quad e = e_1 \geqq e_2 \geqq \cdots \geqq e_r.$

Any representation of V as a direct sum of T-cyclic subspaces has the same number of component subspaces and the same set (36) of T-orders.

Proof. The existence of the direct sum decomposition will be established by induction on the dimension n of V. In case $n = 1$, V is itself a cyclic subspace, and the result is immediate.

For $n > 1$, we have $p(T)^e = O$, but $p(T)^{e-1} \neq O$; hence V contains a vector α_1 with $\alpha_1 p(T)^{e-1} \neq O$. The T-order of α_1 is therefore $p(x)^e$, and α_1 generates a T-cyclic subspace Z_1. Since Z_1 is invariant under T, T induces a linear transformation T' on $V' = V/Z_1$. Since evidently $p(T')^e = O'$, the minimal polynomial of T' on V' is a divisor of $p(x)^e$, and we can use induction on $d[V/Z_1] = d[V] - d[Z_1]$ to decompose V/Z_1 into a direct sum of T'-cyclic subspaces $Z_2', \cdots, Z_r'$, for which the T'-orders are

$$p(x)^{e_2'}, \cdots, p(x)^{e_r'}, \quad e \geqq e_2' \geqq \cdots \geqq e_r'.$$

Lemma 1. *If* α_i' *generates the* T'-*cyclic subspace* $Z_i', i = 2, \cdots, r,$ *then the coset* α_i' *contains a representative* α_i *whose* T-*order is the* T'-*order of* α_i'.

The proof depends on the fact that the T-order $p(x)^e$ of α_1 is a multiple of the T-order of every element of V. In particular let $p(x)^d$ be the T'-order of α_i', so that for any representative $\eta = \eta_i$ of α_i', $\eta p(T)^d = \alpha_1 f(T)$ is in the T-cyclic subspace generated by α_1. Then

$$0 = \eta p(T)^e = \alpha_1 f(T) p(T)^{e-d}.$$

Since α_1 has the T-order $p(x)^e$, this implies that $p(x)^e \,|\, f(x) p(x)^{e-d}$, and hence that $f(x) = g(x) p(x)^d$ for some polynomial $g(x)$. We shall now show that $\alpha_i = \eta - \alpha_1 g(T)$ has a T-order $p(x)^d$ equal to the T'-order of α_i', as required. Since the T-order of α_i is a multiple of the T'-order $p(x)^d$ of $\alpha_i' = \alpha_i + Z_1$, it is sufficient to note that

$$[\eta - \alpha_1 g(T)] p(T)^d = \eta p(T)^d - \alpha_1 f(T) = 0.$$

Having proved Lemma 1, we let Z_i be the T-cyclic subspace generated by α_i. Then $d[Z_i] = d[Z_i']$, since both dimensions are equal to the degree of the common T-order $p(x)^{e_i}$ of α_i'. Hence

(37) $$d[V] - d[Z_1] = d[V/Z_1] = d[Z_2] + \cdots + d[Z_r].$$

By choosing bases, it follows that the subspaces $Z_1, \cdots, Z_r$ span V; hence by (37) and the Corollary to Theorem 20, it follows that V is the direct sum $V = Z_1 \oplus \cdots \oplus Z_r$, as asserted.

It remains to prove the uniqueness of the exponents appearing in any decomposition (36); it will suffice to show that these exponents are determined by T and V. This will be done by the computation of the dimensions of certain subspaces. For example, if d denotes the degree of $p(x)$, then the cyclic subspace Z_i has dimension de_i, and hence the whole space V has dimension $d(e_1 + \cdots + e_r)$. Observe also that for any integer s, the image $Z_i p(T)^s$ of Z_i under $p(T)^s$ is the cyclic subspace generated by $\beta_i = \alpha_i p(T)^s$. It has dimension $d(e_i - s)$ if $e_i > s$, and dimension 0 if $e \leqq s$.

Any vector ξ of V has a unique representation as

$$\xi = \eta_1 + \cdots + \eta_r \qquad (\eta_i \in Z_i; \quad i = 1, \cdots, r).$$

Hence any vector in the range $V_p(T)^s$ of $p(T)^s$ has a unique representation, as

$$(38) \qquad \xi p(T)^s = \eta_1 p(T)^s + \cdots + \eta_r p(T)^s,$$

with components $\eta_i p(T)^s$ in the spaces $Z_i p(T)^s$. The integer s determines an integer t such that

$$e_1 > s, \qquad \cdots, \qquad e_t > s, \qquad e_{t+1} \leqq s$$

(or if $e_r > s, t = r$). Hence, by (38), $Vp(T)^s$ is the direct sum of the cyclic subspaces Z_{β_i} generated by the $\beta_i = \alpha_i p(T)^s$, for $i = 1, \cdots, t$, and its dimension is

$$(39) \qquad d[Vp(T)^s] = d[(e_1 - s) + \cdots + (e_t - s)].$$

The dimensions on the left are determined by V and T; they, in turn, determine the e_i in succession as follows. First take $s = e - 1 = e_1 - 1$, then (39) determines the number of e_i equal to e; next take $s = e - 2$, then (39) determines the number of e_i (if any) equal to $e - 1$, and so forth. This proves the invariance of the exponents $e_1, \cdots, e_r$, and completes the proof of Theorem 23.

Exercises

1. Show that if a vector space V is spanned by the T-cyclic subspaces generated by vectors $\alpha_1, \cdots, \alpha_n$, then the minimal polynomial of T is the l.c.m. of the T-orders of the α_i.
2. Find the minimal polynomial of the matrix B of §8.5, Ex. 3.
3. Prove in detail that $T': V/Z \to V/Z$ is linear if $T: V \to V$ is linear and the subspace Z is invariant under T.
4. Prove, following (37), that $Z_1, \cdots, Z_r$ span V.

10.10. Rational and Jordan Canonical Forms

Using Theorems 20 and 23, it is easy to obtain canonical forms for matrices under similarity. One only needs to give a canonical form for transformations on cyclic subspaces!

One such form is provided by Theorem 21. If A is any $n \times n$ matrix, then a suitable choice of basis in each cyclic subspace represents T_A on that subspace by a companion matrix. Combining all these bases yields a basis of F^n, with respect to which T_A is represented by the direct sum of these companion matrices. The uniqueness assertions of Theorems 20 and 23 show that the set of companion matrices so obtained is uniquely determined by A. We have proved

Theorem 24. *Any matrix A with entries in a field F is similar over F to one and only one direct sum of companion matrices of polynomials*

$$(40) \quad p_i(x)^{e_{i1}}, \cdots, p_i(x)^{e_{ir_i}}, \quad e_{i1} \geqq \cdots \geqq e_{ir_i} > 0, \quad i = 1, \cdots, k,$$

which are powers of monic irreducible polynomials $p_1(x), \cdots, p_k(x)$. The minimal polynomial of A is $m(x) = p_1(x)^{e_{11}} p_2(x)^{e_{21}} \cdots p_k(x)^{e_{k1}}$.

The set of polynomials (40), which is a complete set of invariants of A under similarity (over F) is called the set of *elementary divisors* of A. The representation of A as this direct sum of companion matrices is called the *primary rational canonical form* of A ("Primary" because powers of irreducible polynomials are used and "rational" because the analysis insolves only rational operations in the field F).

Corollary 1. *The characteristic polynomial of an $n \times n$ matrix A is $(-1)^n$ times the product of the elementary divisors of A.*

Proof. It is readily seen that the characteristic polynomial of a direct sum of matrices $B_1, \cdots, B_q$ is the product of the characteristic polynomials of the B_i. But, by Theorem 15, the characteristic polynomial of a companion matrix C_f is $f(x)$, except for sign. These two facts, with the theorem, prove Corollary 1.

Corollary 2. *The eigenvalues of a square matrix are the roots of its minimal polynomial.*

Proof. Since the minimal polynomial $m(x)$ divides the characteristic polynomial, any root of the minimal polynomial is a root of the characteristic polynomial, hence a characteristic root (eigenvalue). Conversely, any root of the characteristic polynomial must, by Corollary 1, be a root of one of the elementary divisors $p_i(x)^{e_{ij}}$, hence by the theorem is a root of $m(x)$.

EXAMPLE. Any 6×6 rational matrix with minimal polynomial $(x^2 + 1)(x + 3)^2$ is similar to one of the following direct sums of companion matrices:

$$C_{(x^2+1)} \oplus C_{(x^2+1)} \oplus C_{(x+3)^2},$$

$$C_{(x^2+1)} \oplus C_{(x+3)^2} \oplus C_{(x+3)^2},$$

$$C_{(x^2+1)} \oplus C_{(x+3)^2} \oplus C_{(x+3)} \oplus C_{(x+3)};$$

in the first case the characteristic polynomial is $(x^2 + 1)^2(x + 3)^2$; in the second and third cases, the characteristic polynomial is $(x^2 + 1)(x + 3)^4$.

Over the field of complex numbers, the only monic irreducible polynomials are the linear polynomials $x - \lambda_i$, with λ_i a scalar. Using this observation, a different canonical form can be constructed for matrices with complex entries or, more generally, for any matrix whose minimal polynomial is a product of powers of linear factors.

In this case, each T-cyclic subspace Z_α in Theorem 23 will have the T-order $(x - \lambda_i)^e$ for some scalar λ_i and positive integer e. Relative to the basis $\alpha, \alpha T, \cdots, \alpha T^{e-1}$ of Z_α, T is represented as in Theorem 24 by the companion matrix of $(x - \lambda_i)^e$. On the other hand, consider the vectors $\beta_1 = \alpha, \beta_2 = \alpha U, \cdots, \beta_e = \alpha U^{e-1}$, where $U = T - \lambda_i I$. Since each β_j is αT^{j-1} plus some linear combination of vectors αT^k with $k < j - 1$, the vectors $\beta_1, \cdots, \beta_e$ also constitute a basis of Z_α. To obtain the effect of T upon the β_j, observe that

$$\beta_j T = \alpha U^{j-1} T = \alpha U^{j-1}(U + \lambda_i I) = \lambda_i \alpha U^{j-1} + \alpha U^j.$$

If $j < e$, this gives $\beta_j T = \lambda_i \beta_j + \beta_{j+1}$; if $j = e$, then $\alpha U^j = 0$ and $\beta_j T = \lambda_i \beta_j$. Now T is represented relative to this basis by the matrix whose rows

$$\begin{pmatrix} \lambda_i & 1 & 0 & 0 \\ 0 & \lambda_i & 1 & 0 \\ 0 & 0 & \lambda_i & 1 \\ 0 & 0 & 0 & \lambda_i \end{pmatrix}$$

are the coordinates of the $\beta_j T$. This is a matrix like that displayed just above, with entries λ_i on the principal diagonal, entries 1 on the diagonal next above the principal diagonal, and all other entries zero. Call such a matrix an *elementary Jordan matrix*.

If we use the above type of basis instead of that leading to the companion matrix in Theorem 24, we obtain

Theorem 25. *If the minimal polynomial for the matrix A over the field F is a product of linear factors*

(41) $$m(x) = (x - \lambda_1)^{e_1}(x - \lambda_2)^{e_2} \cdots (x - \lambda_k)^{e_k},$$

with $\lambda_1, \cdots, \lambda_k$ distinct, then A is similar over F to one and only one direct sum of elementary Jordan matrices, which include at least one $e_i \times e_i$ elementary Jordan matrix belonging to the characteristic root (eigenvalue) λ_i, and no larger elementary Jordan matrix belonging to the characteristic root (eigenvalue) λ_i.

Note that the number of occurrences of λ_i on the diagonal is the multiplicity of λ_i as a root of the characteristic polynomial of A.

The resulting direct sum of elementary Jordan matrices, which is unique except for the order in which these blocks are arranged along the diagonal, is called the *Jordan canonical form* of A. It applies to any matrix over the field of complex numbers. Note that the Jordan canonical form is determined by the set of elementary divisors and, in particular, that if all the e_i in (41) are 1, and only then, the Jordan canonical form is a diagonal matrix with the λ_i as the diagonal entries. Part of the Corollary of Theorem 22 is thus included as a special case.

Corollary. *Any complex matrix is similar to a matrix in Jordan canonical form.*

Exercises

1. Find all possible primary rational canonical forms over the field of rational numbers for the matrices described as follows:
   (a) 5×5, minimal polynomial $(x - 1)^2$.
   (b) 7×7, minimal polynomial $(x^2 - 2)(x - 1)$; characteristic polynomial $(x^2 - 2)^2(x - 1)^3$.
   (c) 8×8, minimal polynomial $(x^2 + 4)^2(x + 8)^2$.
   (d) 6×6, characteristic polynomial $(x^4 - 1)(x^2 - 1)$.
2. Exhibit all possible Jordan canonical forms for matrices with each of the following characteristic polynomials:
   (a) $(x - \lambda_1)^3(x - \lambda_2)^2$,   (b) $(x - \lambda_1)^5(x - \lambda_2)^3$,
   (c) $(x - \lambda_1)(x - \lambda_2)^2(x - \lambda_3)^2$.
3. Express in primary rational canonical form the elementary Jordan matrix displayed in the test.
4. (a) Show that a complex matrix and its transpose necessarily have the same Jordan canonical form.
   (b) Infer that they are always similar.
5. (a) Two of the Pauli "spin matrices" satisfy $ST = -TS$, $S^2 = T^2 = I$, and are hermitian. Prove that $U = iST$ is hermitian and satisfies $TU = -UT$, $U^2 = I$.

(b) Show that, if 2×2, S is similar to $\begin{pmatrix} 1 & 0 \\ 0 & -1 \end{pmatrix}$ and that, with these coordinates, $T = \begin{pmatrix} 0 & b \\ b^{-1} & 0 \end{pmatrix}$ for some b.

★6. Using the methods of §10.9, show that any linear transformation $T: V \to V$ decomposes V into a direct sum of T-cyclic subspaces with T-orders $f_1(x), \cdots, f_r(x)$, where $f_i(x) \mid f_{i-1}(x)$ for $i = 2, \cdots, r$, and $f_1(x)$ is the minimal polynomial of T.

11

Boolean Algebras
and Lattices

11.1. Basic Definition

We will now analyze more closely, from the standpoint of modern algebra, the fundamental notions of "set" (or class) and "subset," briefly introduced in §1.11. Suppose that I is any set, while $X, Y, Z, \cdots$ denote subsets of I. Thus I might be a square, and X, Y, Z three overlapping congruents disks located in I, as in the "Venn diagram" of Figure 1.

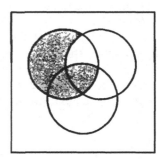

One writes $X \subset Y$ (or $Y \supset X$) whenever X is a subset of Y—i.e., whenever every element of X is in Y. This relation is also expressed by saying that X is "contained" or "included" in Y.

The relation of inclusion is reflexive: trivially, any set X is a subset of itself. It is also transitive: if every element of X is in Y and every element of Y is in Z, then clearly every element of X is in Z. But the inclusion relation is *not* symmetric. On the contrary, if

Figure 1

$X \subset Y$ and $Y \subset X$, then X and Y must contain exactly the same elements, so that $X = Y$.

In summary, the inclusion relation for sets shares with the inequality relation of arithmetic the following properties:

Reflexive: For all X, $X \subset X$.
Antisymmetric: If $X \subset Y$ and $Y \subset X$,   then $X = Y$.
Transitive: If $X \subset Y$ and $Y \subset Z$,   then $X \subset Z$.

It is, however, *not* true that, given two sets X and Y, either $X \subset Y$ or $Y \subset X$.

There are, therefore, four possible ways in which two sets X and Y can be related by inclusion. It may be that $X \subset Y$ and $Y \subset X$, in which case, by antisymmetry, $X = Y$. We can have $X \subset Y$ but not $Y \subset X$, in which case we say that X is *properly contained* in Y, and we write $X < Y$ or $Y > X$. We can have $Y \subset X$ but not $X \subset Y$, in which case X *properly contains* Y. And finally, we have neither $X \subset Y$ nor $Y \subset X$, in which case X and Y are said to be *incomparable*. It is principally the existence of incomparable sets which distinguishes the inclusion relation from the inequality relation between real numbers.

The subsets of a given set I are not only related by inclusion; they can also be combined by two binary operations of "union" and "intersection," analogous to ordinary "plus" and "times." The extent and importance of this analogy were first clearly recognized by the British mathematician George Boole (1815–64), who founded the algebraic theory of sets little more than a century ago.

We define the *intersection* of X and Y (written $X \cap Y$) as the set of all elements in both X and Y; and we define the *union* of X and Y (in symbols, $X \cup Y$) as the set of all elements in either X or Y, or both. The symbols $\cap$ and $\cup$ are called "cap" and "cup," respectively.

Finally, we write X' (read, "the complement of X") to signify the set of all elements not in X. For example, I' is the empty set $\varnothing$, which contains no elements at all! This is because we are considering only *subsets* of I.

The operations of the algebra of classes can be illustrated graphically by means of the Venn diagram of Figure 1. In this diagram, X, Y, and Z are the interiors of the three overlapping disks. Combinations of these regions in the square I can be depicted by shading appropriate areas: thus, Y' is the exterior of Y, and $X \cap (Y' \cup Z)$ is the shaded area.

Exercises

1. The Venn diagrams for X, Y, and Z cut the square into eight nonoverlapping areas. Label each such area by an algebraic combination of X, Y, and Z which represents exactly that area.

2. On a Venn diagram shade each of the following areas:

$$(X' \cap Y) \cup (X \cap Z'), \qquad (X \cup Y)' \cap Z, \qquad (X \cup Y') \cup Z'.$$

3. By shading each of the appropriate areas on a Venn diagram, determine which of the following equations are valid:

    (a) $(X' \cup Y)' = X \cap Y'$,   (b) $X' \cup Y' = (X \cup Y)'$,

    (c) $(X \cup Y) \cap Z = (X \cup Z) \cap Y$,   (d) $X \cup .(Y \cap Z)' = (X \cup Y') \cap Z'$.

11.2. Laws: Analogy with Arithmetic

The analogy between the algebra of sets and ordinary arithmetic will now be described in some detail, and used to define Boolean algebra. The analogy between $\cap$, $\cup$ and ordinary $\cdot$, $+$ is in part described by the following laws, whose truth is obvious.

Idempotent: $\quad X \cap X = X \quad$ and $\quad X \cup X = X.$
Commutative: $\quad X \cap Y = Y \cap X \quad$ and $\quad X \cup Y = Y \cup X.$
Associative: $\quad X \cap (Y \cap Z) = (X \cap Y) \cap Z \quad$ and
$\qquad\qquad\quad X \cup (Y \cup Z) = (X \cup Y) \cup Z.$
Distributive: $\quad X \cap (Y \cup Z) = (X \cap Y) \cup (X \cap Z) \quad$ and
$\qquad\qquad\quad X \cup (Y \cap Z) = (X \cup Y) \cap (X \cup Z).$

Clearly, all of these except for the idempotent laws and the second distributive law correspond to familiar properties of $+$ and $\cdot$, as postulated in Chap 1.

Intersection and union are related to each other and to inclusion by a fundamental law of

Consistency: $\quad$ The three conditions $X \subset Y$, $X \cap Y = X$, and $X \cup Y = Y$ are mutually equivalent.

Further, the void (empty) set being denoted by $\varnothing$, we have the following special properties of $\varnothing$ and I,

Universal bounds: $\quad \varnothing \subset X \subset I \quad$ for all $X.$
Intersection: $\qquad \varnothing \cap X = \varnothing \quad$ and $\quad I \cap X = X.$
Union: $\qquad\qquad \varnothing \cup X = X \quad$ and $\quad I \cup X = I.$

The first three intersection and union properties are analogous to properties of 0 and 1 in ordinary arithmetic.

Finally, complementation is related to intersection and union by three new laws.

Complementarity: $\quad X \cap X' = \varnothing \quad$ and $\quad X \cup X' = I.$
Dualization: $\qquad (X \cap Y)' = X' \cup Y' \quad$ and $\quad (X \cup Y)' = X' \cap Y'.$
Involution: $\qquad\; (X')' = X.$

The first and third laws correspond to laws of ordinary arithmetic, if X' is interpreted as $1 - X$ and $XX = X$ is assumed.

The truth of the above laws can be established in various ways. First, one can test them in particular examples, thus verifying them by "induc-

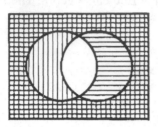

Figure 2

tion." Appropriate examples are furnished by the Venn diagrams. If X and Y are the respective interiors of the left- and right-hand circles in Figure 2, then the area X' is shaded by horizontal lines and Y' by vertical lines. The cross-hatched area is then just the intersection $X' \cap Y'$; the figure shows at once that this area is the complement of the sum $X \cup Y$, as asserted by the second dualization law. Such an argument is convincing to our common sense, but it is not permissible tech-

nically, since only deductive proofs are allowed in mathematical reasoning.

Second, we can consider separately each of the possible cases for an element of I: first, an element in X and in Y; second, an element in X but not in Y; and so on. For example, an element of the first type is in $X \cap Y$, hence not in $(X \cap Y)'$ and not in $X' \cup Y'$, while an element of the second type is in $(X \cap Y)'$ and in Y', hence in $X' \cup Y'$. By looking at the other two cases also, one sees that $(X \cap Y)'$ and $X' \cup Y'$ have the same elements, as in the first dualization law. Note that for two classes X and Y, all four possible cases for an element are represented by points in the four areas of the Venn diagram; while for three classes there are eight cases and eight areas (Figure 1).

Third, we can use the verbal definitions of the operations cup and cap to reformulate the laws. Consider the distributive law. Here

"b in $X \cap (Y \cup Z)$"   means   "b is both in X and in Y or Z,"
"b in $(X \cap Y) \cup (X \cap Z)$"   means   "$b$ is either in both X and Y or in both X and Z."

A little reflection convinces one that these two statements are equivalent, according to the ordinary usage of the connectives "and" and "either $\cdots$ or." This verification of the distributive law may indicate how the laws of the algebra of classes are paraphrases of the properties of the words "and," "or," and "not." If one assumes these properties as basic, as one normally does in mathematical reasoning, one can then prove from them all the above laws for classes.

Exercises

1. Use Venn diagrams to verify the distributive laws.
2. Use the method of subdivision into cases to verify the associative, commutative, and consistency laws.
3. Reformulate the laws of complementarity, dualization, and involution in terms of "and," "or," and "not," as in the third method described above.
4. (a) In treating an algebraic expression in four sets by considering all possible cases for their elements, how many cases occur?
   ★(b) Draw a diagram for four sets which shows every one of these possible cases as a region.
   (c) Show that no such diagram can be made, in which the four given sets are discs.
5. Show that the intersection and union properties of $\varnothing$ and I may be derived from the universal bounds property and the consistency principle.
6. Of the six implications obtained by replacing "equivalent" by "if" and "only if" in the consistency postulate, show that four hold for real numbers x, y if $0 \leqq x, y \leqq 1$.
7. In Ex. 6, which of the intersection and union properties of $\varnothing$, I fails if $\varnothing$ is replaced by the number 0, and I by 1? Which of the properties of complements fails if X' is replaced by $1 - x$ and Y' by $1 - y$?
8. Prove that for arbitrary subsets X, Y of I, $X \subset Y$ if and only if $X' \cup Y = I$.

11.3. Boolean Algebra

We will not concern ourselves further with deriving the preceding algebraic laws from the fundamental principles of logic. Instead, we will simply assume the most basic of these laws as *postulates*, as was done in Chap. 1 for the laws of arithmetic, and then deduce from these postulates as many interesting consequences as possible. Accordingly, we now lay down our basic definition, using a slightly different notation to emphasize the fact that the postulates assumed may apply to other things than sets.

Definition. *A Boolean algebra is a set B of elements* $a, b, c, \cdots$ *with the following properties:*
   (i) *B has two binary operations,* $\wedge$ (wedge) *and* $\vee$ (vee), *which satisfy the idempotent laws*

$$a \wedge a = a \vee a = a,$$

the commutative laws

$$a \wedge b = b \wedge a, \qquad a \vee b = b \vee a,$$

and the associative laws

$$a \wedge (b \wedge c) = (a \wedge b) \wedge c, \qquad a \vee (b \vee c) = (a \vee b) \vee c.$$

(ii) *These operations satisfy the absorption laws*

$$a \wedge (a \vee b) = a \vee (a \wedge b) = a,$$

(iii) *These operations are mutually distributive:*

$$a \wedge (b \vee c) = (a \wedge b) \vee (a \wedge c), \qquad a \vee (b \wedge c) = (a \vee b) \wedge (a \vee c),$$

(iv) *B contains universal bounds O, I which satisfy*

$$O \wedge a = O, \qquad O \vee a = a, \qquad I \wedge a = a, \qquad I \vee a = I,$$

(v) *B has a unary operation $a \to a'$ of* complementation, *which obeys the laws*

$$a \wedge a' = O, \qquad a \vee a' = I.$$

It is understood that the preceding laws are assumed to hold for all $a, b, c \in B$.

Using this definition, the conclusions of §§11.1–11.2 can be summarized in the following statement.

Theorem 1. *Under intersection, union, and complement, the subsets of any set I form a Boolean algebra.*

To illustrate more selectively the significance of the preceding postulates, we now describe examples in which some of them hold, but not all.

EXAMPLE 1. Let L have as "elements" the subspaces of the n-dimensional Euclidean vector space of §7.10. Define $S \wedge T = S \cap T$ as the intersection of S and T, $S \vee T = S + T$ as their linear sum, O as the null vector $\mathbf{0}$, I as the whole space, and S' as the orthogonal complement $S^\perp$ of the subspace S.

Then postulates (i), (ii), (iv), and (v) are satisfied, although the distributive laws (iii) are not. (Let S, T, U be the subspaces of the plane spanned by $(1, 0), (0, 1), (1, 1)$, respectively, for example.)

EXAMPLE 2. Let L have as "elements" the normal subgroups $M, N, \cdots$ of a finite group G. Let $M \wedge N = M \cap N$ be the intersection of M and N, while $M \vee N = MN$ is the set of all products xy ($x \in M, y \in N$). Then $M \wedge N$ and $M \vee N$ are normal subgroups of G. If O denotes the group identity 1, and I is G itself, then postulates (i), (ii), and (iv) are satisfied, although in general (iii) and (v) are not.

Since the systems constructed in Examples 1 and 2 satisfy postulates (i) and (ii), they are lattices in the following sense.

Definition. *A* lattice *is a set L of elements, with two binary operations*† *$\wedge$ and $\vee$ which are idempotent, commutative, and associative and which satisfy the absorption law (ii). If in addition the distributive laws (iii) hold, then L is called a* distributive lattice.

For example, if the void set $\varnothing$ is included, and sets of zero area are neglected, then the set of all polygonal domains is a distributive lattice, under intersection and union. Again, the set of positive integers is a distributive lattice, with $m \wedge n$ the greatest common divisor of m and n, and $m \vee n$ their least common multiple.

The various laws postulated above have many interesting algebraic consequences, of which we will now derive a few of the simplest.

The effect of the associative and commutative laws has already been studied in §1.5. The associative law means essentially that we can form multiple intersections or unions without using parentheses; the commutative law, that we can permute terms in any way we like in an expression involving only vees or only wedges.

In conjunction with the above laws, the effect of the idempotent laws is clearly to permit elimination of repeated occurrences of the same term—all but one of the occurrences of a given term can be deleted. In summary, we have

Lemma 1. *Let f and g be two expressions formed from the letters $a_1, \cdots, a_n$ using only vees $\vee$ and using all these letters (possibly with some repetitions). Then the idempotent, commutative, and associative laws imply that $f = g$. The same holds for expressions involving only wedges $\wedge$.*

If N is the *set* of subscripts $i = 1, \cdots, n$, we can without ambiguity write

$$\bigvee_N a_i \quad \text{or} \quad \bigvee_{i=1}^n a_i \quad \text{and} \quad \bigwedge_N a_i \quad \text{or} \quad \bigwedge_{i=1}^n a_i$$

---

† The wedge operation $\wedge$ is also called *meet*, and the vee operation $\vee$ called *join*; we will use these names interchangeably.

to denote the join and meet of all the a_i, respectively. These notations are analogous to the Σ, $\prod$ notations of algebra.

Again, starting from the commutative, associative, and distributive laws, we can derive by induction, just as in §1.5, generalized distributive laws such as the following:

$$x \wedge (y_1 \vee \cdots \vee y_n) = (x \wedge y_1) \vee \cdots \vee (x \wedge y_n),$$

$$x \vee (y_1 \wedge \cdots \wedge y_n) = (x \vee y_1) \wedge \cdots \wedge (x \vee y_n),$$

$$(x_1 \vee \cdots \vee x_m) \wedge (y_1 \vee \cdots \vee y_n)$$
$$= (x_1 \wedge y_1) \vee (x_1 \wedge y_2) \vee \cdots \vee (x_m \wedge y_n).$$

Exercises

1. Use induction to prove in detail that

    (a) $x \wedge \bigvee\limits_{i=1}^{n} y_i = \bigvee\limits_{i=1}^{n} (x \wedge y_i)$ in any distributive lattice,

    (b) $\left(\bigvee\limits_{i=1}^{n} x_i \right)' = \bigwedge\limits_{i=1}^{n} x_i'$ in any Boolean algebra.

★**2.** Prove in detail by induction that in any distributive lattice,

$$\left(\bigvee_{i=1}^{m} x_i \right) \wedge \left(\bigvee_{i=1}^{m} y_j \right) = \bigvee_{i=1}^{m} \left\{ \bigvee_{j=1}^{n} (x_i \wedge y_j) \right\}.$$

3. Prove in detail that Example 1 defines a lattice which is not distributive if $n > 1$.

4. Prove that Example 2 defines a lattice which is distributive if G is cyclic.

5. Prove that the lattice of all subgroups of the four-group is not distributive.

11.4. Deduction of Other Basic Laws

We now show that the postulates for Boolean algebra listed above imply the other basic formulas of the algebra of classes which were discussed in §§11.1–11.2. For instance, they imply the *uniqueness* of O and I, which we did not postulate.

Lemma 2. *In any Boolean algebra, each of the identities $a \wedge x = a$ and $a \vee x = x$ (for all x) implies that $a = O$; dually, each of the identities $a \vee x = a$ and $a \wedge x = x$ implies that $a = I$.*

Proof. If $a \wedge x = a$ for all x, then in particular $a \wedge O = a$; but $a \wedge O = O$ by (iii); hence $a = O$. Likewise, if $a \vee x = x$ for all x, then $a \vee O = O$; but $a \vee O = a$ by (iii); hence again $a = O$. The proof of the unicity of I is similar.

Lemma 3. *For elements a, b of any lattice, $a \wedge b = a$ holds if and only if $a \vee b = b$.*

Proof. If $a \vee b = b$, then by the absorption law (ii) $a \wedge b = a \wedge (a \vee b) = a$. Conversely, if $a \wedge b = a$, then by the same law $a \vee b = (a \wedge b) \vee b$. Hence, by the commutative laws, $a \vee b = b \vee (b \wedge a) = b$, where the last step uses (ii) again.

Corollary. *In the definition of a Boolean algebra, conditions (iv) can be replaced by either of the following postulates:*
(iv') For all x, $x \wedge O = O$ and $x \vee I = I$
(iv'') For all x, $O \vee x = x$ and $I \wedge x = x$.

The definition of a Boolean algebra given above did not mention the inclusion relation, even though this is the most fundamental concept of all. We shall now define this relation and deduce its basic properties from the postulates stated above. The proof restates the law of consistency, of which a part was already proved as Lemma 2 above.

Definition. *Define $a \leqq b$ to mean that $a \wedge b = a$—or equivalently (Lemma 2), that $a \vee b = b$.*

Lemma 4. *The relation $a \leqq b$ is reflexive, antisymmetric, and transitive in any lattice.*

Proof. Since $a \wedge a = a$, $a \leqq a$ for all a, proving the reflexive law. Again, $a \leqq b$ and $b \leqq a$ imply

$$a = a \wedge b = b \wedge a = b,$$

which proves the antisymmetric law. Finally, $a \leqq b$ and $b \leqq c$ imply $a = a \wedge b = a \wedge (b \wedge c) = (a \wedge b) \wedge c = a \wedge c$, whence $a \leqq c$. This proves the transitive law.   Q.E.D.

The power of the absorption laws was exhibited above in the proofs of Lemmas 2 and 3. Actually, the absorption, commutative, and associative laws imply the idempotent laws: the latter are redundant in the definition of a lattice, for one absorption law is $x = x \wedge (x \vee z)$ for all x, z. Setting $z = x \wedge y$, we infer $x = x \wedge (x \vee (x \wedge y))$ for all x, y. Applying the dual absorption law $x \vee (x \wedge y) = x$, we conclude $x = x \wedge x$ (one idempotent law). The proof that $x = x \vee x$ is similar, interchanging $\wedge$ and $\vee$.

Lemma 5. *In any distributive lattice, $a \vee x = a \vee y$ and $a \wedge x = a \wedge y$ together imply $x = y$.*

Proof. By substitution of equals for equals, and the absorption and distributive laws, successively, we have

$$x = x \wedge (x \vee a) = x \wedge (y \vee a)$$
$$= (x \wedge y) \vee (x \wedge a) = (y \wedge x) \vee (y \wedge a)$$
$$= y \wedge (x \vee a) = y \wedge (y \vee a) = y.$$

Now recall that the operation $a \mapsto a'$ of complementation satisfies

$$a \wedge a' = O \qquad \text{and} \qquad a \vee a' = I.$$

But any element x with $a \wedge x = O$ and $a \vee x = I$ must, by Lemma 5, satisfy $x = a'$. In other words, the complement a' is *uniquely* determined by the complementation laws (v) in the definition of a Boolean algebra. We now show that the remaining properties of set-complements also hold in any Boolean algebra.

Lemma 6. *In any Boolean algebra, we have*

(1) $(x')' = x,$ $(x \wedge y)' = x' \vee y',$ *and* $(x \vee y)' = x' \wedge y'.$

Proof. The statement that x' is a complement of x implies by the commutative law that x is a complement of x', since $x' \wedge x = x \wedge x' = O$ and $x' \vee x = x \vee x' = I$. But we have just seen that complements are unique; hence x is *the* unique complement of x', and $(x')' = x$. Again, by the distributive laws,

$$(x \wedge y) \wedge (x' \vee y') = (x \wedge y \wedge x') \vee (x \wedge y \wedge y')$$
$$= ((x \wedge x') \wedge y) \vee (x \wedge O)$$
$$= (O \wedge y) \vee O = O \vee O = O$$
$$(x \wedge y) \vee (x' \vee y') = (x \vee x' \vee y') \wedge (y \vee x' \vee y')$$
$$= (I \vee y') \wedge (y \vee y' \vee x')$$
$$= I \wedge (I \vee x') = I.$$

This shows that $x' \vee y'$ is *a* complement of $x \wedge y$. Hence, again by the uniqueness of complements, $x' \lor y' = (x \land y)'$ is *the* complement of $x \land y$. The identity $(x \land y)' = x' \land y'$ can be proved similarly.

Corollary. *To find the complement of an expression built up from primed and unprimed letters by iterated vees and wedges (but not using primed parentheses), interchange* $\lor$ *and* $\land$ *throughout, prime each unprimed letter and unprime each primed letter.*

Thus, the complement of $(x' \land y) \lor (z \land w')$ is, by this rule, $(x \lor y') \land (z' \lor w)$.

Proof. If the number n of letters in the given expression f (counting repetitions) is 1, then the lemma is true, since $(x)' = x'$ and $(x')' = x$. Otherwise, since no parentheses are primed, we can write the expression as $f = a \land b$ or $f = a \lor b$—giving, respectively, $f' = a' \lor b'$ or $f' = a' \land b'$. But the expressions a and b contain fewer letters than does f; hence by induction on n we can assume the lemma to be true for them. Substituting in the expressions $f' = a' \lor b'$ or $f' = a' \land b'$, we get the desired formula for the complement.

Exercises

1. Prove that the idempotent law $x \lor x = x$ follows: (a) from the commutative, associative, and absorption laws, (b) the absorption law alone.

Exercises 2–10 refer to Boolean algebras.

2. Prove in detail that $(x \lor y)' = x' \land y'$.

3. Simplify the following Boolean expressions:
   (a) $(x' \land y')'$,   (b) $(a \lor b) \lor (c \lor a) \lor (b \lor c)$,
   (c) $(x \land y) \lor (z \land x) \lor (x' \lor y')'$.

4. Prove that $(x \land y) \lor (x \land y') \lor (x' \land y) \lor (x' \land y') = I$.
   Interpret in terms of the two-circle analog of Venn's diagram.

5. Prove that $x = y$ if and only if $(x \land y') \lor (x' \land y) = O$.

6. Prove Poretzky's law: Given x and t, $x = O$ if and only if

$$t = (x \land t') \lor (x' \land t).$$

7. Prove that
   (a) $y \leq x'$ if and only if $x \land y = O$,   (b) $y \geq x'$ if and only if $x \lor y = I$.

8. Find complements of the following expressions:
   (a) $x \lor y \lor z'$,   (b) $(x \lor y' \lor z') \land (x \lor (y \lor z'))$,
   (c) $x \lor (y \land (z \lor w'))$,   (d) $(x' \lor y)' \land (x \lor y')$.

9. Apply the argument of the corollary to Lemma 5 to the expression $(x' \land y \land z') \lor (x \land y')$, justifying each step.

10. Prove that $(x \lor y) \land (x' \lor z) = (x' \land y) \lor (x \land z)$.

11. Prove that $(x \wedge y) \vee (y \wedge z) \vee (z \wedge y) = (x \vee y) \wedge (y \vee z) \wedge (z \vee x)$ in any distributive lattice.
12. An element a of a lattice L with universal bounds O, I is said to be *complemented* when, for some $x \in L$, $a \wedge x = O$ and $a \vee x = I$. Show that if a and b are complemented elements of a distributive lattice, the same is true of $a \wedge b$ and $a \vee b$.

11.5. Canonical Forms of Boolean Polynomials

Various expressions built up from the $\wedge$, $\vee$, and $'$ operations have been studied in the preceding section. Such expressions are called "Boolean polynomials" (or "Boolean functions"); the analogy with ordinary polynomials (Chap. 3) is obvious.

We now define a *subalgebra* of a Boolean algebra B as a nonvoid subset S of B which contains, with any two elements x and y, also $x \wedge y$, $x \vee y$, and x' (and hence $O = x \wedge x'$ and I). Given an arbitrary nonvoid subset X of B, the set of all values $p(x_1, \cdots, x_n)$ of elements $x_i \in X$ is clearly the smallest subalgebra of B which contains X. As in the case of groups, this subalgebra is said to be *generated* by X. For example, the subalgebra generated by any one element x consists of the four elements x, x', O, I.

This is a special instance of the following surprising fact: *the number of different Boolean polynomials in n variables $x_1, \cdots, x_n$ is 2^{2^n}*. This we now show, illustrating the argument by the polynomial

$$f(x, y, z) = [x \vee z \vee (y \vee z)']' \vee (y \wedge x).$$

First, if any prime occurs outside any parenthesis in the polynomial, it may be moved inside by an application of the dualization law, as in Lemma 5 of §11.4. When all the primes have been moved all the way inside, the polynomial becomes an expression involving only vees and wedges acting on primed and unprimed letters. Thus, in our example:

$$f = [x' \wedge z' \wedge (y \vee z)] \vee (y \wedge x).$$

Secondly, if any $\wedge$ stands outside a parenthesis which contains a $\vee$, then the $\wedge$ can be moved inside by applying the distributive law, as in $c \wedge (a \vee b) = (c \wedge a) \vee (c \wedge b)$. There results a polynomial in which all meets $\wedge$ are formed before any join $\vee$; that is, the expression is a join of terms $T_1, \cdots, T_k$ in which each T_k is a meet of primed and unprimed letters. In the example above,

$$f = (x' \wedge z' \wedge y) \vee (x' \wedge z' \wedge z) \vee (y \wedge x).$$

Thirdly, certain expressions can be shortened or omitted. If a letter "c" appears twice in one term, one occurrence can be omitted, since $c \wedge c = c$. If c appears both primed and unprimed, then the whole term is O, since $c \wedge a \wedge c' = O$ for all a; hence it can be omitted, since $O \vee b = b$ for all b. Thus, above

$$f = (x' \wedge z' \wedge y) \vee (y \wedge x).$$

Now if some term T_k fails to contain a letter c, we can write

$$T_k = T_k \wedge I = T_k \wedge (c \vee c') = (T_k \wedge c) \vee (T_k \wedge c'),$$

replacing T_k by two terms, in each of which c occurs exactly once. Thus, in our example:

$$f = (x' \wedge z' \wedge y) \vee (y \wedge x \wedge z) \vee (y \wedge x \wedge z').$$

Finally, the letters appearing in each term can be rearranged so as to appear in their natural order, thus

$$f = (x' \wedge y \wedge z') \vee (x \wedge y \wedge z) \vee (x \wedge y \wedge z').$$

This is called the *disjunctive canonical form* for f; we have proved the following lemma.

Lemma. *Any Boolean polynomial in $x_1, \cdots, x_n$ can be reduced either to O or to some join of terms T_k of the form*

(2) $\qquad T_k = q_1 \wedge q_2 \wedge \cdots \wedge q_n \qquad$ (*each* $q_j = x_j$ *or* x_j'),

that is, to disjunctive canonical form.

Since there are two alternatives for each q_j, we see that there are exactly 2^n possible T_k. Thus when $n = 3$, our process reduces any Boolean polynomial to O or to some join of the terms

(3)
$$
\begin{array}{llll}
x \wedge y \wedge z, & x' \wedge y \wedge z, & x \wedge y' \wedge z, & x \wedge y \wedge z', \\
x \wedge y' \wedge z', & x' \wedge y \wedge z', & x' \wedge y' \wedge z, & x' \wedge y' \wedge z'.
\end{array}
$$

It is no accident that these eight polynomials represent the eight regions into which the three circles of Figure 1 divide the square. This means geometrically that any Boolean combination of the three circles X, Y, and Z will be the union of some selection of the eight regions of the diagram.

The ultimate terms, such as those listed in (3), will be called minimal Boolean polynomials. In other words, a *minimal* polynomial $M(x_1, \cdots, x_n)$ in n variables $x_1, \cdots, x_n$ is a meet $\bigwedge_{i=1}^{n} q_i$ of n elements in which each ith element q_i is either x_i or x_i'. We have proved

Theorem 2. *Any given Boolean polynomial in $x_1, \cdots, x_n$ is equal to O or to the join of a set S of minimal polynomials.*

Now assign to each M the n-digit binary number $\eta(M) = y_1 y_2 \cdots y_n$, where the digit y_i is 1 or 0 according as q_i is x_i or x_i' in $M = \bigwedge_{i=1}^{n} q_i$ above. Then the function $\eta: M \mapsto \eta(M)$ is a bijection from the set of minimal polynomials in $x_1, \cdots, x_n$ to the set I of all 2^n n-digit binary numbers. Thus in (3), the $\eta(M)$ as listed are

$$111, 011, 101, 110, 100, 010, 001, 000.$$

Alternatively, $\eta(M)$ can be thought of as the vector $\eta = (y_1, y_2, \cdots, y_n) \in \mathbf{Z}_2{}^n$, and each Boolean polynomial $\bigvee_S M_\eta(x_1, \cdots, x_n)$ as corresponding to a set of these vectors.

If $S_i \subset I$ consists of those binary numbers with $y_i = 1$, then S_i' will consists of those with $y_i = 0$. Hence (cf. Ex. 9 below) the set representing a given *minimal* polynomial $M(S_1, \cdots, S_n)$ consists of a single binary number $\alpha(M) = a_1 a_2 \cdots a_n$, whose ith digit a_i is 0 or 1 according as S_i is primed or unprimed in M. Different minimal polynomials $M = M_\alpha$ are clearly represented by different binary numbers; therefore, the joins of different sets of M_α represent different subsets of I. This proves the following result.

Corollary. *There are just 2^{2^n} different Boolean functions of n variables.*

We can now replace haphazard manipulation of Boolean polynomials by a systematic procedure. The truth or falsity of any purported equation $E_1 = E_2$ in Boolean algebra can be settled definitely, simply by reducing each side to disjunctive canonical form.

Exercises

1. Reduce each of the following expressions to canonical form:
   (a) $(x \vee y) \wedge (z' \wedge y)'$,   (b) $(x \vee y) \wedge (y \vee z) \wedge (x \vee z)$.

2. Test each of the following proposed equalities by reducing both sides to (disjunctive) canonical form:

   (a) $[x \wedge (y \vee z)']' = (x \wedge y)' \vee (x \wedge z)$,

   (b) $x = (x' \vee y')' \vee [z \vee (x \vee y)']$.

3. Show that every Boolean polynomial has a dual canonical form which is a "meet" of certain "prime" polynomials. Describe these prime polynomials carefully, and show that they are complements of minimal polynomials. How is the result analogous to the theorem that every ordinary polynomial over a field has a unique representation as a product of irreducible polynomials?

4. Use the canonical form of Ex. 3 to test the equality of Ex. 2(a).

5. Prove that the canonical form of $f(x, y)$ is

$$f(x, y) = [f(I, I) \wedge x \wedge y] \vee [f(I, O) \wedge x \wedge y']$$
$$\vee [f(O, I) \wedge x' \wedge y] \vee [f(O, O) \wedge x' \wedge y'].$$

6. Prove that the meet of any two distinct minimal polynomials is O.

7. Expand $I = (x_1 \vee x_1') \wedge \cdots \wedge (x_n \vee x_n')$ by the generalized distributive law to show that I is the join of the set of all minimal Boolean polynomials.

8. Prove from Ex. 7 and $x_i = x_i \wedge I$ that each x_i is the join of all those minimal polynomials with ith term x_i.

9. (a) Let $\bigvee_A M_\alpha$ denote the join of all minimal polynomials in the set A. Prove:

$$\left(\bigvee_A M_\alpha\right) \vee \left(\bigvee_B M_\beta\right) = \bigvee_{A \cup B} M_\gamma, \qquad \left(\bigvee_A M_\alpha\right) \wedge \left(\bigvee_B M_\beta\right) = \bigvee_{A \cap B} M_\gamma$$

   (b) Show that the preceding formulas are valid also if we define the join $\bigvee_\phi M_\alpha$ of the void set of minimal polynomials to be O.

★10. Using Exs. 7 and 9, prove $\left(\bigvee_A M_\alpha\right)' = \bigvee_{A'} M_\alpha$. (*Hint:* Use Lemma 5 of §11.4.)

★11. Using only Exs. 8–10, give an independent proof that every Boolean polynomial can be written as a join of minimal polynomials.

11.6. Partial Orderings

Little use has been made above of the reflexive, antisymmetric, and transitive laws of inclusion. Yet these are the most fundamental laws of all; thus they apply to many systems which are not Boolean algebras.

For example, they clearly hold for the system of all subsets of a set which are distinguished by any special property (writing either $\subset$ or $\leqq$). Thus, they hold for the subgroups (or the normal subgroups!) of any group, the subfields of any field, the subspaces of any linear space, and so on—even though these do not form Boolean algebras. They also hold for the less-than-or-equal-to relation $x \leqq y$ between real numbers, for the divisibility relation $x \mid y$ between positive integers, and so on.

These examples suggest the abstract concept of a "partial ordering." By this is meant any reflexive, antisymmetric, and transitive relation.

Definition. *A partially ordered set is a set P with a binary relation $\leqq$, which is reflexive, antisymmetric, and transitive.*

For any relation $a \leqq b$ (read, "b includes a") of this type, we may define $a < b$ to mean that $a \leqq b$ but $a \neq b$, while b may be said to *cover* a if $a < b$, and if $a < x < b$ is possible for no x.

The following lemma shows that any lattice can be considered as a partially ordered set (the full significance of this will be explained in the next section). It is a consequence of the absorption law.

Lemma. *In any lattice, the relation $x \leqq y$ defined to mean $x \wedge y = x$ is a partial ordering; it is equivalent to $x \vee y = y$.*

Partially ordered sets with a finite number of elements can be conveniently represented by diagrams. Each element of the system is represented by a small circle so placed that the circle for a is above that for b if $a > b$. One then draws a descending line from a to b in case a covers b. One can reconstruct the relation $a \geqq b$ from the diagram, for $a > b$ if and only if it is possible to climb from b to a along ascending line segments in the diagram.

For example, in Figure 3 the first diagram represents the system of all subgroups of the four-group; the second, the Boolean algebra of all subsets of a set of three points; the third, the numbers 1, 2, 4, 8 under the divisibility relation. The others have been constructed at random, and show how one can construct abstract partially ordered sets simply by drawing diagrams. Figure 3, §6.7, is a diagram for the partially ordered set of all the subgroups of the group of the square.

It is clear that in any partially ordered set, the relation $\geqq$ is also reflexive, antisymmetric, and transitive (simply read the postulates from right to left to see this). Therefore any statement which can be proved

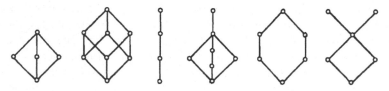

Figure 3

from the postulates defining partially ordered sets by using the relation $a \leqq b$ could be established by exactly the same train of reasoning, if everywhere $a \leqq b$ were replaced by the converse relation $a \geqq b$, and vice versa. This is the

Duality Principle. *Any theorem which is true in every partially ordered set remains true if the symbols* $\leqq$ *and* $\geqq$ *are interchanged through the statement of the theorem.*

It is to be emphasized that this principle is not a theorem about partially ordered sets in the usual sense, but is a theorem about theorems. As such, it belongs to the domain of "metamathematics."

Exercises

1. Show in detail how the second diagram of Figure 3 does represent the algebra of all subsets of a set I of three points.
2. Draw a diagram for each of the following partially ordered sets:
   (a) the Boolean algebra of all subsets of a set of four points,
   (b) the set of all subgroups of a cyclic group of order 12,
   (c) the set of all subgroups of the quaternion group,
   (d) the integers 1, 2, 3, 4, 6, 8, 12, 24 under divisibility,
   (e) the set of all subgroups of a cyclic group of order 54,
   (f) the set of all ideals of the ring $\mathbf{Z}_{40}$ of integers modulo 40.
3. Show that the partially ordered set of parts (d), (e), (f) and Ex. 2 are all "isomorphic" in a suitably defined sense.
4. Which of the following sets are partially ordered sets?
   (a) all subfields of the field $\mathbf{R}$ of real numbers, under the inclusion relation,
   (b) all pairs of numbers (a, b) if $(a, b) \leqq (a', b')$ means that $a \leqq a'$ and $b \leqq b'$,
   (c) all pairs of real numbers if $(a, b) \leqq (a', b')$ means that either $a < a'$ or $a = a'$ and $b \leqq b'$,
   (d) all pairs of real numbers if $(a, b) \leqq (a', b')$ means that $a \leqq a'$ and $b \geqq b'$,
   (e) all subdomains of a given integral domain, under the inclusion relation,
   (f) all polynomials in $F[x]$ if $f(x) \leqq g(x)$ means that $f(x)$ divides $g(x)$.
5. Consider a system of elements with a relation $a < b$ which is transitive and irreflexive ($a < a$ is never true). If $a \leqq b$ is defined to mean that either $a < b$ or $a = b$, prove that one gets a partially ordered set.
6. Prove the lemma stated in the text.
7. (a) Show that in Example 1 of §11.3 the lattice L is defined by the relation of set-inclusion between subspaces.
   (b) Make and prove a similar statement about Example 2 of §11.3.
   (c) If $m \mid n$ is used to define a partial ordering of the positive integers, what do $\wedge$ and $\vee$ signify?

11.7. Lattices

The consistency principle shows how to define inclusion in terms of join or meet; we shall now show that, conversely, one can define join and meet in terms of inclusion. Namely, $x \vee y$ is the least thing which contains both x and y, while $x \wedge y$ is the greatest thing contained in both x and y. This observation is due to C. S. Peirce; we shall state it more precisely as follows.

By a "lower bound" to a set X of elements of a partially ordered set P is meant an element a satisfying $a \leqq x$ for all $x \in X$. By a "greatest lower bound" (g.l.b.) is meant, as in Chap. 4, a lower bound including every other lower bound: a lower bound c such that $c \geqq a$ for any other lower bound a. Clearly, g.l.b.'s are unique if they exist—for if a and b are both g.l.b.'s of the same set X, then $a \geqq b$ and $b \geqq a$, whence $a = b$.

Dually, we can define "upper bounds" and "least upper bounds" (l.u.b.), and prove the uniqueness of the latter when they exist. We are here applying our metamethematical Duality Principle! Hence it is legitimate to speak of *the* g.l.b. and *the* l.u.b. of a set of elements, whenever these bounds exist.

Lemma 1. *In any lattice, the meet $x \wedge y$ and the join $x \vee y$ are the g.l.b. and l.u.b., respectively, of the set consisting of the two elements x and y.*

Proof. Since $x \wedge x \wedge y = x \wedge y$ and $y \wedge x \wedge y = x \wedge y$, the consistency principle shows that $x \wedge y$ is a lower bound of x and y. It is a greatest lower bound, since $z \leqq x$ and $z \leqq y$ imply $z = x \wedge z = x \wedge (y \wedge z)$, and so $z \leqq x \wedge y$, again by the consistency principle. The proof is completed by duality.

This shows that any lattice is a partially ordered set having the "lattice property" that any two elements have a g.l.b. and a l.u.b. We will now show that this property completely characterizes lattices.

Theorem 3. *Let L be any partially ordered set in which any two elements x, y have a g.l.b. $x \wedge y$ and a l.u.b. $x \vee y$. Then L is a lattice under the operations $\wedge$, $\vee$, in which $a \leqq b$ if and only if $a \wedge b = a$ (or, equivalently, $a \vee b = b$).*

Proof. It suffices to prove the idempotent, commutative, associative, and absorption laws, together with the consistency principle. Moreover, by the Duality Principle, it suffices to prove each of the first three laws for g.l.b. The commutative law is obvious from the symmetry of the definition; the associative law, since both $x \wedge (y \wedge z)$ and $(x \wedge y) \wedge z$ are g.l.b.'s of the three elements x, y, and z. The idempotent law is trivial, by substitution in the definition. For the consistency principle, assume first that $x \leqq y$; then any z with $z \leqq x$ and $z \leqq y$ satisfies $z \leqq x$, while $x \leqq x$ and $x \leqq y$, so that x satisfies the definition of the g.l.b. $x \wedge y$. Conversely, if $x = x \wedge y$, then x is a lower bound of y, so that $x \leqq y$; this proves the consistency principle. The absorption law now follows by a proof similar to that of Lemma 3 in §11.4.

The distributive laws were not mentioned above, since they do not hold in all lattices. For example, they do not hold if x, y, z are chosen to be the three subgroups of order 2 in the four-group (Figure 3, first diagram). However, two related inequalities do hold.

Theorem 4. *In any lattice, the semidistributive laws*

$$x \wedge (y \vee z) \geqq (x \wedge y) \vee (x \wedge z), \qquad x \vee (y \wedge z) \leqq (x \vee y) \wedge (x \vee z)$$

hold. Moreover, either distributive law implies its dual.

Proof. The labor of proof is cut in half by the Duality Principle. As regards the first semidistributive law, note that both terms on the left have both terms on the right for lower bounds; hence the g.l.b. of x and $y \vee z$ is an upper bound both to $x \wedge y$ and to $x \wedge z$, and hence to their l.u.b. $(x \wedge y) \vee (x \wedge z)$. Finally, assuming the first distributive law of §11.3, (iii), we get by expansion

$$(x \vee y) \wedge (x \vee z) = ((x \vee y) \wedge x) \vee ((x \vee y) \wedge z)$$
$$= x \vee (x \wedge z) \vee (y \wedge z) = x \vee (y \wedge z),$$

which is the other distributive law of §11.3, (iii). The proof is completed by the Duality Principle.

It is a corollary of the preceding theorems that, to prove that the algebra of classes is a Boolean algebra, we need only know that (i) set-inclusion is reflexive, antisymmetric, and transitive; (ii) the union of two sets is the least set which contains both, and dually for the intersection; (iii) $S \cap (T \cup U) = (S \cap T) \cup (S \cap U)$ identically; (iv) each set S has a "complement" S' satisfying $S \cap S' = O$, $S \cup S' = I$. This also proves

Theorem 5. *A Boolean algebra is a distributive lattice which contains elements O and I with $O \leqq a \leqq I$ for all a, and in which each a has a complement a' satisfying $a \wedge a' = O$, $a \vee a' = I$.*

Boolean algebras can also be described by many other systems of postulates. One such is indicated by Ex. 13 below.

Exercises

1. Which of the diagrams of Figure 3 represent lattices?
2. Draw two new diagrams for partially ordered sets which are not lattices.
3. Which of the examples of Ex. 4, §11.6, represent lattices?
4. Show that if $b \leqq c$ in a lattice L, then $a \wedge b \leqq a \wedge c$ and $a \vee b \leqq a \vee c$ for all $a \in L$.
5. State and prove a Duality Principle for Boolean algebras.
6. Illustrate the Duality Principle by writing out the detailed proof of the second half of Lemma 1 from the proof given for the first half.
7. Show that a lattice having only a finite number of elements has elements O and I satisfying $O \leqq x \leqq I$ for all elements x.
★8. Show that a finite partially ordered set with O and I is a lattice whenever between any elements a_1, a_2, b_1, b_2 with $a_i \leqq b_j$ $(i, j = 1, 2)$ there is an element c such that $a_i \leqq c \leqq b_j$ for all i and j.
9. A *chain* is a simply ordered set (i.e., a partially ordered set in which any x and y have either $x \geqq y$ or $y \geqq x$).
   (a) Prove that every chain is a distributive lattice.
   (b) Prove that a lattice is a chain if and only if all of its subsets are sublattices.
★10. A lattice is called *modular* if and only if $x \geqq z$ always implies $x \wedge (y \vee z) = (x \wedge y) \vee z$.
   (a) Prove that every distributive lattice is modular.
   (b) Construct a diagram for a lattice of five elements which is not modular.
   (c) Prove that each of the following is a modular lattice: (i) all subspaces of a vector space, (ii) all subgroups of an Abelian group, (iii) all normal subgroups of any group.
   (d) Show that in a modular lattice, $x \leqq z$ always implies $x \vee (y \wedge z) = (x \vee y) \wedge z$. Hence infer that the Duality Principle holds for modular lattices.
★11. In any Boolean algebra the symmetric difference of two elements x and y is defined by $x + y = (x \wedge y') \vee (x' \wedge y)$.
   (a) What does this mean if x and y are classes? Draw a figure.
   (b) Show that $x + y$ is associative, commutative, and has a zero element.
   (c) With the symmetric difference as a sum and the meet as a product, show that every Boolean algebra is a commutative ring.
12. (a) Show that if, in the vector space $\mathbf{Z}_2{}^n$, we define multiplication by $\xi\eta = (x_1 y_1, \cdots, x_n y_n)$, then $\mathbf{Z}_2{}^n$ becomes a commutative ring in which $\xi^2 = \xi$ for all ξ.
   ★(b) Show that this ring is a Boolean algebra under the operations $\xi \wedge \eta = \xi\eta$, $\xi \vee \eta = \xi + \eta + \xi\eta$, $\xi' = (1, 1, \cdots, 1) - \xi$.
★13. If L is a lattice with universal bounds O and I, in which each element a has a complement a' with the properties

$$x \leqq a' \quad \text{if and only if} \quad a \wedge x = O,$$

$$y \geqq a' \quad \text{if and only if} \quad a \vee y = I.$$

prove that L is a Boolean algebra. (*Hint:* To prove the first distributive law it suffices to prove that $e \equiv [a \wedge (b \vee c)] \wedge [(a \wedge b) \vee (a \wedge c)]' = O$. Write e as a meet and consider the individual terms.)

11.8. Representation by Sets

The main conclusion of §11.5 is that the postulates which were assumed for Boolean algebra imply all true identities for the algebra of sets with respect to intersection, union, and complement. In fact, a suitable family of subsets $S_1, \cdots, S_n$ of a particular set $\mathbf{Z}_2{}^n$ was shown to have the property that $p(S_1, \cdots, S_n) = q(S_1, \cdots, S_n)$ for two Boolean polynomials p, q if and only if these polynomials had the same disjunctive canonical form. The Boolean algebra consisting of all these disjunctive canonical forms, for given n, is called the *free* Boolean algebra with n generators.

We will now prove a stronger result, showing in passing that the postulates used to define distributive lattices completely characterize the properties of intersections and unions of sets. For this purpose, we will need concepts of homomorphism and isomorphism analogous to those already used for groups.

Definition. *A function* $f: L \to M$ *from a lattice* L *to a lattice* M *is called a* homomorphism *when* $f(x \wedge y) = f(x) \wedge f(y)$ *and* $f(x \vee y) = f(x) \vee f(y)$ *for all* $x, y \in L$. *A homomorphism which is one-one and onto is called an* isomorphism.

For example, the Boolean algebra generated by the circles X, Y, Z of the Venn diagram (Figure 1) is isomorphic with the algebra of all subsets of $\mathbf{Z}_2{}^3$, the function being defined as in §11.5.

Lemma 1. *An isomorphism* $f: A \leftrightarrow B$ *between two Boolean algebras (regarded as lattices) necessarily carries the universal bounds* O, I *and complements in* A *into the corresponding bounds and complements in* B.

Proof. Clearly, $O \wedge x = O$ for all $x \in A$ implies $f(O) \wedge f(x) = f(O \wedge x) = f(O)$ for all $f(x) \in B$; hence $f(O)$ is a universal lower bound in B; the proof that $f(I) = I$ is similar. Therefore $x \wedge x' = O$ in A implies

$$f(x) \wedge f(x') = f(x \wedge x') = f(O) = O \quad \text{in } B,$$

and dually $f(x) \vee f(x') = I$ in B, which proves that $f(x') = [f(x)]'$, completing the proof of Lemma 1.

Definition. A ring *of sets is a family of subsets of a set I which contains with any two subsets S and T also their intersections $S \cap T$ and their union $S \cup T$; a field of sets is a ring of sets which contains I, the empty set $\varnothing$, and with any set S also its complement S'.*

In other words, a field of subsets of I is just a Boolean subalgebra of the Boolean algebra A of all subsets of I; a ring of subsets is just a *sublattice* of A, considered as a distributive lattice. We will prove that every *finite* distributive lattice is isomorphic with a ring of sets and every finite Boolean algebra is isomorphic with the field of *all* subsets of some (finite) set. These may be regarded as partial analogs of Cayley's theorem for groups.

In proving these converses of Theorem 1, we will also want the following concepts.

Definition. An *element $a > O$ of a lattice L is* join-irreducible *if $x \vee y = a$ implies $x = a$ or $y = a$; it is* meet-irreducible *if $a < I$ and $x \wedge y = a$ implies $x = a$ or $y = a$; an element p is an* atom *if $p > O$, and there is no element x such that $p > x > O$.*

Lemma 2. *In a Boolean algebra, an element is join-irreducible if and only if it is an atom.*

Proof. If p is an atom, then $p = x \vee y$ implies $x = p$ or $x = O$; in the second case $p = O \vee y = y$; hence p is join-irreducible. Conversely, if a is not an atom or O, then $a > x > O$ for some x. Therefore

$$a = a \wedge I = a \wedge (x \vee x') = (a \wedge x) \vee (a \wedge x') = x \vee (a \wedge x'),$$

where $x < a$. Since $a \wedge x' \leq a$, and $a \wedge x' = a$ would imply $x = a \wedge x = a \wedge x' \wedge x = O$, $a \wedge x' < a$ also; hence a is join-reducible.

Now, for each element a of any finite lattice L, let $S(a)$ be the set of all join-irreducible elements $p_k \leq a$ in L, and consider the mapping $a \mapsto S(a)$. We have

Lemma 3. *In a finite lattice L, every element a satisfies $a = \bigvee_{S(a)} p_k$.*

Proof. For $a = O$, the result is immediate, since $S(O) = \varnothing$, the void set, and O is the least upper bound of the void set. For any other $a \in L$, we use the Second Principle of Finite Induction, letting $P(n)$ be the proposition that Lemma 3 holds if the number of elements $x \leq a$ in L is n. Trivially, $P(n)$ is true if a is join-irreducible. But if a is not join-irreducible and not O, then $a = x \vee y$, where $x < a$ and $y < a$, whence $n(x) < n(a)$ and $n(y) < n(a)$. By induction on n, it follows that x and y are joins of join-irreducible elements: $x = \bigvee_X p_\xi$ and $y = \bigvee_Y q_\eta$. Hence $a = \bigvee p_\xi \vee \bigvee q_\eta$ is a join of join-irreducible elements.

Lemma 4. *In any finite lattice* L, *the mapping* $a \mapsto S(a)$ *carries meets in* L *into set-theoretic intersections:* $S(a \wedge b) = S(a) \cap S(b)$.

Proof. By definition of $a \wedge b$, $p \leq a \wedge b$ if and only if both $p \leq a$ and $p \leq b$.

Lemma 5. *In a finite* distributive *lattice* L, *the correspondence of Lemma 4 carries joins in* L *into set-theoretic unions*: $S(a \vee b) = S(a) \cup S(b)$.

Proof. A given join-irreducible p is contained in $a \vee b$ if and only if

$$p = p \wedge (a \vee b) = (p \wedge a) \vee (p \wedge b).$$

If p is join-irreducible, this implies either $p \wedge a = p$ (i.e., $p \leq a$) or $p \wedge b = p$ ($p \leq b$). This shows that $S(a \vee b)$ contains p if and only if $S(a)$ contains p or $S(b)$ contains p. But the converse is obvious in any lattice. Q.E.D.

Lemmas 4 and 5 show that the mapping $a \mapsto S(a)$ is a homomorphism from L onto a ring $\Re$ of subsets of the set I of join-irreducible elements of L. Lemma 3 shows that it is, moreover, *one-one* from L onto $\Re$. This proves

Theorem 6. *Any finite distributive* L *is isomorphic with a ring of sets.*

When L is a finite Boolean algebra, Lemma 2 tells us that each $a \in L$ is the join of the atoms $p \leq a$. Also, by Lemmas 4 and 5, for any $a \in L$:

$$S(a) \cap S(a') = S(a \wedge a') = S(O) = \varnothing, \quad \text{and}$$

$$S(a) \cup S(a') = S(a \vee a') = S(I) = J,$$

the set of all atoms (join-irreducibles) of L. That is, $[S(a)]' = S(a')$, and so the function $a \mapsto S(a)$ is an isomorphism.

We have shown that the mapping $a \mapsto T(a)$ is an isomorphism from any Boolean algebra L to a field $\mathfrak{F}$ of subsets of atoms of L. We now show that $\mathfrak{F}$ contains *all* sets of atoms of L, proving

Theorem 7. *Any finite Boolean algebra L is isomorphic with the Boolean algebra of all sets of its atoms.*

Completion of proof. It only remains to show that if S and T are distinct sets of atoms $p_\sigma, p_\tau, \cdots$ of L, then $\bigvee_S p_\sigma \neq \bigvee_T p_\tau$. But this is a corollary of the following result.

Lemma 6. *If an atom $q \leq \bigvee_S p_\sigma$, then $q \in S$.*

For, assuming Lemma 6, $\bigvee_S p_\sigma$ contains the atoms in S and no others.

Proof of lemma. By the generalized distributive law,

$$q = q \wedge \bigvee_S p_\sigma = \bigvee_S (q \wedge p_\sigma).$$

Since q is join-irreducible, it follows that some one $q \wedge p_\sigma = q$, whence $O < p_\sigma \leq q$. Since q is an atom, this implies $p_\sigma = q$.

Exercises

1. If two finite sets I and J have the same number of elements, show that the algebra of all subsets of I is isomorphic to the algebra of all subsets of J.
2. Prove that for every positive integer n there is a Boolean algebra with 2^n elements.
3. Show that the Boolean algebra of all subsets of a class of n elements has exactly $n!$ automorphisms.
4. (a) Find a lattice homomorphism $f: A \to B$ from a Boolean algebra A onto a Boolean algebra B which does not preserve universal bounds or complements.
   (b) Show that such an f preserves complements if and only if it preserves universal bounds.
5. (a) Show that the set $\mathbf{Z}^+$ of all positive integers is a lattice under the partial ordering $m \leq n$ if and only if $m \mid n$.
   (b) Show that this lattice is distributive.
   (c) Identify its join-irreducibles.
6. Show that if the join-irreducibles of a finite distributive lattice L are a chain C, then L itself is a chain. How many more elements does L have than C?

12

Transfinite Arithmetic

12.1. Numbers and Sets

The present chapter will be concerned with the relationship between numbers and sets. This is the *cardinal* approach to the positive integers, as contrasted with the *ordinal* approach exemplified by the Peano postulates of §2.6, which regard position in the familiar sequence "one, two, three, four, · · ·" as basic. Developed with care, this cardinal approach enables one to define numbers in terms of sets, thereby reducing the totality of undefined terms which must be assumed in mathematics. But to carry out this program requires too great a reshuffling of basic ideas to fit neatly into the present book.

Instead, therefore, we shall assume the reader to be familiar with both the positive integers and the concept of a set, and shall proceed from there. Our object will be to extend the cardinal approach so as to give a precise definition of *infinite* cardinal numbers, which play a basic role in modern mathematics. Using this definition, we shall show how to add, multiply, and raise to powers arbitrary cardinal numbers, showing in the process that these operations have most (though not all) of the properties possessed by the corresponding operations on positive integers.

The source of the relationship between numbers and sets is the following definition.

Definition. *Let n be any positive integer. A set S will be said to have* cardinal number *n (in symbols,† $o(S) = n$) if and only if there exists a bijection between the elements of S and the integers* $1, 2, 3, \cdots, n$.

---

† An *empty* set is sometimes said to have the cardinal number *zero*.

This definition means that the elements of S can be labeled $s_1, s_2, s_3, \cdots, s_n$, where s_k is the element of S corresponding to the integer k. In other words, one can *count* the elements of S by counting up to n, counting each element once and only once. It is a corollary that if two sets S and T have the same cardinal number, then there is a bijection between them—namely, the correspondence $s_1 \leftrightarrow t_1, \cdots, s_n \leftrightarrow t_n$. But what is not obvious, a priori, is the fact that the same set cannot have two different cardinal numbers—that, by recounting in a different order, one will not arrive at a different total number of elements. We shall now prove this fact, stating first a somewhat more general result.

Theorem 1. *Let m and n be positive integers. There exists a bijection between the set $1, \cdots, m$ and a proper subset of the set $1, \cdots, n$ if and only if $m < n$.*

Proof. If $m < n$, then the bijections $1 \leftrightarrow 1, 2 \leftrightarrow 2, \cdots, m \leftrightarrow m$ is of the desired sort. This half of the proposition is obvious, but the converse must be analyzed more carefully.

The converse is trivial if $m = 1$, since 1 is the least positive integer; hence we can use induction on m. But now suppose there were a bijection $1 \leftrightarrow f(1), \cdots, m \leftrightarrow f(m)$ between $1, \cdots, m$ and a proper subset S of the integers $1, \cdots, n$. Define a new bijection $i \leftrightarrow g(i)$ $[i = 1, \cdots, m - 1]$ as follows:

(1) $g(i) = f(i)$ unless $f(i) = n$; $g(i) = f(m)$ if $f(i) = n$.

Since $f(i) = n$ for at most one i, the correspondence $i \leftrightarrow g(i)$ would be one-one between the integers $1, \cdots, m - 1$ and certain of the integers $1, \cdots, n - 1$.

By hypothesis, the set S of all integers $f(i)$ is a *proper* subset of the set $1, \cdots, n$; this means that S does not contain all the integers $1, \cdots, n$. Let us select the first positive integer $k \leqq n$ which is not in S, so that $f(i)$ is never k for $i = 1, \cdots, m$. If $k < n$, the definition (1) shows that no $g(i)$ equals k; if $k = n$, $f(i) = n$ is never true, so no $g(i)$ equals $f(m)$. In either event, the integers $g(1), \cdots, g(m - 1)$ fail to include all the integers $1, \cdots, n - 1$, so $i \leftrightarrow g(i)$ is one-one between the integers $1, \cdots, m - 1$ and a *proper* subset of the integers $1, \cdots, n - 1$. Now, by mathematical induction, we can assume $m - 1 < n - 1$—whence, adding one to both sides, $m < n$.

Corollary 1. *There exists a bijection between the set $\{1, \cdots, m\}$ and a subset of the set $\{1, \cdots, n\}$ if and only if $m \leqq n$.*

Proof.   If $m \leqq n$, the bijection $1 \leftrightarrow 1, \cdots, m \leftrightarrow m$ is of the desired sort. Conversely, if $i \leftrightarrow f(i)$ is a bijection between $\{1, \cdots, m\}$ and certain of the integers $1, \cdots, n$, it is a bijection between $\{1, \cdots, m\}$ and a *proper* subset of $\{1, \cdots, n, n + 1\}$. Hence by Theorem 1, $m < n + 1$, so $m \leqq n$.

Corollary 2.   *If there exists a bijection between the set* $\{1, \cdots, m\}$ *and the set* $\{1, \cdots, n\}$, *then* $m = n$.

For, by Corollary 1, $m \leqq n$ and $n \leqq m$—whence $m = n$. This shows that the same set cannot have two different positive integers for cardinal numbers.

Corollary 3.   *If S is a proper subset of the set* $\{1, \cdots, n\}$, *there is no bijection between the set* $\{1, \cdots, n\}$ *and the set S.*

Proof.   If there were such a bijection, Theorem 1 would prove $n < n$, a contradiction.

The preceding results immediately imply the following. Let S and T be any two sets whose cardinal numbers are positive integers m and n. Then $m \leqq n$ if and only if there is a bijection between S and a subset of T; $m = n$ if and only if there is a bijection between S and all of T.

Exercises

1. If a set S has cardinal n, and if t is a particular element of S, show that there exists a bijection between S and $1, \cdots, n$ in which t corresponds to n.
2. If a set S has cardinal n, show that the deletion of a single element from S leaves a set S^* of cardinal $n - 1$.
3. Prove Corollary 1 directly by the method used in the proof of Theorem 1.
4. Do the same for Corollary 3.

12.2. Countable Sets

A set is called *finite* if and only if its elements can be counted in the usual way. We shall formulate this more precisely as follows.

Definition.   *A nonempty set S is called* finite *if and only if its cardinal number is a positive integer. A set which is not empty or finite is called* infinite.

For example, the set $\mathbf{Z}^+$ of all positive integers is infinite. (It is not hard to prove this, using Theorem 1.) We shall now introduce the idea that infinite sets may also be considered to have cardinal numbers.

Definition. *A set S is said to be* countable *or* denumerable *or to have the cardinal number* $\mathbf{d}$ *(in symbols,*† $o(S) = \mathbf{d}$*) if it is bijective with the set of all positive integers.*

This is equivalent to the requirement that it be possible to enumerate all the elements of S in an ordinary infinite sequence: $s_1, s_2, s_3, \cdots, s_n, \cdots$, so that each element of S appears once and only once. If another set T is bijective with a countable set S, it follows that T is itself countable.

Theorem 2 (Paradox of Galileo). *Any denumerable set has a bijection onto a proper subset of itself.*

Proof. All the elements of the set (say S) can by hypothesis be written $s_1, s_2, s_3, \cdots$, with the different positive integers as subscripts. The bijection $s_1 \leftrightarrow s_2, s_2 \leftrightarrow s_3, \cdots, s_i \leftrightarrow s_{i+1}, \cdots$ is one-one between the set S and the set obtained from S by deleting s_1. Q.E.D.

It may be shown that $\mathbf{d}$ ("countable infinity") is the smallest infinite cardinal number. More precisely, this is

Theorem 3. *Any infinite set contains a countable subset.*

Proof. Let S be the infinite set; choose for s_1 any element in it. From $S - \{s_1\}$, then choose a second element s_2; from $S - \{s_1, s_2\}$ a third element s_3, and so on. Since S is infinite, $S - \{s_1, s_2, \cdots, s_n\}$ can never be empty; hence we can always choose an s_{n+1} in it,‡ and the process can never stop until we have constructed an infinite sequence of different elements of S.

Corollary (Dedekind–Peirce). *A set S is infinite if and only if it has a bijection with a proper subset of itself.*

Proof. If S is a finite set of cardinal number n, then S is bijective with $1, \cdots, n$, so Corollary 3 to Theorem 1 asserts that S cannot have a

---

† The Hebrew letter $\aleph_0$ (aleph-nought) is often used instead of $\mathbf{d}$.

‡ This construction uses a basic principle of set theory known as the Axiom of Choice: that given any set S, there exists a "choice function" γ which chooses from any nonempty set $T \subset S$ an element $\gamma(T) \in T$.

bijection with a proper part of itself. Conversely, let S be any infinite set; it will contain a countable subset U of elements $u_1, u_2, u_3, \cdots$. The function associating each element u_i in U with its successor u_{i+1}, and each element of S not in U with itself, is a bijection from S to a proper part of itself.   Q.E.D.

In practice, surprisingly many infinite sets turn out to be countable (to have the cardinal number **d**). Examples are given by

Theorem 4.   *The set* **Z** *of all integers is countable; the set* **Q** *of all rational numbers is countable.*

$$
\begin{array}{cccc}
1/1 & 2/1 & 3/1 & \vdots \\
\downarrow & \uparrow & \uparrow & \vdots \\
1/2 \longrightarrow & 2/2 & 3/2 & \vdots \\
& & \uparrow & \vdots \\
1/3 \longrightarrow & 2/3 \longrightarrow & 3/3 & \vdots
\end{array}
$$

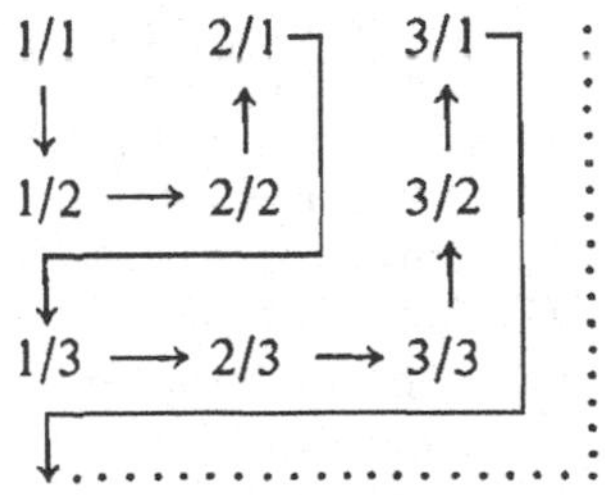

Figure 1

Proof.   The correspondence $n \leftrightarrow 2n + 1$ $(n = 0, 1, 2, \cdots)$ $(-n) \leftrightarrow 2n$ $(n = 1, 2, 3, \cdots)$ is one-one between the set $0, -1, +1, -2, +2, \cdots$ of all integers and the set $1, 2, 3, 4, 5, \cdots$ of positive integers. This proves the first assertion.

We shall next prove that the set $\mathbf{Q}^+$ of all positive rational numbers is countable. To do this, we first arrange the quotients of positive integers in an infinite square, as in Figure 1. By going around the borders of smaller squares in order, we can then arrange all such quotients in the following ordinary infinite sequence. The first term is $1/1$; the successor of $n/1$ is $1/(n + 1)$; the successor of m/n is $(m + 1)/n$ if $m < n$, and is $m/(n - 1)$ if $m \geqq n > 1$. Delete from this sequence all fractions which are not in their lowest terms (or equivalently, which are equal to other quotients of integers previously enumerated). The resulting subsequence enumerates the positive rational numbers in an ordinary sequence, establishing a bijection $m/n \leftrightarrow k$ between $\mathbf{Q}^+$ and $\mathbf{Z}^+$. But this can be easily extended to a bijection $m/n \leftrightarrow k, 0 \leftrightarrow 0, -(m/n) \leftrightarrow -k$ between the set $\mathbf{Q}$ of all rational numbers and the set $\mathbf{Z}$ of all integers. Since $\mathbf{Z}$ is countable, it follows that $\mathbf{Q}$ is.

Exercises

1. Show that the set of all integral multiples of 7 is countable.
2. Show that the set of all vectors in a finite-dimensional space $\mathbf{Q}^n$ over the field of rationals is countable.
3. Prove directly that a bijection between the set of all positive integers and a finite set is impossible.
4. If $S = T \cup U$, where T and U are countable, prove that S is countable.

5. If $S = T \cup U$, where S is countable and T is finite, prove that U is countable.
6. Show that every subset of a countable set is either finite or countable.
7. Show that the number of decimals ending in an infinite sequence which consists exclusively of 9's is countable.
8. Establish specific bijections between the set of all integers and three different proper subsets of itself.
9. Show that in Figure 1, m/n is the $((n-1)^2 + m)$th term if $m \leqq n$ and the $(m^2 - n + 1)$st if $m > n$.
10. Prove that the field $\mathbf{Q}(\sqrt{2})$ is countable (cf. §2.1).
11. Prove that every group contains a subgroup which is countable or finite.
12. Exhibit a bijection between the field of real numbers and a proper subset thereof.
13. Prove that the set of all numbers of the form $r + r'\sqrt{-1}$ $(r, r' \in \mathbf{Q})$ is countable.
★14. Prove that the ring $\mathbf{Q}[x]$ of all polynomials with rational coefficients is countable.

12.3. Other Cardinal Numbers

Not all infinite classes are countable: there is more than one "infinite" cardinal number. For instance.

Theorem 5 (Cantor). *The set* **R** *of all real numbers is not countable.*

Proof. We use the so-called "diagonal process." Suppose there were an enumeration $x_1, x_2, x_3, \cdots$ of all real numbers. List the decimal expansions of these numbers after the decimal point in their enumerated order in a square array as in Figure 2. From the digits along the diagonal of this array construct a new real number b between 0 and 1 as follows. Where a_{nn} is the nth diagonal term, let the bth digit b_n in the decimal expansion of b be $a_{nn} - 1$ if $a_{nn} \neq 0$ and 1 if $a_{nn} = 0$. Then $b = .b_1 b_2 b_3 b_4 \ldots$ is the decimal expansion of a real number b which differs from the nth number x_n of the enumeration in at least the nth decimal place. Thus, b is equal to no x_n, contradicting our supposition that the enumeration included *all* real numbers.

$$x_1 = \cdots .a_{11}a_{12}a_{13}a_{14} \cdots$$
$$x_2 = \cdots .a_{21}a_{22}a_{23}a_{24} \cdots$$
$$x_3 = \cdots .a_{31}a_{32}a_{33}a_{34} \cdots$$
$$x_4 = \cdots .a_{41}a_{42}a_{43}a_{44} \cdots$$

Figure 2

Remark. This proof is complicated by the circumstance that certain numbers, such as $1.000 = 0.999 \cdots$, may have *two* different decimal expansions, one ending in an infinite succession of 9's, the other in a succession of 0's. The trouble may be avoided by assuming that the former type of expansion (with 9's) is never used for the decimals $x_1, x_2, \cdots$ in the original enumeration. The construction of b never yields a digit $b_n = 9$; hence the decimal expansion of b is in proper form to be compared with the expansions x_n.

Definition. *A set S which is bijective to the set* **R** *of all real numbers will be said to have the cardinal number* **c** *of the continuum (in symbols, $o(S) = $* **c**$)$.

In practice, most of the sets occurring in geometry and analysis have one of the cardinal numbers **d** or **c**. This can be shown case by case, using special constructions. But in the long run it is much easier to prove first a general principle due to E. Schroeder and F. Bernstein. The formulation of this principle involves the general concept of a cardinal number, and so we proceed to define this concept now.

Definition. *The* cardinal number *of a set S is the class of all sets*† *which have a bijection onto S; the cardinal number of S is denoted by $o(S)$.*

It follows that two sets S and T have the *same* cardinal number (or are *cardinally equivalent*) if and only if there exists a bijection between them. We denote this by the symbolic equation $o(S) = o(T)$.

In virtue of the last sentence of §12.1, the concept of *inequality* between cardinal numbers can be defined in a way which is consistent with the ordinary notion of inequality between positive integers.

Definition. *We shall say that a set S is cardinally majorizable by a set T—and write $o(S) \leqq o(T)$—whenever there is an injection from S to T.*

Theorem 6 (Schroeder–Bernstein). *If $o(S) \leqq o(T)$ and $o(T) \leqq o(S)$, then $o(S) = o(T)$.*

In words, if there is an injection from S to T and another from T to S, then there is a bijection between all of S and all of T. (The converse is trivial.)

Proof.   Let $s \mapsto s\tau$ be the given injection from S to T, and let $t \mapsto t\sigma$ be the given injection from T to a subset of S. Each element s of S is the image $t\sigma$ of at most one element $t = s\sigma^{-1}$ of T; this (if it exists) in turn has at most one *parent* $s' = t\tau^{-1} = s\sigma^{-1}\tau^{-1}$ in S, and so on. Tracing

---

† This concept is like that of a "chemical element," which is likewise an abstraction, referring to all atoms having a specified nuclear charge (i.e., a specified structure).

back in this way the *ancestry* of each element of S (and also of T) as far as possible, we see that there are three alternative cases: (a) elements whose ancestry goes back forever, perhaps periodically (see Ex. 13), (b)

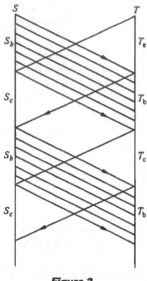

Figure 3

elements descended from a *parentless* ancestor in S, (c) elements descended from a *parentless* ancestor in T. Corresponding to these cases, we divide S into subsets S_a, S_b, S_c, and T into subsets T_a, T_b, T_c. Moreover, the category containing any element of S or T contains all its *ancestors* and *descendants*.

Indeed, σ (also τ!) is clearly bijective between S_a and T_a—each element of S_a is the image under σ of one and only one element of T_a, while each element t of T_a is the parent of one and only one element $t\sigma$ of S_a. Similarly, τ (but not σ!) is bijective between S_b and T_b, while σ (but not τ!) bijective between S_c and T_c. Combining these three bijections, $S_a \leftrightarrow T_a$, $S_b \leftrightarrow T_b$, and $S_c \leftrightarrow T_c$, we get a bijection between all of S and all of T. Q.E.D.

The situation is illustrated by Figure 3, which does not show the elements of infinite ancestry. The sets S and T are represented by points on two vertical lines, where τ is represented by the arrows slanting down to the right and σ by those slanting to the left, while the bijection of S_b to T_b is indicated by the lines without arrowheads.

Theorem 7. *The line segment S_1: $0 < x < 1$, the unit square S_2: $0 < x, y < 1$ in the plane, and the unit cube S_3: $0 < x, y, z < 1$ in space, all have the cardinal number* **c.**

Proof. The function $x \mapsto e^x = y$ (with inverse $y \mapsto \log_e y$) is one-one between $-\infty < x < +\infty$ and $0 < y < +\infty$; the function $y \mapsto y/(1 + y) = z$ with inverse $z \mapsto z/(1 - z)$ is bijective between $0 < y < +\infty$ and $0 < z < 1$. Hence the function $x \mapsto e^x/(1 + e^x) = z$ is bijective from $-\infty < x < +\infty$ to $0 < z < 1$. This proves the first assertion.

To prove the second, consider the mapping

(2) $$(.x_1x_2x_3 \cdots, .y_1y_2y_3 \cdots) \mapsto .x_1y_1x_2y_2x_3y_3 \cdots$$

between ordered couples of real numbers between 0 and 1, written in decimal form, and single real numbers between 0 and 1. It is injective (although not continuous) between the square S_2 and a subset of the line segment S_1—if it were not that decimals consisting exclusively of 9's after one point were excluded, it would be bijective between S_2 and all of S_1. This proves $o(S_2) \leqq o(S_1)$. But $o(S_1) \leqq o(S_2)$ (by the obvious mapping $x \mapsto (x, 1/2)$); hence by Theorem 6, $o(S_2) = o(S_1)$, which is c. A similar mapping

$$(3) \qquad (.x_1x_2x_3 \cdots, .y_1y_2y_3 \cdots, .z_1z_2z_3 \cdots) \mapsto (.x_1y_1z_1x_2y_2z_2 \cdots)$$

shows that $o(S_3) \leqq o(S_1)$, whence, similarly, $o(S_3) = c$.   Q.E.D.

Further examples of sets of cardinal number c are listed in the exercises.

Exercises

1. Why is the Schroeder–Bernstein theorem trivial in case the set T_c is void? What about S_c in this case?
2. Determine explicitly the sets $S_a, S_b, S_c, T_a, T_b, T_c$ when S and T are the intervals $-1 \leqq s \leqq 1/2$ and $-1 \leqq t \leqq 1/2$, τ is the injection $s \mapsto s^3$ and σ the injection $t \mapsto t^3$.
3. The same exercise, if S is the set of positive integers, T that of nonnegative integers, σ is $s \mapsto s$, τ is $t \mapsto t + 1$.
4. Prove: Any subset of real n-dimensional space which contains a continuous arc has cardinal number c.
5. Prove that if there is a surjection from S to all of a second set T, then $o(T) \leqq o(S)$.
6. Prove, conversely, that if $o(T) \leqq o(S)$, then there is a surjection of S onto T. (You may assume the Axiom of Choice.)
7. Do the properties "reflexive," "symmetric," and "transitive" apply to the relation of cardinal equivalence ($o(S) = o(T)$)? Do they apply to the relation $o(S) \leqq o(T)$? Give reasons.
8. If $o(S) \leqq o(T)$ and $o(U) = o(S)$, prove that $o(U) \leqq o(T)$.
★9. Prove: There are c $n \times n$ matrices with quaternion coefficients.
★10. Establish an explicit bijection between the set of all real numbers between 0 and 1, inclusive, and the set of all unlimited decimals $.a_1a_2a_3 \cdots$.
★11. The same exercise, for the intervals $0 < x < 1$ and $0 \leqq x \leqq 10$.
★12. Without using the Schroeder–Bernstein theorem, prove directly that
    (a) the set of nonnegative real numbers has cardinality c;
    (b) the set of those positive real numbers which are not integers has cardinality c.
★13. Determine explicitly the sets $S_a, S_b, S_c, T_a, T_b, T_c$ when $S = N \cup \{a, b, c\}$, $T = N \cup \{a, b, c\}$, $\sigma(n) = n + 1$ and cyclic on $\{a, b, c\}$, and $\tau(n) = n + 2$ and identity on $\{a, b, c\}$.

★12.4. Addition and Multiplication of Cardinals

Infinite cardinal numbers can be added and multiplied just like finite ones with preservation of all laws except the cancellation laws.

If m and n are positive integers, one may construct a set with cardinal number $m + n$ by starting with a set S' of cardinal m (say the set $1, \cdots, m$) and adding to it a disjoint set S'' of cardinal n (say the set $m + 1, m + 2, \cdots, m + n$). The union $S' \cup S''$ then has cardinal $m + n$. Similarly, the class of all couples (i, j), where i runs through the integers $1, \cdots, m$ and j through the integers $1, \cdots, n$ (e.g., the subscripts of an $m \times n$ matrix) has the cardinal number mn. We shall not prove these familiar facts; instead, we shall point out that they suggest the following extension of the operations of ordinary addition and multiplication to infinite cardinal numbers.

Definition. *Let α and β be arbitrary cardinal numbers. Then $\alpha + \beta$ is the cardinal number of those sets which are sums of disjoint subsets having α and β elements, respectively, and $\alpha\beta$ is the cardinal number of the set of all couples (x, y), where x runs through a set of α elements and y through a set of β elements.*

Addition is single-valued, for if S and T are sums of disjoint subsets S' and S'' respectively T' and T'', and there are bijections between S' and T' and between S'' and T'', then one can combine these into a bijection between all of S and all of T. Similarly, multiplication is single-valued. Indeed, most of the laws of ordinary arithmetic apply to infinite as well as to finite numbers.†

Theorem 8. *Addition and multiplication are commutative and associative; multiplication is distributive over addition; 1 is an identity.*

Proof. The commutative and associative laws of addition are corollaries of the laws of Boolean algebra. The commutative law of multiplication follows, since the function $(x, y) \mapsto (y, x)$ is bijective between the set of all couples $(x, y)[x \in S, y \in T]$ and that of all couples (y, x) $[y \in T, x \in S]$, whatever the sets S and T. The associative law of multiplication follows from an obvious bijection from the set of all triples $((x, y), z)[x \in S, y \in T, z \in U]$ to that of all triples $(x, (y, z))[x \in S,$

---

† Unfortunately, this fact loses much of its interest in the light of the theorem (which we shall not prove) that the sum or product of any two infinite cardinal numbers is simply the greater of the two. Transfinite exponentiation (§12.5) is much more interesting.

$y \in T, z \in U$], where S, T, and U are any sets. Finally, if T and U are disjoint, $o(S)(o(T) + o(U))$ is clearly the cardinal number of the set of all couples (x, w) [$x \in S$, w in T or U]; while $o(S)o(T) + o(S)o(U)$ is that of the set of all couples (x, y) [$x \in S, y \in T$] plus all couples (x, z) [$x \in S, z \in U$]. There is an obvious bijection between these two sets, proving the distributive law. The proof that $1 \cdot \alpha = \alpha$ for any cardinal number α is trivial.

Theorem 9. *The cancellation laws of addition and of multiplication do not hold for infinite cardinal numbers.*

Proof. The proof of Theorem 2 shows that $\mathbf{d} = \mathbf{d} + 1$. But this implies $\mathbf{d} + 1 = (\mathbf{d} + 1) + 1 = \mathbf{d} + 2$, although $1 \neq 2$—which violates the cancellation law of addition. Again, the set $\mathbf{Z}^+$ of positive integers is divisible into the disjoint subsets of even and odd integers, and these are countable, whence $\mathbf{d} + \mathbf{d} = \mathbf{d}$. Hence by Theorem 8, $(1 + 1)\mathbf{d} = 1 \cdot \mathbf{d}$ or $2\mathbf{d} = 1\mathbf{d}$—yet $2 \neq 1$. Q.E.D.

Actually, the equations $\alpha = \alpha + 1$ and $\alpha = \alpha + \alpha$ hold for all infinite cardinal numbers, but we shall not prove this.

It is a corollary that the system of finite and infinite cardinal numbers cannot be embedded in any system in which subtraction and division are possible; can you prove it?

Exercises

1. Prove in detail (using Boolean algebra) that addition of cardinal numbers is commutative and associative.
2. Prove $\alpha = \alpha + 1$ for any finite cardinal number α. (*Hint:* Use Theorem 3.)
3. Prove $\mathbf{d} + \mathbf{d} + \mathbf{d} = \mathbf{ddd} = \mathbf{d}$. (*Hint:* See Figure 1.)
4. (a) If n is a finite cardinal, show that $\mathbf{d} + n = \mathbf{d}$.
   (b) Show likewise that $\mathbf{d}n = \mathbf{d}$.
5. Prove $\mathbf{c} + \mathbf{d} = \mathbf{c}$ without using Theorem 6.
6. Prove $\mathbf{c} + \mathbf{c} = \mathbf{c} \cdot \mathbf{c} = \mathbf{c}$ without using §12.5.
7. Prove $\mathbf{dc} = \mathbf{c}$.
8. Prove the last statement in §12.4.
9. (a) Prove that if $x \geq \mathbf{d}$, then $x + \mathbf{d} = x$.
   (b) Prove that if $x + \mathbf{d} = \mathbf{c}$, then $x = \mathbf{c}$.
★10. For a *denumerable* group G consider the proof of the Lagrange theorem (Chap. 6) on the orders of possible finite subgroups S of G.
   (a) Show that the proof puts no restriction on the order of S.
   (b) Show that there may exist subgroups of any given finite order in a denumerable G.

★12.5. Exponentiation

If S and T are finite sets with cardinalities $m = o(S)$ and $n = o(T)$, then the ordinary power $n^m = o(T)^{o(S)}$ can be described as the number of functions from the set S to T. For any such function $x \mapsto y$ determines a function $y = f(x)$ which prescribes for each argument x in S a value y in T. To count the number of different such abstract functions f, observe that the first element x of S has just $o(T)$ possible images; for each of these, there are $o(T)$ choices for the image y of the second element of S, and so on—so that the number of ways of choosing all $o(S)$ images is $o(T)$ multipied by itself $o(S)$ times, or $o(T)^{o(S)}$.

This combinatory characterization of $o(T)^{o(S)}$ can be applied to infinite cardinal numbers.

Definition. *Let α and β be arbitrary cardinal numbers, not 0. Then β^α is the number of functions from a class of α elements to a class of β elements.*

We omit the essentially trivial proof that this defines a univalent operation: that if $\alpha = \alpha'$ and $\beta = \beta'$, then $\beta^\alpha = \beta'^{\alpha'}$.

Theorem 10. $c = 2^d$.

Proof. Each real number x between 0 and 1 has a dyadic expansion $.x_1 x_2 x_3 \cdots$ as an infinite seqence of x_i equal to 0 or 1. Distinct real numbers x and y have different expansions (§4.3); hence the function $f(x) = (x_1, x_2, x_3, \cdots)$ is one-one. But the number of such sequences is by definition the number of functions from a countable domain (namely, the set of all **d** places in the sequence) to a domain of two elements (namely, 0 and 1). We infer that there are at most 2^d real numbers between 0 and 1—hence by Theorem 7, that $c \leqq 2^d$.

On the other hand, each infinite decimal composed exclusively of (say) 3's and 7's represents a different real number—hence $2^d \leqq c$. Now, using Theorem 6, we get $c = 2^d$.

Theorem 11. *The following laws on exponentiation hold for arbitrary cardinal numbers α, β, and γ: (i) $\alpha^\beta \alpha^\gamma = \alpha^{\beta+\gamma}$; (ii) $(\alpha\beta)^\gamma = \alpha^\gamma \beta^\gamma$; (iii) $(\alpha^\beta)^\gamma = \alpha^{\beta\gamma}$; (iv) $\alpha^1 = \alpha$ and $1^\alpha = 1$.*

Proof. The proofs of the two parts of (iv) are trivial. To prove identities (i)–(iii), we suppose that S, T, and U are sets of α, β, and γ elements, respectively, with T and U disjoint.

Proof of (*i*). Consider the functions $h(v)$ from a set V, in which T and U are complementary subsets, to the set S. By definition, the number of such functions is $\alpha^{\beta+\gamma}$. On the other hand, each such function determines and is determined by a pair $(f(t), g(u))$ of independent functions, one from T to S and the other from U to S. The number of these is by definition $\alpha^{\beta}\alpha^{\gamma}$.

Proof of (*ii*). Consider the functions $h(u)$, assigning to each $u \in U$ a pair $(s, t) = (f(u), g(u))$ of arbitrary values in S and T, respectively. The number of such functions is $(\alpha\beta)^{\gamma}$, by definition. But it is also the number $\alpha^{\gamma}\beta^{\gamma}$ of pairs of functions $f(u), g(u)$—one from U to S and the other from U to T.

Proof of (*iii*). Consider the functions $f(t, u)$ of two variables $t \in T$ and $u \in U$ with values in S; their number is by definition $\alpha^{\beta\gamma}$. But every $f(t, u)$ associates with each fixed u a rule $f_u(t)$, assigning to each t a value $f_u(t) = f(t, u)$ in S. Conversely, each mapping $u \mapsto f_u$ defines a function $f(y, u) = f_u(t)$ of the variables t and u. Since the number of f_u is by definition α^{β}, the number of $f(t, u)$ is $(\alpha^{\beta})^{\gamma}$. Q.E.D.

Theorems 10 and 11 allow one to infer a number of equations involving **c** from corresponding equations on **d**. Thus,

$$\mathbf{c}^2 = (2^{\mathbf{d}})^2 = 2^{2\mathbf{d}} = 2^{\mathbf{d}} = \mathbf{c},$$

$$2\mathbf{c} = 2^1 2^{\mathbf{d}} = 2^{\mathbf{d}+1} = 2^{\mathbf{d}} = \mathbf{c},$$

$$\mathbf{c}^{\mathbf{d}} = (2^{\mathbf{d}})^{\mathbf{d}} = 2^{\mathbf{d}^2} = 2^{\mathbf{d}} = \mathbf{c} \qquad \text{(cf. Theorem 4)}.$$

Using these results, Ex. 1 below, and Theorem 6, we obtain easily such rules as $\mathbf{d}^{\mathbf{d}} = \mathbf{c}$, $n^{\mathbf{d}} = \mathbf{c}$ for any $n > 1$, and so on.

We shall conclude by proving a generalization of Theorem 5.

Theorem 12. *For any cardinal number α, $\alpha < 2^{\alpha}$.*

Explanation. By this notation is intended the assertion that $\alpha \leqq 2^{\alpha}$, and yet $\alpha \neq 2^{\alpha}$.

Proof. Let S be any set of cardinal number α. Then 2^{α} is by definition the number of functions $f(x), g(x), \cdots$ with domain S and with values 0 and 1. By defining $f_x(y) = 0$ if $x \neq y$ and $f_x(x) = 1$, we get a bijection $x \leftrightarrow f_x$ between S and a special set of functions from S to the set $(0, 1)$. This proves that $\alpha \leqq 2^{\alpha}$.

Conversely, let there be given any bijection $x \leftrightarrow g_x$ between S and functions with the domain S and with values 0 and 1. Construct a new function $h(x)$: $h(x) = 0$ if $g_x(x) = 1$ and $h(x) = 1$ if $g_x(x) = 0$. This defines a function with domain S and with values 0 and 1; moreover, by construction, $h(x) \neq g_x(x)$ for all g_x. We conclude that h is different from every g_x, and so, that there exists no bijection between S and the set of *all* functions with domain S and with values 0 and 1. In symbols, $\alpha \neq 2^\alpha$.

Exercises

1. Show that if $\alpha \leq \beta$, then for all γ:
    (a) $\alpha + \gamma \leq \beta + \gamma$,   (b) $\alpha\gamma \leq \beta\gamma$,   (c) $\alpha^\gamma \leq \beta^\gamma$,   (d) $\gamma^\alpha \leq \gamma^\beta$.
2. Prove $c^c = 2^c$. (*Hint:* Use Ex. 7 of §12.4.)
3. If a set S has cardinal α, prove that the set of all possible subsets of S has cardinal 2^α. (*Hint:* Each subset $T \leq S$ determines a so-called *characteristic function* $f_T(x)$, with $f_T(x) = 1$ if $x \in T$, $f_T(x) = 0$ otherwise.)
4. Show that the number of subsets of the square is equal to the number of all real functions of a real variable.
★5. What is the number of (a) finite and (b) countable sets of real numbers?
★6. How many sets of real numbers are there whose cardinal number is c?
7. Show that the conventions $0^0 = 1$ and $0^\alpha = 0$ for all $\alpha > 0$ are consistent with laws (i)–(iv) of Theorem 11.

13

Rings and Ideals

13.1. Rings

In this chapter, we shall take up the study of general rings and their homomorphisms, showing how the latter are associated with ideals. We shall then apply the concept of ideals to the geometry of algebraic curves and surfaces, and (in Chap. 14) to the factorization theory of algebraic numbers. Our basic postulates will be as follows.

Definition. *A ring A is a system of elements which is an Abelian group under an operation of addition, and is closed under an associative operation of multiplication which is distributive with respect to addition. Thus, for all a, b, c in the ring A,*

(1) $a(bc) = (ab)c, \qquad a(b + c) = ab + ac, \qquad (a + b)c = ac + bc.$

We shall also assume that every ring A has a unity $1 \neq 0$, *such that $1a = a1 = a$ for all $a \in A$.*

Rings include all the integral domains and other *commutative* rings studied in Chapters 1–3, such as Z_m (the integers modulo m), and $A[x], A[x, y]$, the rings of polynomials with coefficients in any given commutative ring A. They also include *noncommutative* rings, such as the quaternion ring of §8.11. The set $M_n(F)$ of all $n \times n$ matrices over any given field F is a ring under $A + B$ and AB, which is also noncommutative if $n > 1$.

If A and B are any two rings, the set of all pairs (a, b), with a in A and b in B, becomes a ring under the two operations defined by

(2)
$$(a_1, b_1) + (a_2, b_2) = (a_1 + a_2, b_1 + b_2),$$
$$(a_1, b_1)(a_2, b_2) = (a_1 a_2, b_1 b_2).$$

The resulting ring $A \oplus B$ is called the *direct sum* of A and B. Thus if $\mathbf{Q}$ is the rational field, $\mathbf{Z}$ the domain of integers, and Q the quaternion ring, then $\mathbf{Q} \oplus \mathbf{Z} \oplus Q$ is a ring. This bizarre example gives some indication of the enormous variety of rings!

Much of the theory of commutative rings extends to the noncommutative case. Thus the definition of isomorphism of rings given in §1.12 applies whether or not $ab = ba$; so does the definition of subring given in §3.3. Moreover, much of the discussion of commutative rings applies to any ring. Thus one can prove that a subset S of a ring A is a subring if and only if $1 \in S$, while b and c in S imply that $b - c$ and bc are in S; see also Ex. 1.

Linear Algebras.† Matrices and quaternions are important examples of a class of rings having an additional vector space structure. Such rings were originally constructed as "hypercomplex number systems" more extensive than $\mathbf{C}$; today, they are usually called linear associative algebras.

Definition. *A linear algebra* over a field F is a set $\mathfrak{A}$ which is a *finite-dimensional vector space over F and which admits an associative and bilinear multiplication,*

(3) $\qquad\qquad\qquad \alpha(\beta\gamma) = (\alpha\beta)\gamma \qquad\qquad$ (associative),

(4) $\quad \alpha(c\beta + d\gamma) = c(\alpha\beta) + d(\alpha\gamma), \qquad (c\alpha + d\beta)\gamma = c(\alpha\gamma) + d(\beta\gamma)$
$\qquad\qquad\qquad\qquad\qquad\qquad\qquad\qquad\qquad\qquad\qquad$ (bilinear),

where these laws are to hold for all scalars c and d in F and for all α, β, γ in $\mathfrak{A}$. The order *of $\mathfrak{A}$ is its dimension as a vector space. $\mathfrak{A}$ has a* unity *element 1 if $1\alpha = \alpha = \alpha 1$ for all α in $\mathfrak{A}$. The algebra is called a* division *algebra if, in addition, it contains with every $\alpha \neq 0$ an α^{-1} for which $\alpha^{-1}\alpha = 1$.*

In particular, every linear algebra is a ring.

---

† The material on linear algebras has been included primarily as a source of examples and because of its intrinsic interest; it and §13.6 can be omitted without loss of continuity.

A celebrated theorem of Frobenius (1878) states that the quaternions constitute the only noncommutative division algebra over the field of real numbers.

EXAMPLE 1. Construct over the real numbers an algebra of "dual numbers" which has two basis elements δ and ϵ, which multiply according to the rules $\delta\epsilon = \epsilon\delta = \delta$, $\delta^2 = 0$, $\epsilon^2 = \epsilon$. From these rules the product of any two elements of A can be found, for

$$(a\delta + b\epsilon)(c\delta + d\epsilon) = ac\delta^2 + ad\delta\epsilon + bc\epsilon\delta + bd\epsilon^2$$
$$= (ad + bc)\delta + bd\epsilon.$$

The requisite postulates, such as the associative law for multiplication, may be verified. This example, like the quaternions, shows how an algebra may be defined by giving a suitable multiplication table for the basis elements.

EXAMPLE 2. The total matrix algebra $M_n(F)$ of all $n \times n$ matrices over F has as a basis the matrices E_{ij}, which have entry 1 in the i, j position and zeros elsewhere. The multiplication table for the basis elements is $E_{ij}E_{jk} = E_{ik}, E_{ij}E_{kl} = O \, (j \neq k)$.

EXAMPLE 3. Let G be any finite group, with elements $\alpha_1, \cdots, \alpha_n$ and multiplication $\alpha_i\alpha_j = \alpha_k$. If F is any field, there exists a linear algebra $\mathfrak{A}$ over F which has the elements of G for a basis, and in which multiplication is determined by bilinearity from the group table for G,

$$(x_1\alpha_1 + \cdots + x_n\alpha_n)(y_1\alpha_1 + \cdots + y_n\alpha_n) = \sum_{i,j} (x_iy_j)(\alpha_i\alpha_j).$$

This algebra is known as the *group algebra* of G over F.

In particular, the group algebra of the cyclic group of order two with generator α has the basis $1 = \alpha^2$ and α, and the multiplication

$$(x \cdot 1 + y\alpha)(u \cdot 1 + v\alpha) = (xu + yv)1 + (xv + yu)\alpha.$$

Relative to the basis $\beta = (1 + \alpha)/2$, $\gamma = (1 - \alpha)/2$, it has the multiplication table $\beta^2 = \beta$, $\gamma^2 = \gamma$, $\beta\gamma = \gamma\beta = 0$.

EXAMPLE 4. The set of all those $2n \times 2n$ matrices which have $n \times n$ blocks of zeros, at the upper right and the lower left, forms an algebra which is a subring of $M_{2n}(F)$. It is the direct sum of two copies of $M_n(F)$.

We now prove an analogue for matrices of Cayley's theorem (§6.5, Theorem 8). First, we define two algebras $\mathfrak{A}$ and $\mathfrak{A}'$ over the same field F to be *isomorphic* when there is a bijection $\alpha \leftrightarrow \alpha'$ between their elements that preserves all three operations:

$$(5) \qquad (\alpha + \beta)' = \alpha' + \beta', \qquad (c\alpha)' = c\alpha', \qquad (\alpha\beta)' = \alpha'\beta',$$

for all $\alpha, \beta \in \mathfrak{A}$ and all $c \in F$.

Theorem 1. *Every linear associative algebra of order n with a unity is isomorphic to an algebra of n × n matrices.*

Proof. The algebra $\mathfrak{A}$ is a vector space of elements ξ. Associate with each element α in $\mathfrak{A}$ the transformation T obtained by right multiplication as $\xi T = \xi\alpha$ for any ξ in $\mathfrak{A}$. Since multiplication is bilinear as in (4), T is a linear transformation. Since a unity 1 is present, $1\alpha = 1\beta$ implies $\alpha = \beta$, so distinct elements α and β induce distinct transformations T and U. Moreover, the algebra postulates give

$$\xi(\alpha + \beta) = \xi\alpha + \xi\beta, \qquad \xi(c\alpha) = c(\xi\alpha), \qquad \xi(\alpha\beta) = (\xi\alpha)\beta,$$

so the corresponding transformations are $\alpha + \beta \mapsto T + U$, $c\alpha \mapsto cT$, $\alpha\beta \mapsto TU$. This means that the correspondence $\alpha \mapsto T$ is an isomorphism of the given algebra to an algebra of linear transformations on $\mathfrak{A}$. The transformations in turn are represented isomorphically by matrices, hence the statement of the theorem.

Exercises

1. (a) In any ring, prove that $(-a)(-b) = ab$ and that $-(-a) = a$.
  (b) Prove that $a0 = 0a$ for all a, and that the unity 1 is unique.
2. Prove that the direct sum defined by (2) is actually a ring.
3. Prove that the direct sum of two integral domains is not an integral domain.
4. Define the direct sum of n given rings, and prove it a ring.
5. Prove that the direct sum of two linear algebras over a field F can be made into a linear algebra over F, after suitable definition of scalar multiplication.
6. Prove the statement of the text characterizing a subring S.
7. Show that the zero element 0 of a linear algebra satisfies $\xi \cdot 0 = 0 = 0 \cdot \xi$ for all ξ.
8. Is the algebra of dual numbers a division algebra? Justify.

9. Show that the following systems are linear algebras:
(a) a vector space V_n, with $\alpha \cdot \beta = 0$ for all α and β,
(b) all $n \times n$ triangular matrices (entries all 0 below the diagonal).
10. Show that if P is any invertible $n \times n$ matrix over F, then $A \mapsto P^{-1}AP$ is an automorphism of $M_n(F)$. Generalize.
11. Prove that an $n \times n$ matrix A which commutes with every $n \times n$ matrix is necessarily a scalar matrix. (*Hint:* A commutes with each E_{ij}.)
12. If $\mathfrak{A}$ is an algebra, show that the set $\mathfrak{Z}$ of all those elements z in $\mathfrak{A}$ which commute with every element of $\mathfrak{A}$ is a subalgebra of $\mathfrak{A}$. (It is called the *center* of $\mathfrak{A}$.)

13.2. Homomorphisms

Given two rings A and A', the correspondence $a \rightarrow aH$ is called a *homomorphism* of A to A' if aH is a uniquely defined element of A' for each element a of A, and if, for all a and b in A,

$$(6) \quad (a + b)H = aH + bH, \quad (ab)H = (aH)(bH), \quad 1H = 1'.$$

In brief, just as in the commutative case of §3.3, a homomorphism is a mapping which preserves unity, sums, and products. As with groups, a homomorphism onto is also called an *epimorphism*.

A homomorphism H from the ring A to A' is certainly a homomorphism of the additive group of A to that of A'. Therefore, H has the properties, proved in §6.11 for groups,

$$(7) \quad 0H = 0', \quad (-a)H = -(aH), \quad (a-b)H = aH - bH.$$

Here $0'$ is the zero element of the ring A', that is, the identity element of the additive group of A'.

The familiar correspondence $a \mapsto a_m$, which carries each integer a into its residue class modulo m, is a homomorphism of the ring $\mathbf{Z}$ of integers to $\mathbf{Z}_m$. If $f(x)$ is any polynomial with coefficients in an integral domain D, the correspondence $f(x) \mapsto f(b)$ found by "substituting" for x a fixed element b of D is a homomorphism of the polynomial domain $D[x]$ to D, for the rules for adding and multiplying polynomial forms in an indeterminate x certainly apply to the corresponding polynomial expressions in b. If $\mathbf{Q}[x]$ is the ring of polynomials with rational coefficients, the correspondence $f(x) \mapsto f(\sqrt{2})$ is an epimorphism of the polynomial ring $\mathbf{Q}[x]$ onto the field of all numbers $a + b\sqrt{2}$ (see the discussion in §2.1). The direct sum $A \oplus B$ of two rings A and B is mapped epimorphically on the summand B by the correspondence $(a, b) \mapsto b$; this correspondence preserves sums and products by the very definition (2) of the operations in a direct sum.

To describe a particular homomorphism explicitly, one would naturally ask when two elements a and b of the first ring have the same image in the second. By the rule (7), this can happen only when their difference has the image $(a - b)H = 0'$. Hence we search for the set of elements mapped by H on the zero element $0'$ of A'. For example, the homomorphism $\mathbf{Z} \rightarrow \mathbf{Z}_m$ maps onto zero all multiples km of the modulus m. The set of all these multiples is closed under subtraction, and also under multiplication by any integer of $\mathbf{Z}$ whatever. Similarly, the homomorphism $f(x) \mapsto f(b)$ maps onto zero all polynomials divisible by $(x - b)$, and no others. The set S of all these polynomials is also closed under subtraction and under multiplication by all members of $D[x]$ (whether in S or not). These two examples suggest the following definition and theorem (cf. §3.8).

Definition. *An* ideal *C in a ring A is a nonvoid subset of A with the properties*
  (i) *c_1 and c_2 in C imply that $c_1 - c_2$ is in C;*
  (ii) *c in C and a in A imply that ac and ca are in C.*

Theorem 2. *In any homomorphism H of a ring A, the set of all elements mapped on zero is an ideal in A.*

To prove Theorem 2 in general, let C be the set of all elements c in A with $cH = 0'$, where $0'$ is the zero element of the image A'. Then, for any a whatever in A, $(ac)H = (aH)(cH) = (aH)0' = 0'$ and $(ca)H = (cH)(aH) = 0'$, which proves (ii). Moreover, $c_1 H = c_2 H = 0'$ gives by (7)

$$(c_1 - c_2)H = c_1 H - c_2 H = 0' - 0' = 0',$$

hence property (i).

This result suggests that ideals in a ring are analogous to normal subgroups in a group. To express this analogy, we call the set of all elements mapped on zero by a homomorphism H the *kernel* of H, and we say that a ring B is an *epimorphic image* of a ring A under the homomorphism H when H is surjective (an epimorphism), so that every element $b \in B$ is the image aH of some $\alpha \in A$ under H.

Theorem 3. *An epimorphic image of a ring A is determined up to isomorphism by its kernel.*

Proof. We have to show that if H and K are epimorphisms of A onto rings A' and A'', respectively, and if $aH = 0'$ if and only if $aK = 0''$, then A' and A'' are isomorphic. It is natural to let an element $a' \in A'$ correspond to $a'' \in A''$ if and only if these two elements have a common antecedent a in A, so

$$a' \leftrightarrow a'' \quad \text{when} \quad aH = a', \quad aK = a'',$$

for some a. This correspondence is one-one: under it each a' in A' corresponds to one and only one a'' in A''. To see this, note first that each a' in A' has at least one antecedent a in A and hence corresponds to at least one $a'' = aK$ in A''. Second, if $a' \leftrightarrow a''$ and $a' \leftrightarrow b''$, then

$$aH = a', \quad aK = a'', \quad bH = a', \quad bK = b''$$

for some a, b in A, whence $(a - b)H = a' - a' = 0'$, implying that $0'' = (a - b)K = a'' - b''$ by hypothesis. The correspondence also preserves sums and products, for if $a' \leftrightarrow a''$ and $b' \leftrightarrow b''$, then

$$a' + b' = (a + b)H \leftrightarrow (a + b)K = a'' + b''$$
$$a'b' = (ab)H \leftrightarrow (ab)K = a''b''$$

where a is a common antecedent of a' and a'', and b one for b' and b''.

The two properties (i) and (ii) of an ideal have several immediate consequences. Any ideal C contains some element c, hence (i) shows $c - c = 0$ to be in C. Therefore $0 - c = -c$ is also in C for any c in C. By property (i), we find that the sum $c_1 + c_2 = c_1 - (-c_2)$ of any two elements of C lies in C. Thus, since $1 \in A$, a nonvoid subset C of A is an ideal of A if and only if every linear combination $a_1 c_1 \pm a_2 c_2$ and $c_1 a_1 \pm c_2 a_2$ lies in C, for c_1 and c_2 in C and coefficients a_1 and a_2 in A. In particular, an ideal of A need not be a subring of A, since it may not contain the unity of A. The whole ring A and the subset (0) consisting of 0 alone are always ideals in any ring A. They are called *improper ideals* of A. Any other ideal is called *proper*. Correspondingly, a *proper epimorphism* of a ring A is one whose kernel is a proper ideal, so that the epimorphism is not an isomorphism (mapping only (0) on $0'$).

Theorem 4. *A division ring has no proper epimorphic images.*

Proof. It suffices to show that a division ring D can have no proper ideals. Let C be any ideal in D which is not the ideal (0), and which thus contains an element $c \neq 0$. By (ii), C then contains $1 = c^{-1}c$ and, by (ii) again, C contains any element $a = a \cdot 1$ of the whole division ring. Therefore C is improper, as asserted.

If b is an element in a *commutative* ring A, the set (b) of all multiples xb of b, for variable x in A, is an ideal, for properties (i) and (ii) may be verified. This ideal (b) is known as a *principal ideal*; it is the smallest ideal of A containing b. We recall that, by Theorem 6 of §1.7, every ideal in the domain $\mathbf{Z}$ of integers is principal. By Theorem 11 of §3.8, the same is true in the domain $F[x]$ of polynomials in one indeterminate over any field F.

In the ring $\mathbf{Q}[x, y]$ of polynomials in two variables with rational coefficients, the set C of all polynomials with constant term zero is an ideal. It is not a principal ideal, for the two polynomials x and y both lie in C and cannot both be multiples of one and the same polynomial $f(x, y)$. Though this ideal C is not generated by any single polynomial $f(x, y)$, all its elements can be represented by linear combinations $xg(x, y) + yh(x, y)$ with polynomial coefficients, so the whole ideal is given by the linear combinations of *two* generating elements x and y.

Consider now the ideal generated by any given finite set of elements in a commutative ring A. If an ideal C contains elements $c_1, c_2, \cdots, c_m$, then it must contain all linear combinations $\sum_i x_i c_i$ of these elements with coefficients x_i in A. But the set

$$(8) \qquad (c_1, c_2, \cdots, c_m) = [\text{all elements } \sum_i x_i c_i \text{ for } x_i \text{ in } A]$$

is itself an ideal, for

$$\sum_i x_i c_i - \sum_i y_i c_i = \sum_i (x_i - y_i) c_i \qquad \text{and} \qquad a\left(\sum_i x_i c_i\right) = \sum_i (a x_i) c_i,$$

so the set has the properties (i) and (ii) requisite for an ideal. Since A has a unity element 1, each c_i is necessarily one of the elements $c_i = 0 \cdot c_1 + \cdots + 0 \cdot c_{i-1} + 1 \cdot c_i + 0 \cdot c_{i+1} + \cdots + 0 \cdot c_m$ in this set (8). Therefore, the set $(c_1, \cdots, c_m)$, defined by (8), is an ideal of A containing the c_i and contained in every ideal containing all the c_i. It is called the ideal with the *basis* $c_1, \cdots, c_m$. (Such basis elements do *not* resemble bases of vector spaces because $x_1 c_1 + \cdots + x_m c_m = 0$ need *not* imply $c_1 = \cdots = c_m = 0$.)

In most familiar integral domains, every ideal has a finite basis, but there exist domains where this is not the case.

Exercises

1. Which of the following mappings are homomorphisms, and why? If the mapping is a homomorphism, describe the ideal mapped into zero.
   (a) $a \mapsto 5a$, a an integer in $\mathbf{Z}$;

(b) $f(x) \mapsto f(\omega)$, $f(x)$ a polynomial in $\mathbf{Q}[x]$, ω a cube root of unity;

(c) $f(x, y) \mapsto f(t, t)$,   mapping $F[x, y]$ into $F[t]$ (x, y, t indeterminates).

2. Show that every homomorphic image of a commutative ring is commutative.

3. In the ring $\mathbf{Q}[x, y]$ of polynomials $f(x, y) = a + b_1 x + b_2 y + c_1 x^2 + c_2 xy + c_3 y^2 + \cdots$, which of the following sets of polynomials are ideals? If the set is an ideal, find a basis for it.
   (a) all $f(x, y)$ with constant term zero ($a = 0$),
   (b) all $f(x, y)$ not involving x ($b_1 = c_1 = c_2 = \cdots = 0$),
   (c) all polynomials without a quadratic term ($c_1 = c_2 = c_3 = 0$).

4. (a) Find all ideals in $\mathbf{Z}_6$.   (b) Find all homomorphic images of $\mathbf{Z}_6$.

5. Prove in detail that the only proper epimorphic images of $\mathbf{Z}$ are the rings $\mathbf{Z}/(m) = \mathbf{Z}_m$ defined in Chap. 1.

6. (a) Find all ideals in $\mathbf{Z}_m$ for every m.
   (b) Find all epimorphic images of $\mathbf{Z}_m$.

7. Find all ideals in the direct sum of two fields. Generalize.

8. Find all ideals in the direct sum $\mathbf{Z} \oplus \mathbf{Z}$, where $\mathbf{Z}$ is the ring of integers.

★9. If C_1 and C_2 are ideals in rings A_1 and A_2, prove that $C_1 \oplus C_2$ is an ideal in the direct sum $A_1 \oplus A_2$, and that every ideal in the direct sum has this form.

10. In an integral domain show that $(a) = (b)$ if and only if a and b are associates (§ 3.6).

11. If A is a commutative ring in which every ideal is principal, prove that any two elements a and b in A have a g.c.d. which has an expression $d = ra + sb$.

★12. Let A be a ring containing a field F with the same unity element (e.g., A might be a ring of polynomials over F). Prove that every proper homomorphic image of A contains a subfield isomorphic to F.

★13. Let $\mathbf{Z}_p^*$ be the ring of all rational numbers m/n with denominator relatively prime to a given prime p. Prove that every proper ideal in $\mathbf{Z}_p^*$ has the form (p^k) for some positive integer k.

13.3 Quotient-rings

For every homomorphism of a ring there is a corresponding ideal of elements mapped on zero. Conversely, given an ideal, we shall now construct a corresponding homomorphic image. An ideal C in a ring A is a subgroup of the additive group of A. Each element a in A belongs to a coset, often called the *residue class* $a' = a + C$, which consists of all sums $a + c$ for variable c in C. Two elements a_1 and a_2 belong to the same coset if and only if their difference lies in the ideal C. Since addition is commutative, C is a normal subgroup of the additive group A, so the cosets of C form an Abelian quotient-group, in which the sum of two cosets is a third coset found by adding representative elements, as

(9) $$(a_1 + C) + (a_2 + C) = (a_1 + a_2) + C.$$

This sum was shown in §6.13 to be independent of the choice of the elements a_1 and a_2 in the given cosets.

To construct the product of two cosets, choose any element $a_1 + c_1$ in the first and any element $a_2 + c_2$ in the second. The product

$$(a_1 + c_1)(a_2 + c_2) = a_1 a_2 + (a_1 c_2 + c_1 a_2 + c_1 c_2) = a_1 a_2 + c'$$

is always an element in the coset $a_1 a_2 + C$, for by property (ii) of an ideal the terms $a_1 c_2$, $c_1 a_2$, and $c_1 c_2$ lie in the ideal C. Therefore all products of elements in the first coset by elements in the second lie in a single coset; this product coset is

(10) $$(a_1 + C)(a_2 + C) = a_1 a_2 + C.$$

The associative and the distributive laws follow at once from the corresponding laws in A, and the coset which contains 1 acts like a unity, so the cosets of C in A form a ring.

The correspondence $a \mapsto a' = a + C$ which carries each element of A into its coset is an epimorphism by the very definitions (9) and (10) of the operations on cosets. In the epimorphic image, the zero element is the coset $0 + C$, so the elements of C are mapped upon zero. These results may be summarized as follows:

Theorem 5. *Under the definitions (9) and (10), the cosets of any ideal C in a ring A form a ring, called the* quotient-ring† *A/C. The function $a \mapsto a + C$ which carries each element of A into the coset containing it is an epimorphism of A onto the quotient-ring A/C, and the kernel of this epimorphism is the given ideal C.*

Corollary 1. *If A is commutative, so is A/C.*

The relation of ideals to homomorphisms is now complete. In particular, the uniqueness assertion of Theorem 3 can be restated thus:

Corollary 2. *If an epimorphism H maps A onto A' and has the kernel C, then A' is isomorphic to the quotient-ring A/C.*

---

† The ring A/C is also often called a residue class ring, since its elements are the residue classes (cosets) of C in A.

The ring $\mathbf{Z}_m$ of integers modulo m can now be described as the quotient-ring $\mathbf{Z}/(m)$. Conversely, with this example in mind, one often writes $a \equiv b$ (C), and says that a and b are *congruent* modulo an ideal of a ring R, when $(a - b) \in C$.

Every property of a quotient-ring is reflected in a corresponding property of its generating ideal C. To illustrate this principle, call an ideal $C < A$ *maximal*† when the only ideals of A containing C are C and the ring A itself. Call an ideal P in A *prime* when every product ab which is in P has at least one factor, a or b, in P.

In *commutative* rings, prime ideals play a special role. Thus, in the ring $\mathbf{Z}$ of integers, a (principal) ideal (p) is a prime ideal if and only if p is a prime number, for a product ab of two integers is a multiple of p if and only if one of the factors is a multiple of p, when p is a prime but not otherwise.

Theorem 6. *If A is a commutative ring, the quotient-ring A/C is an integral domain if and only if C is a prime ideal, and is a field if and only if C is a maximal ideal in A.*

Proof. The commutative ring A/C is an integral domain if and only if it has no divisor of zero (§1.2, Theorem 1). This requirement reads formally

$$(11) \qquad a'b' = 0 \qquad \text{only if} \qquad a' = 0 \quad \text{or} \quad b' = 0,$$

where a' and b' are cosets of elements a and b in A. Now a coset a' of C is zero if and only if a is in the ideal C, so the requirement above may be translated by

$$(12) \qquad ab \text{ in } C \qquad \text{only if} \qquad a \text{ is in } C \quad \text{or} \quad b \text{ is in } C.$$

This is exactly the definition of a prime ideal C.

Suppose next that C is maximal, and let b be any element of A not in C. Then the set of all elements $c + bx$, for any c in C and any x in A, can be shown to be an ideal. This ideal contains C and contains an element b not in C; since C is maximal, it must be the whole ring A. In particular, the unity 1 is in the ideal, so for some $a, 1 = c + ba$. In terms of cosets this equation reads $1' = b'a'$. Thus, for any coset $b' = b + C \neq C$, we have found a reciprocal coset $a' = a + C$, which is to say that the commutative ring of cosets is a field. Conversely, if A/C is a field, one may prove C maximal (Ex. 10). Q.E.D.

---

†"Maximal" is sometimes replaced by the term "divisorless."

Since every field is an integral domain, Theorem 6 implies that every maximal ideal is prime. Conversely, however, a prime ideal need not be a maximal ideal. For example, consider the homomorphism $f(x, y) \mapsto f(0, y)$ which maps the domain $F[x, y]$ of all polynomials in x and y with coefficients in a field on the smaller domain $F[y]$. The ideal thereby mapped onto 0 is the principal ideal (x) of all polynomials which are multiples of x. Since the image ring $F[y]$ is indeed a domain, this ideal (x) is a prime ideal, as one can also verify directly. But $F[y]$ is not a field, so (x) cannot be maximal. It is in fact contained in the larger ideal (x, y), which consists of all polynomials with constant term zero.

Exercises

1. Prove the associative and distributive laws for the multiplication of cosets.
2. Let congruence modulo an ideal $C \le A$ be defined so that $a \equiv b \pmod{C}$ if and only if $a - b$ is in C. Prove that congruences can be added and multiplied, and show that a coset of C consists of mutually congruent elements.
3. Prove in detail Corollary 1, Theorem 5.
4. Find all prime ideals in the ring $\mathbf{Z}$ of integers.
5. Find all prime ideals and all maximal ideals in the ring $F[x]$ of polynomials over F.
★6. Prove without using Theorem 6 that every maximal ideal of an integral domain is prime.
★7. Find a prime ideal which is not maximal in the domain $\mathbf{Z}[x]$ of all polynomials with integral coefficients.
8. Show that, in the domain $\mathbf{Z}[\omega]$ of all numbers $a + b\omega$ (a, b integers, ω an imaginary cube root of unity), (2) is a prime ideal. Describe $\mathbf{Z}[\omega]/(2)$.
9. In the polynomial ring $\mathbf{Q}[x, y]$, which of the following ideals are prime and which are maximal?
   (a) (x^2)
   (b) $(x - 2, y - 3)$,
   (c) $(y - 3)$,
   (d) $(x^2 + 1)$,
   (e) $(x^2 - 1)$,
   (f) $(x^2 + 1, y - 3)$.
10. Prove that if a quotient-ring A/C is a field, then C is maximal.
11. Find a familiar ring isomorphic to each of the following quotient rings A/C:
   (a) $A = \mathbf{Q}[x]$, $C = (x - 2)$;
   (b) $A = \mathbf{Q}[x]$, $C = (x^2 + 1)$;
   (c) $A = \mathbf{Q}[x, y]$, $C = (x, y - 1)$;
   (d) $A = \mathbf{Z}[x]$, $C = (3, x)$;
   (e) $A = \mathbf{Z}_p^*$, $C = (p)$, as in Ex. 13 of §13.2.
12. (The "Second Isomorphism Theorem.") Let $C > D$ be two ideals in a ring A.
   (a) Prove that the quotient C/D is an ideal in A/D.
   (b) Prove that A/C is isomorphic to $(A/D)/(C/D)$. (*Hint:* The product of two homomorphisms is a homomorphism.)

★13.4. Algebra of Ideals

Inclusion between ideals is closely related to divisibility between numbers. In the ring $\mathbf{Z}$ of integers $n \mid m$ means that $m = an$, hence that every multiple of m is a multiple of n. The multiples of n constitute the principal ideal (n), so the condition $n \mid m$ means that (m) is contained in (n). Conversely, $(m) \subset (n)$ means in particular that m is in (n), hence that $m = an$. Therefore

$$(m) \subset (n) \qquad \text{if and only if} \qquad n \mid m.$$

More generally, in any commutative ring R, $(b) \subset (a)$ implies that $b = ax$ for some $x \in R$—that is, that $a \mid b$. Conversely, if $a \mid b$, then $b = ax$ for some $x \in R$ and so $by = axy \in (a)$ for all $by \in (b)$, whence $(b) \subset (a)$. This proves

Theorem 7. *In a commutative ring R,*

$$(13) \qquad\qquad (b) \subset (a) \qquad \textit{if and only if} \qquad a \mid b.$$

But beware! The "bigger" number corresponds to the "smaller" ideal; for instance, the ideal (6) of all multiples of 6 is properly contained in the ideal (2) of all even integers.

The g.c.d. and l.c.m. also have ideal-theoretic interpretations. The least common multiple m of integers n and k is a multiple of n and k which is a divisor of every other common multiple. The set (m) of all multiples of m is thus the set of all common multiples of n and k, so is just the set of elements common to the principal ideals (n) and (k). This situation can be generalized to arbitrary ideals in arbitrary (not necessarily commutative) rings, as follows.

The *intersection* $B \cap C$ of any two ideals B and C of a ring A may be shown to be an ideal. If D is any other ideal of A, the ideal $B \cap C$ has the three properties

$$B \cap C \subset B, \qquad B \cap C \subset C, \qquad \text{and}$$

$$D \subset B \quad \text{and} \quad D \subset C \quad \text{imply} \quad D \subset B \cap C.$$

The intersection is thus the g.l.b. of B and C in the sense of lattice theory.

Dual to the intersection is the *sum* of two ideals. If B and C are ideals in A, one may verify that the set

$$(14) \qquad B + C = [\text{all sums } b + c \quad \text{for } b \text{ in } B, \ c \text{ in } C]$$

is an ideal in A. Since any ideal containing B and C must contain all sums $b + c$, this ideal $B + C$ contains B and C and is contained in every ideal containing B and C. Thus $B + C$ is a l.u.b. or *join* in the sense of lattice theory.

Theorem 8. *The ideals in a ring A form a lattice under the ordinary inclusion relation with the join given by the sum $B + C$ of (14) and the meet by the intersection $B \cap C$.*

If the integers m and n have d as g.c.d., then the ideal sum $(m) + (n)$ is just the principal ideal (d). For, by (13), $(d) \supset (m)$ and $(d) \supset (n)$; since d has a representation $d = rm + sn$, any ideal containing m and n must needs contain d and so all of (d). Therefore, (d) is the join of (m) and (n); that is, $(d) = (m) + (n)$.

The preceding observation can be generalized as follows:

Lemma. *In a commutative ring R, the sum $(b) + (c)$ of two principal ideals is itself a principal ideal (d) if and only if d is a greatest common divisor of b and c.*

We leave the proof to the reader.

In general, if ideals B and C in a commutative ring are generated by bases

$$(15) \qquad B = (b_1, \cdots, b_m), \qquad C = (c_1, \cdots, c_n),$$

then we have, for any $b + c \in B + C$,

$$b + c = \sum_i x_i b_i + \sum_j y_j c_j,$$

as in (8). That is, $B + C$ is generated by b's and c's, so that

$$(16) \qquad (b_1, \cdots, b_m) + (c_1, \cdots, c_n) = (b_1, \cdots, b_m, c_1, \cdots, c_n).$$

This rule, in combination with natural transformations of bases, may be used to compute greatest common divisors of integers explicitly. For example,

$$(336) + (270) = (336, 270) = (336 - 270, 270) = (66, 270)$$
$$= (66, 270 - 4 \times 66) = (66, 6) = (6),$$

so that the g.c.d. of 336 and 270 is 6.

In any commutative ring, one can also define the *product* $B \cdot C$ of any two ideals B and C,

(17) $B \cdot C = [\text{all sums } b_1 c_1 + \cdots + b_m c_m \text{ for } b_i \text{ in } B, \ c_j \text{ in } C]$.

This set is in fact an ideal; it is generated by all the products bc with one factor in B and another in C, so is the smallest ideal containing all these products. In particular, the product of two principal ideals (b) and (c) is simply the principal ideal (bc) generated by the product of the given elements b and c. More generally, if ideals B and C are determined by bases as in (12), any product bc has the form

$$bc = \left(\sum_i x_i b_i \right) \left(\sum_j y_j c_j \right) = \sum_{i,j} (x_i y_j)(b_i c_j).$$

Hence the product ideal BC has the basis

(18) $BC = (b_1 c_1, b_1 c_2, \cdots, b_m c_{n-1}, b_m c_n)$.

Such products are useful for algebraic number theory (§14.10).

Exercises

1. Prove in detail that $B \cap C$ and $B + C$ are always ideals.
2. Prove that the product BC of (17) is an ideal.
3. Draw a lattice diagram for all the ideals in $\mathbf{Z}_{24}$.
4. If $f(x)$ and $g(x)$ are polynomials over a field, and $d(x)$ is their g.c.d., prove that $(f(x)) + (g(x)) = (d(x))$.
5. Compute by ideal bases the g.c.d.'s $(280, 396)$ and $(8624, 12825)$.
6. Prove that every ideal in the ring $\mathbf{Z}$ of integers can be represented uniquely as a product of prime ideals.
7. Prove the following rules for transforming a basis of an ideal:
$$(c_1, c_2, \cdots, c_m) = (c_1 + x c_2, c_2, \cdots, c_m),$$
$$(x c_1, c_1, c_2, \cdots, c_m) = (c_1, c_2, \cdots, c_m).$$
8. Simplify the bases of the following ideals in $R[x, y]$:
$$(x^2 + y, 3y, 4x^3 + x^2), \qquad (x^2 + 3xy + y^2, 2x^2 - y^2, x^2 + 6xy, x^3 + y^2).$$
9. (a) In any commutative ring, show that $BC \subseteq B \cap C$.
   (b) Give an example to show that $BC < B \cap C$ is possible.
   (c) Prove that $B(C + D) = BC + BD$.
★10. Prove that the lattice of ideals in any ring is modular in the sense of Ex. 10, §11.7.

11. In a commutative ring A, let $B : C$ denote the set of all elements x such that xc is in B whenever c is in C.
    (a) If B and C are ideals, prove that $B : C$ is also an ideal in A. (It is called the "ideal quotient.")
    (b) Show that $(B_1 \cap B_2) : C = (B_1 : C) \cap (B_2 : C)$.
    (c) Prove that $B : C$ is the l.u.b. (join) of all ideals X with $CX \subset B$.
12. Prove that if a ring R contains ideals B and C with $B \cap C = 0$, $B + C = R$, then R is isomorphic to the direct sum of B and C.

13.5. Polynomial Ideals

The notion of an ideal is fundamental in modern algebraic geometry. The reason for this soon becomes apparent if one considers algebraic curves in three dimensions.

Generally, in the n-dimensional vector space F^n, an (affine) *algebraic variety* is defined as the set. V of all points $(x_1, \cdots, x_n)$ satisfying a suitable finite system of polynomial equations

$$(19) \qquad f_1(x_1, \cdots, x_n) = 0, \qquad \cdots, \qquad f_m(x_1, \cdots, x_n) = 0.$$

For example, in $\mathbf{R}^3$, the circle C of radius 2 lying in the plane parallel to the (x, y)-plane and two units above it in space is usually described analytically as the set of points (x, y, z) in space satisfying the simultaneous equations

$$(20) \qquad x^2 + y^2 - 4 = 0, \qquad z - 2 = 0.$$

These describe the curve C as the intersection of a circular cylinder and a plane. But C can be described with equal accuracy as the intersection of a sphere with the plane $z = 2$, by the equivalent simultaneous equations

$$(21) \qquad x^2 + y^2 + z^2 - 8 = 0, \qquad z - 2 = 0.$$

Still another description is possible, by the equations

$$(22) \qquad x^2 + y^2 - 4 = 0, \qquad x^2 + y^2 - 2z = 0.$$

These describe C as the intersection of a circular cylinder with the paraboloid of revolution $x^2 + y^2 = 2z$.

One can avoid the preceding ambiguity by describing C in terms of *all* the polynomial equations which its points satisfy. But if $f(x, y, z)$ and $g(x, y, z)$ are any two polynomials whose values are identically zero on C, then their sum and difference also vanish identically on C. So, likewise, does any multiple $a(x, y, z)f(x, y, z)$ of $f(x, y, z)$ by any polynomial $a(x, y, z)$ whatsoever. This means that the *set of all polynomials whose values are identically zero on C is an ideal.* This ideal then, and not any special pair of its elements, is the ultimate description of C. We will now show that the set of all such equations is an ideal.

Theorem 9. *In F^n, the set $J(S)$ of all polynomials which vanish identically on a given set S is an ideal in $F[x_1, \cdots, x_n]$.*

For, if $p(x_1, \cdots, x_n)$ vanishes at a given point, then so do all multiples of p, while if p and q vanish there, so do $p \pm q$. The same is true of polynomials which vanish identically on a given set; in fact $J(S)$ is just the intersection of the ideals $J(\xi)$ of polynomials which vanish at the different points $\xi \in S$.

Thus, in the case of the circle C discussed above, $J(C)$ is the ideal of all linear combinations

$$(23) \qquad h(x, y, z) = a(x, y, z)(x^2 + y^2 - 4) + b(x, y, z)(z - 2),$$

with polynomial coefficients $a(x, y, z)$ and $b(x, y, z)$. That is, $J(C)$ is simply the ideal $(x^2 + y^2 - 4, z - 2)$ with *basis* $x^2 + y^2 - 4$ and $z - 2$. The polynomials of (21) generate the same ideal, for these polynomials are linear combinations of those of (20), while those of (20) can conversely be obtained by combination of the polynomials of (21). The polynomial ideal determined by this curve thus has various bases,

$$(24) \qquad (x^2 + y^2 - 4, z - 2) = (x^2 + y^2 + z^2 - 8, z - 2)$$
$$= (x^2 + y^2 - 2z, z - 2).$$

The quotient ring $\mathbf{R}[x, y, z]/(x^2 + y^2 - 4, z - 2)$ has an important meaning. Namely, it is isomorphic with the ring of all *functions* on C (cf. §3.2) which are definable as *polynomials* in the variables x, y, z. It is clearly isomorphic with $\mathbf{R}[x, y]/(x^2 + y^2 - 1)$, and hence to the ring of all *trigonometric polynomials* $p(\cos \theta, \sin \theta)$ with the usual rules of identification. This quotient ring is called the ring of *polynomial functions* on C, and its extension to a field is called the field of *rational functions on C*.

The *twisted cubic* $C_3 \colon x = t, y = t^2, z = t^3$ is an algebraic curve which (unlike C) can be defined parametrically by polynomial functions of the parameter t. Evidently, a given point (x, y, z) lies on C_3 if and only if $y = x^2$ and $z = x^3$. Hence C_3 is the algebraic curve defined in $\mathbf{R}^3$ by the ideal $M = (y - x^2, z - x^3)$.

By definition, a polynomial $p(x, y, z)$ vanishes identically on C_3 if and only if $p(t, t^2, t^3) = 0$ for all $t \in \mathbf{R}$. Now consider the homomorphism†

(25) $f(x, y, z) \mapsto f(t, t^2, t^3)$ (t an indeterminate).

Clearly, $y = x^2$ and $z = x^3$ for all points on C_3, which shows that $y - x^2$ and $z - x^3$ will lie in our ideal M. But, conversely, observe that the substitution $y = y' + x^2, z = z' + x^3$ will turn any polynomial $f(x, y, z)$ into a polynomial $f'(x, y', z')$, and that in this form the homomorphism (25) is

(25′) $f'(x, y', z') \mapsto f'(t, 0, 0)$.

This correspondence maps onto 0 every term of f' which contains y' or z', and no others, so the polynomials mapped onto zero are simply those which are linear combinations $g(x, y, z)y' + h(x, y, z)z'$. Therefore, our ideal M is exactly the ideal $(y', z') = (y - x^2, z - x^3)$ with basis $y' = y - x^2, z' = z - x^3$. This expresses C_3 as the intersection of a parabolic cylinder and another cylinder. In the further analysis of C_3, the quotient-ring $\mathbf{R}[x, y, z]/M$ plays an important role. The mapping (25) shows that this quotient-ring is isomorphic to the polynomial ring $\mathbf{R}[t]$.

The sum of two ideals has a simple geometric interpretation. For example, in $\mathbf{R}[x, y, z]$ the principal ideal $(z - 2)$ represents the plane $z = 2$, because all the polynomials $f(x, y, z)(z - 2)$ of this ideal vanish whenever x, y, and z are replaced by the coordinates of a point on the plane $z = 2$. Similarly, the principal ideal $(x^2 + y^2 - 4)$ defines a cylinder of radius 2 with the z-axis as its axis. The sum of these two ideals is $(x^2 + y^2 - 4, z - 2)$, according to the rule (16). We have just seen that this sum (23) represents the circle which is the *intersection* of the plane and the cylinder. In fact, it is obvious that the locus corresponding to the sum of two ideals is the intersection of the loci determined by the ideals separately.

Conversely, any ideal J in the polynomial ring $\mathbf{R}[x_1, \cdots, x_n]$ determines a corresponding locus, which consists of all points $(a_1, \cdots, a_n)$ of n-space such that $f(a_1, \cdots, a_n) = 0$ for each polynomial $f \in J$. Hilbert's Basis Theorem asserts that J has a finite basis $f_1, \cdots, f_m$, so that the corresponding locus V is indeed an algebraic variety. However, the ideal $J(V)$ of this variety may be larger than the given ideal J (cf. Ex. 3 below).

---

† *Caution.* The fact that (25) defines a homomorphism is not obvious; to prove it requires an extension of Theorem 1 of §3.1.

Exercises

1. Find the ideal belonging to the curve with the parametric equations $x = t + 1$, $y = t^3$, $z = t^4 + t^2$ in $\mathbf{R}^3$.

2. Show that any ideal $(ax + by + cz, a'x + b'y + c'z)$ generated by two linearly independent linear polynomials determines a line in $\mathbf{R}^3$.

3. (a) Show that the ideals (x, y) and (x^2, xy, y^2) in $\mathbf{R}[x, y, z]$ determine the same algebraic variety.
   (b) Show that any ideal and its square determine the same locus.

4. Show in detail that the set of polynomials in $\mathbf{R}[x_1, \cdots, x_n]$ vanishing identically on any locus C is an ideal.

5. (a) What is the locus determined by $xy = 0$ in three-dimensional space?
   (b) Prove that the locus determined by the *product* of two principal ideals is the *union* of the loci determined by the two ideals individually.
   (c) Generalize to arbitrary ideals. (*Hint:* If a point in the locus of the product is not in the locus determined by the first factor, it fails to make zero at least one of the polynomials in the first ideal.)
   (d) What is the locus determined by the intersection of two ideals?

6. (a) Compute the inverse of the "birational" transformation
$$T: x' = x, \quad y' = y - x^2, \quad z' = z + y + x^3.$$
   (b) Prove that the set of all substitutions of the form $x' = x$, $y' = y + p(x)$, $z = z' + q(x, y)$ (p, q polynomials) is a group.
   (c) Show that each such substitution induces an *automorphism* on the ring $\mathbf{R}[x, y, z]$.

7. (a) If H is an ideal in a commutative ring A, the *radical* of H is the set $\sqrt{H}$ of all x in A with some power x^m in H. Prove that $\sqrt{H}$ is an ideal.
   (b) If H is an ideal in the polynomial ring $\mathbf{C}[x, y, z]$, V the corresponding locus, prove that $J(V)$ contains $\sqrt{H}$. (The Hilbert Nullstellensatz asserts that $J(V) = \sqrt{H}$.)

8. Describe the locus determined by $x^2 + y^2 = 0$ (a) in $\mathbf{R}^3$ and (b) in $\mathbf{C}^3$.

★13.6. Ideals in Linear Algebras

In a noncommutative ring one may consider "one-sided" ideals. A *left ideal L* in a ring A is a subset of A such that $x - y$ and ax lie in L whenever x and y are in L and a is in A. A *right ideal* may be similarly defined. In contrast to these notions, an ideal in our previous sense is called a *two-sided* ideal. For example, in the ring M_2 of all 2×2 matrices, the matrices in which the first column is all zero form a left ideal but do not form a two-sided ideal.

These concepts may be profitably applied to a linear algebra A with a unity element 1; as observed in §13.1, any such linear algebra A is a ring. In this case, any left ideal L or right ideal R is closed also with respect to scalar multiplication. Thus, if ξ is any element in L and c any scalar, then L contains $c\xi$, because $c\xi = (c \cdot 1)\xi$ is the product of an element in L by some element $c \cdot 1$ in A. If A is regarded as a linear space over its field F of scalars, any left (or right) ideal of A is thus a subspace.

A linear algebra is said to be *simple* if it has no proper (two-sided) ideals. Thus, a simple algebra has no proper homomorphic images.

Theorem 10. *The algebra of all $n \times n$ matrices over a field is simple.*

Proof. This algebra M_n has as a basis the n^2 matrices E_{ij}, which have entry 1 in the (i, j) position and zero elsewhere. A proper ideal B in M_n would contain at least one nonzero matrix $A = \sum_i a^{ij}E_{ij}$, with a coefficient $a_{rs} \neq 0$. Each matrix

$$(26) \qquad (a_{rs})^{-1}E_{kr}AE_{sk} = (a_{rs})^{-1}\sum_{i,j} E_{kr}E_{ij}E_{sk}a_{ij} = E_{kk}$$

then lies in B. Consequently, the identity matrix $I = \sum_k E_{kk}$ is in B, so B must be the whole algebra, and is improper.   Q.E.D.

Wedderburn (1908) proved a celebrated converse of Theorem 10. This converse asserts that, in particular, every simple algebra over the field $\mathbf{C}$ of complex numbers is isomorphic to the algebra of all $n \times n$ matrices over $\mathbf{C}$. To handle the general case, one needs the concept of a *division algebra*. By this is meant a linear algebra which is a division ring. Using the fundamental theorem of algebra, one can prove that the only division algebra over the complex field $\mathbf{C}$ is $\mathbf{C}$ itself. A famous theorem of Frobenius asserts that the only division algebras over the real field $\mathbf{R}$ are $\mathbf{R}$, $\mathbf{C}$, and the algebra of quaternions (§8.11).

One can construct a total matrix algebra $M_n(D)$ of any order n over any division ring D, as follows. To add or multiply two $n \times n$ matrices with coefficients in a division algebra D, apply the ordinary rules,

$$(27) \qquad \|a_{ij}\| + \|b_{ij}\| = \|a_{ij} + b_{ij}\|, \qquad c\|a_{ij}\| = .\|ca_{ij}\|,$$

$$\|a_{ij}\| \cdot \|b_{ij}\| = \left\| \sum_{k=1}^{n} a_{ik}b_{kj} \right\|.$$

Wedderburn's result is that if F is *any* field, the most general simple algebra A over F is obtained as follows. Take any division algebra D over F and any positive integer n. Then A consists of all $n \times n$ matrices with coefficients in D.

Exercises

1. Prove that every division algebra is simple.
2. Find all right ideals in a division algebra.
3. Discuss the algebra of left ideals in a ring, describing sums, intersections, and principal left ideals.
4. Show that every quotient-ring of a linear algebra over F is itself a linear algebra.
5. (a) If S is a subspace of the vector space F^n, prove that the set of all matrices with rows in S is a left ideal of $M_n(F)$.
   ★(b) Show that every left ideal C of $M_n(F)$ is one of those described in part (a). (*Hint:* Show that every row of a matrix of C is the first row of a matrix in C which has its remaining rows all zero. Use the methods of §§7.6–7.7.)
★6. Extend Theorem 10 to the total matrix algebra $M_n(D)$ over an arbitrary division ring D.

13.7. The Characteristic of a Ring

Any ring R can be considered as an additive (Abelian) group. The cyclic subgroup generated by any $a \in R$ consists of the mth powers of a, where m ranges over the integers. In additive notation we write $m \times a$ for the mth "power" of a. Thus, if m is positive integer,

$$(28) \qquad m \times a = a + a + \cdots + a \qquad (m \text{ summands});$$

if $m = 0, 0 \times a = 0$; while if $m = -n$ is negative,

$$(29) \qquad (-n) \times a = n \times (-a)$$
$$= (-a) + (-a) + \cdots + (-a) \qquad (n \text{ summands}).$$

We call $m \times a$ the mth *natural multiple* of a; it is defined for any $m \in \mathbf{Z}$ and $a \in R$.

These natural multiples of elements in a domain D have all the properties which have been proved in §6.6, in the multiplicative notation, for powers in any commutative group; hence,

$$(30) \qquad \begin{aligned} (m \times a) + (n \times a) &= (m + n) \times a, \\ m \times (n \times a) &= (mn) \times a, \qquad \text{and} \end{aligned}$$

$$(31) \qquad \begin{aligned} m \times (a + b) &= m \times a + m \times b, \\ m \times (-a) &= (-m) \times a. \end{aligned}$$

There are further properties which result from the distributive law. One general distributive law (see §1.5) is

$$(a + a + \cdots + a)b = ab + ab + \cdots + ab \qquad (m \text{ summands}).$$

In terms of natural multiples, this becomes

(32) $$(m \times a)b = m \times (ab) = a(m \times b).$$

This also holds for $m = 0$ and for negative m, for with $m = -n$ the definition (29) gives

$$(-n) \times ab = n \times (-ab) = [n \times (-a)]b = [(-n) \times a]b.$$

The rule $(a + \cdots + a)(b + \cdots + b) = ab + \cdots + ab$ is another general distributive law. It may be reformulated as

(33) $$(m \times a)(n \times b) = (mn) \times (ab).$$

This also is valid for all integers m and n, positive, negative, or zero.

Setting $a = 1$, the unity (multiplicative identity) of R, (32) shows that $m \times b$ is just $(m \times 1)b$, the product of b with the mth natural multiple of 1. Moreover, setting $a = 1$ in (30), we see that the mapping $m \mapsto m \times 1$ from $\mathbf{Z}$ into R preserves sums. Finally, setting $a = b = 1$ in (33), we obtain

(33') $$(m \times 1)(n \times 1) = (mn) \times (1 \cdot 1) = (mn) \times 1;$$

the mapping preserves products. This proves

Theorem 11. *The mapping* $m \mapsto m \times 1$ *is a* homomorphism *from the ring* $\mathbf{Z}$ *into R for any ring R.*

Corollary 1. *The set of natural multiples of 1 in any ring R is a subring isomorphic to $\mathbf{Z}$ or to $\mathbf{Z}_m$ for some integer $m > 1$.*

Definition. *The characteristic of a ring R is the number m of distinct natural multiples $m \times 1$ of its unity element 1.*

Corollary 2. *In the additive group of an integral domain D, all nonzero elements have the same order—namely, the characteristic of D.*

Proof. For all nonzero $b \in D$, $m \times b = 0$ if and only if $(m \times 1)b = 0$, which is equivalent by the cancellation law to $m \times 1 = 0$.   Q.E.D.

The domain **Z** of all integers has characteristic† ∞, while the domain **Z**$_p$ has characteristic p. These are the only characteristics possible:

Theorem 12. *The characteristic of an integral domain is either ∞ or a positive prime p.*

To prove this, suppose, to the contrary, that some domain D had a finite characteristic which was composite, as $m = rs$. Then by (33'), the ring unity 1 of D satisfies

$$0 = m \times 1 = (rs) \times 1 = (r \times 1) \cdot (s \times 1).$$

By the cancellation law, either $r \times 1 = 0$ or $s \times 1 = 0$. Hence the characteristic must be a divisor or r or of s, and not m, as assumed.

Corollary. *In any domain, the additive subgroup generated by the unity element is a subdomain isomorphic to* **Z** *or to* **Z**$_p$.

The binomial formula (9) of §1.5 illustrates the value of natural multiples. In any commutative ring R, the expansion

$$(a + b)^2 = a^2 + ab + ba + b^2 = a^2 + 2 \times (ab) + b^2$$

has a middle term which is, properly speaking, a natural multiple $2 \times (ab)$. More generally, the proof by induction given in §1.5 of the binomial formula (9) there involves the binomial coefficients as natural *multiples*, and so we can write

(34)
$$(a + b)^n = a^n + \binom{n}{1} \times (a^{n-1}b) + \binom{n}{2} \times (a^{n-2}b^2) + \cdots + \binom{n}{n} \times b^n,$$

where the coefficients $\binom{n}{i}$ are natural integers given by the formulas

(35) $$\binom{n}{i} = [n!]/[(n - i)!i!], \qquad i = 0, 1, \cdots, n,$$

and where $n! = n(n - 1) \cdots 3 \cdot 2 \cdot 1$ and $0! = 1$.

---

† Most writers use "characteristic 0" in place of "characteristic ∞."

Theorem 13. *In any commutative ring R of prime characteristic p, the correspondence $a \mapsto a^p$ is a homomorphism.*

Proof. By (6), we are required to prove that $1^p = 1$, that $(ab)^p = a^p b^p$, and that $(a \pm b)^p = a^p \pm b^p$ for all $a, b \in R$. The first two equations hold in every commutative ring. As for the third, set $n = p$ in formulas (34) and (35). Since p is a prime, it is not divisible by any of the factors of $i!$ or $(p - i)!$ for $0 < i < p$. Hence all the binomial coefficients in (34) with $0 < i < p$ are multiples of p. But the ring R has characteristic p; hence all terms in (34) with factor $\binom{p}{i}$, $0 < i < p$, drop out. There follows the identity

(36) $$(a \pm b)^p = a^p \pm b^p,$$

completing the proof.

Corollary. *In a finite field F of characteristic p the correspondence $a \mapsto a^p$ is an automorphism.*

Proof. Since $a^p = 0$ implies $a = 0$ in F, the kernel of the homomorphism $a \mapsto a^p$ is 0, and the homomorphism is one-one. Since F is finite, this implies that $a \mapsto a^p$ is also onto, hence an automorphism.

Exercises

1. Show that the natural multiple $m \times a$ can be defined for positive m by the "recursion formulas" $1 \times a = a$, $(m + 1) \times a = m \times a + a$.
2. Prove by induction the rules (30) and (32) for positive natural multiples.
3. Obtain Fermat's theorem (§1.9, Theorem 18) as a corollary of Theorem 13.
4. What can you say about the characteristic of an ordered integral domain?
5. (a) Show that $\alpha : a \mapsto a^p$ is one-one (a monomorphism) in any integral domain D of characteristic p.
   (b) Show that if $D = Z_p(x)$, then the image of α is a proper subdomain of D.
   (c) Show that a finite field must have a proper automorphism unless it is one of the "prime" fields Z_p.

13.8. Characteristics of Fields

Since a field is defined as an integral domain in which division (except by zero) is possible, the discussion of characteristics applies at once to fields. If a field F has characteristic p, then by Theorem 12 the additive subgroup of F generated by its unity element is a subfield and is isomorphic to the finite field composed of the integers modulo p. If a field F has characteristic ∞, then by Theorem 12 the subgroup generated by the unity element 1 consists of all multiples $m \times 1$, and so the subfield generated by c is composed of all the quotients $(m \times 1)/(n \times 1)$, with $n \neq 0$. This subfield is the field of quotients of the subdomain of all multiples $m \times 1$. As such, by Theorem 7 of §2.2, it is isomorphic to the field of rational numbers, which is the field of quotients of the domain of the integers $m \leftrightarrow m \times 1$. Indeed, the map $(m \times 1)/(n \times 1) \leftrightarrow m/n$ is an isomorphism between the subfield generated by 1 and the field of rational numbers. This proves the following result (cf. Corollary 2 of Theorem 18, §2.6):

Theorem 14. *In a field of characteristic* ∞, *the subfield generated by the unity element is isomorphic to the field* **Q** *of all rational numbers.*

The isomorphism $(m \times 1)/(n \times 1) \leftrightarrow m/n$ preserves all four rational operations in such a field F. In dealing with a single field F, it is thus possible (and convenient) to identify each quotient $(m \times 1)/(n \times 1)$ with its corresponding rational number m/n. With this convention, each field of characteristic ∞ may be said to contain all the rational numbers m/n, with $n \neq 0$. By a similar convention every field of characteristic p may be said to contain the field Z_p. In this sense, every field is an extension of one of the minimal fields (so-called *prime* fields) **Q** and Z_p. Therefore it is natural to begin a systematic classification of fields with a survey of the ways of extending a given field. Such a survey will be made in the next chapter.

Exercises

1. Let F_4 be any field with exactly four elements.
   (a) Show that F_4 has characteristic 2.
   (b) Show that both elements not in the prime subfield Z_2 of F_4 satisfy $x^2 = x + 1$.
   (c) Using this fact, show that F_4 is isomorphic to the field $Z[\omega]/(2)$ of Ex. 8, §13.3.
2. Find all automorphisms of the field F_4 of Ex. 1.
3. Show that the conventional formula for the solution of a quadratic equation applies to any field of characteristic not 2.
4. Over which fields is the usual formula (§5.5) for solving a cubic equation valid?

14

Algebraic Number Fields

14.1. Algebraic and Transcendental Extensions

The remaining two chapters are concerned with solutions of polynomial equations $p(x) = 0$ over a general field F and their properties. It will be shown that any such equation can be solved in a suitable *extension* of F, by which is meant a field K containing F as a subfield. Thus $p(x) = 0$ always has one root in the quotient-field $F[x]/(p)$ of the polynomial ring $F[x]$ by the principal ideal of multiples of p.

After describing general properties of such extensions, we will study specifically the field of all "algebraic numbers" obtained by extending the rational field $\mathbf{Q}$ in this way. A brief introduction is given to algebraic number theory, through the problem of proving unique factorization theorems for "integers" in certain quadratic extensions $\mathbf{Q}[x]/(x^2 - r) = \mathbf{q}(\sqrt{r}), r \in \mathbf{Z}$. For instance, the Gaussian integers $m + n\sqrt{-1}$ (the case $r = -1$) can be uniquely factored into Gaussian primes.

The simplest kind of extension K of a field F is that consisting of rational expressions $p(c)/q(c) = (\sum a_k c^k)/(\sum b_l c^l)$ of a single element $c \in K$ with coefficients a_i, b_j in F. For example, the complex numbers $a + bi$ are generated by the reals and the single complex number i, while the field $\mathbf{Q}(x)$ of all rational forms (with rational coefficients) in an indeterminate x is generated by the field $\mathbf{Q}$ and the element x. A single field may be generated in several different ways. For example, the field $\mathbf{Q}(\sqrt{2})$ is generated by a root $\sqrt{2}$ of the equation $x^2 - 2$ and consists of all real numbers $a + b\sqrt{2}$ with rational coefficients a and b (see the example in §2.1). A different equation $x^2 + 4x + 2 = 0$ has a root $-2 + \sqrt{2}$ which generates the same field $\mathbf{Q}(\sqrt{2})$, for any number in the field can be expressed in terms of this new generator as

$$a + b\sqrt{2} = (a + 2b) + b(-2 + \sqrt{2}).$$

The usual process of completing the square, applied to this equation, gives $x^2 + 4x + 2 = (x + 2)^2 - 2 = 0$, so that $y = x + 2$ satisfies a new equation $y^2 - 2 = 0$ with a root generating the same field. The use of a transformations of variables to simplify an equation thus corresponds to the choice of a new generator for the corresponding field.

Let us describe in general the subfield *generated* by a given element in *any* extension K of a field F. Let K be a given field, F a subfield of K, and c an element of K. Consider those elements of K which are given by polynomial expressions of the form

(1) $f(c) = a_0 + a_1 c + a_2 c^2 + \cdots + a_n c^n$ (each a_i in F).

Any subdomain of K containing F and c necessarily contains all such elements $f(c)$. Conversely, the set of all such polynomials is closed under addition, subtraction, and multiplication. Therefore these expressions (1) constitute the subdomain of K generated by F and c. This subdomain is conventionally denoted by $F[c]$, with *square* brackets.

If $f(c)$ and $g(c) \neq 0$ are polynomial expressions like (1), their quotient $f(c)/g(c)$ is an element of K, called a *rational expression* in c with coefficients in F. The set of all such quotients is a subfield; it is the field *generated* by F and c and is conventionally denoted by $F(c)$, with *round* brackets.

A field K is called a *simple extension* of its subfield F if K is generated over F by a single element c, so that $K = F(c)$. The fields $\mathbf{Q}(\sqrt{2})$, $\mathbf{Q}(\sqrt[3]{5})$, and $\mathbf{Q}(\omega)$ discussed in §2.1 are all instances of simple extensions. It can be proved that any extension of F whatever is obtainable by a finite or (well-ordered) transfinite sequence of simple extensions.

Over the field of rational numbers, some complex numbers, such as i, $\sqrt{2}$, $\sqrt[3]{5}$, $\sqrt{-3}$, satisfy polynomial equations with rational coefficients. There are other numbers, like π and $e = 2.71828\cdots$, which can be shown to satisfy no such equations (except trivial ones). The latter numbers are called "transcendental." This important dichotomy applies to elements over any field.

Definition. *Let K be any field, and F any subfield of K. An element c of K will be called* algebraic *over F if c satisfies a polynomial equation with* *coefficients not all zero in F,*

(2) $a_0 + a_1c + a_2c^2 + \cdots + a_nc^n = 0$ (a_i *in F, not all* 0).

An element c of K which is not algebraic over F is called transcendental *over F.*

A simple extension $K = F(c)$ is said to be algebraic or transcendental over F, according as the generating element c is algebraic or transcendental over F. The structure of a simple transcendental extension is especially easy to describe.

Theorem 1. *If c is transcendental over F, the subfield F(c) generated by F and c is isomorphic to the field F(x) of all rational forms in an indeterminate x, with coefficients in F. The isomorphism may be so chosen that a $\mapsto$ a for each a in F, and c $\mapsto$ x.*

Proof. The extension $F(c)$ clearly contains F and all the rational expressions $f(c)/g(c)$ with coefficients in F. If two polynomial expressions $f_1(c)$ and $f_2(c)$ are equal in $F(c)$, their coefficients must be equal term by term, because otherwise the difference $f_1(e) - f_2(c)$ would yield a polynomial equation for c with coefficients not all zero, contrary to the assumption that c is transcendental over F. Therefore the correspondence $f(c) \leftrightarrow f(x)$ is a bijection between the domain $F[c]$ and the domain $F[x]$ of polynomial forms in an indeterminate x. By the rules for operating with polynomials, this correspondence is an isomorphism. It may be extended by Theorem 6 of §2.2 to give the isomorphism $f(c)/g(c) \leftrightarrow f(x)/g(x)$ between $F(c)$ and $F(x)$.

Exercises

1. Identify each of the following complex numbers as algebraic or transcendental over the field $\mathbf{Q}$ of rational numbers and cite your reasons: $\sqrt{7}$, $\sqrt[3]{5}$, π^2, $e + 3$ (where $e = 2.71828 \cdots$), $i + 3$, $e^{2\pi i}$, $\sqrt{2} + i$.
2. Show that if x is algebraic (over F), then so are x^2 and $x + 3$, and conversely.
3. What numbers in $\mathbf{Q}(\sqrt{5})$ generate the whole field?
4. (a) If d is an integer which is not a square, describe the field $\mathbf{Q}(\sqrt{d})$.
   (b) Find those elements in $\mathbf{Q}(\sqrt{d})$ which generate the whole field.
   (c) Express each such element as a root of a quadratic equation with coefficients in $\mathbf{Q}$.

14.2. Elements Algebraic over a Field

We next investigate the nature of *simple* algebraic extensions of a field *F*, generated by *F* and a single element *u* algebraic over *F*. By definition, this element must satisfy over *F* a polynomial equation of degree at least one. The same element *u* may satisfy many different equations; for example, $\sqrt{2}$ is a root of $x^2 - 2 = 0$, $x^3 - 2x = 0$, $x^4 - 4 = 0$, and so on. But it is the root of just one irreducible and monic polynomial equation (see also Ex. 6 below).

Theorem 2. *If an element u of an extension K of a field F is algebraic over F, then u is a zero of one and only one monic polynomial p(x) that is irreducible in the polynomial domain F[x]. If h is another polynomial in F[x], then h(u) = 0 if and only if h is a multiple of p in the domain F[x], that is, if and only if h is in the principal ideal (p) of F[x].*

Proof. The polynomials $h \in F[x]$ with $h(u) = 0$ constitute an ideal in $F[x]$; this ideal is just the kernel of the homomorphism $\phi_u : F[x] \to K$ defined by the "evaluation map" $p \mapsto p(u)$ that assigns to each polynomial *p* its value at $u \in K$. Like all ideals of $F[x]$, this ideal is principal (§3.8, Theorem 11), and so consists of all multiples of any one of its members of least degree. Just one of these is monic; call it *p*. This *p* is irreducible, for otherwise it could be factored as $p = fg$, where *f* and *g* are polynomials of smaller degree, which would imply $f(u)g(u) = p(u) = 0$, so either $f(u) = 0$ *or* $g(u) = 0$ contrary to the choice of *p* as a polynomial of least degree with $p(u) = 0$. The proof is complete.

Definition. *The* minimal polynomial *of an element u algebraic over a field F is the (unique) monic irreducible polynomial* $p \in F[x]$ *with* $p(u) = 0$; *the* degree $n = [u:F]$ *of u over F is the degree of this polynomial.*

Corollary. *If the element u has degree n over a field F, then one has* $a_0 + a_1 u + \cdots + a_{n-1}u^{n-1} = 0$ *for coefficients* a_i *in F if and only if* $a_0 = a_1 = \cdots = a_{n-1} = 0$.

We are now in a position to describe the subfield of *K* generated by *F* and our algebraic element *u*. This subfield $F(u)$ clearly contains the sub-domain $F[u]$ of all elements expressible as polynomials $f(u)$ with coefficients in *F*. Moreover, the mapping $f(x) \mapsto f(u)$ will be shown to be an isomorphism $\phi' : F[x]/(p) \to F(u)$ of fields between the quotient-ring $F[x]/(p)$ and $F(u)$.

The rest of this section will be concerned with this result. From the formulas for adding and multiplying polynomials, it is evident that ϕ' is an epimorphism from $F[x]$ to the subdomain $F[u]$. But actually, the domain $F[u]$ is a subfield. Indeed, let us find an inverse for any element $f(u) \neq 0$ in $F[u]$. The statement that $f(u) \neq 0$ means that u is not a root of $f(x)$, hence by Theorem 2 that $f(x)$ is not a multiple of the irreducible polynomial $p(x)$, hence that $f(x)$ and $p(x)$ are relatively prime. Therefore we can write

$$(3) \qquad\qquad 1 = t(x)f(x) + s(x)p(x)$$

for suitable polynomials $t(x)$ and $s(x)$ in $F[x]$. The corresponding equation in $F[u]$ is $1 = t(u)f(u)$. This states that the nonzero element $f(u)$ of $F[u]$ does have a reciprocal $t(u)$ which is also a polynomial† in u, and shows that $F[u]$ is a subfield of K.

Since, conversely, every subfield of K which contains F and u evidently contains every polynomial $f(u)$ in $F[u]$, we see that $F[u]$ is the subfield of K generated by F and u. We have proved

Theorem 3. *Let K be any field, and u an element of K algebraic over the subfield F of K; let $p(x)$ be the monic irreducible polynomial over F of which u is the root. Then the mapping $\phi': f(x) \mapsto f(u)$ from the polynomial domain $F[x]$ to $F(u)$ is an epimorphism with kernel $(p(x))$.*

Combining this result with Corollary 2 of Theorem 5, §13.3, we have an immediate corollary.

Theorem 4. *In Theorem 3, $F(u)$ is isomorphic to the quotient-ring $F[x]/(p)$, where p is the monic irreducible polynomial of u over F.*

The quotient-ring $F[x]/(p)$ can be described very simply. Each polynomial $f(x) \in F[x]$ is congruent modulo (p) to its *remainder* $r(x) = f(x) - a(x)p(x)$ when divided by $p(x)$, and this is a unique polynomial

$$(4) \qquad\qquad r(x) = r_0 + r_1 x + \cdots + r_{n-1} x^{n-1}$$

of degree less than n. To add or subtract two such polynomials, just do the same to their coefficients. To multiply them, form their polynomial product as in §3.1, (3'), and compute the remainder under division by $p(x)$.

Thus, in the special case of the extension $\mathbf{Q}(\sqrt{2})$ of the rational field

---

† For example, in $\mathbf{Q}[\sqrt{3}]$, $1 + \sqrt{3}$ has the multiplicative inverse, found by "rationalization of the denominator" as $1/(1 + \sqrt{3}) = (1 - \sqrt{3})/(1 + \sqrt{3})(1 - \sqrt{3}) = -\frac{1}{2} + \frac{1}{2}\sqrt{3}$.

$F = \mathbf{Q}$ by $u = \sqrt{2}$, we have $p(x) = x^2 - 2$. Hence any element of $\mathbf{Q}(u)$ can be written as $a + b\sqrt{2}$ with rational a, b, and

$$(a + b\sqrt{2})(c + d\sqrt{2}) = a^2 + (ad + bc)\sqrt{2} + bd(\sqrt{2})^2$$
$$= (a^2 + 2bd) + (ad + bc)\sqrt{2}$$

Formula (4) reveals the quotient-ring $F[x]/(p)$ as an n-dimensional *vector space* over F; it is the quotient space of the infinite-dimensional vector space $F[x]$ by the subspace of multiples of $p(x)$. Note also that multiplication is *bilinear* (linear in each factor). Hence the algebraic extension $F(x)/(p)$ can also be considered as a commutative *linear algebra* over F, in the sense of §13.1.

Exercises

1. Find five different polynomial equations for $\sqrt{3}$ and show explicitly that they are all multiples of the monic irreducible equation for $\sqrt{3}$ (over the field $\mathbf{Q}$).
2. In the simple $\mathbf{Q}(u)$ generated by a root u of the irreducible equation $u^3 - 6u^2 + 9u + 3 = 0$, express each of the following elements in terms of the elements $1, u, u^2$, as in (4): u^4, u^5, $3u^5 - u^4 + 2$, $1/(u + 1)$, $1/(u^2 - 6u + 8)$.
3. In the simple extension $\mathbf{Q}(u)$ generated by a root u of $x^5 + 2x + 2 = 0$, express each of the following elements in the form (4): $(u^3 + 2)(u^3 + 3u)$, $u^4(u^4 + 3u^2 + 7u + 5)$, $1/u$, $(u + 2)/(u^2 + 3)$.
4. Represent the complex number field as a quotient-ring from the domain $\mathbf{R}[x]$ of all polynomials with real coefficients.
5. Represent the field $\mathbf{Q}(\sqrt{2})$ as a quotient-ring from the domain $\mathbf{Q}[x]$ of polynomials with rational coefficients.
6. Prove directly from the relevant definitions: If u is algebraic over F, then the monic polynomial of least degree with root u is irreducible over F.
7. Prove from the relevant definitions: If u is any element of a field K, and F any subfield of K, then the set of all polynomials $g(x)$, with coefficients in F, of which u is a root is an *ideal* of $F[x]$.

14.3. Adjunction of Roots

So far we have assumed as given an extension K of a field F, and have characterized the subfield of K generated by F and a given $u \in K$ in terms of the minimal (i.e., monic irreducible) polynomial p over F such that $p(u) = 0$. Alternatively, we can start just with F and an irreducible polynomial p and *construct* a larger field containing a root of $p(x) = 0$.

This "constructive" approach generalizes the procedure used in Chapter 5 to construct the complex field C from the real field R by adjoining an "imaginary" root of the equation $x^2 + 1 = 0$. The characterizations of Theorems 3 and 4 show how to achieve the same result in general.

Theorem 5. *If F is a field and p a polynomial irreducible over F, there exists a field $K \cong F[x]/(p)$ which is a simple algebraic extension of F generated by a root u of $p(x)$.*

Proof. Since $p(x)$ is irreducible, the principal ideal (p) is maximal in $F[x]$. Hence the quotient-ring $F[x]/(p)$ is a field, by §13.3, Theorem 6. It contains F and the residue class $x + (p)$ containing x, which satisfies $p(x) = 0$ in $F[x]/(p)$.

This simple extension is unique, up to isomorphism:

Theorem 6. *If the fields $F(u)$ and $F(v)$ are simple algebraic extensions of the same field F, generated respectively by roots u and v of the same polynomial p irreducible over F, then $F(u)$ and $F(v)$ are isomorphic. Specifically, there is exactly one isomorphism of $F(u)$ to $F(v)$ in which u corresponds to v and each element of F to itself.*

Proof. Take the composite $\phi_u^{-1}\phi_v$ of the isomorphisms

$$F(u) \xleftarrow{\phi_u} F[x]/(p) \xrightarrow{\phi_v} F(v)$$

provided in Theorem 3.

Theorem 5 may be used to construct various finite fields. For example, start with the field Z_3 of integers modulo 3. The polynomial $x^2 - x - 1$ has none of the three elements 0, 1, or 2 as a zero; hence it is irreducible in $Z_3[x]$. Therefore the quotient-ring $Z_3[x]/(x^2 - x - 1)$ is a field K generated by its subfield Z_3 and the coset, call it u, of x. Moreover, since $[u:F] = 2$, every element of this field K can be written uniquely as $a + bu$, with $a, b \in Z_3$, so K has exactly nine elements.

This field can also be constructed directly without using the concept of a quotient-ring. It consists of just nine elements of the form $a + bu$. The sum of two of them is given by the rule

$$(a + bu) + (c + du) = (a + c) + (b + d)u.$$

To compute the product of two elements of this type, we "multiply out" in the natural fashion and then simplify by the proposed equation

$u^2 = u + 1$. The result is

$$(a + bu)(c + du) = ac + (ad + bc)u + bdu^2$$
$$= (ac + bd) + (ad + bc + bd)u.$$

One can verify in detail that the nine elements $a + bu$ ($a, b \in Z_3$) under these two operations satisfy all the postulates for a field. In particular, the inverses of the nonzero elements are given by

1	2	u	$2u$	$1 + u$	$1 + 2u$	$2 + u$	$2 + 2u$
1	2	$2 + u$	$1 + 2u$	$2 + 2u$	$2u$	u	$1 + u$

By its construction, this field is clearly the field $Z_3(u)$ generated by u from the field Z_3 of residue classes. It is one of the simplest examples of a finite field (see §15.3).

The preceding adjunction process may be applied to any base field F whatever. If F is the field $\mathbf{R}$ of all real numbers, and $p(x)$ the polynomial $x^2 + 1$ irreducible over $\mathbf{R}$, then the construction yields a field $\mathbf{R}(u)$ generated by a quantity u with $u^2 = -1$. This quantity u behaves like $i = \sqrt{-1}$, and the field $\mathbf{R}(u)$ is actually isomorphic to the field $\mathbf{C}$ of complex numbers; we thus have a slight variant of the construction used in Chap. 5 to obtain the complex numbers from the real numbers.

If F is the field Z_p of integers modulo p, and if $p(x)$ is some irreducible polynomial over F, the construction above will yield a field consisting of elements $a_0 + a_1u + \cdots + a_{n-1}u^{n-1}$. There are only a finite number p of choices for each coefficient a_i; hence the field constructed is a finite field of p^n elements, where n is the degree of the polynomial p.

One can construct *algebraic function fields* in the same way. Thus, let $F = \mathbf{C}(z)$ be the field of all rational complex functions; let it be desired to adjoin to F a function $t(z)$ such that $t^2 = (z^2 - 1)(z^2 - 4)$. We can consider the polynomial $p(t) = f(z, t) = t^2 - (z^2 - 1)(z^2 - 4)$ as an irreducible quadratic polynomial in t with coefficients in $\mathbf{C}(z)$. The quotient-ring $K = F[t]/p(t)$ is then a field containing all rational functions *and* the algebraic function t. One can study $t(z)$ as an element of K, without having to construct a Riemann surface for it (it is two-valued). The field K is called an *elliptic function field* because it is generated by the integrand of an elliptic integral,

$$\int \sqrt{(z^2 - 1)(z^2 - 4)} \, dz.$$

If Theorem 6 is applied to an ordinary polynomial such as $x^3 - 5$, irreducible over the field $\mathbf{Q}$ of rationals, it can refer equally to the extension of $\mathbf{Q}$ by the positive $\sqrt[3]{5}$ or to $\mathbf{Q}(\omega\sqrt[3]{5})$ where $\omega = (-1 + \sqrt{3}i)/2$ is a complex cube root of unity. It shows that these two fields $\mathbf{Q}(\sqrt[3]{5})$ and $\mathbf{Q}(\omega\sqrt[3]{5})$ are algebraically indistinguishable because they are isomorphic.

This isomorphism means, roughly speaking, that any two roots of an irreducible polynomial $p(x)$ have the same behavior, and that all the algebraic properties of a root u may be derived from the irreducible equation which it satisfies. There are many examples of such an isomorphism. For instance, the field $\mathbf{C} = \mathbf{R}(i)$ of complex numbers is generated over the field $\mathbf{R}$ of real numbers by either of the two roots $\pm i$ of the equation $x^2 + 1 = 0$. Hence there is by Theorem 6 an automorphism of $\mathbf{C}$ carrying i into $-i$. This automorphism is just the correspondence $a + bi \leftrightarrow a - bi$ between a number and its ordinary complex conjugate.

Exercises

1. Exhibit an automorphism not the identity of each of the following fields: $\mathbf{Q}(\sqrt{2})$, $\mathbf{Q}(\sqrt{-3})$, $\mathbf{Q}(i)$.

2. Exhibit a nonreal field of complex numbers isomorphic to each of the real fields $\mathbf{Q}(\sqrt[3]{5})$, $\mathbf{Q}(\sqrt[4]{2})$.

3. Prove that $x^3 + x - 1$ is irreducible over the field $\mathbf{Z}_5$ of integers modulo 5. If a root of this polynomial is adjoined to $\mathbf{Z}_5$, how many elements has the resulting field?

4. (a) Find polynomials of degrees 2 and 3 irreducible over the field of integers modulo 2.
   (b) Construct addition and multiplication tables for a field with four elements.

5. (a) Show that the field of nine elements constructed in the text has characteristic 3.
   (b) Exhibit explicitly the isomorphism $a \leftrightarrow a^3$ for this field.

6. (a) Find all the irreducible quadratic polynomials over the field $\mathbf{Z}_3$.
   (b) Prove that any two fields with nine elements are isomorphic. (*Hint:* First show that every element in such a finite field is quadratic over $\mathbf{Z}_3$.)

7. Prove that the polynomial $t^2 - (x^2 - 1)(x^2 - 4)$ is irreducible in t over the field $\mathbf{C}(x)$. (*Hint:* Use the results of §3.9.)

8. Prove that the elliptic function field $\mathbf{C}(x, y)$ of the text can be generated over $\mathbf{C}(x)$ by a root z of the equation $t^2 = (x^2 - 4)/(x^2 - 1)$.

9. If $g(t)$ is a reducible polynomial, which elements in the quotient-ring $F[t]/(g(t))$ actually have inverses?

10. Use Theorem 6 of §13.3 to give another proof that $F[t]/(p(t))$ is a field.

14.4. Degrees and Finite Extensions

In a simple extension $F(u)$ generated by an element u of degree n, every element w has by formula (4) a unique representation as

$$(5) \qquad w = a_0 + a_1 u + \cdots + a_{n-1} u^{n-1},$$

with coefficients in F. This unique representation closely resembles the representation of a vector in terms of the vectors of a "basis" $1, u, \cdots, u^{n-1}$. This suggests an application of vector space concepts. Indeed, any extension K of a field F may be considered as a *vector space* over F: simply ignore the multiplication of elements of K, and use as operations of the vector space the addition of two elements of K and the multiplication (a "scalar" multiplication) of an element of K by an element of F. All the vector space postulates are satisfied by this addition and scalar multiplication. If this vector space K has a finite dimension, then the field K is called a *finite extension* of F, and the dimension n of the vector space is known as the *degree* $n = [K:F]$ of the extension.

For example, the complex field $\mathbf{C} = \mathbf{R}(i)$ is a two-dimensional vector space over the real subfield $\mathbf{R}$ (as in §5.2); the field $\mathbf{Q}(\sqrt[3]{5})$ generated by the rational numbers and a cube root of 5 is a three-dimensional vector space over the rational subfield $\mathbf{Q}$, and so on. In general, Theorem 4 on simple algebraic extensions may be restated in terms of dimensions as follows.

Theorem 7. *The degree of an algebraic element u over a field F is equal to the dimension of the extension $F(u)$, regarded as a vector space over F. This vector space has a basis* $1, u, \cdots, u^{n-1}$.

In §14.5 we shall show how the vector space approach may be used to analyze extensions of a field F obtained by adjoining several different algebraic elements. But before discussing such "multiple" extensions we shall first see how the vector space approach enables one to compare the irreducible equations satisfied by different elements in the same simple algebraic extension $F(u)$ over F.

A fundamential fact about vector spaces is the invariance of the dimension (any two bases of a space have the same number of elements). This fact may be applied to the special case of finite extensions of fields, as follows,

Corollary. *If two algebraic elements u and v over a field F generate the same extension $F(u) = F(v)$, then u and v have the same degree over F.*

A simple algebraic extension is finite, and, conversely, every finite extension consists of algebraic elements.

Theorem 8. *Every element w of a finite extension K of F is algebraic over F and satisfies an equation irreducible over F of degree at most n, where n = $[K:F]$ is the degree of the given extension.*

Proof. The $n + 1$ powers $1, w, w^2, \cdots, w^n$ of the given element w are elements of the n-dimensional vector space K, hence must be linearly dependent over F (§7.4, Theorem 5, Corollary 2). There must, therefore, be a linear relation $b_0 + b_1 w + \cdots + b_n w^n = 0$ with not all coefficients zero. Interpreted as a polynomial, this relation implies that w is algebraic over F.

Corollary. *Every element of a simple algebraic extension $F(u)$ is algebraic over F.*

This important conclusion assures us that a transcendental element would never appear in a simple algebraic extension.

In working with a particular simple algebraic extension $F(u)$, the irreducible polynomial $p(x)$ for u must be used systematically, for by Theorem 2 an element $g(u)$ in the extension is zero if and only if the polynomial $g(x)$ is divisible by $p(x)$. Suppose, for instance, that $\mathbf{Q}(u)$ is an extension of degree 3 over the field $\mathbf{Q}$ of rationals, generated by a root u of $x^3 - 2x + 2$. This polynomial is irreducible by the Eisenstein irreducibility criterion (§3.10). The element $w = u^2 - u$ in this extension $\mathbf{Q}(u)$ must satisfy some polynomial equation of degree at most 3. To find this equation, express the powers $w^2 = u^4 - 2u^3 + u^2$ and $w^3 = u^6 - 3u^5 + 3u^4 - u^3$ linearly in terms of 1, u, and u^2, as in Theorem 4. This is done by applying repeatedly the given equation $u^3 = 2u - 2$. This gives

$$w = u^2 - u, \qquad w^2 = 3u^2 - 6u + 4, \qquad w^3 = 16u^2 - 28u + 18.$$

To obtain the linear relation which must hold between 1, w, w^2, and w^3, one may solve the equations for w and w^2 linearly to get u and u^2, as

(6) $\qquad u = -w^2/3 + w + 4/3, \qquad u^2 = -w^2/3 + 2w + 4/3.$

These, substituted in the expression for w^3, give the desired equation

$$w^3 - 4w^2 - 4w - 2 = 0.$$

This equation is irreducible over **Q**, by the Eisenstein theorem. Alternatively, one may argue by equation (6) that u is in $\mathbf{Q}(w)$, so that $\mathbf{Q}(u) = \mathbf{Q}(w)$ and u and w generate the same extension, and by the Corollary to Theorem 7 have the same degree 3 over **Q**. This means that any equation of degree 3 for w must be irreducible.

Exercises

1. Each of the following numbers is in a simple algebraic extension of **Q**, hence is algebraic over **Q**. Find in each case the monic irreducible equation satisfied by the number. (a) $2 + \sqrt{3}$, (b) $\sqrt[4]{5} + \sqrt{5}$, (c) $\sqrt[3]{2} + \sqrt[3]{4}$, (d) $u^2 - 1$, where u satisfies $u^3 = 2u + 2$, (e) $u^2 + u$, where u satisfies $u^3 = -3u^2 + 3$.

2. Prove that every finite extension of the field **R** of real numbers either is **R** itself or is isomorphic to the field **C** of complex numbers.

3. Prove that the field of all complex numbers has no proper finite extensions.

4. (a) If K is an extension of degree 2 of the field **Q** of rationals, prove that $K = \mathbf{Q}(\sqrt{d})$, where d is an integer which is not a square and which has no factors which are squares of integers.

   (b) How much of this result remains true if **Q** is replaced by a field F of characteristic ∞? by a field F of any characteristic?

5. Is the field $F(x)$ of rational forms in the indeterminate x a finite extension of F? Why?

6. Prove that the number of elements in a finite field of characteristic p is a power of p.

7. (a) Prove that there are exactly $(p^2 - p)/2$ monic irreducible quadratic polynomials over the field $\mathbf{Z}_p$ of integers modulo p (exception: $p = 5$).

   (b) Prove that for each p there exists a field of characteristic p with p^2 elements.

★8. Prove that, unless $p \equiv 2 \pmod 3$, there are exactly $(p^2 - p)/3$ monic irreducible cubic polynomials over the field $\mathbf{Z}_p$ of integers modulo p.

★9. Let F be any field contained in an integral domain D. Prove:

   (a) D is a vector space over F.

   (b) If, as a vector space, D has a finite dimension over F, then D is a field.

14.5. Iterated Algebraic Extensions

Finite extensions of a field F may be built up by repeated simple extensions. If F has characteristic ∞, one may prove that any such iterated extension can be obtained as a simple extension; that is, it is generated over F by a suitably chosen single element. We shall omit this proof and discuss the properties of iterated extensions directly. In general, if K is any extension of F containing elements $c_1, c_2, \cdots, c_r$, the symbol $F(c_1, c_2, \cdots, c_r)$ denotes the subfield of K generated by $c_1, \cdots, c_r$ and the elements of F (the subfield consisting of all elements rationally expressible in terms of $c_1, \cdots, c_r$ over F). Alternatively, such a multiple extension may be obtained by iterated simple extensions; thus, $F(c_1, c_2)$ is the simple extension $L(c_2)$ of the simple extension $L = F(c_1)$.

Iterated algebraic extensions may arise in the solution of equations, where it is often useful to introduce appropriate auxiliary equations. For example, the equation $x^4 - 2x^2 + 9 = 0$ may be written as

$$x^4 - 2x^2 + 9 = (x^4 - 6x^2 + 9) + 4x^2 = (x^2 - 3)^2 + 4x^2 = 0.$$

The equation, therefore, is $[(x^2 - 3)/2x]^2 = -1$. This formula indicates that any field which contains a root u of the given equation also contains a root $i = (u^2 - 3)/2u$ of the equation $y^2 = -1$. If we adjoin the auxiliary quantity i to the field Q of rationals, the original equation becomes reducible over $Q(i)$, for

$$x^4 - 2x^2 + 9 = (x^2 - 3 + 2xi)(x^2 - 3 - 2xi).$$

By the usual formula, the factor $x^2 - 3 - 2ix$ has a root $u = i + \sqrt{2}$. The original equation thus has a root in the field $K = Q(i, \sqrt{2})$. This field K could have been obtained by adjoining to Q first $\sqrt{2}$, then i. The intermediate field $Q(\sqrt{2})$ consists of real numbers, hence cannot contain i. The quadratic equation $y^2 + 1 = 0$ for i must therefore remain irreducible over the real field $Q(\sqrt{2})$, so that the extension $Q(\sqrt{2}, i)$ has over $Q(\sqrt{2})$ a degree 2 and a basis of two elements 1 and i. The field $Q(\sqrt{2})$ in turn has a basis $1, \sqrt{2}$ over Q. Therefore any element w in the whole field $Q(\sqrt{2}, i)$ can be expressed as

$$(7) \qquad w = (a + b\sqrt{2}) + (c + d\sqrt{2})i = a + b\sqrt{2} + ci + d\sqrt{2}i,$$

with rational coefficients a, b, c, and d. The four elements $1, \sqrt{2}, i, \sqrt{2}i$ thus form a basis for the whole extension $K = Q(\sqrt{2}, i)$ over Q. This method of compounding bases can be stated in general, as follows:

Theorem 9. *If the elements $u_1, \cdots, u_n$ form a basis for a finite extension K of F, while $w_1, \cdots, w_m$ constitute a basis for an extension L of K, then the mn products $u_i w_j$ for $i = 1, \cdots, n$ and $j = 1, \cdots, m$ form a basis for L over F.*

Proof. Any element y in L can be represented as a linear combination $y = \sum_j r_j w_j$ of the given basis, with coefficients r_j in K. Each coefficient r_j is in turn some combination $r_j = \sum_i a_{ij} u_i$ of the basis elements of K, with each a_{ij} in F. On substitution of these values,

$$y = \sum_j \sum_i a_{ij} u_i w_j$$

appears as a linear combination of the suggested elements $u_i w_j$, with coefficients in F. The same type of successive argument proves that these mn elements are linearly independent over F, hence do constitute a basis for K. Q.E.D.

Many consequences flow from Theorem 9. In the first place, one may state the result without reference to the particular bases used, as follows:

Corollary 1. *If K is a finite extension of F and L a finite extension of K, then L is a finite extension of F, and its degree is*

(8) $$[L:F] = [L:K][K:F] \qquad (L \supset K \supset F).$$

Corollary 2. *If K is is a a finite extension of degree $n = [K:F]$ over F, every element u of K has over F a degree which is a divisor of n.*

Proof. The element u generates a simple extension $F(u)$; hence by (8), $n = [K:F(u)][F(u):F]$, where the second factor is the degree of u under consideration.

Corollary 3. *An element u of a finite extension $K \supset F$ generates the whole extension if and only if $[K:F] = [u:F]$.*

Proof. If u satisfies over F an irreducible equation of degree $[K:F]$, then u generates a subfield $F(u)$ of degree n over F. By (8) this subfield must include all of K.

Corollary 4. *If $K = F(y_1, y_2, \cdots, y_r)$ is a field generated by r quantities y_i, where each successive y_i is algebraic over the field $F(y_1, \cdots, y_{i-1})$ generated by the preceding $i-1$ quantities, then K is a finite extension of F, and every element in K is algebraic over F.*

Proof. Every degree $[F(y_1, \cdots, y_{i-1}, y_i):F(y_1, \cdots, y_{i-1})]$ is finite; hence by Corollary 1 the whole degree $[K:F]$ is finite. By Theorem 8, every element in K is then algebraic over F.

Corollary 5. *If $p(x)$ is an irreducible cubic polynomial over a field F, and if K is an extension of F of degree 2^m, then $p(x)$ is irreducible over K.*

This corollary means in particular that an irreducible cubic equation could never be solved by successive square roots, for the adjunction of a square root to a field F either will give no extension at all or will give an extension of degree 2, so that the extension $K = F(\sqrt{a}, \sqrt{b}, \sqrt{c}, \cdots)$ obtained by any number of square roots will have as degree some power 2^m of 2. By Corollary 5, this extension will never contain a root of the given irreducible cubic.

For a proof, suppose $p(x)$ reducible over the field K of degree 2^m. Then the cubic $p(x)$ must have at least one linear factor $x - u$, so that K contains a root u of $p(x)$. But such an element u of degree 3 over F cannot be contained in a field K of degree 2^m over F, by Corollary 2. This proves $p(x)$ irreducible.

This corollary is the algebraic basis of the theorem that it is impossible to solve the classical problem of duplicating a general cube or trisecting a general angle by ruler and compass alone. Any such construction problem may be reduced to analytic terms. The data of the problem consist of a number of points and lines. Relative to some set of axes, the coordinates of these points (and the ratios of the coefficients in the equations for these lines) are a set of real numbers which generates a certain field F of real numbers. Each step in a ruler and compass construction provides certain new points and lines. It can be shown† that the corresponding new field of numbers is either F itself or a quadratic extension of F. Hence repeated constructions yield a set of points and lines corresponding to a field K of degree 2^m over F.

Consider now the duplication of the cube. The data consist of a pair of coordinate axes, a unit segment along one of these axes, and a cube with this segment as side. The problem is to construct another cube of double the volume. The side of this new cube will satisfy the equation $x^3 - 2 = 0$. By Eisenstein's theorem this equation is irreducible over the field $\mathbf{Q}$ of rationals (the field associated with the data). Over any field K corresponding to a ruler and compass construction, the polynomial $x^3 - 2$ will still be irreducible, by Corollary 5. Hence by these methods it is impossible to construct (say along the x-axis) a segment which is the side of the duplicated cube.

The trisection problem is treated in similar fashion; the essential device consists in writing the trigonometric equation for the cosine of one third of an angle in terms of the cosine of the whole angle. For most angles, this will again give an irreducible cubic equation.

---

† This depends essentially on the fact that the equation of a circle (compass) is quadratic and the equation of a straight line (ruler) is linear.

Exercises

1. In Theorem 9, prove in detail that the mn elements $u_i w_j$ are independent over F.

2. Prove that the equation $x^4 - 2x^2 + 9$ treated in the text is irreducible over **Q**. (*Hint:* Use the degree of $\mathbf{Q}(\sqrt{2}, i)$.)

3. If $p(x)$ is a polynomial of degree q and is irreducible over F and if K is a finite extension of F of degree relatively prime to q, prove that $p(x)$ is irreducible over K.

4. Determine the degree of each of the following multiple extensions of the field **Q** of rational numbers. Give reasons.
   (a) $\mathbf{Q}(\sqrt{3}, i)$,   (b) $\mathbf{Q}(\sqrt[3]{5}, \sqrt{-2})$,   (c) $\mathbf{Q}(\sqrt{18}, \sqrt[4]{2})$,
   (d) $\mathbf{Q}(\sqrt{8}, 3 + \sqrt{50})$,   (e) $\mathbf{Q}(\sqrt[3]{2}, u)$,   where $u^4 + 6u + 2 = 0$,
   (f) $\mathbf{Q}(\sqrt{3}, \sqrt{-5}, \sqrt{7})$,   (g) $\mathbf{Q}(\sqrt{3}, \sqrt{2})$.

5. Give a basis over **Q** for each field of Ex. 4.

6. Determine whether the polynomial given is irreducible over the field indicated. Give reasons.
   (a) $x^2 + 3$, over $\mathbf{Q}(\sqrt{7})$;   (b) $x^2 + 1$, over $\mathbf{Q}(\sqrt{-2})$;
   (c) $x^3 + 8x - 2$, over $\mathbf{Q}(\sqrt{2})$;
   (d) $x^5 + 3x^3 - 9x - 6$, over $\mathbf{Q}(\sqrt{7}, \sqrt{5}, 1 + i)$.

7. Determine in each of the following cases whether the number u given generates the given extension of the field **Q** of rational numbers. In each case, prove your answer correct.
   (a) $u = \sqrt[3]{7}$, in $\mathbf{Q}(\sqrt[3]{7})$;   (b) $u = \sqrt{2} + \sqrt{5}$, in $\mathbf{Q}(\sqrt{2}, \sqrt{5})$;
   (c) $u = 2 + \sqrt[3]{9}$, in $\mathbf{Q}(\sqrt[3]{3})$;   (d) $u = \sqrt{2} - 1/(1 + \sqrt{2})$, in $\mathbf{Q}(\sqrt{2})$;
   (e) $u = v^2 + v + 1$, in $\mathbf{Q}(v)$,   where $v^3 + 5v - 5 = 0$.

8. Is $c = \pi^6 + 5\pi^3 + 2\pi - 14$ transcendental or algebraic over the field **Q** of rational numbers? Why?

9. If K is an extension of F of prime degree, prove that any element in K but not in F generates all of K over F.

10. (a) Find the cubic equation which gives $\cos\theta$ in terms of $\cos 3\theta$.
    (b) Show that this equation is irreducible over **Q** when $3\theta = 60°$ (this means that an angle of $60°$ cannot be trisected with ruler and compass).

14.6. Algebraic Numbers

An *algebraic number* u is a complex number which satisfies a polynomial equation with rational coefficients not all zero.

$$(9) \quad a_0 + a_1 u + a_2 u^2 + \cdots + a_n u^n = 0 \quad (a_i \text{ in } \mathbf{Q}, \text{ not all } a_i = 0).$$

In other words, an algebraic number is any complex number which is algebraic over the field **Q** of rationals. In discussing extensions of fields, we have repeatedly used examples of algebraic numbers, such as $i\sqrt{-2}$, $\sqrt[5]{3}$, or ω.

Theorem 10. *The set of all algebraic numbers is countable.*

The verification of this statement requires that we describe a method of enumerating or of listing all algebraic numbers. First, we list all the equations which they satisfy. Observe that an equation (9) for an algebraic number can be multiplied through by a common denominator for its rational coefficients; there results an equation with integral coefficients not all zero, in which the first coefficient may be assumed to be positive. We know that the possible integral coefficients of these polynomials can be enumerated, for example, as $0, +1, -1, +2, -2, +3, -3, \cdots$. The linear polynomials with integral coefficients can be displayed in an array, such as:

$$
\begin{array}{llllll}
x, & x+1, & x-1, & x+2, & x-2, & x+3, \ldots \\
-x, & -x+1, & -x-1, & -x+2, & -x-2, & -x+3, \ldots \\
2x, & 2x+1, & 2x-1, & 2x+2, & 2x-2, & 2x+3, \ldots \\
-2x, & -2x+1, & -2x-1, & -2x+2, & -2x-2, & \ldots
\end{array}
$$

One can then make a single list including them all by taking in succession as indicated the diagonals of the above array. The result is the list

$$x, -x, x+1, x-1, -x+1, 2x, -2x, 2x+1, -x-1, \cdots.$$

We then find a rectangular array of quadratic polynomials by simply adjoining the various second-degree terms mx^2 to each element in this list. From this array we again obtain a list of all quadratic polynomials, and so on for higher degrees. When this is done for every degree, there results an array of lists, in which the nth row is the list of all polynomials of degree n. Take again the diagonal development of this list, and we get a list of all polynomials. In this list replace every polynomial by its roots and drop out any duplications. The result is a list of all the roots of polynomials with integral coefficients; that is, it is the required enumeration of all algebraic numbers.

A consequence is that the real algebraic numbers are countable. But Cantor's diagonal process proves (§12.3, Theorem 5) that the set of *all* real numbers is *not* countable. Hence this set must be larger than the set of algebraic real numbers. This argument gives an indirect proof of the existence of a transcendental real number. The result we state as follows:

Corollary. *Not every real number is algebraic.*

Cantor's argument for this result was at first rejected by many mathematicians, since it did not exhibit any specific transcendental real numbers. His argument is now generally accepted, but it is possible to give more explicit proofs of this corollary (see Exs. 10–13 below).

Theorem 11. *The set of all algebraic numbers is a field.*

Proof.   We need only demonstrate that the sum, product, difference, and quotient of any two algebraic numbers u and $v \neq 0$ are again algebraic numbers. But all these combinations are contained in the subfield $Q(u, v)$ of the field of complex numbers generated by u and v. Since u is algebraic over Q, $Q(u)$ is a finite extension of Q; since v is algebraic over $Q(u)$, $Q(u, v)$ is finite over $Q(u)$. Hence by Theorem 9, $Q(u, v)$ is a finite extension of Q, so each of its elements is an algebraic number (Theorem 8).   Q.E.D.

A field F is called *algebraically complete*† if every polynomial equation with coefficients in F has a root in F. Over such a field F every polynomial $f(x)$ has a root c, hence has a linear factor $x - c$. Consequently, the only irreducible polynomials over F are linear, and every polynomial over an algebraically complete field F can be written as a product of linear factors (as in formula (11), §5.3). Furthermore, there can be no simple algebraic extension of F except F itself. We conclude that a field F is algebraically complete if and only if F has no proper simple algebraic extensions. The fundamental theorem of algebra (§5.3, Theorem 5) asserts that the field of all complex numbers is algebraically complete.

Theorem 12. *The field A of all algebraic numbers is algebraically complete.*

Proof.   Take a polynomial equation $x^n + u_{n-1}x^{n-1} + \cdots + u_0 = 0$ whose coefficients are algebraic numbers u_i in A. These coefficients generate an extension $K = Q(u_0, u_1, \cdots, u_{n-1})$ which is a finite extension of the field Q of rationals, by Corollary 4 to Theorem 9. Any complex root r of the given equation is algebraic over the field K, so that $K(r)$ is a finite extension of K and hence of Q. The element r of this extension is then algebraic over Q, by Theorem 8. This means that the root r is an algebraic number, in the field A, so A is algebraically complete.   Q E D

We now have the field Q embedded in the algebraically complete field A of all algebraic numbers, and the field R of real numbers embedded in

---

† Instead of "algebraically complete," some sources use "algebraically closed." The term "complete" seems preferable, in view of the topological analogy.

the algebraically complete field C of complex numbers. These results are special cases of a general theorem, which states that any field F whatever has an extension A which is algebraically complete and in which every element is algebraic over F (cf. §15.1, Appendix).

The theory of algebraic numbers has been elaborately developed. It concerns chiefly fields K of algebraic numbers which are finite extensions of the field Q. Such a field is known as an *algebraic number field*. We consider next the arithmetic properties of such a field.

Exercises

1. Illustrate Theorem 11 by finding an equation with rational coefficients for each of the following algebraic numbers:
   (a) $\sqrt{2} + \sqrt{-3}$,   (b) $\sqrt{-1} + \sqrt[3]{5}$,   (c) $(\sqrt{7})(\sqrt[3]{2})$,
   (d) $\sqrt{7}/(1 + \sqrt{2})$,   (e) $u\sqrt{-2}$, where $u^3 + 7u - 14 = 0$.

2. (a) If u and v are algebraic numbers of degrees m and n, respectively (over Q), prove that the degree of $u + v$ never exceeds mn.
   (b) What about the degree of u/v?
   (c) If t is transcendental and u algebraic, prove that $t + u$ and tu are transcendental, provided, in the latter case, that $u \neq 0$.

3. Illustrate Theorem 12 by finding an equation with rational coefficients for a root of each of the following equations:
   (a) $x^2 + 3x + \sqrt{2} = 0$,   (b) $x^2 + \sqrt{3}x - \sqrt{-1} = 0$,
   (c) $x^3 - \sqrt{3}x + 1 + \sqrt[3]{2} = 0$,
   (d) $x^2 + u + 2 = 0$,   where u is a root of $u^3 + 5u^2 - 10u + 5 = 0$.

4. Give the first sixteen terms in the list of all quadratic polynomials, as in the proof of Theorem 10.

5. Prove that the set of all algebraic numbers of a fixed degree is countable, without using Theorem 10.

6. Prove that any finite extension of a countable field is countable.

7. Show that the set A of all elements of a field F which are algebraic over any countable subfield S of F is countable.

8. (a) Show that there exists a real number transcendental over $Q(\pi)$.
   (b) Show that there exist countably many algebraically independent real numbers, using Ex. 7 and the definition of §3.4.

9. Show that the proof given for Theorem 10 implicitly uses the following formulas of transfinite arithmetic:
   (a) There are $\mathbf{d}^{n+1} = \mathbf{d}$ polynomials of degree n.
   (b) There are $\mathbf{d} + \mathbf{d} + \cdots + \mathbf{d} + \cdots$ (to $\mathbf{d}$ terms) $= \mathbf{d}^2$ polynomials of all degrees.

★10. (a) If u is any fixed real number, show by factorization of $x^i - u^i$ that there is a constant $N(j)$, such that $|x^j - u^j| \leq N|x - u|$ whenever $|x - u| < 1$.
   (b) If $f(x)$ is any polynomial with real coefficients, u any real number, show

that there is a constant M depending on f and u, such that $|f(x) - f(u)| \leq M|x - u|$ whenever $|x - u| < 1$.

★11. Let the real algebraic number u satisfy the polynomial equation $f(x) = 0$ of degree r with *integral* coefficients. If m and n are integers such that $|m/n - u| < 1/Mn^r$, where M is the constant of Ex. 10, show that $f(m/n) = 0$. (*Hint:* By Ex. 10, $|f(m/n)| < 1/n^r$, while $f(m/n)$ is a rational number of denominator n^r.)

★12. If u is a real number for which an infinite sequence of distinct rational fractions m_k/n_k can be found, such that $|u - (m_k/n_k)| < 1/kn_k^k$ for all k, show that u is transcendental. (*Hint:* If the degree of u were r, Ex. 11 would give $f(m_k/n_k) = 0$ for all sufficiently large k.)

★13. Numbers satisfying the hypothesis of Ex. 12 are called "Liouville (transcendental) numbers."

 (a) Show $\sum_{k=1}^{\infty} 10^{-k!} = 0.110001 \cdots$ is a Liouville number.

 (b) Exhibit two other Liouville numbers.

14.7. Gaussian Integers

A *Gaussian integer* is a complex number $\alpha = a + bi$ whose components a, b are both integers. Any such Gaussian integer satisfies a monic equation $\alpha^2 - 2a\alpha + (a^2 + b^2) = 0$ with integral coefficients; hence it is an algebraic number. The sum, difference, and product of two such integers is again such an integer, hence the Gaussian integers form an integral domain $\mathbf{Z}[i]$. In this domain questions of divisibility and decomposition into primes (irreducibles) may be considered.

It is convenient to introduce the "norm" of any complex number σ (integral or not). If $\sigma = r + si$. the *norm* $N(\sigma)$ is the product of σ by its conjugate $\sigma^* = r - si$:

(10) $$N(\sigma) = \sigma\sigma^* = (r + si)(r - si) = r^2 + s^2.$$

This norm is always nonnegative and is the square of the absolute value of σ. For any two numbers σ and τ, one has

(11) $$N(\sigma\tau) = N(\sigma)N(\tau).$$

This equation means that the correspondence $\sigma \mapsto N(\sigma)$ preserves products; in other words, it is a homomorphic mapping of the multiplicative group of nonzero numbers σ on a multiplicative group of real numbers. In particular, the norm of a Gaussian integer is a (rational) integer.

Recall now the general concepts involving divisibility (§3.6). A *unit* of $\mathbf{Z}[i]$ is a Gaussian integer $\alpha \neq 0$ with a reciprocal α^{-1} which is also a

Gaussian integer. Then $\alpha\alpha^{-1} = 1$, so that $N(\alpha\alpha^{-1}) = N(\alpha)N(\alpha^{-1}) = 1$, and the norm of a unit α must be $N(\alpha) = 1$. Inspection of (10) shows that the only possible units are ± 1 and $\pm i$. Two integers are *associate* in $\mathbf{Z}[i]$ if each divides the other. Hence the only associates of α in $\mathbf{Z}[i]$ are $\pm\alpha$ and $\pm i\alpha$.

The rational prime number 5 has in $\mathbf{Z}[i]$ four different decompositions

$$(12) \qquad \begin{aligned} 5 &= (1 + 2i)(1 - 2i) = (2i - 1)(-2i - 1) \\ &= (2 + i)(2 - i) = (i - 2)(-i - 2). \end{aligned}$$

These decompositions are not essentially different; for instance, $(2 + i) = i(1 - 2i)$ and $2 - i = -i(1 + 2i)$, and in each of the other cases corresponding factors are associates. Each factor in (12) is *prime* (irreducible). For example, if $2 + i$ had a factorization $2 + i = \alpha\beta$, then $N(2 + i) = 5 = N(\alpha)N(\beta)$, so that $N(\alpha)$ (or $N(\beta)$) would be 1, hence α (or β) would be a unit. The factors (12) give essentially the only way of decomposing 5, for in any decomposition $5 = \gamma\delta$, $N(5) = 25 = N(\gamma)N(\delta)$, so each factor which is not a unit must have norm 5. By trial one finds that the only integers of norm 5 are those used in (12).

On the other hand, the rational prime 3 is prime in $\mathbf{Z}[i]$. Suppose $3 = \alpha\beta$; then $N(\alpha)N(\beta) = 9$ and $N(\alpha)|9$. If $N(\alpha) = 1$, α is a unit, while if $N(\alpha) = N(a + bi) = 3$, then $a^2 + b^2 = 3$, which is impossible for integers a and b. Hence 3 has no proper factor α in the domain of Gaussian integers.

A unique factorization theorem can be proved for the Gaussian integers by developing first a division algorithm, analogous to that used for ordinary integers and for polynomials.

Theorem 13. *For given Gaussian integers α and $\beta \neq 0$ there exist Gaussian integers γ and ρ with*

$$(13) \qquad \alpha = \beta\gamma + \rho, \qquad N(\rho) < N(\beta).$$

Proof. Start with the quotient $\alpha/\beta = r + si$ and select integers r' and s' as close as possible to the rational numbers r and s. Then

$$\alpha/\beta = (r' + s'i) + [(r - r') + (s - s')i] = \gamma + \sigma, \qquad \gamma = r' + s'i,$$

where $|r - r'| \leq 1/2, |s - s'| \leq 1/2$, so that

$$N(\sigma) = (r - r')^2 + (s - s')^2 \leq 1/4 + 1/4 < 1.$$

The equation may now be written as $\alpha = \beta\gamma + \beta\sigma$, where α and $\beta\gamma$, and hence $\beta\sigma$, are integers, and where $N(\beta\sigma) = N(\beta)N(\sigma) < N(\beta)$.   Q.E.D.

Lemma 1. *Two Gaussian integers α_1 and α_2 have a greatest common divisor δ which is a Gaussian integer expressible in the form $\delta = \beta_1\alpha_1 + \beta_2\alpha_2$ where β_1 and β_2 are Gaussian integers.*

Proof. By repeated divisions, one may construct a Euclidean algorithm, much as in the case of rational integers (§1.7). The successive remainders ρ of (13) decrease in norm, hence the algorithm eventually reaches an end. The last remainder not zero is the desired greatest common divisor.   Q.E.D.

A more sophisticated proof starts with the ideal (α_1, α_2) generated by α_1 and α_2 in the ring $\mathbf{Z}[i]$. Among the elements of this ideal choose one, δ, of minimum norm, and write $\alpha_1 = \delta\gamma_1 + \rho_1$, $\alpha_2 = \delta\gamma_2 + \rho_2$, as in (13). The remainders ρ_i lie in the ideal and have norm less than δ, hence must be zero. Therefore $\alpha_1 = \delta\gamma_1$, $\alpha_2 = \delta\gamma_2$, so δ is a common divisor. Since δ is in the ideal, it has the form $\delta = \beta_1\alpha_1 + \beta_2\alpha_2$, hence it is a multiple of every common divisor of α_1 and α_2. Therefore δ is the required g.c.d.

The rest of the treatment of the decomposition of Gaussian integers proceeds exactly as in the case of rational integers (§§1.7–1.8) and of polynomials (§3.5 and §3.8); hence we state only the important stages. A Gaussian integer π is said to be *prime* if it is not 0 or a unit and if its only factors in $\mathbf{Z}[i]$ are units and associates of π. One proves

Lemma 2. *If π is prime, then $\pi | \alpha\beta$ implies that $\pi | \alpha$ or that $\pi | \beta$.*

Theorem 14. *Every Gaussian integer α can be expressed as a product $\alpha = \pi_1 \cdots \pi_n$ of prime Gaussian integers. This representation is essentially unique, in the sense that any other decomposition of α into primes has the same number of factors and can be so rearranged that correspondingly placed factors are associates.*

In order appropriately to generalize these notions, we first investigate the irreducible polynomial equations satisfied by Gaussian integers. If $\alpha = a + bi$ is a Gaussian integer which is not a rational integer, then $b \neq 0$, and α must satisfy an irreducible quadratic equation. This is

$$[x - (a + bi)][x - (a - bi)] = x^2 - 2ax + (a^2 + b^2) = 0;$$

it is a monic irreducible equation with rational integers as coefficients. Conversely, it may be shown that if a number $r + si$ in the field $\mathbf{Q}(i)$

satisfies a monic irreducible equation with integral coefficients, then this number is a Gaussian integer.† This gives

Theorem 15. *A number in the field* $\mathbf{Q}(i)$ *is a Gaussian integer if and only if the monic irreducible equation which it satisfies over* $\mathbf{Q}$ *has integers as coefficients.*

Exercises

1. Find the decomposition into primes of the following Gaussian integers: 5, $3 + i, 6i, 11, 1 - 7i$.
2. Find the g.c.d. of each of the following pairs of Gaussian integers α_1 and α_2 and express it as $\beta_1\alpha_1 + \beta_2\alpha_2$:
   (a) $3 + 6i$ and $12 - 3i$, $\qquad$ (b) $5 + 3i$ and $13 + 18i$.
3. Find all possible factorizations of 13 into Gaussian integers, and show explicitly that any two factorizations differ only by associates.
4. Prove that every ideal of Gaussian integers is principal.
5. (a) Prove Lemma 1, using a Euclidean algorithm.
   (b) Prove Lemma 2.
6. Prove Theorem 14 from Lemma 2.
7. (a) Prove that a rational prime p is prime in $\mathbf{Z}[i]$ if and only if the equation $x^2 + y^2 = p$ has no solution in integers x and y.
   (b) Show that any rational prime of the form $p = 4n + 3$ is prime in $\mathbf{Z}[i]$.
★8. (a) Prove that the quotient-ring $\mathbf{Z}[x]/(p, x^2 + 1)$ is isomorphic to both $\mathbf{Z}[i]/(p)$ and $\mathbf{Z}_p[x]/(x^2 + 1)$.
   (b) Prove that the first is an integral domain if and only if p is prime in $\mathbf{Z}[i]$; while the second is an integral domain if and only if $x^2 \equiv -1 \pmod{p}$ has no solution in $\mathbf{Z}$.
   (c) Assuming that the multiplicative group $\bmod\, p$ is cyclic (§15.3, Theorem 6), show that if $p = 4n + 1$, $x^2 \equiv -1 \pmod{p}$ has a solution in $\mathbf{Z}$.
   (d) Conclude that $p = 4n + 1$ cannot be a prime in $\mathbf{Z}[i]$.

Exs. 9–13 all refer to the domain $\mathbf{Z}[\sqrt{-2}]$ of numbers $a + b\sqrt{-2}$, where a and b are integers.
9. Define a norm as $N(a + b\sqrt{-2}) = a^2 + 2b^2$ and exhibit its properties.
10. Prove a division algorithm in the domain $\mathbf{Z}[\sqrt{-2}]$.
11. Prove the existence of greatest common divisors in $\mathbf{Z}[\sqrt{-2}]$.
12. State and prove the unique decomposition theorem for $\mathbf{Z}[\sqrt{-2}]$.
13. Factor the following numbers in $\mathbf{Z}[\sqrt{-2}]$: $5, 1 + 3\sqrt{-2}, 2 + \sqrt{-2}$.
14. (a) Find a unit different from ± 1 in $\mathbf{Z}[\sqrt{2}]$.
   (b) Show that there is an infinite number of distinct units in $\mathbf{Z}[\sqrt{2}]$. (*Hint:* Use powers of one unit.)

---

† The proof is given in a slightly more general case in the next section (Theorem 16).

14.8. Algebraic Integers

In general, an algebraic number u is said to be an *algebraic integer* if the *monic* irreducible equation satisfied by u over the field of rationals has integers as coefficients; so that

$$(14)\quad p(u) = a_0 + a_1 u + \cdots + a_{n-1} u^{n-1} + u^n = 0, \qquad a_i \text{ integers,}$$

where $p(x)$ is irreducible over $\mathbf{Q}$. The irreducible equation satisfied by a rational number m/n is just the linear equation $x - m/n = 0$. Therefore a rational number is an algebraic integer if and only if it is an integer in the ordinary sense. Such an (ordinary) integer of $\mathbf{Z}$ may be called a *rational integer* to distinguish it from other algebraic integers. An algebraic number $u \neq 0$ is called a *unit* if both u and u^{-1} are algebraic integers.

In testing whether a given algebraic number is an integer, it is not necessary to appeal to an irreducible equation, by virtue of the following result:

Theorem 16. *A number is an algebraic integer if and only if it satisfies over $\mathbf{Q}$ a monic polynomial equation with integral coefficients.*

Proof. Suppose that u is a root of some monic polynomial $f(x)$ with integral coefficients. Over $\mathbf{Q}$, u also satisfies an irreducible polynomial $p(x)$, which may be taken to have integral coefficients. Any common divisor of these coefficients may be removed, so we can assume that the coefficients of $p(x)$ have 1 as g.c.d. This amounts to saying that $p(x)$ is primitive, in the sense of §3.9, in the domain $\mathbf{Z}[x]$ of all polynomials with integral coefficients. The given polynomial $f(x)$ is monic, hence is also primitive. By Theorem 2 we know that the polynomial $f(x)$ with root u must be divisible, in $\mathbf{Q}[x]$, by the irreducible polynomial $p(x)$ for u, so $f(x) = q(x)p(x)$. Since f and p are primitive, Lemma 3, §3.9, asserts that the quotient $q(x)$ also has integral coefficients. The leading coefficient 1 in $f(x)$ is then the product of the leading coefficients in q and p; hence $\pm p(x)$ is monic, which means that u is integral according to the definition (14). Q.E.D.

A number may be an algebraic integer even if it doesn't look the part; for example, $u = (1 + \sqrt{5})/2$ looks like a fraction but satisfies an equation,

$$(x - (1 + \sqrt{5})/2)(x - (1 - \sqrt{5})/2) = x^2 - x - 1 = 0,$$

which is monic and has integral coefficients. This suggests a systematic search for those numbers in quadratic fields which are algebraic integers. Any field K of degree 2 over the field $\mathbf{Q}$ of rationals can be expressed as a simple algebraic extension $K = \mathbf{Q}(\sqrt{d})$. Without loss of generality, one may assume that d is an integer and that it has no factor (except 1) which is the square of an integer. This is the case to be considered:

Theorem 17. *If $d \neq 1$ is an integer with no square factors, then in case $d \equiv 2$ or $d \equiv 3$ (mod 4), the algebraic integers in $\mathbf{Q}(\sqrt{d})$ are the numbers $a + b\sqrt{d}$, with (rational) integers a and b as coefficients. However, if $d \equiv 1$ (mod 4), the integers of $\mathbf{Q}(\sqrt{d})$ are the numbers $a + b(1 + \sqrt{d})/2$, with a and b rational integers.*

Proof. As a preliminary, observe that $a \equiv 1$ (mod 2) means that $a = 1 + 2r$, hence that $a^2 = 1 + 4r + 4r^2 \equiv 1$ (mod 4). In other words,

(15) $\qquad a \equiv 1 \,(\text{mod}\, 2) \qquad \text{implies} \qquad a^2 \equiv 1 \,(\text{mod}\, 4),$

(16) $\qquad a \equiv 0 \,(\text{mod}\, 2) \qquad \text{implies} \qquad a^2 \equiv 0 \,(\text{mod}\, 4),$

so a square is always congruent to 0 or 1, modulo 4.

Any number u in $\mathbf{Q}(\sqrt{d})$ may be expressed as $u = (a + b\sqrt{d})/c$, where the integers a, b, and c have no factor in common. We assume $b \neq 0$ to exclude the trivial case of a rational number. The monic irreducible quadratic equation for u is then

(17) $\quad [x - (a + b\sqrt{d})/c][x - (a - b\sqrt{d})/c]$
$$= x^2 - (2a/c)x + (a^2 - db^2)/c^2 = 0.$$

If u is an algebraic integer, these coefficients $2a/c$ and $(a^2 - db^2)/c^2$ must also be integers. Therefore, $4a^2/c^2$, $(4a^2 - 4db^2)/c^2$, and $4db^2/c^2$ must all be integers, so that $c \mid 2a$ and $c^2 \mid 4db^2$. Since d was assumed to contain no square factors, any prime $p \neq 2$ contained in c must divide both a and b^2, contrary to the arrangement that a, b, c have no factor (except ± 1) in common. For similar reasons $4 \mid c$ is impossible, so the only choices for c are $c = 1$ and $c = 2$.

Consider now the case $d \equiv 2$ or $d \equiv 3$ (mod 4), with $c = 2$. In this case the last coefficient $(a^2 - db^2)/4$ of (17) must be integral, so $a^2 \equiv db^2$ (mod 4). If $b \equiv 1$ (mod 2), then $b^2 \equiv 1$ (mod 4), and $a^2 \equiv db^2 \equiv 2$ or 3 (mod 4), a contradiction to the rules (15) and (16). If $b \equiv 0$ (mod 2), then $a^2 \equiv 0$ (mod 4), and $a \equiv 0$ (mod 2), so that a, b, and c have a common factor 2. In either event we conclude that c is 1, and that all the integers of $\mathbf{Q}(\sqrt{d})$ are of the form $a + b\sqrt{d}$. Conversely, the monic equation (17) for a number of this form does have integral coefficients.

The remaining case $d \equiv 1 \pmod 4$ is given a similar treatment, except that $a \equiv b \equiv 1 \pmod 2$ turns out to be possible.

Corollary. *In any field of degree* 2 *over* **Q** *the set of all algebraic integers is an integral domain.*

Proof. Sums, differences, and products of integers, represented as in Theorem 17, are again integers of this form. Q.E.D.

The next task is that of generalizing this corollary to any algebraic number field.

Exercises

1. Prove that every root of unity is an algebraic integer.
2. (a) Find all integers and all units in $\mathbf{Q}(\omega)$, where ω is a complex cube root of unity.
   (b) Prove that every unit in $\mathbf{Q}(\omega)$ is a root of unity.
3. Complete the proof of the second case of Theorem 17 ($d \equiv 1 \pmod 4$).
4. (a) Prove that any algebraic number can be written as a quotient u/b, where u is an algebraic integer and b a rational integer (i.e., an integer of **Z**).
   (b) Prove that any field K of algebraic numbers is the field of quotients of the domain of all algebraic integers in K.
★5. Find all the integers in $\mathbf{Q}(\sqrt{2}, i)$.

14.9. Sums and Products of Integers

This section is devoted to the proof of the following result:

Theorem 18. *The set of all algebraic integers is an integral domain.*

The following specialization is an immediate consequence:

Corollary. *In any field* K *of algebraic numbers, the algebraic integers form an integral domain.*

An instructive proof of Theorem 18 depends on an analysis of the additive groups generated by algebraic integers. If $v_1, \cdots, v_n$ are any algebraic numbers, we let $G = [v_1, \cdots, v_n]$ denote the subgroup† generated by these numbers in the additive group of all complex numbers. This

† Such an additive group is sometimes called a **Z**-*module* because its elements can be multiplied by "scalars" from **Z**.

group G simply consists of all numbers representable in the form

(18) $u = a_1 v_1 + a_2 v_2 + \cdots + a_n v_n$ (a_i rational integers).

Recall that the natural multiple $av = a \times v$ is simply a "power" of v in the additive cyclic subgroup generated by v.

Lemma 1. *Any subgroup S of the group $G = [v_1, \cdots, v_n]$ can also be generated by n or fewer numbers.*

Proof. For each index k let G_k be the subgroup $[v_k, \cdots, v_n]$ generated by the last $n - k + 1$ generators of G, so that G_k consists of all sums of the form $a_k v_k + \cdots + a_n v_n$. Among the elements of G_k which lie in the given subgroup S, select an element

(19) $w_k = c_k v_k + c_{k+1} v_{k+1} + \cdots + c_n v_n,$

in which the first coefficient c_k has the least positive value possible. (If in every element the coefficient of v_k is zero, set $w_k = 0$.) If $w = b_k v_k + \cdots$ is any other element of S in G_k, its first coefficient b_k may be written $b_k = q_k c_k + r_k$, with a nonnegative remainder $r_k < c_k$. The difference $w - q_k w_k = r_k v_k + \cdots$ then lies in the groups G_k and S and has a nonnegative first coefficient r_k less than the minimum c_k. Therefore $r_k = 0$, and any element w of S in G_k gives an element $w' = w - q_k w_k$ in G_{k+1}.

The n selected elements $w_1, \cdots, w_n$ generate the whole group S, for given any any element w in S, one may find q_1 so that $w - q_1 w_1$ depends only on $v_2, \cdots, v_n$, and then some q_2 so that $w - q_1 w_1 - q_2 w_2$ depends only on $v_3, \cdots, v_n$, and so on; at the end $w = \sum q_i w_i$.   Q.E.D.

Lemma 2. *A number u is an algebraic integer if and only if the additive group generated by all the powers $1, u, u^2, u^3, \cdots$ of u can be generated by a finite number of elements.*

Proof. If u is an integer, it satisfies a monic equation (14) of degree n with integral coefficients. This equation expresses u^n as an element in the group $G = [1, u, \cdots, u^{n-1}]$ generated by n smaller powers of u. By iteration, the same equation may be used to express any higher power of u as an element of this group. Therefore u satisfies the criterion of Lemma 2.

Conversely, suppose that the group G generated by $1, u, u^2, \cdots$ can be generated by any n numbers $v_1, \cdots, v_n$ of G. The product of u by any element $\sum a_j u^j$ of G is still an element $\sum a_j u^{j+1}$ of G, so each of the products uv_i must lie in G and must be expressible in terms of the generators as $uv_i = \sum_j a_{ij}v_j$, where the a_{ij} are integers. These expressions give n homogeneous equations in the v's, of the form

$$(a_{11} - u)v_1 + \quad a_{12}v_2 + \cdots + \quad a_{1n}v_n = 0,$$
$$a_{21}v_1 + (a_{22} - u)v_2 + \cdots + \quad a_{2n}v_n = 0,$$
$$\cdots$$
$$a_{n1}v_1 + \quad a_{n2}v_2 + \cdots + (a_{nn} - u)v_n = 0.$$

This system of equations has a set of solutions $v_1, v_2, \cdots, v_n$ not all zero, so the matrix of coefficients must be linearly dependent (§7.7, Theorem 13, Corollary). The matrix of coefficients may be written as $A - uI$, where $A = \|a_{ij}\|$. Since it is singular, its determinant is zero, so

$$(20) \qquad |A - uI| = (-1)^n u^n + b_{n-1}u^{n-1} + \cdots + b_n = 0,$$

where the coefficients b_i are certain polynomials in the integers a_{ij} and are thus themselves integers. This equation (20) means† that u is an algebraic integer, as required in the lemma.

The conclusion of Lemma 2 may be reformulated thus:

Corollary. *If all the positive powers of an algebraic number u lie in an additive group generated by a finite set of numbers $y_1, \cdots, y_n$, then u is an algebraic integer.*

Proof. The group S generated by $1, u, u^2, \cdots$ is a subgroup of the group generated by $1, y_1, \cdots, y_n$. Hence by Lemma 1, this subgroup S can be generated by a finite number of its members, and therefore, by Lemma 2, the number u is an algebraic integer. Q.E.D.

Return now to the proof of Theorem 18. If u and v are algebraic integers, we are to show that $u + v$ and uv are integers. The hypothesis means that all powers u^k and v^k can be expressed in terms of a finite number of powers $1, u, \cdots, u^{n-1}$ and $1, v, \cdots, v^{r-1}$. Therefore every power $(uv)^k = u^k v^k$ and $(u + v)^k$ lies in the additive group generated by the products $1, u, uv, uv^2, \cdots, u^{n-1}v^{r-1}$. By the corollary it follows that uv and $u + v$ are algebraic integers, as required for the theorem.

Exercises

1. Show explicitly that each of the following numbers is an algebraic integer by displaying an appropriate monic equation with integral coefficients:
   (a) $\sqrt{2} + \sqrt{3}$,   (b) $i + \omega$,   (c) $\sqrt{7} + (1 + \sqrt{5})/2$.

† Note that (20) is simply the characteristic polynomial of A, in the sense of Chap. 10.

2. (a) If numbers $v_1, \cdots, v_n$ are linearly independent over $\mathbf{Q}$, prove that any subgroup S of finite index in $G = [v_1, \cdots, v_n]$ also can be generated by n linearly independent numbers $w_1, \cdots, w_n$.

  (b) Show that any such subgroup S is (group) isomorphic to the whole group G.

3. If numbers $v_1, \cdots, v_n$ are linearly independent over $\mathbf{Q}$, show how the basis found in Lemma 1 for a subgroup S of $G = [v_1, \cdots, v_n]$ may be used to compute the index of S in G. (*Hint:* Find a representative for each coset of S.)

★4. Show that a group $G = [v_1, \cdots, v_n]$ has no infinite ascending chain of distinct subgroups; i.e., show that, given an infinite sequence of subgroups $S_1 \subseteqq S_2 \subseteqq S_3 \subseteqq \cdots \subseteqq G$, there is an index m for which $S_m = S_{m+1} = S_{m+2} = \cdots$. (*Hint:* Apply Lemma 1 to the join of the groups S_k.)

5. (a) Show that every module contained in the domain $\mathbf{Z}$ of ordinary integers is an ideal of $\mathbf{Z}$.

  (b) Exhibit a module contained in the domain $\mathbf{Z}[i]$ of Gaussian integers which is not an ideal of $\mathbf{Z}[i]$.

★6. If an algebraic number u satisfies a monic polynomial equation in which the other coefficients are *algebraic* integers, prove that u is also an algebraic integer.

14.10. Factorization of Quadratic Integers

To illustrate the factorization theory of algebraic integers, we consider in more detail the simplest case, that of *quadratic integers*. That is, we consider factorizations of the integers of $\mathbf{Q}(\sqrt{d})$, as characterized in Theorem 17. The basic tool for this purpose is the concept of *norm*.

The formula for the norm depends on the field, but the idea is the same in all cases, even for algebraic number fields of higher degrees. The norm is defined essentially by means of the automorphisms of the field. The quadratic field $\mathbf{Q}(\sqrt{d})$ has by Theorem 6 an automorphism $u = a + b\sqrt{d} \leftrightarrow \bar{u} = a - b\sqrt{d}$ which carries each number u into its "conjugate" $\bar{u}$

Definition. *The norm $N(u)$ of a number $u = a + b\sqrt{d}$ of $\mathbf{Q}(\sqrt{d})$ is the product $u\bar{u}$ of u by its conjugate $\bar{u}$,*

$$(21) \qquad N(u) = u\bar{u} = (a + b\sqrt{d})(a - b\sqrt{d}).$$

Since the correspondence $u \leftrightarrow \bar{u}$ is an isomorphism, $\overline{uv} = \bar{u} \cdot \bar{v}$, hence

$$(22) \qquad N(uv) = N(u)N(v).$$

The norm thus transfers any factorization $w = uv$ of an integer in the field into a factorization $N(w) = N(u)N(v)$ of a rational integer $N(w)$. (The norm of an algebraic integer is a rational integer; see Ex. 1.)

The properties of the norm depend basically on whether d is positive or negative—i.e., on whether $Q(\sqrt{d})$ is a *real* or *complex* quadratic field. If $d < 0$, then $N(u)$ is simply $|u|^2$, the square of the absolute value of u, and it is positive unless $u = 0$. Whereas if $d > 0$, then $N(u) = a^2 - b^2 d$ may be positive or negative. This difference shows up in the group U of the *units* of $Q(\sqrt{d})$, as we shall now see.

Lemma 1. *An integer* $u \in Q(\sqrt{d})$ *is a* unit *if and only if* $N(u) = \pm 1$.

Proof. Trivially, $N(1) = 1$; moreover, $N(u)$ is necessarily a rational integer. Hence if $uv = 1$ for some other integer $v \in Q(\sqrt{d})$, then $N(u)N(v) = N(uv) = 1$, whence $N(u) = \pm 1$. Conversely, if $N(u) = u\bar{u} = \pm 1$, then $u(\pm\bar{u}) = 1$ and u is a unit of $Q(\sqrt{d})$.

A similar argument applies to algebraic number fields generally.

Combining Lemma 1 with Theorem 17, one can determine the units of any *complex* quadratic number field $Q(\sqrt{-d})$, $d > 0$ a square-free integer. The integers of $Q(\sqrt{-d})$ then have the form $u = m + n\alpha\,(m, n \in J)$, where

$$\alpha = \begin{cases} \sqrt{-d} & \text{if } d \not\equiv 3 \ (\text{mod } 4) \\[2mm] \dfrac{1 + \sqrt{-d}}{2} & \text{if } d \equiv 3 \ (mod\ 4). \end{cases}$$

Correspondingly, the norm of u satisfies

$$N(u) = \begin{cases} m^2 + n^2 d & \text{if } d \not\equiv 3 \ (\text{mod } 4) \\[2mm] \left(m + \dfrac{n}{2}\right)^2 + \dfrac{n^2 d}{4} & \text{if } d \equiv 3 \ (\text{mod } 4). \end{cases}$$

If $d \not\equiv 3 \ (\text{mod } 4)$ and $d > 1$, $m^2 + n^2 d \leq 1$ is possible only if $m = \pm 1, n = 0$. Likewise, if $d \equiv 3 \ (\text{mod } 4)$ and $d > 3$, then $d \geq 7$ and $N(u) \geq 7n^2/4 > 1$ unless $n = 0$. Hence, again, the only units of $Q(\sqrt{-d})$ are ± 1. This proves

Theorem 19. *The only complex quadratic number fields having units other than ± 1 are* $Q(\sqrt{-1})$ *and* $Q(\sqrt{-3})$.

The units of $Q(\sqrt{-1})$ are ± 1 and $\pm i$; those of $Q(\sqrt{-3})$ are the powers of $\omega = (1 + \sqrt{-3})/2$, which is a primitive *sixth* root of unity.

Real quadratic number fields have infinitely many units. For example, $1 + \sqrt{2}$ is a unit of $\mathbf{Q}(\sqrt{2})$, since $N(1 + \sqrt{2}) = -1$. Hence so are all the powers $(1 + \sqrt{2})^{\pm k}$ of $(1 + \sqrt{2})$.

Though factorization into primes is unique for many rings of quadratic integers, this is not the case in $\mathbf{Q}(\sqrt{-5})$. For example, consider the factorizations of the number 6:

(23) $$6 = 2 \cdot 3 = (1 + \sqrt{-5})(1 - \sqrt{-5}).$$

If two integers u and v of $\mathbf{Q}(\sqrt{-5})$ satisfy $uv = 6$, then $N(u)N(v) = N(6) = 36$. A proper factor u of 6 will thus have a norm which is a proper factor of $2^2 3^2$, so only the cases $N(u) = 2, 3, 4, 6, 9, 12, 18$ require investigation. Since, in these cases, $N(v) = 18, 12, 9, 6, 4, 3, 2$, respectively, it suffices to consider $N(u) = 2, 3, 4, 6$. One easily sees from $N(m + n\sqrt{-5}) = m^2 + 5n^2$ that all possible factors are listed in (23).

One can rescue the unique factorization theorem in the preceding example by considering products of *ideals*, as in §13.4, instead of products of numbers. One finds that the principal ideals (2), (3), $(1 + \sqrt{-5})$, and $(1 - \sqrt{-5})$ are *not* prime ideals. The relevant prime ideals are the ideals $P = (2, 1 + \sqrt{-5})$, $Q = (3, 1 + \sqrt{-5})$, and $Q' = (3, 1 + \sqrt{-5})$, as described by their bases in $\mathbf{Z}[\sqrt{-5}]$. These ideals are not principal; moreover

$$P^2 = (4, 2 + 2\sqrt{-5}, 6) = (2),$$
$$QQ' = (9, 3 + 3\sqrt{-5}, 6) = (3),$$

This shows that the ideals (2) and (3) are not prime.

To show that P is a prime ideal in $\mathbf{Z}[\sqrt{-5}]$, we observe that $(m + n\sqrt{-5}) \in P$ if and only if $m + n \equiv 0 \pmod{2}$. Therefore $\mathbf{Z}[\sqrt{-5}]/P$ contains only two elements and is the field $\mathbf{Z}_2$. Hence, as in §13.3, Theorem 6, P is a prime ideal. Similarly, $\mathbf{Z}[\sqrt{-5}]/Q \cong \mathbf{Z}[\sqrt{-5}]/Q' \cong \mathbf{Z}_3$, and so Q and Q' are prime ideals.

In conclusion, we have shown that the ideal (6) of $\mathbf{Z}[\sqrt{-5}]$ has the *unique* factorization $(6) = P^2 QQ'$ into prime ideals.

This unique ideal decomposition which we have derived in the domain $\mathbf{Z}[\sqrt{-5}]$ serves merely to indicate how the notion of an ideal may be used systematically to reestablish the unique decomposition theorem in domains of algebraic integers where the ordinary factorization is not unique. By a further development one may establish the "fundamental theorem of ideal theory": *In the domain D of all algebraic integers in an algebraic number field K, every ideal can be represented uniquely, except for order, as a product of prime ideals. In particular, every integer u of the domain generates a principal ideal (u) which has such a unique factorization.*

Exercises

1. (a) In any quadratic field, show that the norm of an integer is an integer.
   (b) If $u = a + b\sqrt{d}$ is not rational, show that $N(u)$ is the constant term in the monic irreducible equation satisfied by u.
2. Find all units in $\mathbf{Q}[\sqrt{-7}]$.
3. Prove that the number of units in a quadratic field $\mathbf{Q}(\sqrt{-d})$ with d positive is finite, and show that every unit is a root of unity.
★4. Prove that the roots of unity which lie in any given algebraic number field form a cyclic group.
5. State and prove a division algorithm for $\mathbf{Z}[\omega]$, where $\omega = (-1 + \sqrt{-3})/2$. (*Hint*: The integral multiples of any β divide the complex plane into equilateral triangles.)
★6. Let D be any integral domain in which a norm $N(\alpha)$ is defined, where (i) $N(\alpha)$ is a positive integer if $\alpha \neq 0$; (ii) $N(\alpha\beta) = N(\alpha)N(\beta)$; (iii) given α and $\beta \neq 0$, γ and ζ exist such that $\alpha = \beta\gamma + \zeta$, and $N(\zeta) < N(\beta)$.
   (a) Prove D is a unique factorization domain.
   (b) Prove every ideal in D is principal.

15

Galois Theory

15.1. Root Fields for Equations

Classically, algebraists tried to solve real and (later) complex polynomial equations by explicit formulas. Their efforts produced the solutions "by radicals" of the general quadratic, cubic, and quartic equations which we derived in Chap. 5. But repeated attempts to obtain similar formulas which would solve general quintic (fifth-degree) equations proved fruitless.

The reason for this was finally discovered by Évariste Galois, who showed that an equation is solvable by radicals if and only if the group of automorphisms associated with it is "solvable" in a purely group-theoretic sense. The automorphisms in question are those automorphisms of the extension field generated by *all* the roots of the equation, which leave fixed all the coefficients of the equation. This final chapter presents the most essential arguments of Galois in modern form, beginning with an examination of the extension field generated by all the roots of a given polynomial $p(x)$ over a given field F. This is the so-called "root field" of $p(x)$, which we now define formally.

Definition. *An extension N of F is a* root field *of a polynomial $f(x)$ of degree $n \geq 1$ with coefficients in F when (i) $f(x)$ can be factored into linear factors $f(x) = c(x - u_1) \cdots (x - u_n)$ in N; (ii) N is generated over F by the roots of $f(x)$, as $N = F(u_1, \cdots, u_n)$.*

If $f(x) = ax^2 + bx + c$ $(a \neq 0)$ is a *quadratic* polynomial over F with the conjugate roots† $u_j = (b \pm \sqrt{b^2 - 4ac})/2a$, $j = 1, 2$, the simple

---

† By a "root" of a polynomial f, we mean, of course, a number x such that $f(x) = 0$; such an x is also called a "zero" of $f(x)$.

extension $K = F(u_1) \cong F[x]/(f(x))$ of F generated by one root u_1 of $f(x) = 0$ is already the root field of f over F. This is true because $u_2 = c/au_1$, whence $f(x) = a(x - u_1)(x - u_2)$ can be factored into linear factors over in $K = F(u_1)$.

However, this is not generally true of irreducible *cubic* polynomials. Thus the root field N of $x^2 - 5$ over **Q** is $Q(\sqrt[3]{5}, \omega\sqrt[3]{5}, \omega^2\sqrt[3]{5}) = Q(\sqrt[3]{5}, \omega)$, where $\omega = (-1 + \sqrt{3}i)/2$ is a complex cube root of unity. The real extension field $Q(\sqrt[3]{5}) \cong Q[x]/(x^3 - 5)$ of the rational field generated by the *real* cube root of 5 is of degree three over **Q**, while the smallest extension of **Q** containing *all* cube roots of 5 is $N = Q(\sqrt[3]{5}, \omega)$. This is of degree two over $Q(\sqrt[3]{5})$, since ω satisfies the cyclotomic equation $\omega^2 + \omega + 1 = 0$. Considered as a vector space over **Q**, the root field N of $x^3 - 5$ thus has the basis $(1, \sqrt[3]{5}, \sqrt[3]{25}, \omega, \omega\sqrt[3]{5}, \omega\sqrt[3]{25})$, and is an extension of **Q** of degree six.

A general existence assertion for root fields may be obtained by using the known existence of simple algebraic extensions, as follows:

Theorem 1. *Any polynomial over any field has a root field.*

For a polynomial of first degree, the root field is just the base field F; hence we may use induction on the degree n of $f(x)$. Suppose the theorem true for all fields F and for all polynomials of degree $n - 1$, and let $p(x)$ be a factor, irreducible over F, of the given polynomial $f(x)$. By Theorem 5 of §14.3 there exists a simple extension $K = F(u)$ generated by a root u of $p(x)$. Over K, $f(x)$ has a root u and hence a factor $x - u$, so $f(x) = (x - u)g(x)$. The quotient $g(x)$ is a polynomial of degree $n - 1$ over K, and the induction assumption provides a root field N over K generated by $n - 1$ roots of $g(x)$. This field N is a root field for $f(x)$.

It will be proved in the next section (Theorem 2) that all root fields of a given polynomial f over a given base field F are isomorphic, so that it is legitimate to speak of *the* root field of f over F.

Appendix. Theorem 1 can be used to construct, *purely algebraically*, an algebraically complete extension of any finite or countable field F, as follows. The number of polynomials of degree n over F is finite or countable, being $\mathbf{d}^{n+1} = \mathbf{d}$ ($\mathbf{d}$ = countable infinity) if F is countable. Hence the number of *all* polynomials over F is countable (cf. Ex. 14, §12.2), and we can arrange these polynomials in a sequence $p_1(x), p_2(x), p_3(x), \cdots$.

Now let F_1 be the root field of $p_1(x)$ over F; let F_2 be the root field of $p_2(x)$ over $F_1; \cdots$; and generally, let F_n be the root field of $p_n(x)$ over F_{n-1}. Finally, let F^* be the set of all elements that appear in one of the F_n—and hence in all its successors. If a and b are any two elements of

F^*, they must both be in some F_n and hence in all its successors. Therefore $a + b$, ab, and (for $b \neq 0$) a/b must also have the same value in F_n and all its successors, which shows that F^* is a field.

To show that F^* is algebraically complete, let $g(x)$ be any polynomial over F^*; all the coefficients of $g(x)$ will be in some F_n, and so algebraic over F. Using Theorem 9, §14.5, one can then find a nonzero multiple $h(x)$ of $g(x)$ with coefficients in F (see Ex. 5 below). But $h(x)$ can certainly be factored into linear factors in *its* root field F_m over an appropriate F_{m-1}—hence so can its divisor $g(x)$. Hence $g(x)$ can also be factored into linear factors over the larger field F^*, which is therefore an algebraically complete field of characteristic p. Furthermore, every element of F^* is algebraic over F.

Using general well-ordered sets and so-called transfinite induction instead of sequences, the above line of argument can be modified† so as to apply to any field F. The modification establishes the following important partial generalization of the Fundamental Theorem of Algebra. *Any field F has an algebraically complete extension.*

15.2. Uniqueness Theorem

We now prove the uniqueness (up to isomorphism) of the root field of Theorem 1.

Theorem 2. *Any two root fields N and N' of a given polynomial $f(x)$ over F are isomorphic. The isomorphism of N to N' may be so chosen as to leave the elements of F fixed.*

Proof. The assertion that the root field is unique is essentially a straightforward consequence of the fact that two different roots of the same irreducible polynomial generate isomorphic simple extensions (Theorem 6, §14.3). Specifically, two root fields $N = F(u_1, \cdots, u_n)$ and $N' = F(u_1', \cdots, u_n')$ of an irreducible $p(x)$ contain isomorphic simple extensions $F(u_1)$ and $F(u_1')$ generated by roots u_1 and u_1' of $p(x)$. Hence there is an isomorphism T of $F(u_1)$ to $F(u_1')$; it remains only to extend appropriately this isomorphism to the whole root field. The basic procedure for such an extension is given by

Lemma 1. *If an isomorphism S between fields F and F' carries the coefficients of an irreducible polynomial $p(x)$ into the corresponding coeffi-*

† A detailed proof appears in B. L. van der Waerden, *Moderne Algebra*, Part I. Berlin, 1930 (in some but not all editions).

cients of a polynomial $p'(x)$ over F', and if $F(u)$ and $F'(u')$ are simple extensions generated, respectively, by roots u and u' of these polynomials, then S can be extended to an isomorphism S^* of $F(u)$ to $F'(u')$, in which $uS^* = u'$.

Proof. Exactly as in the discussion of Theorem 6, §14.3, the desired extension S^* is given explicitly by the formula

(1) $$(a_0 + a_1 u + \cdots + a_{n-1} u^{n-1}) S^*$$
$$= a_0 S + (a_1 S) u' + \cdots + (a_{n-1} S)(u')^{n-1}$$

for all a_i in F, where n is the degree of u over F.

Lemma 2. *If an isomorphism S of F to F' carries $f(x)$ into a polynomial $f'(x)$ and if $N \supset F$ and $N' \supset F'$ are, respectively, root fields of $f(x)$ and $f'(x)$, the isomorphism S can be extended to an isomorphism of N to N'.*

This will be established by induction on the degree $m = [N:F]$. For $m = 1$ it is trivial, since S is then already extended to N; hence take $m > 1$ and assume the lemma true for all root fields N of degree less than m over some F. Since $m > 1$, not all roots of $f(x)$ lie in F, so there is at least one irreducible factor $p(x)$ in $f(x)$ of degree $d > 1$. Let u be a root of $p(x)$ in N, while $p'(x)$ is the factor of $f'(x)$ corresponding to $p(x)$ under the given isomorphism S. The root field N' then contains a root u' of $p'(x)$, and by Lemma 1 the given S can be extended to an isomorphism S^*, with

(2) $uS^* = u'$, $[F(u)]S^* = F'(u')$, $p(u) = 0$, $p'(u') = 0$.

Since N is generated over F by the roots of $f(x)$, N is certainly generated over the larger field $F(u)$ by these roots, so N is a root field of $f(x)$ over $F(u)$, with a degree m/d. For the same reason, N' is a root field of $f'(x)$ over $F'(u')$. Since $m/d < m$, the induction assumption of our lemma therefore asserts that the isomorphism S^* of (2) can be extended from $F(u)$ to N. This proves Lemma 2.

In case the two root fields N and N' are both extensions of the same base field F, and S is the identity mapping of F on itself, Lemma 2 shows that N is isomorphic to N', thereby proving Theorem 2.

Exercises

1. Find the degrees of the root fields of the following polynomials over $\mathbf{Q}$:
   (a) $x^3 - x^2 - x - 2 = 0$,   (b) $x^3 - 2 = 0$,
   (c) $x^4 - 7 = 0$,   (d) $(x^2 - 2)(x^2 - 5) = 0$.

2. Prove: The root field of a polynomial of degree n over a field F has at most the degree $n!$ over F.
3. (a) If ζ is a primitive nth root of unity, prove that $\mathbf{Q}(\zeta)$ is the root field of $x^n - 1 = 0$ over $\mathbf{Q}$.
   (b) Compute its degree for $n = 3, 4, 5, 6$.
4. Prove that any algebraically complete field of characteristic p contains a subfield isomorphic to the field constructed in the Appendix of §15.1.
★5. Let $g(x) = a_0 + a_1 x + \cdots + a_n x^n$ have coefficients algebraic over a field F; prove that $g(x)$ is a divisor of some nonzero $h(x)$ with coefficients in F. (*Hint:* Form a root field of $g(x)$ over $F(a_0, \cdots, a_n)$; factor $g(x)$ into linear factors $(x - u_i)$ in this root field; the u_i will be algebraic over F with irreducible equations $h_i(x)$; set $h(x) = \prod h_i(x)$.)
6. Let $p \in \mathbf{Q}[x]$ be any monic polynomial with rational coefficients, and let $z_1, \cdots, z_n$ be its complex roots. Show that $\mathbf{Q}(z_1, \cdots, z_n)$ is the root field of p over $\mathbf{Q}$.

15.3. Finite Fields

By systematically using the properties of root fields, one can obtain a complete treatment of all fields with a finite number of elements (finite fields). Since a field of characteristic ∞ always contains an infinite subfield isomorphic to the rationals (Theorem 14, §13.8), every finite field F has a prime characteristic p. Without loss of generality, we can assume that F contains the field $\mathbf{Z}_p$ of integers modulo p (see Theorem 12, Corollary, §13.7). The finite field F is then a finite extension of $\mathbf{Z}_p$ and so has a basis $u_1, \cdots, u_n$ over $\mathbf{Z}_p$. Every element in F has a unique expression as a linear combination $\sum a_i u_i$. Each coefficient here can be chosen in $\mathbf{Z}_p$ in exactly p ways, so there are p^n elements in F all told. This proves

Theorem 3. *The number q of elements in a finite field is a power p^n of its characteristic.*

In a finite field F with $q = p^n$ elements, the non-zero elements form a multiplicative group of order $q - 1$. The order of every element in this group is then a divisor of $q - 1$, so that every element satisfies the equation $x^{q-1} = 1$. Therefore all q elements $a_1, a_2, \cdots, a_q$ of F (including zero) satisfy the equation

$$(3) \qquad\qquad x^q - x = 0, \qquad q = p^n.$$

Hence the product $(x - a_1)(x - a_2) \cdots (x - a_q)$ is a divisor of $x^q - x$, being a product of relatively prime polynomials each dividing $x^q - x$.

Since it, like $x^q - x$, is monic and of degree q, we conclude that

(4) $$x^q - x = (x - a_1)(x - a_2) \cdots (x - a_q).$$

Therefore *F is the root field of $x^q - x$ over* $\mathbf{Z}_p$. Any other finite field F' with the same number of elements is the root field of the same equation; hence is isomorphic to F by the uniqueness of the root field (Theorem 2). This argument proves

Theorem 4. *Any two finite fields with the same number of elements are isomorphic.*

Next consider the question: which finite fields really exist? To exhibit a finite field one would naturally form the root field N of the polynomial $x^q - x$ over $\mathbf{Z}_p$. We now prove that the desired root field consists precisely of the roots of this polynomial.

Lemma. *The polynomial $x^q - x$ has q distinct linear factors in its root field N.*

The proof will be by contradiction. If $x^q - x$ had a multiple factor $(x - u)$, we could write $x^q - x = (x - u)^2 g(x)$. Comparing formal derivatives (§3.1, Ex. 7), we would have

$$(x^q - x)' = q \times x^{q-1} - 1 = -1$$

$$[(x - u)^2 g(x)]' = (x - u)[2g(x) + (x - u)g'(x)],$$

whence $(x - u)$ would be a divisor of -1, a contradiction. This proves the lemma.

On the other hand, the sum of any two of the roots $u_1, \cdots, u_q$ of $x^q - x$ is a root, for $(a \pm b)^p = a^p \pm b^p$ in any field of characteristic p, so that if $a^{p^n} = a$ and $b^{p^n} = b$, then

$$(a \pm b)^{p^n} = a^{p^n} \pm b^{p^n} = a \pm b.$$

The product ab is also a root, for $(ab)^{p^n} = a^{p^n} b^{p^n} = ab$, and a similar result holds for a quotient. The set of all q roots of $x^q - x$ is therefore a subfield of the root field N; since this subfield contains all the roots, it must actually be the whole root field N. This means that we have constructed a field with q elements, hence

Theorem 5. *For any prime p and any positive integer n, there exists a finite field with $p^n = q$ elements: the root field of $x^q = x$ over $\mathbf{Z}_p$.*

By Theorems 4 and 5 there is one and essentially only one field with p^n elements. This field is sometimes called the *Galois field GF(p^n)*. The structure of the multiplicative group of this field can be described completely, as follows.

Theorem 6. *In any finite field F, the multiplicative group of all nonzero elements is cyclic.*

Proof. Each nonzero element in F is a $(q-1)$st root of unity, in the sense that it satisfies the equation $x^{q-1} = 1$, where q is the number of elements in F. To prove the group cyclic, we must find in F a "primitive" $(q-1)$st root of unity, which has no lower power equal to 1; the powers of the primitive root will then exhaust the group. To this end, write $q-1$ as a product of powers of distinct primes

$$q - 1 = p_1^{e_1} p_2^{e_2} \cdots p_r^{e_r} \qquad (0 < p_1 < p_2 < \cdots < p_r).$$

For each $P = p_i$, $P^e \mid (q-1)$, so the roots of $x^{P^e} = 1$ are all roots of $x^{q-1} = 1$, hence all lie in F. Of all the P^e distinct roots of this equation $x^{P^e} = 1$, exactly P^{e-1} satisfy the equation $x^{P^{e-1}} = 1$; therefore F contains at least one root $c = c_i$ of $x^{P^e} = 1$ which does not satisfy $x^{P^{e-1}} = 1$. This element c_i thus has order $p_i^{e_i}$ in the multiplicative group of F. The product $c_1 c_2 \cdots c_r$ is an element of order $q-1$ (cf. Ex. 8 below), as desired.

Theorem 7. *Every finite field of characteristic p has an automorphism $a \leftrightarrow a^p$.*

Proof. From the general discussion of fields of characteristic p, we know that the correspondence $a \mapsto a^p$ maps F isomorphically into the set of pth powers (§13.7, Theorem 13). Since this correspondence is one-one, the q elements a give exactly q pth powers, which must then include the whole field F. Therefore $a \mapsto a^p$ maps F on all of F.

Corollary. *In a finite field of characteristic p, every element has a pth root.*

Some additional properties of finite fields are stated in the exercises.

Exercises

1. Prove that there exists an irreducible polynomial of every positive degree over $\mathbf{Z}_p$.

2. Prove that every finite field containing $\mathbf{Z}_p$ is a *simple* extension of $\mathbf{Z}_p$.

3. Prove that every finite extension of a finite field is a simple extension.

4. (a) Using degrees, show that any subfield of $GF(p^n)$ has p^m elements, where $m \mid n$.

   (b) If $m \mid n$, prove that $(p^m - 1) \mid (p^n - 1)$.

   (c) Use (b) to show that, if $m \mid n$, then $GF(p^n)$ has a subfield with p^m elements.

5. Show that the lattice of all subfields of the Galois field of order p^n is isomorphic to the lattice of all positive divisors of n.

6. In $GF(p^n)$ show that the automorphism $a \leftrightarrow a^p$ has order n.

7. If m is relatively prime to the characteristic p of F, show that there exists a primitive mth root of unity over F. (*Hint*: Apply the method used for Theorem 6. Does this apply to a field of characteristic ∞?)

8. Prove: in an Abelian group the product $c_1 c_2 \cdots c_r$ of elements c_i whose orders are powers $p_i^{e_i}$ of distinct primes has order exactly $p_1^{e_1} \cdots p_r^{e_r} = h$. (*Hint*: Show the order divides h, but fails to divide h/p_i for any i.)

9. (a) Show from first principles that the multiplicative group of nonzero integers mod p (in $\mathbf{Z}_p$) is cyclic.

   (b) Let ζ be a primitive pth root of unity over the field $\mathbf{Q}$ of rational numbers. Use (a) to prove that the Galois group of $\mathbf{Q}(\zeta)$ over $\mathbf{Q}$ is cyclic of order $p - 1$.

10. (a) Show that in any finite field of order $q = p^n$, the set S of perfect squares has cardinality $(q + 1)/2$ at least.

    (b) Infer that $S \cap (a - S)$ cannot be void for any $a \in S$.

    (c) Conclude that every element is a sum of two squares.

15.4. The Galois Group

Groups can be used to express the symmetry not only of geometric figures but also of algebraic systems. For example, the field $\mathbf{C}$ of complex numbers has, relative to the real numbers, two symmetries; one is the identity and the other is the isomorphism $a + bi \leftrightarrow a - bi$, which maps each number on its complex conjugate. Such an isomorphism of a field onto itself is known as an automorphism. In general, an *automorphism* T of a field K is a bijection $a \leftrightarrow aT$ of the set K with itself such that sums and products are preserved, in the sense that for all a and b in K,

(5) $$(a + b)T = aT + bT, \qquad (ab)T = (aT)(bT).$$

The composite ST of two automorphisms S and T is also an automorphism, and the inverse of an automorphism is again an automorphism. Hence

Theorem 8. *The set of all automorphisms of a field K is a group under composition.*

Let K be an extension of F and consider those automorphisms T such that $aT = a$ for every a in F. These are the automorphisms which leave F elementwise invariant; in the whole group of automorphisms of K, they form a subgroup called the *automorphism group* of K over F. Thus the automorphism group of **C** over **R** consists of the two automorphisms $a + bi \mapsto a + bi$ and $a + bi \mapsto a - bi$.

Definition. *The* automorphism group *of a field K over a subfield F is the group of those automorphisms of K which leave every element of F invariant.*

The most important special case is the automorphism group of a field of algebraic numbers over the field **Q** of rationals, but before we consider specific examples, let us determine the possible images of an algebraic number under an automorphism.

Theorem 9. *Any automorphism T of a finite extension K over F maps each element u of K on a conjugate uT of u over F.*

This theorem asserts that u and its image uT both satisfy the same irreducible equation over F. To prove it, let the given element u, which is algebraic over F, satisfy a monic irreducible polynomial equation $p(x) = x^n + b_{n-1}x^{n-1} + \cdots + b_0$ with coefficients in F. The automorphism T preserves all rational relations, by (5), and leaves each b_i fixed; hence $p(u) = 0$ gives

$$(u^n + b_{n-1}u^{n-1} + \cdots + b_0)T$$
$$= (uT)^n + b_{n-1}(uT)^{n-1} + \cdots + b_1(uT) + b_0 = 0.$$

This equation states that uT is also a root of $p(x)$, hence that uT is a conjugate of u.

EXAMPLE 1.   Consider the field $K = \mathbf{Q}(\sqrt{2}, i)$ of degree four over the field of rationals† generated by $\sqrt{2}$ and $i = \sqrt{-1}$. Over the intermediate field $F = \mathbf{Q}(i)$ the whole field K is an extension of degree two, generated by either of the conjugate roots $\pm\sqrt{2}$ of $x^2 = 2$. By Theorem 6, §14.3, there is an automorphism S of K carrying $\sqrt{2}$ into $-\sqrt{2}$, and leaving the elements of $\mathbf{Q}(i)$ fixed. That is, the conjugate roots $\sqrt{2}$ and

---

† As in §14.5, one may observe that this field is the root field of $x^4 - 2x^2 + 9$.

$-\sqrt{2}$ are algebraically indistinguishable. The effect of S on any element u of K is

(6) $\qquad (a + b\sqrt{2} + ci + d\sqrt{2}i)S = a - b\sqrt{2} + ci - d\sqrt{2}i,$

where we have written each element of K in terms of the basis $1, \sqrt{2}, i, \sqrt{2}i$ (cf. §14.5). By a similar argument, there is an automorphism T leaving the members of $\mathbf{Q}(\sqrt{2})$ fixed and carrying i into $-i$. Then

(7) $\qquad (a + b\sqrt{2} + ci + d\sqrt{2}i)T = a + b\sqrt{2} - ci - d\sqrt{2}i,$

so T simply maps each number on its complex conjugate. The product ST is still a third automorphism of K. The effect of these automorphisms on $\sqrt{2}$ and i may be tabulated as

$$S: \begin{cases} \sqrt{2} \mapsto -\sqrt{2} \\ \phantom{\sqrt{2}} i \mapsto i, \end{cases} \qquad T: \begin{cases} \sqrt{2} \mapsto \sqrt{2} \\ \phantom{\sqrt{2}} i \mapsto -i, \end{cases}$$

$$ST: \begin{cases} \sqrt{2} \mapsto -\sqrt{2} \\ \phantom{\sqrt{2}} i \mapsto -i, \end{cases} \qquad I: \begin{cases} \sqrt{2} \mapsto \sqrt{2} \\ \phantom{\sqrt{2}} i \mapsto i. \end{cases}$$

We assert that I, S, T, and ST are the only automorphisms of K over $\mathbf{Q}$. By Theorem 9, any other automorphism U must carry $\sqrt{2}$ into a conjugate $\pm\sqrt{2}$, and i into a conjugate $\pm i$. These are exactly the four possibilities tabulated above for I, S, T, and ST. Hence U must agree with one of these four automorphisms in its effect upon the generators $\sqrt{2}$ and i and, therefore, in its effect upon the whole field. Thus $U = I$, S, T, or ST.

The multiplication table for these automorphisms can be found directly from the tabulation of the effects on $\sqrt{2}$ and i displayed above. It is

(8) $\qquad\qquad S^2 = I, \qquad T^2 = I, \qquad ST = TS.$

This is exactly like the multiplication table for the elements of the four group (§6.7), so we conclude that the automorphism group of $\mathbf{Q}(\sqrt{2}, i)$ is isomorphic to the four group $\{I, S, T, ST\}$.

Definition. *If $N = F(u_1, \cdots, u_n)$ is the root field of a polynomial $f(x) = (x - u_1) \cdots (x - u_n)$, then the automorphism group of N over F is known as the* Galois group *of the equation $f(x) = 0$ or as the Galois group of the field N over F.*

To describe explicitly the automorphisms T of a particular Galois group, one proceeds as follows. Let N be the root field of $f(x)$ over F. Then T maps roots of $f(x)$ onto roots of $f(x)$ (Theorem 9), and distinct roots onto distinct roots. Hence T effects a permutation ϕ of the distinct roots $u_1, \cdots, u_k$ of $f(x)$, so that

$$(9) \qquad u_1 T = u_{1\phi}, \cdots, u_k T = u_{k\phi}, \qquad k \leqq n.$$

On the other hand, every element w in the root field is expressible as a polynomial $w = h(u_1, \cdots, u_k)$, with coefficients in F. Since T leaves these coefficients fixed, the properties (9) of T give

$$[h(u_1, \cdots, u_k)]T = h(u_1 T, \cdots, u_k T) = h(u_{1\phi}, \cdots, u_{k\phi}).$$

This formula asserts that the effect of T on w is entirely determined by the effect of T on the roots, or that T is uniquely determined by the permutation (9). Since the product of two permutations is obtained by applying the corresponding automorphisms in succession, the permutations (9) form a group isomorphic to the group of automorphisms. The permutations (9) include only those permutations which do preserve all polynomial identities between the roots and so can correspond to automorphisms. The results so established may be summarized as follows:

Theorem 10. *Let $f(x)$ be any polynomial of degree n over F which has exactly k distinct roots $u_1, \cdots, u_k$ in a root field $N = F(u_1, \cdots, u_k)$. Then each automorphism T of the Galois group G of $f(x)$ induces a permutation $u_i \leftrightarrow u_i T$ on the distinct roots of $f(x)$, and T is completely determined by this permutation.*

Corollary 1. *The Galois group of any polynomial is isomorphic to a group of permutations of its roots.*

Corollary 2. *The Galois group of a polynomial of degree n has order dividing n!.*

EXAMPLE 2. The equation $x^4 - 3 = 0$ is irreducible over the field $\mathbf{Q}$, by Eisenstein's Theorem, and has the four distinct roots $r, ir, -r, -ir$, where $i = \sqrt{-1}$, and $r = \sqrt[4]{3}$ is the real, positive fourth root of 3. The root field $N = \mathbf{Q}(r, ir, -r, -ir)$ may be generated as $N = \mathbf{Q}(r, i)$. Since r is of degree four over $\mathbf{Q}$ and since i is complex, hence of degree two over the real field $\mathbf{Q}(r)$, the whole root field N has degree eight over $\mathbf{Q}$. By Theorem 9, §14.5, this extension N has a basis of 8 elements $1, r, r^2, r^3, i, ir, ir^2, ir^3$. Since every element in N can be expressed as a linear combination of these basis elements, with rational coefficients, the effect of an automorphism T will be completely determined once rT and iT are known.

Several automorphisms of N may be readily constructed. Since N is an extension of degree two over the real field $\mathbf{Q}(r)$, it has an automorphism T which maps each number of N on its complex conjugate; hence $rT = r$, $iT = -i$. On the other hand, N is an extension of degree four of the subfield $\mathbf{Q}(i)$, generated by the element r. By Theorem 6, §14.3, N has an automorphism S mapping r into its conjugate ir, so $rS = ir$, $iS = i$. It follows that S^2 is an automorphism with $rS^2 = i^2 r$, $iS^2 = i$, while $rS^3 = -ir$, $iS^3 = i$. By further combinations of S and T, one finds for N eight automorphisms, with the following effects upon the generators i and r:

	I	S	S^2	S^3	T	TS	TS^2	TS^3
r mapped into	r	ir	$-r$	$-ir$	r	ir	$-r$	$-ir$
i mapped into	i	i	i	i	$-i$	$-i$	$-i$	$-i$

One may also compute that $TS^3 = ST$, $S^4 = T^2 = I$, so that these eight automorphisms form a group, isomorphic with the group of the square (§6.4). These automorphisms constitute the whole Galois group, for any automorphism must map i into one of its conjugates $\pm i$, and r into a conjugate $\pm r$ or $\pm ir$; the table above includes all eight possible combinations of these effects.

Many concepts of group theory can be applied to such a Galois group G. Thus G contains the subgroup $H = [I, S, S^2, S^3]$ generated by S and the smaller subgroup $L = [I, S^2]$ generated by S^2. Each automorphism of the subgroup H leaves i fixed, and hence leaves fixed every element in the subfield $\mathbf{Q}(i)$. The smaller subgroup L consists of those automorphisms which leave fixed everything in the larger subfield $\mathbf{Q}(i, r^2)$. In this sense, the descending sequence of subgroups $G \supset H \supset L \supset I$ corresponds to an ascending sequence of subfields $\mathbf{Q} \subset \mathbf{Q}(i) \subset \mathbf{Q}(i, \sqrt{3}) \subset \mathbf{Q}(i, r)$. Such an ascending sequence of subfields gives a method of solving the given equation by successively adjoining the roots of simpler equations $x^2 = -1$, $y^2 = 3$, $z^2 = \sqrt{3}$. This example illustrates the significance of the subgroups of a Galois group for the solution of an equation by radicals.

Homomorphisms of Galois groups arise naturally. Each automorphism U of the group G above carries i into $\pm i$, hence carries each element of the field $\mathbf{Q}(i)$ into some element of the same field. This means that U induces an automorphism U^* of $\mathbf{Q}(i)$, where U^* is defined for an element w in $\mathbf{Q}(i)$ by the identity $wU^* = wU$. The correspondence $U \mapsto U^*$ is a homomorphism mapping the group G of all automorphisms U of N on the group G^* of automorphisms of $\mathbf{Q}(i)$. But G^* has only two elements: the identity I^* and the automorphism interchanging i and $-i$. Furthermore, $U^* = I^*$ if and only if U leaves $\mathbf{Q}(i)$ elementwise fixed; that is, if and only if U is in the subgroup $H = [1, S, S^2, S^3]$. Hence $U \mapsto U^*$ is that epimorphism of G whose kernel is H, and the group G^* is therefore isomorphic to the quotient-group G/H.

Exercises

1. Draw a lattice diagram for the system of all subfields of $\mathbf{Q}(i, r)$.
2. Prove that $x^4 - 3$ is irreducible over $\mathbf{Q}$ by showing that none of the linear or quadratic factors of $x^4 - 3$ have coefficients in $\mathbf{Q}$.
3. Represent each automorphism of the Galois group of $x^4 - 3$ as a permutation of the roots.
4. (a) Prove that $x^4 - 3$ is irreducible over $\mathbf{Q}(i)$.
   (b) Describe the Galois group of $x^4 - 3$ over $\mathbf{Q}(i)$.
5. Show from first principles that the following permutation of the roots of $x^4 - 3$ cannot possibly correspond to an automorphism: $r \mapsto ir, ir \mapsto -ir, -r \mapsto r, -ir \mapsto -r$.
6. Let $F = \mathbf{Q}(\omega)$ be the field generated by a complex cube root ω of unity. Discuss the Galois group $x^3 - 2$ over F, including a determination of the degree of the root field, a description of the Galois group in purely group-theoretic language, and a representation of each automorphism as a permutation.
7. Do the same for $x^5 - 7$ over $\mathbf{Q}(\zeta)$, ζ a primitive fifth root of unity.
8. Prove that the Galois group of a finite field is cyclic.
9. If ζ is a primitive nth root of unity, prove that the Galois group of $\mathbf{Q}(\zeta)$ is Abelian. (*Hint:* Any automorphism has the form $\zeta \mapsto \zeta^e$.)
10. (a) If K is an extension of $\mathbf{Q}$, prove that every automorphism of K leaves each element of $\mathbf{Q}$ fixed.
    (b) State and prove a similar result for fields of characteristic p.

15.5.   Separable and Inseparable Polynomials

The general discussion of Galois groups is complicated by the presence of so-called *inseparable* irreducible polynomials—or elements which are algebraic of degree n but have *fewer* than n conjugates. This complication occurs for some fields of characteristic p, and can be illustrated by a simple example.

Let $K = \mathbf{Z}_p(u)$ denote a simple transcendental extension of the field $\mathbf{Z}_p$ of integers mod p, and let F denote the subfield $\mathbf{Z}_p(u^p)$ of K generated by $u^p = t$. Thus, F consists of all rational forms in an element t transcendental over $\mathbf{Z}_p$. Over F the original element u satisfies an equation $f(x) = x^p - t = 0$. This polynomial $f(x)$ is actually irreducible over $F = \mathbf{Z}_p(t)$, for if f were reducible over $\mathbf{Z}_p(t)$, it would, by Gauss's Lemma (§3.9), be reducible over the domain $\mathbf{Z}_p[t]$ of polynomials in t; but such a factorization $f(x) = g(x, t)h(x, t)$ is impossible, since $f(x) = x^p - t$ is linear in t. Therefore the root u of $f(x)$ has degree p over F. But $f(x)$ has over K the factorization

$$(10) \qquad f(x) = x^p - u^p = (x - u)^p.$$

Hence it has only *one* root, and u (although of degree $p > 1$) has no conjugates except itself.

We can describe the situation in the following terms:

Definition. *A polynomial $f(x)$ of degree n is* separable *over a field F if it has n distinct roots in some root field $N \geq F$; otherwise, $f(x)$ is* inseparable. *A finite extension $K \geq F$ is called* separable *over F if every element in K satisfies over F a separable polynomial equation.*

There is an easy test for the separability or inseparability of a given polynomial $f(x) = a_0 + a_1 x + \cdots + a_n x^n$. Namely, first define the *formal derivative $f'(x)$* of $f(x)$ by the formula (cf. §3.1, Ex. 7)

$$(11) \qquad f'(x) = a_1 + (2 \times a_2)x + \cdots + (n \times a_n)x^{n-1},$$

where $n \times a_n$ denotes the *nth natural multiple* of a_n (see §13.7). If the coefficients are in the field of real numbers, this derivative agrees with the ordinary derivative as found by calculus. From the formal definition (11), without any use of limits, one can deduce many of the laws for differentiation, such as

$$(f + g)' = f' + g', \qquad (fg)' = fg' + gf', \qquad (f^m)' = mf^{m-1}f',$$

and so on.

Now factor $f(x)$ into powers of distinct linear factors over any root field N,

$$(12) \qquad f(x) = c(x - u_1)^{e_1} \cdots (x - u_k)^{e_k} \qquad (c \neq 0).$$

Differentiating both sides of (12) formally, we see that $f'(x)$ is the sum of $ce_1(x - u_1)^{e_1-1}(x - u_2)^{e_2} \cdots (x - u_k)^{e_k}$ and $(k - 1)$ terms each containing $(x - u_1)^{e_1}$ as a factor. Hence if $e_1 > 1, (x - u_1)$ divides $f'(x)$, while if

$e_1 = 1$, then it does not. Repeating the argument for $e_2, \cdots, e_k$, we find that $f(x)$ and $f'(x)$ have a common factor unless $e_1 = e_2 = \cdots = e_k = 1$, that is, unless $f(x)$ is separable; hence the polynomial $f(x)$ is separable when factored over N if and only if $f(x)$ and its formal derivative $f'(x)$ are relatively prime.

But the g.c.d. of $f(x)$ and $f'(x)$ can be computed as in Chap. 3 directly by the Euclidean algorithm in $F[x]$; it is not altered if F is extended to a larger field. We infer

Theorem 11. *Let $f(x)$ be any polynomial over a field F; compute by the Euclidean algorithm the (monic) greatest common divisor $d(x)$ of $f(x)$ and its formal derivative $f'(x)$. If $d(x) = 1$, then $f(x)$ is separable; otherwise, $f(x)$ is inseparable.*

If $f(x)$ is irreducible, then the g.c.d. $(f(x), g(x))$ is 1 unless $f(x)$ divides $g(x)$, and $f(x)$ cannot divide any polynomial of lower degree except 0. Hence

Corollary 1. *An irreducible polynomial is separable unless its formal derivative is 0.*

Corollary 2. *Any irreducible polynomial over a field of characteristic ∞ is separable.*

For $f'(x) = n \times a_n x^{n-1} + \cdots \neq 0$ if $n > 0$ and $a_n \neq 0$.

It is a further corollary that if F is of characteristic ∞, then the root field of any irreducible polynomial $f(x)$ of degree n contains exactly n distinct conjugate roots of $f(x)$. Furthermore, any algebraic element over a field of characteristic ∞ satisfies an equation which is irreducible and hence separable, so that any algebraic extension of such a field is separable in the sense of the definition above.

The result of Corollary 2 does not hold for fields of prime characteristic. For example, the irreducible polynomial $f(x) = x^p - t$ mentioned at the beginning of the section has a formal derivative $(x^p - t)' = p \times x^{p-1} = 0$.

Exercises

1. Without using Theorem 11, show that the roots of an irreducible quadratic polynomial over $\mathbf{Q}$ are distinct.
2. Let $f(x)$ be a polynomial with rational coefficients, while $d(x)$ is the g.c.d. of $f(x)$ and $f'(x)$. Prove that $f(x)/d(x)$ is a polynomial which has the same roots as $f(x)$, but which has no multiple roots.

3. (a) Show that if $f'(x) = 0$, then $f(x)$ is inseparable over any field F.

★(b) Show that if $f'(x) = 0$ over $\mathbf{Z}_p$, then $f(x) = [g(x)]^p$ for some $g(x)$.

4. Show that $x^3 - 2u$ is inseparable over $\mathbf{Z}_3(u)$. Show that the Galois group of its root field is the identity.

5. Use Theorem 11 to show that $x^q - x$ is separable over $\mathbf{Z}_p$, if $q = p^n$.

6. (a) If $f(x)$ is a polynomial over a field F of characteristic p with $f'(x) = 0$, show that $f(x)$ can be written in the form $a_0 + a_1 x^p + \cdots + a_n x^{np}$.

(b) Show that if F is finite, $f(x) = [g(x)]^p$ for suitable $g(x)$.

(c) Use part (b) to show that every irreducible polynomial over a finite field is separable.

15.6. Properties of the Galois Group

The root fields and Galois groups of separable polynomials have two especially elegant properties, which we now state as theorems.

Theorem 12. *The order of the Galois group of a separable polynomial over F is exactly the degree $[N:F]$ of its root field.*

In the second example of §15.4 we have already seen that this is the case for the root field of $x^4 = 3$.

Theorem 13. *In the root field $N \supset F$ of a separable polynomial, the elements left invariant by every automorphism of the Galois group of N over F are exactly the elements of F.*

This theorem gives us some positive information about the Galois group G, for it asserts that for each element a in N but not in F there is in G an automorphism T with $aT \neq a$.

For the proof of Theorem 12, refer back to Lemma 2 of §15.2, which concerned the extensibility of isomorphisms between fields. Note that in this lemma (unlike in §15.5) $f'(x)$ does not signify the derivative of $f(x)$.

Lemma. *If the polynomial $f(x)$ of Lemma 2 in §15.2 is separable, S can be extended to N in exactly $m = [N:F]$ different ways.*

This result can be proved by mathematical induction on m. Any extension T of the given isomorphism S of F to F' will map the root u used in (2) into some one of the roots u' of $p'(x)$; hence every possible extension of S is yielded by one of our constructions. Since $f(x)$ is separable, its factor $p(x)$ of degree d will have exactly d distinct roots u'.

These d choices of u' give exactly d choices for S^* in (2). By the induction assumption, each such S^* can then be extended to N in $m/d = [N:F(u)]$ different ways, so there are all told $d(m/d) = m$ extensions, as asserted.

If $f(x) = f'(x)$ is separable of degree m and we set $N = N'$ in Lemma 2 of §15.2, our new lemma asserts that the identity automorphism I of F can be extended in exactly m different ways to an automorphism of N. But these automorphisms constitute the Galois group of N over F, proving Theorem 12.

Finally, to prove Theorem 13, let G be the Galois group of the root field N of a separable polynomial over F, while K is the set of all elements of N invariant· under every automorphism of G. One shows easily that K is a field, and that $K \supset F$. Hence every automorphism in G is an extension to N of the identity automorphism I of K. Since N is a root field over K, there are by our lemma only $[N:K]$ such extensions, while by Theorem 12 there are $[N:F]$ automorphisms, all told. Hence $[N:K] = [N:F]$. Since $K \supset F$, this implies that $K = F$, as asserted in Theorem 13.

Still another consequence of the extension lemmas is the fact that a root field is always "normal" in the following sense.

Definition. *A finite extension N of a field F is said to be normal over F if every polynomial $p(x)$ irreducible over F which has one root in N has all its roots in N.* .

In other words, every polynomial $p(x)$ which is irreducible over F, and has a root in N, can be factored into linear factors over N.

Theorem 14. *A finite extension of F is normal over F if and only if it is the root field of some polynomial over F.*

Proof. If N is normal over F, choose any element u of N not in F and find the irreducible equation $p(x) = 0$ satisfied by u. By the definition of normality, N contains all roots of $p(x)$, hence contains the root field M of $p(x)$. If there are elements of N not in M, one of these elements v satisfies an irreducible equation $q(x) = 0$, and M is contained in the larger root field of $p(x)q(x)$, and so on. Since the degree of N is finite, one of the successive root fields so obtained must be the whole field N.

Conversely, the root field N of any $f(x)$ is normal. Suppose that there is some polynomial $p(x)$ irreducible over F which has one but not all of its roots in N. Let w be a root of $p(x)$ in N, and adjoin to N another root w' which is not in N. The simple extension $F(w)$ is isomorphic to $F(w')$ by a correspondence T with $wT = w'$. The field N is a root field for $f(x)$ over $F(w)$; on the other hand, $N' = N(w')$ is generated by roots of $f(x)$ over $F(w')$, hence is a root field for $f(x)$ over $F(w')$. Hence, by Lemma 2, §15.2, the correspondence T can be extended to an isomorphism of N to N'. Since T leaves the elements of the base field F fixed, these isomorphic fields N and N' must have the same degree over F. But we assumed that $N' = N(w')$ is a proper extension of N, so that its degree over F is larger than that of N. This contradiction proves the theorem.

If the first half of this proof is applied to a separable extension (one in which each element satisfies a separable equation), all the polynomials $p(x), q(x)$ used are separable. This proves the

Corollary. *Every finite, normal, and separable extension of F is the root field of a separable polynomial.*

In particular, every finite and normal extension N of the field $\mathbf{Q}$ of rational numbers is automatically separable (Theorem 11, Corollary 2), hence is the root field of some separable polynomial. The order of the automorphism group of N over $\mathbf{Q}$ is therefore exactly the degree $[N:\mathbf{Q}]$.

The Galois group may be used to treat properties of symmetric polynomials, as defined in §6.10.

Theorem 15. *Let $N = F(u_1, \cdots, u_n)$ be the field generated by all n roots $u_1, \cdots, u_n$ of a separable polynomial $f(x)$ of degree n, and let $g(x_1, \cdots, x_n)$ be any polynomial form over F symmetric in n indeterminates x_i. The element $w = g(u_1, \cdots, u_n)$ of N then lies in the base field F.*

Proof. Any automorphism T of the Galois group G of N effects a permutation $u_i \mapsto u_i T$ of the roots of $f(x)$, by Theorem 10. The symmetry of $g(x_1, \cdots, x_n)$ means that it is unaltered by any permutation of the indeterminates; hence

$$w \mapsto wT = g(u_1 T, \cdots, u_n T) = g(u_1, \cdots, u_n) = w.$$

Since w is altered by no automorphism T, w lies in F, by Theorem 13.

Corollary. *Any polynomial (over F) symmetric in n indeterminates $x_1, \cdots, x_n$ can be expressed as a rational† function (over F) in the n*

---

† Cf. Theorem 19 of §6.10, which states a stronger result.

elementary symmetric functions

$$\sigma_1 = x_1 + x_2 + \cdots + x_n,$$

$$\sigma_2 = x_1 x_2 + x_1 x_3 + \cdots + x_{n-1} x_n,$$

(13)
$$\vdots$$

$$\sigma_n = x_1 x_2 \cdots x_n.$$

To simplify the formulas, we write out the proof for the case $n = 3$ only. Over F the elementary symmetric functions $\sigma_1, \sigma_2,$ and σ_3 generate a field $K = F(\sigma_1, \sigma_2, \sigma_3)$. The field $N = F(x_1, x_2, x_3)$ generated by the three original indeterminates is a finite extension of K; in fact, the generators x_i of N are the roots of a cubic polynomial

$$f(x) = (x - x_1)(x - x_2)(x - x_3) = x^3 - \sigma_1 x^2 + \sigma_2 x - \sigma_3,$$

with coefficients which prove to be exactly the given symmetric functions (13). Introduce the Galois group G of the root field N over K. By Theorem 10, every automorphism induces a permutation of the x_i; hence by Theorem 15, any symmetric polynomial of the x_i lies in the base field K. Since $K = F(\sigma_1, \sigma_2, \sigma_3)$, it follows that such a symmetric polynomial is a rational function of $\sigma_1, \sigma_2, \sigma_3$.

Exercises

1. In the proof of the corollary of Theorem 15, show that the Galois group of $N = K(x_1, x_2, x_3)$ over K is exactly the symmetric group on three letters.
2. Express $x_1^3 + x_2^3 + x_3^3$ in terms of the elementary symmetric functions. (Cf. also §6.10, Exs. 7 and 8.)
3. (a) Show that there exist a field K and a subfield F, such that the Galois group of K over F is the symmetric group of degree n.
   (b) Show that in (a) K may be chosen as a subfield of the field of real numbers. (*Hint:* Use n algebraically independent real numbers.)
4. If a polynomial of degree n has n roots $x_1, \cdots, x_n$, its *discriminant* is $D = \Pi(x_i - x_j)^2$, where the product is taken over all pairs of subscripts with $i < j$.
   (a) Show that the discriminant of a polynomial with rational coefficients is a rational number.
   (b) For a quadratic polynomial, express D explicitly as a rational function of the coefficients.
   ★(c) The same problem for a cubic polynomial.
5. Show that if K is normal over F, and $F \subset L \subset K$, then K is normal over L.

15.7. Subgroups and Subfields

If H is any set of automorphisms of a field N, the elements a of N left invariant by all the automorphisms of H (such that $aT = a$ for each T in H) form a *subfield* of N. In particular, this is true if N is the root field of any polynomial over any base field F, and H is any subgroup of the Galois group of N over F.

Theorem 16. *If H is any finite group of automorphisms of a field N, while K is the subfield of all elements invariant under H, the degree $[N:K]$ of N over K is at most the order of H.*

Proof.† If H has order n, it will suffice to show that any $n + 1$ elements $c_1, \cdots, c_{n+1}$ of N are linearly dependent over K. From the n elements T of H we construct a system of n homogeneous linear equations

$$y_1(c_1 T) + y_2(c_2 T) + \cdots + y_{n+1}(c_{n+1} T) = 0$$

in $n + 1$ unknowns y_i. Such a system always has in N a solution different from $y_1 = y_2 = \cdots = y_{n+1} = 0$, by Theorem 10 of §2.3. Now pick the smallest integer m such that the n equations

$$(14) \qquad y_1(c_1 T) + y_2(c_2 T) + \cdots + y_m(c_m T) = 0, \qquad T \in H,$$

still have such a solution. This solution $y_1, \cdots, y_m$ consists of elements of N and is unique to within a constant factor, for if there were two nonproportional solutions, a suitable linear combination would give a solution of the system with $m - 1$ unknowns. Without loss of generality, we can also assume $y_1 = 1$.

Now apply any automorphism S in H to the left side of (14). Since $TS = T'$ runs over all the elements of H, the result is a system

$$(y_1 S)(c_1 T') + (y_2)(c_2 T') + \cdots + (y_m S)(c_m T') = 0, \qquad T' \in H,$$

identical with (14) except for the arrangement of equations. Therefore $y_1 S, \cdots, y_m S$ is also a solution of (14), and so by the uniqueness of the solution is $t y_1, \cdots, t y_m$, where t is a factor of proportionality. However, since $y_1 = 1$ and S is an automorphism, $y_1 S = 1$ also and so $t = 1$. We conclude that $y_i S = y_i$ for every $i = 1, \cdots, m$ and every S in H, which

---

† This proof, which involves the idea of looking on a Galois group simply as a finite group of automorphisms, with no explicit reference to a base field, is due to Professor Artin.

means that the coefficients y_i lie in the subfield K of invariant elements. Equation (14) with $T = I$ now asserts that the elements $c_1, \cdots, c_m$ are linearly dependent over the field K. This proves the theorem.

On the basis of this result, we can establish, at least for separable polynomials, a correspondence between the subgroups of a Galois group and the subfields of the corresponding root field. This correspondence provides a systematic way of reducing questions about fields related to a given equation to parallel questions about subgroups of (finite) Galois groups.

Theorem 17 (Fundamental Theorem of Galois Theory). *If G is the Galois group for the root field N of a separable polynomial $f(x)$ over F, then there is a bijection $H \leftrightarrow K$ between the subgroups H of G and those subfields K of N which contain F. If K is given, the corresponding subgroup $H = H(K)$ consists of all automorphisms in G which leave each element of K fixed; if H is given, the corresponding subfield $K = K(H)$ consists of all elements in N left invariant by every automorphism of the subgroup H. For each K, the subgroup $H(K)$ is the Galois group of N over K, and its order is the degree $[N:K]$.*

Proof. For a given K, $H(K)$ is described thus:

(15) T is in $H(K)$ if and only if $bT = b$ for all b in K.

If S and T have this property, so does the product ST, so the set $H(K)$ is a subgroup. The field N is a root field for $f(x)$ over K, and every automorphism of N over K is certainly an automorphism of N over F leaving every element of K fixed, hence is in the subgroup $H(K)$. Therefore $H(K)$ is by definition the Galois group of N over K. If Theorem 12 is applied to this Galois group, it shows that the order of $H(K)$ is exactly the degree of N over K.

Two different intermediate fields K_1 and K_2 determine distinct subgroups $H(K_1)$ and $H(K_2)$. To prove this, choose any a in K_1 but not in K_2, and apply Theorem 13 to the group $H(K_2)$ of N over K_2. It asserts that $H(K_2)$ contains some T with $aT \neq a$. Since a is in K_1, this automorphism T does not lie in the group $H(K_1)$, so $H(K_1) \neq H(K_2)$.

We know now that $K \mapsto H(K)$ is a bijection between *all* of the subfields of N and *some* of the subgroups of G. In order to establish a bijection between *all* subfields and *all* subgroups we must show that every subgroup appears as an $H(K)$. Let H be a subgroup of order h and $K = K(H)$ be defined as in the statement of Theorem 17:

(16) b is in $K(H)$ if and only if $bS = b$ for all S in H.

According to Theorem 16, $[N:K] \leqq h$. By comparing (15) with (16), one sees that the subgroup $H(K)$ corresponding to $K = K(H)$ certainly includes the group H originally given, while by Theorem 12 the order of $H(K)$ is $[N:K]$. Since $[N:K] \leqq h$, this means that the order of the group $H(K)$ does not exceed the order of its subgroup H. Therefore $H(K) = H$, as asserted. This completes the proof.

The set of all fields K between N and F is a lattice relative to the ordinary relation of inclusion between subfields. If K_1 and K_2 are two subfields, their g.l.b. or meet in this lattice is the intersection $K_1 \cap K_2$, which consists of all elements common to K_1 and K_2, while their l.u.b. or join is $K_1 \vee K_2$, *the subfield of N generated by the elements of K_1 and K_2 jointly.* For instance, if $K_1 = F(v_1)$ and $K_2 = F(v_2)$ are simple extensions, their join is the multiple extension $F(v_1, v_2)$.

Theorem 18. *The lattice of all subfields $K_1, K_2, \cdots$ is mapped by the correspondence $K \mapsto H(K)$ of Theorem 17 onto the lattice of all subgroups of G, in such a way that*

$$(17) \qquad K_1 \subset K_2 \quad \text{implies} \quad H(K_1) \supset H(K_2),$$

$$(18) \qquad H(K_1 \vee K_2) = H(K_1) \cap H(K_2),$$

$$(19) \qquad H(K_1 \cap K_2) = H(K_1) \vee H(K_2).$$

In particular, the subgroup consisting of the identity alone corresponds to the whole normal field N.

These results state that the correspondence inverts the inclusion relation and carries any meet into the (dual) join and conversely. Any bijection between two lattices which has these properties is called a *dual isomorphism*.

To prove the theorem, observe first that the definition (15) of the group belonging to a field K shows that for a larger subfield the corresponding group must leave more elements invariant, hence will be smaller. This gives (17). The meet and the join are defined purely in terms of the inclusion relation (see §11.7); hence by the Duality Principle, a bijection which inverts inclusion must interchange these two, as is asserted in (18) and (19).

We omit the proof of the following further result.

Theorem 19. *A field K, with $N \supset K \supset F$, is a normal field over F if and only if the corresponding group $H(K)$ is a normal subgroup of the Galois group G of N. If K is normal, the Galois group of K over F is isomorphic to the quotient-group $G/H(K)$.*

The conclusions of this theorem have already been illustrated in a special case by the example at the end of §15.4.

Exercises

1. (a) Prove that if H is any set of automorphisms of a field N, the elements of N left invariant by all the automorphisms in H form a subfield K of N.
   (b) Show that N is normal over this subfield K.
2. Exhibit completely the subgroup-subfield correspondence for the field $Q(\sqrt{2}, i)$ over Q.
3. Do the same for the root field of $x^4 - 3$, as discussed in §15.4.
4. Prove that the index of $H(K)$ in G is the degree of K over F.
5. If N is the root field of a separable polynomial $f(x)$ over F, prove that the number of fields between N and F is finite.
6. Prove that the fields K between N and F form a lattice.
7. If K is a finite extension of a field F of characteristic ∞, prove that the number of fields between K and F is finite.
★8. Prove Theorem 19.
★9. Two subfields K_1 and K_2 are called *conjugate* in the situation of Theorem 17, if there exists an automorphism T of N over F carrying K_1 into K_2. Prove that this is the case if and only if $T^{-1}H(K_1)T = H(K_2)$ (i.e., if and only if $H(K_1)$ and $H(K_2)$ are "conjugate" subgroups of G).

15.8. Irreducible Cubic Equations

Galois theory can be applied to show the impossibility of resolving various classical problems about the solution of equations by radicals. As a simple example of this technique, we shall consider the famous "irreducible case" of cubic equations with real roots.

A cubic equation may be taken in the form (see §5.5, (17))

$$(20) \qquad f(y) = y^3 + py + q = (y - y_1)(y - y_2)(y - y_3),$$

with real coefficients p and q and with three real or complex roots y_1, y_2, and y_3. The coefficients p and q may be expressed as symmetric functions of the roots, for on multiplying out (20), one finds

$$(21) \qquad 0 = y_1 + y_2 + y_3, \qquad p = y_1y_2 + y_1y_3 + y_2y_3, \qquad q = -y_1y_2y_3.$$

It is important to introduce the discriminant D of the cubic, defined by the formula

$$(22) \qquad D = [(y_1 - y_2)(y_1 - y_3)(y_2 - y_3)]^2.$$

The permutation of any two roots does not alter D, so that D is a polynomial symmetric in y_1, y_2, and y_3. By Theorem 15 it follows that D is expressible as a quantity in the field $F = \mathbf{Q}(p, q)$ generated by the coefficients. This expression is, as in §5.5, (24),

$$(23) \qquad\qquad D = -4p^3 - 27q^2;$$

this equation is a polynomial identity in y_1, y_2, and y_3 and may be checked by straightforward use of the equations (21) and (22).

Theorem 20. *A real cubic equation with a positive discriminant has three real roots; if $D = 0$, at least two roots are equal; while if $D < 0$, two roots are imaginary.*

This may be verified simply by observing how the various types of roots affect the formula (22) for D. If all roots are real, D is clearly positive, while $D = 0$ if two roots are equal. Suppose, finally, that one root $y_1 = a + bi$ is an imaginary number ($b \neq 0$). The complex conjugate $y_2 = a - bi$ must then also be a root (§5.4), while the third root is real. In (22), $y_1 - y_2 = (a + bi) - (a - bi) = 2bi$ is a pure imaginary, while

$$(y_1 - y_3)(y_2 - y_3) = (y_1 - y_3)(y_1{}^* - y_3) = (y_1 - y_3)(y_1 - y_3)^*$$

is a real number. The discriminant D is therefore negative. This gives exactly the alternatives listed in the theorem.

Theorem 21. *If the cubic polynomial (20) is irreducible over $F = \mathbf{Q}(p, q)$, has roots y_1, y_2, y_3, and discriminant D, then its root field $F(y_1, y_2, y_3)$ is $F(\sqrt{D}, y_1)$.*

Proof. By the definition (22) of D, the root field certainly contains $\sqrt{D}$; hence it remains only to prove that the roots y_2 and y_3 are contained in $K = F(\sqrt{D}, y_1)$. In this field K the cubic has a linear factor $(y - y_1)$, so the remaining quadratic factor

$$(24) \qquad (y - y_2)(y - y_3) = y^2 - (y_2 + y_3)y + y_2 y_3$$

also has its coefficients in K. By substitution in (24), $(y_1 - y_2)(y_1 - y_3)$ is in K, so that

$$y_2 - y_3 = \pm\sqrt{D}/(y_1 - y_2)(y_1 - y_3)$$

is in K. But the coefficient $y_2 + y_3$ of (24) is also in K. If both $y_2 + y_3$ and $y_2 - y_3$ are in K, so are y_2 and y_3. This proves the theorem.

Consider now a cubic which is irreducible over its coefficient field but which has three real roots. Formula (19) of §5.5 gives the roots as $y = z - p/3z$, where

$$z^3 = -q/2 + \sqrt{q^2/4 + p^3/27} = -q/2 + \sqrt{-D/108}.$$

(We have used the expression (23) for D.) Since the roots are real, D is positive (Theorem 20); hence the square root in these formulas is an imaginary number. The formula thus gives the *real* roots y in terms of *complex* numbers!

For many years this was regarded as a serious blemish in this set of formulas, and mathematicians endeavoured to find for the real roots of the cubic other formulas which would involve only real radicals (square roots, cube roots, or higher roots). This search was in vain, by reason of the following theorem.

Theorem 22. *If a cubic polynomial has real roots and is irreducible over the field $F = Q(p, q)$ generated by its coefficients, then there is no rational formula for a root of the cubic in terms of real radicals over F.*

Before proving this, we discuss more thoroughly the properties of a radical $\sqrt[m]{a} = a^{1/m}$. If m is composite, with $m = rs$, then $a^{1/m} = (a^{1/r})^{1/s}$, and so on, so that any radical may be obtained by a succession of radicals with prime exponents. In the latter case we can determine the degree of the field obtained by adjoining a radical.

Lemma. *A polynomial $x^r - a$ of prime degree r over a real field† K is either irreducible over K or has a root in K.*

Proof. Adjoin to K a primitive rth root of unity ζ and then a root u of $x^r - a$. The resulting extension $K(\zeta, u)$ contains the r roots $u, \zeta u, \zeta^2 u, \cdots, \zeta^{r-1} u$ of the polynomial $x^r - a$, hence is the root field of this polynomial, which has the factorization

$$(x^r - a) = (x - u)(x - \zeta u)(x - \zeta^2 u) \cdots (x - \zeta^{r-1} u).$$

Suppose that $x^r - a$ has over F a proper factor $g(x)$ of positive degree $m < r$. This factor $g(x)$ is then a product of m of the linear factors of $x^r - a$ over $K(\zeta, u)$, so that the constant term b in $g(x)$ is a product of m

---

† A real field is any field with elements which are real numbers. This lemma is true for any field, though the proof must be slightly modified if K has characteristic r.

roots $\zeta^i u$. Therefore $b = \zeta^k u^m$, for some integer k, and

$$b^r = (\zeta^k u^m)^r = (\zeta^r)^k (u^r)^m = (u^r)^m = a^m.$$

From this we can find in K an rth root of a, for $m < r$ is relatively prime to r and there exist integers s and t with $sm + tr = 1$ (§1.7, (13)), so that

$$b^{sr} = a^{sm} = a^{1-tr} = a/a^{tr},$$

and $a = (b^s a^t)^r$. The assumed reducibility of $x^r - a$ over K thus yields a root $b^s a^t$ of $x^r - a$ in K.   Q.E.D.

We can now prove Theorem 22. To do this, suppose the conclusion false. Then some root of the cubic can be expressed by real radicals, which is to say that a root y_1 lies in some field $L = F(\sqrt[r]{a}, \sqrt[s]{b}, \cdots)$ generated over F by real radicals. Since D is positive, the real radical $\sqrt{D}$ adjoined to this field gives another real field $K = L(\sqrt{D})$. By Theorem 21 the roots of the cubic will *all* lie in this field, so they all can be expressed by formulas involving real radicals. The field K is obtained by a finite number of radicals. If $\sqrt{D}$ is adjoined first, this amounts to saying that K is the last of a finite chain of fields

(25) $$F \subset K_1 \subset K_2 \subset K_3 \subseteq \cdots \subset K_n = K,$$

where

(26) $$K_1 = F(\sqrt{D}), \qquad K_{i+1} = K_i(a_i^{1/r_i}), \qquad i = 1, \cdots, n-1,$$

with each a_i in K_i and each r_i a prime. By dropping out extra fields, one may assume that the real root a_i^{1/r_i} is not in the field K_i; by the lemma this means that $x^{r_i} - a_i$ is irreducible over K_i and hence that the degree of K_{i+1} is $[K_{i+1}:K_i] = r_i$.

By assumption, the roots of the cubic lie in K; they do not lie in F or in $F(\sqrt{D})$, since the cubic is irreducible over F. In the chain (25) there is then a first field K_{j+1} which contains a root of the cubic, say the root y_1. Over the previous field K_j the given cubic must be irreducible, for otherwise it would have a linear factor $(y - y_i)$ over K_j, contrary to the fact that K_j contains none of the y_i. The extension

(27) $$K_{j+1} = K_j(a^{1/r}), \qquad a = a_j, \qquad r = r_j,$$

then has degree r and contains an element y_1 of degree three over K_j. By Theorem 9, Corollary 2, §14.5, $3 \mid r$, so the prime r must be 3; we are dealing in (27) with a *cube* root $\sqrt[3]{a}$. The field K_{j+1} is generated over K_j by y_1, contains $\sqrt{D}$, and hence by Theorem 21 contains all roots of the cubic. Therefore K_{j+1} is the root field of the given cubic over K_j. As a root field it is normal by Theorem 14; since it contains *one* root $a^{1/3}$ of the polynomial $x^3 - a$ irreducible over K_j, it must therefore contain *all* roots of this polynomial. The other roots are $\omega a^{1/3}$ and $\omega^2 a^{1/3}$, so K_{j+1} also contains ω, a complex cube root of unity. This violates the assumption that $K_{j+1} \subset K$ is a real field. The proof is complete.

Exercises

1. Verify formula (23) for the discriminant.
2. Express the roots of the cubic explicitly in terms of y_1 and $\sqrt{D}$, after the method of Theorem 21.
★3. How much of the discussion of cubic equations applies to cubics over $\mathbf{Z}_3$?
4. Prove: A polynomial $x^n - a$ which has a factor of degree prime to n over a field F of characteristic ∞ has a root in F.
5. Prove: If F is a field of characteristic ∞ containing all nth roots of unity, then the degree $[F(a^{1/n}):F]$ is a divisor of n.
6. Consider the Galois group G of the irreducible cubic (20) over $F = \mathbf{Q}(p, q)$. Prove that if D is the square of a number of F, then G is the alternating group on three letters, and that otherwise it is the symmetric group.

15.9. Insolvability of Quintic Equations

Throughout the present section, F will denote a subfield of the field of complex numbers which contains all roots of unity, and K will denote a variable finite extension of F.

Suppose $K = F(a^{1/r})$ is generated by F and a single rth root $a^{1/r}$ of an element $a \in F$, where r is a prime. The other roots of $x^r = a$ are, as in Chap. 5, $\zeta a^{1/r}, \cdots, \zeta^{r-1} a^{1/r}$, where ζ is a primitive rth root of unity, and so is in F. Therefore K is the root field of $x^r = a$ over F, and hence is normal over F. Unless $K = F$, the polynomial $x^r - a$ is irreducible over F, by the lemma of §15.8, so there is an automorphism S of K carrying the root $a^{1/r}$ into the root $\zeta a^{1/r}$. The powers $I, S, S^2, \cdots, S^{r-1}$ of this automorphism carry $a^{1/r}$ respectively into each of the roots of the equation $x^r = a$; hence these powers include all the automorphisms of K over F. We conclude that the *Galois group of K over F is cyclic.*

More generally, suppose K is normal over F and can be obtained from F by a sequence of simple extensions, each involving only the adjunction of an n_ith root to the preceding extension of F. This means that there exists a sequence of intermediate fields K_i,

$$(28) \qquad\qquad F = K_0 \subset K_1 \subset K_2 \subset \cdots \subset K_s = K,$$

such that $K_i = K_{i-1}(x_i)$, where $x_i^{n_i} \in K_{i-1}$. Without loss of generality, we can assume each n_i is prime. Such a K we shall call an *extension of F by radicals*. Since K is normal, it is the root field of a polynomial $f(x)$ over F, and so the root field of the same $f(x)$ over K_1—and so (Theorem 14) normal over K_1. But K_1 is normal over F by the preceding paragraph. Consequently, every automorphism of K over F induces an automorphism of K_1 over F, and the multiplication of automorphisms is the same. Further, by Lemma 2 of §15.2, every automorphism of K_1 over F can be extended to one of K over F. Hence the correspondence is an *epimorphism* from the Galois group of K over F to that of K_1 over F, like that described at the end of §15.4. Under this epimorphism, moreover, the elements inducing the identity automorphism on K_1 over F are by definition just the automorphisms of K over K_1. This shows that the Galois group $G(K/F)$ of K over F is mapped epimorphically onto $G(K_1/F)$. The latter is therefore *isomorphic* to the quotient-group $G(K/F)/G(K/K_1)$. Combining this with the result of the last paragraph, we infer that $G(K/K_1)$ is a normal subgroup of $G(K/F)$ with cyclic quotient-group $G(K_1/F)$.

Now use induction on s. By definition, K is an extension of K_1 by radicals; as above, it is also normal over K_1. Hence the preceding argument can be reapplied to $G(K/K_1)$ to prove that $G(K/K_2)$ is a normal subgroup of $G(K/K_1)$ with cyclic quotient-group $G(K_2/K_1)$. Repeating this argument s times and denoting the subgroup $G(K/K_i)$ by S_i, we get the following basic result.

Theorem 23. *Let K be any normal extension of F by radicals. Then the Galois group G of K over F contains a sequence of subgroups $S_0 = G \supset S_1 \supset S_2 \supset \cdots \supset S_s$, each normal in the preceding with cyclic quotient-group S_{i-1}/S_i, and with S_s consisting of I alone.*

This states that the Galois group of K over F is *solvable* in the sense of the following definition.

Definition. *A finite group G is* solvable *if and only if it contains a chain of subgroups $S_0 = G \supset S_1 \supset S_2 \supset \cdots \supset S_s = I$ such that for all k, (i) S_k is normal in S_{k-1} and (ii) S_{k-1}/S_k is cyclic.*

A great deal is known about abstract solvable groups; for example, any group whose order is divisible by fewer than three distinct primes is solvable (Burnside); it is even known (Feit–Thompson) that every group of odd order is solvable. We shall, however, content ourselves with the following meager fact.

Lemma 1. *Any epimorphic image G' of a finite solvable group G is itself solvable.*

Proof. Let G have the chain of subgroups S_k as described in the definition of solvability, and let $S_0' = G', S_1', \cdots, S_s' = I'$ be their homomorphic images. Then each S_k' contains, with any x' and y', also $x'y' = (xy)'$ and $x'^{-1} = (x^{-1})'$ (x, y being arbitrary antecedents of x' and y' in S_k), and so is a subgroup of G'. Furthermore, if a is in S_{k-1} and x is in S_k, the normality of S_k in S_{k-1} means that $a^{-1}xa$ is in S_k and hence that $a'^{-1}x'a' = (a^{-1}xa)'$ is in S_k'. Since a' may be any element of S_{k-1}', this proves S_k' normal in S_{k-1}'. Finally, since S_{k-1} consists of the powers $(S_k a)^n = S_k a^n$ of some single coset of S_k (S_{k-1}/S_k being cyclic), S_{k-1}' consists of the powers $S_k'a'^n = (S_k'a')^n$ of the image of this coset, and so is also cyclic. The chain of these subgroups $S_0' \supset S_1' \supset S_2' \supset \cdots \supset S_s'$ thus has the properties which make G' solvable, as required for Lemma 1.   Q.E.D.

Now let us define an equation $f(x) = 0$ with coefficients in F to be *solvable by radicals* over F if its roots lie in an extension K of F obtainable by successive adjunctions of nth roots. This is the case for all quadratic, cubic, and quartic equations, by §5.5. It should be observed that K is not required to be normal, but only to contain the root field N of $f(x)$ over F. However, since any conjugate of an element expressible by radicals is itself expressible by conjugate radicals, the root field N of $f(x)$ must also be contained in a finite extension $K^* \supset K$, normal over F and an extension of F by radicals. This K^* contains N as a normal subfield over F. Hence each automorphism S of K^* over F induces an automorphism S_1 of N over F, and the correspondence $S \mapsto S_1$ is an epimorphism. That is, the Galois group of K^* over F is epimorphic to that of N over F; but the former is solvable (by Theorem 23); hence by Lemma 1, so is the latter. This proves

Theorem 24. *If an equation $f(x) = 0$ with coefficients in F is solvable by radicals, then its Galois group over F is solvable.*

In order to prove that equations of the fifth degree are not always solvable by radicals, we need therefore find only one whose Galois group is not solvable. We shall do this: first we shall prove that the symmetric group of degree five is not solvable, and then we shall exhibit a quintic equation whose Galois group is the symmetric group of degree five.

Theorem 25. *The symmetric group G on n letters is not solvable unless* $n \leq 4$.

Proof. Let $G = S_0 \supset S_1 \supset S_2 \supset \cdots \supset S_s$ be any chain of subgroups, each normal in the preceding with cyclic quotient-group S_{k-1}/S_k; we shall prove by induction on s that S_s *must contain every* 3-cycle (ijk). This will imply that $S_s > I$, and so that G cannot be solvable.

Since $S_0 = G$ contains every 3-cycle, it is sufficient by induction to show that if S_{s-1} contains every 3-cycle, then so does S_s. First, note that if the permutations ϕ and ψ are both in S_{s-1}, then their so-called "commutator" $\gamma = \phi^{-1}\psi^{-1}\phi\psi$ is in S_s. To see this, consider the images ϕ', ψ', and γ' in S_{s-1}/S_s. This quotient-group, being cyclic, is commutative; hence

$$\gamma' = \phi'^{-1}\psi'^{-1}\phi'\psi' = I' \qquad \text{in } S_{s-1}/S_s,$$

which implies $\gamma \in S_s$. But in the special case when $\phi = (ilj)$ and $\psi = (jkm)$, where i, j, k are given and l, m are any two other letters (such letters exist unless $n \leq 4$), we have

$$\gamma = (jli)(mkj)(ilj)(jkm) = (ijk) \in S_s \qquad \text{for all } i, j, k.$$

This proves that S_s contains every 3-cycle, as desired.

Incidentally, it is possible to prove a more explicit form of this theorem. It is known that the alternating group A_n is a normal subgroup of the symmetric group G, so there is a chain beginning $G > A_n$. One may then prove that the alternating group A_n (for $n > 4$) has no normal subgroups whatever except itself and the identity.

Lemma 2. *There is a (real) quintic equation whose Galois group is the symmetric group on five letters.*

Proof. Let A be the field of all algebraic numbers; it will be countable and contain all roots of unity. Hence we can choose in succession, as in §14.6, five algebraically independent real numbers $x_1, \cdots, x_5$ over A. Form the transcendental extension $A(x_1, \cdots, x_5)$. Now let $\sigma_1, \cdots, \sigma_5$ be the elementary symmetric polynomials in the x_i, and let $F = A(\sigma_1, \cdots, \sigma_5)$. As in Theorem 15, the Galois group of the polynomial

$$(29) \qquad f(t) = t^5 - \sigma_1 t^4 + \sigma_2 t^3 - \sigma_3 t^2 + \sigma_4 t - \sigma_5 = 0$$

over F is the symmetric group on the five letters x_i.

It follows from Lemma 2 and Theorem 25 that there exists a (real) quintic equation over a field containing all roots of unity, whose Galois group is not solvable. Now applying Theorem 24, we get our final result.

Theorem 26. *There exists a (real) quintic equation which is not solvable by radicals.*

Exercises

1. Prove that the symmetric group on three letters is solvable.
2. Prove that any finite commutative group is solvable. (*Hint:* Show that it contains a (normal) subgroup of prime index.)
3. Prove that if a finite group G contains a normal subgroup N such that N and G/N are both solvable, then G is solvable.
4. (a) Prove that in the symmetric group on four letters the commutators of 3-cycles form a normal subgroup of order 4.
   (b) Using this and the alternating subgroup, prove that the symmetric group on four letters is solvable.
5. Prove that any finite abstract group G is the Galois group of a suitable equation. (*Hint:* By Cayley's theorem, G is isomorphic with a subgroup of a symmetric group.)
★6. (a) Show that the Galois group of $x^n = a$ is solvable even over a field not containing roots of unity.
   (b) Show that Theorem 24 holds for any F, whether it contains roots of unity or not.
7. Show explicitly that if K is an extension of F by radicals, then there exists an extension K^* of K which is normal over F and which is also an extension of F by radicals. (This fact was used above in the proof of Theorem 24.)
8. If F contains the nth roots of unity, and if $K = F(a^{1/n})$, where a is in K, show that the Galois group of K over F is cyclic, even when n is not prime.
9. If $\mathbf{Q}$ is the rational field, f the special polynomial of (29), show that the Galois group of f over the field $\mathbf{Q}(\sigma_1, \cdots, \sigma_5)$ is still the symmetric group on five letters.
10. Show that if $n > 4$, there exists a real equation of degree n which is not solvable by radicals.

Bibliography

General References

Albert, A. A. (ed.). *Studies in Modern Algebra* (MAA Studies in Mathematics, II).
   Englewood Cliffs, N.J.: Prentice-Hall, 1963.
Artin, E. *Geometric Algebra*. New York: Interscience, 1957.
Birkhoff, G., and T. C. Bartee. *Modern Applied Algebra*. New York: McGraw-
   Hill, 1970.
Godement, Roger. *Cours d'algèbre*. Paris: Hermann, 1963.
Herstein, I. N. *Topics in Algebra*. New York: Wiley, 1964.
Jacobson, N. *Basic Algebra. I. Basic Algebra. II*. San Francisco: Freeman, 1974,
   1976.
Mac Lane, Saunders, and Garrett Birkhoff. *Algebra*. New York: Chelsea, 1988.
Schreier, O., and E. Sperner. *Introduction to Modern Algebra and Matrix Theory*
   (English translation). New York: Chelsea, 1952.
Uspensky, J. V. *Theory of Equations*. New York: McGraw-Hill, 1948.
van der Waerden, B. L. *Modern Algebra*, I, 4th ed., and II, 5th ed. (English
   translation). New York: Ungar, 1966 and 1967.

Number Theory

Hardy, G. H., and E. M. Wright. *An Introduction to the Theory of Numbers*, 4th
   ed. Oxford: Clarendon, 1954.
LeVeque, W. J. *Topics in Number Theory*. 2 vols. Reading, Mass.: Addison-
   Wesley, 1956.
Niven, Ivan, and H. S. Zuckerman. *An Introduction to the Theory of Numbers*.
   New York: Wiley, 1960.
Rademacher, H. *Lectures on Elementary Number Theory*. New York: Wiley, 1964.

Algebraic Number Theory

Lang, S. *Algebraic Numbers*. Reading, Mass.: Addison-Wesley, 1964.
Ribenboim, P. *Algebraic Numbers*. New York: Wiley, 1972.
Weiss, E. *Algebraic Number Theory*. New York: McGraw-Hill, 1963.

Group Theory

Curtis, C. W., and I. Reiner. *Representation Theory of Finite Groups and Associative Algebras*. New York: Interscience, 1962.

Fuchs, L. *Abelian Groups*. Budapest: Hungarian Academy of Sciences, 1958.
Gorenstein, D. *Finite Groups*. New York: Harper & Row, 1968.
Hall, M. *The Theory of Groups*. New York: Macmillan, 1959.
Rotman, J. J. *The Theory of Groups*. Boston: Allyn & Bacon, 1965.

Matrix Theory

Faddaeva, V. N. *Computational Methods of Linear Algebra*. Translated by C. D. Benster. New York: Dover, 1959.
Varga, R. S. *Matrix Iterative Analysis*. Englewood Cliffs, N.J.: Prentice-Hall, 1962.

Galois Theory

Artin, E. *Galois Theory*, 2nd ed. (Notre Dame Mathematical Lecture No. 2). Notre Dame, Ind.: University of Notre Dame Press, 1944.

Linear Algebra and Rings

Jacobson, N. *Lie Algebras*. New York: Wiley, 1962.
Jacobson, N. *The Structure of Rings*, 2nd ed. New York: American Mathematical Society, 1964.
McCoy, N. H. *The Theory of Rings*. New York: Macmillan, 1964.

Algebraic Geometry

Fulton, W. *Algebraic Curves*. New York: Benjamin, 1969.
Jenner, W. E. *Rudiments of Algebraic Geometry*. New York: Oxford University Press, 1963.
Lang, S. *Introduction to Algebraic Geometry*. New York: Interscience, 1958.
Zariski, O., and P. Samuel. *Commutative Algebras*. 2 vols. New York: Van Nostrand, 1958, 1960.

Logic

Kleene, S. C. *Mathematical Logic*. New York: Wiley, 1967.
Mendelson, E. *Introduction to Mathematical Logic*. New York: Van Nostrand, 1964.

Lattice Theory

Abbott, J. C. *Sets, Lattices and Boolean Algebras*. Boston: Allyn & Bacon, 1969.
Birkhoff, Garrett. *Lattice Theory*, 3rd ed. Providence: American Mathematical Society, 1966.

Homological Algebra

Freyd, P. *Abelian Categories.* New York: Harper & Row, 1964.
Jans, J. P. *Rings and Homology.* New York: Holt, 1964.
Mac Lane, Saunders. *Homology.* Berlin: Springer, 1963.
Mac Lane, Saunders. *Categories for the Working Mathematician.* Berlin: Springer, 1971.

Universal Algebra

Cohn, P. M. *Universal Algebra.* New York: Harper & Row, 1965.
Grätzer, G. *Universal Algebra.* New York: Van Nostrand, 1968.
Jonsson, Bjarni. *Topics in Universal Algebra* (Lecture Notes in Mathematics No. 250). Berlin: Springer, 1972.

List of
Special Symbols

A	Matrix (also B, C, etc.)
A^T	Transposed matrix
A^*	Complex conjugate of matrix
$\mathfrak{A}$	Linear algebra
$A_n(F)$	Affine group over F
B	Boolean algebra
$\mathfrak{c}$	Cardinality of $\mathbf{R}$
C	Complex field
D	Integral domain
$D[x]$	Polynomial forms in x, coefficients in D
$D\langle x \rangle$	Polynomial functions in x, coefficients in D
$\mathfrak{d}$	Cardinality $o(R\mathbf{Z})$ of set of positive integers
$\mathbf{E}_n$	Euclidean n-space
E_{ij}	Special matrix, (i,j)-entry 1, others O
$e, 1$	Group identity
F	Field
F^n	Space of n-tuples over F
$F[x]$	Polynomial forms in x, coefficients in F
$F(x)$	Rational forms in x, coefficients in F
G	Group
g.l.b.	Greatest lower bound
i	$\sqrt{-1}$
I	Identity matrix or transformation; greatest element of lattice
j, k	Quaternion units
J	Ideal in ring (also H, L, etc.)
K	Field
$[K:F]$	Degree of K over F
$L_n(F)$	Full linear group over F
l.u.b.	Least upper bound
$M_n(F)$	Total matric algebra over F

O	Zero matrix; least element of lattice
$O_n(F)$	Orthogonal group
$o(S)$	Cardinal number of set S
P	Prime ideal (also nonsingular matrix)
p, q	Positive prime number
$Q(D)$	Field of quotients of domain D
$\mathbf{Q}$	Rational field
R	Ring
$\mathbf{R}$	Real field
S	Set; subgroup; subspace
S'	Set-complement
$S^{\perp}$	Orthogonal complement (of subspace)
T	Linear transformation
T_A	Transformation given by matrix A
$[u : F]$	Degree of u over F
V, W	Vector spaces
V^*	Dual vector space
X	Vector or row matrix
$\mathbf{Z}$	Domain or group of integers
$\mathbf{Z}_n$	Integers mod n
$\mathbf{Z}^+$	Semiring of positive integers
α, β	Vectors
(α, β)	Inner (dot) product of vectors
δ_{ij}	Kronecker delta
ε_i	Unit vector
ϕ, ψ	Transformations; mappings; functions
$\prod$	Product
$\sum$	Summation
ξ, η	Vectors
$\mathbf{0}$	Zero vector
$\varnothing$	Void set
$\cap, \cup$	Intersection, union (of sets)
$\wedge, \vee$	Meet, join (in Boolean algebra, lattice)
$\in$	Is a member of
$\subset$	Is a subset of
$<$	Less than; properly included in
$\leqq$	Inequality

$\perp$	Orthogonal to
$\otimes$	Direct Product
$\oplus$	Direct sum
$\circ$	Binary operator
$\mapsto$	Goes into (for elements)
$\to$	Goes into (for sets)
$*$	Conjugate complex number
∞	Infinity
$\sim$	Associate
$\equiv$	Congruent
$\lvert A \rvert$	Determinant of A
$\lvert a \rvert$	Absolute value
$\lVert a_{ij} \rVert$	Matrix
$a \mid b$	a divides b
(a, b)	Greatest common divisor (g.c.d.)
$[a, b]$	Least common multiple (l.c.m.)

Printed in the United States
by Baker & Taylor Publisher Services